Advanced Membrane Technologies for Wastewater Treatment and Recycling

Advanced Membrane Technologies for Wastewater Treatment and Recycling

Editors

Hongjun Lin
Meijia Zhang

Basel • Beijing • Wuhan • Barcelona • Belgrade • Novi Sad • Cluj • Manchester

Editors
Hongjun Lin
College of Geography and
Environmental Sciences,
Zhejiang Normal University
Jinhua, China

Meijia Zhang
College of Geography and
Environmental Sciences,
Zhejiang Normal University
Jinhua, China

Editorial Office
MDPI
St. Alban-Anlage 66
4052 Basel, Switzerland

This is a reprint of articles from the Special Issue published online in the open access journal *Membranes* (ISSN 2077-0375) (available at: https://www.mdpi.com/journal/membranes/special_issues/Membrane_Technologies_Wastewater).

For citation purposes, cite each article independently as indicated on the article page online and as indicated below:

Lastname, A.A.; Lastname, B.B. Article Title. *Journal Name* **Year**, *Volume Number*, Page Range.

ISBN 978-3-0365-8862-9 (Hbk)
ISBN 978-3-0365-8863-6 (PDF)
doi.org/10.3390/books978-3-0365-8863-6

© 2023 by the authors. Articles in this book are Open Access and distributed under the Creative Commons Attribution (CC BY) license. The book as a whole is distributed by MDPI under the terms and conditions of the Creative Commons Attribution-NonCommercial-NoDerivs (CC BY-NC-ND) license.

Contents

About the Editors vii

Preface ix

Hongjun Lin and Meijia Zhang
Advanced Membrane Technologies for Wastewater Treatment and Recycling
Reprinted from: *Membranes* 2023, *13*, 558, doi:10.3390/membranes13060558 1

Mamdouh S. Alharthi, Omar Bamaga, Hani Abulkhair, Husam Organji, Amer Shaiban, Francesca Macedonio and et al.
Evaluation of a Hybrid Moving Bed Biofilm Membrane Bioreactor and a Direct Contact Membrane Distillation System for Purification of Industrial Wastewater
Reprinted from: *Membranes* 2023, *13*, 16, doi:10.3390/membranes13010016 5

Anna Mittag, Md Musfiqur Rahman, Islam Hafez and Mehdi Tajvidi
Development of Lignin-Containing Cellulose Nanofibrils Coated Paper-Based Filters for Effective Oil-Water Separation
Reprinted from: *Membranes* 2023, *13*, 1, doi:10.3390/membranes13010001 23

Xin Sun, Hana Shiraz, Riccardo Wong, Jingtong Zhang, Jinxin Liu, Jun Lu and et al.
Enhancing the Performance of PVDF/GO Ultrafiltration Membrane via Improving the Dispersion of GO with Homogeniser
Reprinted from: *Membranes* 2022, *12*, 1268, doi:10.3390/membranes12121268 39

Miguel Aguilar-Moreno, Sergi Vinardell, Mònica Reig, Xanel Vecino, César Valderrama and José Luis Cortina
Impact of Sidestream Pre-Treatment on Ammonia Recovery by Membrane Contactors: Experimental and Economic Evaluation
Reprinted from: *Membranes* 2022, *12*, 1251, doi:10.3390/membranes12121251 51

Philipp Jahn, Michael Zelner, Viatcheslav Freger and Mathias Ulbricht
Polystyrene Sulfonate Particles as Building Blocks for Nanofiltration Membranes
Reprinted from: *Membranes* 2022, *12*, 1138, doi:10.3390/membranes12111138 71

João Cunha, Miguel P. da Silva, Maria J. Beira, Marta C. Corvo, Pedro L. Almeida, Pedro J. Sebastião and et al.
Water Molecular Dynamics in the Porous Structures of Ultrafiltration/Nanofiltration Asymmetric Cellulose Acetate–Silica Membranes
Reprinted from: *Membranes* 2022, *12*, 1122, doi:10.3390/membranes12111122 85

Hui Zou, Ying Long, Liguo Shen, Yiming He, Meijia Zhang and Hongjun Lin
Impacts of Calcium Addition on Humic Acid Fouling and the Related Mechanism in Ultrafiltration Process for Water Treatment
Reprinted from: *Membranes* 2022, *12*, 1033, doi:10.3390/membranes12111033 103

M. João Pereira, Manuela Pintado, Carla Brazinha and João Crespo
Recovery of Valuable Aromas from Sardine Cooking Wastewaters by Pervaporation with Fractionated Condensation: Matrix Effect and Model Validation
Reprinted from: *Membranes* 2022, *12*, 988, doi:10.3390/membranes12100988 117

Kai Liu, Jinwen Guo, Yingdong Li, Jinguang Chen and Pingli Li
High-Flux Ultrafiltration Membranes Combining Artificial Water Channels and Covalent Organic Frameworks
Reprinted from: *Membranes* 2022, *12*, 824, doi:10.3390/membranes12090824 125

Hui Zou, Neema Christopher Rutta, Shilei Chen, Meijia Zhang, Hongjun Lin and Baoqiang Liao
Membrane Photobioreactor Applied for Municipal Wastewater Treatment at a High Solids Retention Time: Effects of Microalgae Decay on Treatment Performance and Biomass Properties
Reprinted from: *Membranes* **2022**, *12*, 564, doi:10.3390/membranes12060564 139

Tian Li, Hongjian Yu, Jing Tian, Junxia Liu, Tonghao Yuan, Shaoze Xiao and et al.
PAC-UF Process Improving Surface Water Treatment: PAC Effects and Membrane Fouling Mechanism
Reprinted from: *Membranes* **2022**, *12*, 487, doi:10.3390/membranes12050487 153

Jiarui Guo, Yan Zhang, Fenghua Chen and Yuman Chai
A Membrane with Strong Resistance to Organic and Biological Fouling Using Graphene Oxide and D-Tyrosine as Modifiers
Reprinted from: *Membranes* **2022**, *12*, 486, doi:10.3390/membranes12050486 169

Zirui Wang, Shusu Shen, Linbin Zhang, Abdessamad Ben Hida and Ganwei Zhang
Hydrophilic and Positively Charged Polyvinylidene Fluoride Membranes for Water Treatment with Excellent Anti-Oil and Anti-Biocontamination Properties
Reprinted from: *Membranes* **2022**, *12*, 438, doi:10.3390/membranes12040438 189

Marn Soon Qua, Yan Zhao, Junyou Zhang, Sebastian Hernandez, Aung Thet Paing, Karikalan Mottaiyan and et al.
Novel Sandwich-Structured Hollow Fiber Membrane for High-Efficiency Membrane Distillation and Scale-Up for Pilot Validation
Reprinted from: *Membranes* **2022**, *12*, 423, doi:10.3390/membranes12040423 205

Duowei Lu, Hao Bai and Baoqiang Liao
Comparison between Thermophilic and Mesophilic Membrane-Aerated Biofilm Reactors—A Modeling Study
Reprinted from: *Membranes* **2022**, *12*, 418, doi:10.3390/membranes12040418 223

Zhiyu Yang, Linlin Zang, Tianwei Dou, Yajing Xin, Yanhong Zhang, Dongyu Zhao and et al.
Asymmetric Cellulose/Carbon Nanotubes Membrane with Interconnected Pores Fabricated by Droplet Method for Solar-Driven Interfacial Evaporation and Desalination
Reprinted from: *Membranes* **2022**, *12*, 369, doi:10.3390/membranes12040369 243

Xiaofeng Fang, Shihao Wei, Shuai Liu, Ruo Li, Ziyi Zhang, Yanbiao Liu and et al.
Metal-Coordinated Nanofiltration Membranes Constructed on Metal Ions Blended Support toward Enhanced Dye/Salt Separation and Antifouling Performances
Reprinted from: *Membranes* **2022**, *12*, 340, doi:10.3390/membranes12030340 255

Kun Dong, Xinghui Feng, Wubin Wang, Yuchao Chen, Wei Hu, Haixiang Li and et al.
Simultaneous Partial Nitrification and Denitrification Maintained in Membrane Bioreactor for Nitrogen Removal and Hydrogen Autotrophic Denitrification for Further Treatment
Reprinted from: *Membranes* **2021**, *11*, 911, doi:10.3390/membranes11120911 269

About the Editors

Hongjun Lin

Hongjun Lin, Ph.D., is a Distinguished Professor and serves as the Dean of the College of Geography and Environmental Sciences at Zhejiang Normal University. He holds the distinction of being recognized as a China Highly Cited Chinese Scholar by Elsevier and a Globally Highly Cited Scientist by Clarivate. Over the course of his career, he has overseen more than 20 scientific projects and garnered numerous awards for his contributions. Hongjun Lin's impressive body of work includes over 300 scientific publications, and he holds editorial roles on the Journal Editorial Board/Youth Editorial Board of prestigious international journals such as Chinese Chemical Letters, among others. His ongoing research interest encompasses a wide range of topics, including the preparation of separation membranes, mechanisms related to water treatment, environmental pollutant control, modeling calculations, photo-electrocatalysis, anaerobic ammonia oxidation, granular sludge, algal–bacterial systems, material preparation, industrial wastewater treatment, and artificial neural networks.

Meijia Zhang

Dr. Meijia Zhang earned her Ph.D. in Biotechnology from Lakehead University in 2020 and currently holds the esteemed position of Distinguished Professor at Zhejiang Normal University. Her academic journey has been marked by active involvement in five significant scientific projects, resulting in two notable awards in recognition of her contributions. In recent years, Dr. Zhang has co-authored over 60 research papers featured in esteemed journals. Her research focus centers on cutting-edge areas, including membrane bioreactor technology, the elucidation of membrane fouling mechanisms, and the application of innovative approaches to control fouling, alongside pioneering work in algae wastewater treatment.

Preface

This reprint, entitled "Advanced Membrane Technologies for Wastewater Treatment and Recycling," encompasses a wide spectrum of subjects, including membrane technology, membrane modification, wastewater treatment, membrane bioreactors, membrane fouling, and fouling mechanisms. Its primary focus is to comprehensively explore cutting-edge advancements and forthcoming developments in membrane technologies as applied to wastewater treatment and recycling. Grateful for the substantial support received from numerous scholars regarding this Special Issue, we are pleased to announce the publication of seventeen research articles and one communication, all of which have significantly contributed to advancing our scientific understanding of this field. We extend our heartfelt appreciation to the authors of these articles for their valuable contributions and to the referees for their rigorous review process. This reprint aims to broaden its sphere of influence, raising awareness among a wider readership regarding the latest research advances in membrane technology within the domain of wastewater treatment and recycling.

Hongjur Lin and Meijia Zhang
Editors

Editorial

Advanced Membrane Technologies for Wastewater Treatment and Recycling

Hongjun Lin * and Meijia Zhang *

College of Geography and Environmental Sciences, Zhejiang Normal University, Jinhua 321004, China
* Correspondence: hjlin@zjnu.cn (H.L.); mzhang15@zjnu.edu.cn (M.Z.)

Citation: Lin, H.; Zhang, M. Advanced Membrane Technologies for Wastewater Treatment and Recycling. *Membranes* **2023**, *13*, 558. https://doi.org/10.3390/membranes13060558

Received: 16 May 2023
Accepted: 23 May 2023
Published: 29 May 2023

Copyright: © 2023 by the authors. Licensee MDPI, Basel, Switzerland. This article is an open access article distributed under the terms and conditions of the Creative Commons Attribution (CC BY) license (https://creativecommons.org/licenses/by/4.0/).

In the face of the ever-growing severe problem of water scarcity, wastewater reuse, recycling and resource recovery are increasingly recognized as crucial part of the solution. Compared to other wastewater treatment processes, membrane technology stands out with its distinctive advantages, including simple operation, easy scalability, and chemical-free operations, and has, therefore, been extensively employed in wastewater treatment and recycling. Despite the significant progress in implementing membrane technologies for wastewater treatment and recycling, the application of membranes in wastewater treatment still confronts various unresolved challenges, such as membrane fouling.

This current Special Issue on *Membranes* aims to comprehensively cover state-of-the-art advancements and future developments in the field of membrane technologies applied to wastewater treatment and recycling. Following the proposal for this Special Issue, seventeen research articles and one communication were published, all of which contributed significantly to the scientific understanding of this field. This article presents a concise summary of the research articles included in this Special Issue.

Half of the published papers in this Special Issue focused on membrane fabrication and modification, with new strategies developed to improve membrane separation [1–4], permeability [5–7], and anti-fouling properties [8,9] by fabricating the membrane structure or surface properties. For example, Fang et al. [1] fabricated a loose nanofiltration membrane by integrating blending and interfacial coordination strategies; this resulted in a membrane with good pure water flux, dye rejection, and salt penetration. Liu et al. [5] synthesized a covalent–organic framework (COF) composite membrane by assembling COF layers and the imidazole-quartet water channel, which exhibited excellent performance above 271.7 L m^{-2} h^{-1} bar^{-1} water permeance and above a 99.5% congo red rejection rate. Wang et al. [9] improved the antifouling performance of this membrane by modifying the membrane surface hydrophilicity and porosity by mixing polycationic liquid into the polyvinylidene fluoride (PVDF) membrane.

Of the remaining eight articles, half were related to the application of membrane bioreactors in wastewater treatment. Dong et al. [10] optimized the membrane biofilm reactor and hydrogen-based membrane biofilm reactor (MBR-MBfR) to treat low C/N wastewater and found that proper system functioning was achieved by coupling the partial nitrification-denitrification (PN-D) process in an MBR with further treatment in an MBfR. Aguilar-Moreno et al. [11] demonstrated the economic feasibility of ammonia recovery from anaerobic digestion concentrate using a combination of C/F, aeration, and membrane contactor. Zou et al. [12] reported that the MPBR system could not maintain long-term operations under high SRT for municipal wastewater treatment. Such operating conditions may lead to the decay and deterioration of MPBR's biological performance while improving the antifouling performance of microalgae flocs. Alharthi et al. [13] successfully integrated the moving bed biofilm reactor (MBBR), membrane bioreactor (MBR), and direct contact membrane distillation (DCMD) treatment steps for industrial wastewater treatment, and the results showed that high-quality effluents were obtained by the three-step process.

Cunha et al. [14] characterized water dynamics in cellulose acetate-silica asymmetric membranes by combining multiple methods, including ^{1}H NMR spectroscopy, diffusometry and relaxometry. In the research conducted by Lu, Bai, and Liao [15], mathematical modeling was utilized to examine the impact of temperature (mesophilic versus thermophilic) and oxygen partial pressure on the performance of the membrane-aerated biofilm reactor (MABR), and the results indicated that ThMABR had significant advantages over conventional mesophilic MABR. Zou et al. [16] investigated the role of Ca^{2+} addition in humic acid (HA) fouling and the potential of adding Ca^{2+} for fouling mitigation in the coagulation-ultrafiltration process. The results demonstrated the feasibility of fouling mitigation by adding Ca^{2+} into the ultrafiltration process to treat HA pollutants. Li et al. [17] investigated the water purification effect and membrane fouling mechanism of two types of powdered activated carbon (PAC) that enhanced PVDF ultrafiltration membranes for surface water treatment. The results showed that PAC could effectively enhance membrane filtration performance.

Pereira et al. [18] reported on a study that evaluated the applicability of a previously developed mathematical model to predict the fractionation of aromas from different chemical families in real effluents (sardine cooking wastewaters) and remove off-flavors. Their findings demonstrated that the model simulations were not substantially impacted by the food matrix, which served to validate and expand the applicability of the model.

The field of advanced membrane technologies for wastewater treatment and recycling encompassed a broad range of research topics. This Special Issue presents significant contributions to membrane research, covering membrane fabrication and modification, membrane bioreactor applications in wastewater treatment, and membrane fouling control and mechanisms. In conclusion, the editors appreciate the authors' and reviewers' valuable contributions to this Special Issue. We are also grateful to the editorial staff of "Advanced Membrane Technologies for Wastewater Treatment and Recycling" for their invaluable support.

Author Contributions: Writing—original draft preparation, M.Z.; writing—review and editing, M.Z. and H.L. All authors have read and agreed to the published version of the manuscript.

Funding: This research received no external funding.

Institutional Review Board Statement: Not applicable.

Data Availability Statement: Not applicable.

Acknowledgments: We thank authors of the papers published in this research topic for their valuable contributions and the referees for their rigorous review. We also thank the editorial board of *Membranes*.

Conflicts of Interest: The authors declare no conflict of interest.

References

1. Fang, X.; Wei, S.; Liu, S.; Li, R.; Zhang, Z.; Liu, Y.; Zhang, X.; Lou, M.; Chen, G.; Li, F. Metal-Coordinated Nanofiltration Membranes Constructed on Metal Ions Blended Support toward Enhanced Dye/Salt Separation and Antifouling Performances. *Membranes* **2022**, *12*, 340. [CrossRef] [PubMed]
2. Jahn, P.; Zelner, M.; Freger, V.; Ulbricht, M. Polystyrene Sulfonate Particles as Building Blocks for Nanofiltration Membranes. *Membranes* **2022**, *12*, 1138. [CrossRef] [PubMed]
3. Yang, Z.; Zang, L.; Dou, T.; Xin, Y.; Zhang, Y.; Zhao, D.; Sun, L. Asymmetric Cellulose/Carbon Nanotubes Membrane with Interconnected Pores Fabricated by Droplet Method for Solar-Driven Interfacial Evaporation and Desalination. *Membranes* **2022**, *12*, 369. [CrossRef] [PubMed]
4. Mittag, A.; Rahman, M.M.; Hafez, I.; Tajvidi, M. Development of Lignin-Containing Cellulose Nanofibrils Coated Paper-Based Filters for Effective Oil-Water Separation. *Membranes* **2023**, *13*, 1. [CrossRef] [PubMed]
5. Liu, K.; Guo, J.; Li, Y.; Chen, J.; Li, P. High-Flux Ultrafiltration Membranes Combining Artificial Water Channels and Covalent Organic Frameworks. *Membranes* **2022**, *12*, 824. [CrossRef]
6. Qua, M.S.; Zhao, Y.; Zhang, J.; Hernandez, S.; Paing, A.T.; Mottaiyan, K.; Zuo, J.; Dhalla, A.; Chung, T.-S.; Gudipati, C. Novel Sandwich-Structured Hollow Fiber Membrane for High-Efficiency Membrane Distillation and Scale-Up for Pilot Validation. *Membranes* **2022**, *12*, 423. [CrossRef]

7. Sun, X.; Shiraz, H.; Wong, R.; Zhang, J.; Liu, J.; Lu, J.; Meng, N. Enhancing the Performance of PVDF/GO Ultrafiltration Membrane via Improving the Dispersion of GO with Homogeniser. *Membranes* **2022**, *12*, 1268. [CrossRef] [PubMed]
8. Guo, J.; Zhang, Y.; Chen, F.; Chai, Y. A Membrane with Strong Resistance to Organic and Biological Fouling Using Graphene Oxide and D-Tyrosine as Modifiers. *Membranes* **2022**, *12*, 486. [CrossRef] [PubMed]
9. Wang, Z.; Shen, S.; Zhang, L.; Ben Hida, A.; Zhang, G. Hydrophilic and Positively Charged Polyvinylidene Fluoride Membranes for Water Treatment with Excellent Anti-Oil and Anti-Biocontamination Properties. *Membranes* **2022**, *12*, 438. [CrossRef] [PubMed]
10. Dong, K.; Feng, X.; Wang, W.; Chen, Y.; Hu, W.; Li, H.; Wang, D. Simultaneous Partial Nitrification and Denitrification Maintained in Membrane Bioreactor for Nitrogen Removal and Hydrogen Autotrophic Denitrification for Further Treatment. *Membranes* **2021**, *11*, 911. [CrossRef] [PubMed]
11. Aguilar-Moreno, M.; Vinardell, S.; Reig, M.; Vecino, X.; Valderrama, C.; Cortina, J.L. Impact of Sidestream Pre-Treatment on Ammonia Recovery by Membrane Contactors: Experimental and Economic Evaluation. *Membranes* **2022**, *12*, 1251. [CrossRef] [PubMed]
12. Zou, H.; Rutta, N.C.; Chen, S.; Zhang, M.; Lin, H.; Liao, B. Membrane Photobioreactor Applied for Municipal Wastewater Treatment at a High Solids Retention Time: Effects of Microalgae Decay on Treatment Performance and Biomass Properties. *Membranes* **2022**, *12*, 564. [CrossRef] [PubMed]
13. Alharthi, M.S.; Bamaga, O.; Abulkhair, H.; Organji, H.; Shaiban, A.; Macedonio, F.; Criscuoli, A.; Drioli, E.; Wang, Z.; Cui, Z.; et al. Evaluation of a Hybrid Moving Bed Biofilm Membrane Bioreactor and a Direct Contact Membrane Distillation System for Purification of Industrial Wastewater. *Membranes* **2023**, *13*, 16. [CrossRef] [PubMed]
14. Cunha, J.; da Silva, M.P.; Beira, M.J.; Corvo, M.C.; Almeida, P.L.; Sebastião, P.J.; Figueirinhas, J.L.; de Pinho, M.N. Water Molecular Dynamics in the Porous Structures of Ultrafiltration/Nanofiltration Asymmetric Cellulose Acetate-Silica Membranes. *Membranes* **2022**, *12*, 1122. [CrossRef] [PubMed]
15. Lu, D.; Bai, H.; Liao, B. Comparison between Thermophilic and Mesophilic Membrane-Aerated Biofilm Reactors-A Modeling Study. *Membranes* **2022**, *12*, 418. [CrossRef] [PubMed]
16. Zou, H.; Long, Y.; Shen, L.; He, Y.; Zhang, M.; Lin, H. Impacts of Calcium Addition on Humic Acid Fouling and the Related Mechanism in Ultrafiltration Process for Water Treatment. *Membranes* **2022**, *12*, 1033. [CrossRef] [PubMed]
17. Li, T.; Yu, H.; Tian, J.; Liu, J.; Yuan, T.; Xiao, S.; Chu, H.; Dong, B. PAC-UF Process Improving Surface Water Treatment: PAC Effects and Membrane Fouling Mechanism. *Membranes* **2022**, *12*, 487. [CrossRef] [PubMed]
18. Pereira, M.J.; Pintado, M.; Brazinha, C.; Crespo, J. Recovery of Valuable Aromas from Sardine Cooking Wastewaters by Pervaporation with Fractionated Condensation: Matrix Effect and Model Validation. *Membranes* **2022**, *12*, 988. [CrossRef] [PubMed]

Disclaimer/Publisher's Note: The statements, opinions and data contained in all publications are solely those of the individual author(s) and contributor(s) and not of MDPI and/or the editor(s). MDPI and/or the editor(s) disclaim responsibility for any injury to people or property resulting from any ideas, methods, instructions or products referred to in the content.

Article

Evaluation of a Hybrid Moving Bed Biofilm Membrane Bioreactor and a Direct Contact Membrane Distillation System for Purification of Industrial Wastewater

Mamdouh S. Alharthi [1], Omar Bamaga [2], Hani Abulkhair [1,2,*], Husam Organji [2], Amer Shaiban [2], Francesca Macedonio [3], Alessandra Criscuoli [3], Enrico Drioli [3], Zhaohui Wang [4], Zhaoliang Cui [4,*], Wanqin Jin [4] and Mohammed Albeirutty [1,2,*]

1. Department of Mechanical Engineering, King Abdulaziz University, P.O. Box 80200, Jeddah 21589, Saudi Arabia
2. Center of Excellence in Desalination Technology, King Abdulaziz University, P.O. Box 80200, Jeddah 21589, Saudi Arabia
3. Institute on Membrane Technology, National Research Council of Italy (CNR-ITM), Via P. Bucci 17/C, 87036 Rende, Italy
4. State Key Laboratory of Materials-Oriented Chemical Engineering, College of Chemical Engineering, Nanjing Tech University, Nanjing 211816, China
* Correspondence: haboalkhaoir@kau.edu.sa (H.A.); zcui@njtech.edu.cn (Z.C.); mbeirutty@kau.edu.sa (M.A.)

Abstract: Integrated wastewater treatment processes are accepted as the best option for sustainable and unrestricted onsite water reuse. In this study, moving bed biofilm reactor (MBBR), membrane bioreactor (MBR), and direct contact membrane distillation (DCMD) treatment steps were integrated successively to obtain the combined advantages of these processes for industrial wastewater treatment. The MBBR step acts as the first step in the biological treatment and also mitigates foulant load on the MBR. Similarly, MBR acts as the second step in the biological treatment and serves as a pretreatment prior to the DCMD step. The latter acts as a final treatment to produce high-quality water. A laboratory scale integrated MBBR/MBR/DCMD experimental system was used for assessing the treatment efficiency of primary treated (PTIWW) and secondary treated (STIWW) industrial wastewater in terms of permeate water flux, effluent quality, and membrane fouling. The removal efficiency of total dissolved solids (TDS) and effluent permeate flux of the three-step process (MBBR/MBR/DCMD) were better than the two-step (MBR/DCMD) process. In the three-step process, the average removal efficiency of TDS was 99.85% and 98.16% when treating STIWW and PTIWW, respectively. While in the case of the two-step process, the average removal efficiency of TDS was 93.83% when treating STIWW. Similar trends were observed for effluent permeate flux values which were found, in the case of the three-step process, 62.6% higher than the two-step process, when treating STIWW in both cases. Moreover, the comparison of the quality of the effluents obtained with the analysed configurations with that obtained by Jeddah Industrial Wastewater Treatment Plant proved the higher performance of the proposed membrane processes.

Keywords: industrial wastewater; moving bed biofilm reactor; membrane bioreactor; membrane distillation; hybrid process

1. Introduction

The increased demand for freshwater has led to the increase in withdrawals of limited nonrenewable water resources, leading to water scarcity [1,2]. This scarcity has led to water and wastewater treatment innovations, and the need to follow better environmental practices. Stringent water quality standards have helped the evolution of advanced effluent treatment technologies, thereby preserving water quality [2,3]. The high water demand and the environmental threat added more pressure on managing and recycling water. Water reuse through bioreactors can be considered an additional source of environmental sustainability. Recent trends have made the industries either minimize production or recycle treated wastewater to reduce their effects and follow the concept of zero liquid discharge [3]. The effluents of the industrial wastewater treatment plants are characterized by the levels of their constituents, such as biological oxygen demand (BOD5), chemical oxygen demand (COD), total suspended solids (TSS), total dissolved solids (TDS), oil, and grease. Apart from these, industry-specific effluents can include various organic matters, toxins, heavy metals, phenols, dioxins, and furans. The degree of treatment required depends on the quality of effluent and its characteristics. Treatment can be done in aerobic, anaerobic, and anoxic conditions to obtain the desired quality of treated water [4,5].

The wastewater treatment process is realized by chemical, physical, physio-chemical, and biological or a combination of these methods. Biological methods include biofilters, trickling filters, biological contactors, Activated Sludge Process (ASP), Sequencing Batch Reactor, Membrane Bio Reactor (MBR), and Moving Bed Biofilm Reactor (MBBR). Chemical treatment includes aeration, chlorination, and disinfection methods. Physical treatment includes screening, sedimentation, filtration, and flotation. Physio-chemical processes include coagulation and flocculation. The advanced treatment includes carbon adsorption, absorption, stripping, ion exchange, reverse osmosis, and disinfection. Hybrid treatment plants include a combination of these processes for effective treatment [6]. The selection of the suitable combination depends on the wastewater's characteristics and on the required quality of the obtained effluent.

In fact, a wastewater treatment process typically consists of four steps, namely primary treatment, secondary treatment, tertiary treatment, and advanced treatment. The primary treatment includes screening, neutralization, sedimentation, and flotation processes. It is meant to remove contaminants such as debris, grit, sand, etc. Secondary treatment includes aerobic and anaerobic treatments with sedimentation, and its function is to remove organic contaminants and ammonia. Tertiary treatment includes adsorption, precipitation, and disinfection; it is meant to remove nutrients and pathogens [7].

Membrane bioreactors (MBR) are used primarily in wastewater treatment. MBR involves the use of suspended mixed microbial cultures. MBR technology combines biological processes, such as ASP, with membrane filtration. The most common configuration is called a submerged membrane bioreactor [8]. Membrane bioreactors are configured by the type of separation they are designed for. The separation is carried out by either pressure-driven membranes in side-stream MBR's or vacuum-driven membranes submerged in the reactor. In the side-stream MBRs, the wastewater is pumped through the membrane and returned to the bioreactor for further treatment; whereas, vacuum-driven membranes are submerged in the bioreactor [9]. The major advantage of MBR over conventional systems is its smaller carbon footprint [10]. MBR is up to one-third of the size of conventional ASP systems with the same treatment capacity. Moreover, low sludge production and higher-quality degraded sludge are produced in the MBR system. This advantage of MBR contributes to the better competitiveness of MBRs compared to ASPs [9]. The disadvantages of MBRs are their higher oxygen demand (requiring higher energy input when compared to conventional systems), and the fouling of the membranes (requiring constant monitoring and maintenance) [9].

The moving bed biofilm reactor (MBBR) was developed in Norway between the late 1980s and early 1990s by Odegaard H. et al., 1994 [11]. The MBBR process involves utilizing the entire volume of the reactor space for biomass growth. The process uses carriers that move freely in the reactor, acting as a biomass growth medium. The carriers are retained in the reactor by a sieve arrangement at the outlet of the reactor. MBBR can be used under aerobic, anaerobic, and anoxic conditions. For effective treatment, the biofilm carriers are required to be in motion; under the aerobic condition, the movement of carriers is caused by air movement, whereas in anaerobic and anoxic conditions, a mixer is used for agitation of the carriers. The carriers are made from high-density polyethylene (PEHD) with a density of 0.96 g/cm^3. The biomass grows on the surface of these carriers, and due to the shape of the carriers, they provide a higher surface area for the growth of biomass. The ideal growth of biomass is thin and evenly distributed along the surface area of the carrier, for which turbulence is a crucial factor. Turbulence provides movement of the carriers and maintains a thin biomass layer through the shearing force [12]. Sohail et al. (2020) proposed the integration of MBBR with MBR for the mitigation of MBR membrane fouling [13]. The MBBR has a superior performance in reducing the concentration of suspended solids due to the high biodegradation rate of the organic matter facilitated by the biofilm carriers. They have concluded that the MBBMR was a superior option with respect to effluent rate, operation time, and sludge mass compared to stand alone MBR, MBBR, or ASP.

Membrane Distillation (MD) is a water desalination technology comprised of a hydrophobic membrane that only allows volatile components (and, therefore water vapor) to pass through, not the liquid. MD operates on a temperature difference between the hot feed, which is in contact with the upstream side of the membrane, and the cold condensate, which is in contact with the downstream side of the membrane [14,15]. The configuration can operate at relatively low temperatures of the feed, and has an excellent salts and pollutants rejection efficiency (close to 100%). However, MD is not as energy efficient as reverse osmosis (RO) systems. Jeong et al. [16] conducted a feasibility study of the MD process for the treatment of wastewater from sewage treatment plants for potable water reuse. It was found that MD achieved treated water quality levels as required for drinking purposes, most of the dissolved organic matter was rejected, and a few naturally found amino acids such as tyrosine passed through the membrane. All pharmaceuticals were removed in such a way that their concentrations were below the quantification limits; however, membrane fouling was found to be an issue [16].

The membrane distillation bioreactor (MDBR) combines the thermophilic biological process with the MD process [17]. A vapor-liquid interface is created in the MD process and passes the same through a hydrophobic membrane. The influent wastewater vaporizes at the vapor-liquid interface close to the surface of the hydrophobic membrane, and permeates under the effect of the vapor pressure gradient. Finally, it is condensed and the distillate is removed. Conventional MBR system uses either microfiltration or ultrafiltration microporous membranes to retain the biomass or mixed liquor within the reactor. In comparison, MDBR has an MD membrane to retain the same. MDBR membranes can be placed in a side-stream, or submerged in the bioreactor. MDBRs are operated at 50–60 °C to treat the influent wastewater. The average specific energy consumption (SEC) of the DCMD system was calculated at around 500 kWh/m^3 [18] which is much higher than the SEC of the RO system. The high SEC of the MD process is attributed to the nature of the driving force of the process which implies the necessity of a temperature gradient across the membrane. For this reason, MD modules must have a heat sink on the permeate side to induce a temperature gradient. This makes MD work as a heat exchanger where most of the exergy of feed water is destructed and lost to the permeate side. In cases of the availability of waste heat on the treatment site, which is a common feature of many industries, the waste heat can be utilized suitably for MD operation [19]. Under such conditions, the MD system can produce freshwater without any high energy costs [20]. The MDBR systems, similar to MBR systems, are also prone to fouling. This can, however, be managed or controlled to a certain extent by way of bubbling and cleaning the membranes [21]. MDBR produces a higher quality of effluent than MBR and can be operated on the principle of using waste heat produced by specific industries [21]. Goh et al. [17] noted that the inclusion of biomass in the MDBR system could result in a decline in flux and bio-fouling. In their study, they successfully delayed wetting by 1.7–3.6 times by just lowering the retentate organic and nutrient concentration. Fast flux decline was due to the thermal and mass transfer resistance of the biofilm; however, the same cannot be controlled with periodic membrane cleaning and process optimization. It was concluded that MDBR can be used for the reclamation of industrial wastewater with low volatile organic content and can be feasible if access to waste heat is readily available.

Khaing et al. [22] conducted a study utilizing submerged MDBR for the treatment of petrochemical wastewater. Membranes were found to be thermally stable and could maintain the flux over 5.5 L/(m² h) throughout the study period of 105 days; however, flux decline was found due to inorganic fouling of the membrane. Leyva-Diaz et al. [23] tested a hybrid MBBR-MBR system at two different scales of operation to analyze their effect on municipal wastewater treatment. The configurations were reliable for organic matter removal, with COD removal percentages of 90.97 ± 2.55% and 95.56 ± 2.01% for hybrid MBBR–MBRL and hybrid MBBR–MBRP, respectively. Trapani et al. [24] compared two pilot-scale MB-MBR and MBR systems by increasing salinity in feed wastewater. Pore fouling tendency was noted to be higher in the MBR system. It was concluded that the MB-MBR system performed better and had potential for treatment of high strength or industrial wastewater.

In this study, an MBBR combined with a UF membrane is introduced before a DCMD system to obtain the combined advantages of MBBR, MBR, and DCMD. The process was analysed as a hybrid system for industrial wastewater treatment. Two configurations of the hybrid moving bed biofilm membrane distillation bioreactor (MBBMDBR) system were assessed and compared for purification of primary and secondary treated industrial wastewater in terms of the system permeate water flux, quality, and membrane fouling.

2. Methodology

Three experiments were carried out in this study. The first and second experiments were devoted to evaluating the performance of a hybrid moving bed biofilm reactor (MBBR) combined with a UF membrane followed by a DCMD system for the treatment of primary and secondary wastewater, respectively. In the third experiment, the performance of the submerged MDBR system was assessed for the purification of secondary treated wastewater (Figure 1).

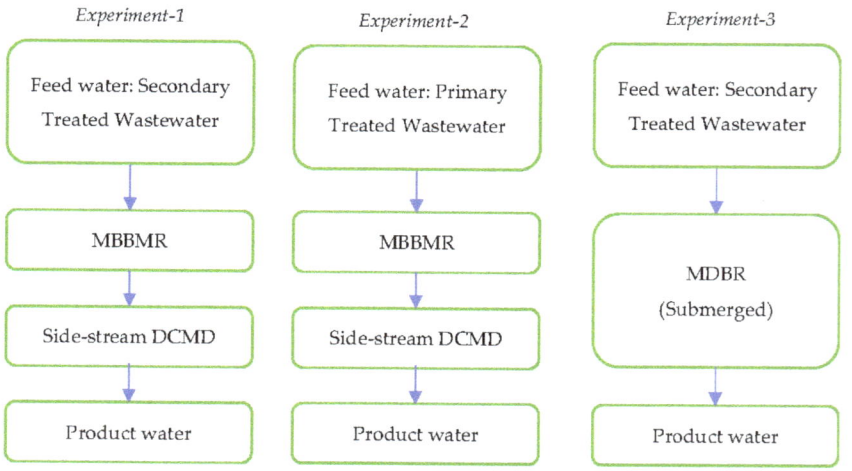

Figure 1. Configurations of the hybrid wastewater treatment and membrane processes tested, and the related experiments.

2.1. Materials

Feed water was collected at two locations from Jeddah Industrial Wastewater Treatment Plant: (a) primary treated water (F1) has undergone primary treatment of grit removal, oil and grease removal, and primary sedimentation, and (b) secondary treated water (F2) was collected from the secondary clarifier after activated sludge treatment. The physicochemical and biological analysis data of the feed wastewater is presented in Table S1 in Supplementary Materials. F1 The influent feed water has a light-yellow tint to it. The major parameters analyzed were TDS (962 mg/L), pH (9.4), TSS (524 mg/L), Turbidity (40 NTU), and TOC (286 ppm). Major anions and cations were Chloride (342 ppm), Sodium (313 ppm), Sulphate (110 ppm), Potassium (31.9 ppm), Calcium (24 ppm), and Magnesium (7.3 ppm). Heavy metal content Aluminum (1.9 ppm), Iron (0.3 ppm), and Zinc (0.16 ppm).

F2 influent feed water has a greenish-yellow tint to it. Major parameters analyzed were TDS (818 mg/L), pH (9.4), TSS (90 mg/L), Turbidity (1.6 NTU), and TOC (11.7 ppm). Major anions and cations were Chloride (304 ppm), Sodium (286 ppm), Sulphate (36.8 ppm), Potassium (24.3 ppm), Calcium (16.6 ppm), and Magnesium (5.5 ppm). Heavy metal content in influent feed water-1 was found to be very low with the highest concentration being aluminum (0.4 ppm).

Hydrophilic PVDF ultrafiltration membranes were procured from Nanjing Tech University (NTU), Nanjing, China, and used to prepare experimental membrane modules for the moving bed biofilm membrane bioreactor MBBMR configuration (Figure 2a). Hydrophobic PVDF membranes (supplied by Econity, South Korea) were used for preparing submerged (Figure 2b) and side-stream (Figure 2c) membrane distillation modules. Specifications of UF modules and MD modules are provided in Tables 1 and 2, respectively.

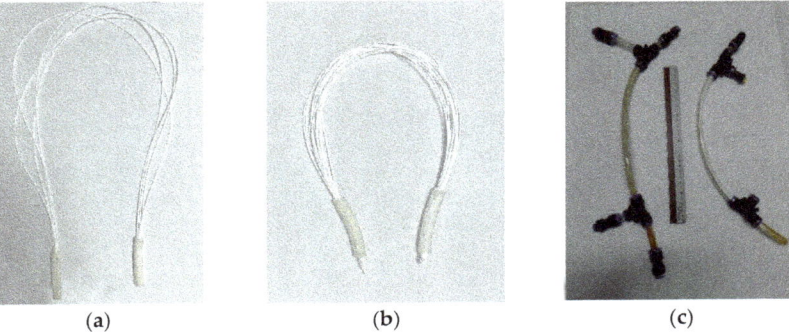

Figure 2. Experimental membrane modules used in the study; (**a**) UF membrane module used in hybrid MBBMR configuration, (**b**) MD membrane module used for submerged MD (Experiment 3), and (**c**) MD membrane module used for side stream MD (Experiment 1 and Experiment 2).

Table 1. Properties of UF modules used in the study.

Properties	Units	UF Membrane
No of membranes in a module	number	4
Length	cm	80
Contact angle [°]	[°]	85
Outer diameter	mm	2.20
Wall thickness	mm	1
Porosity	%	55.6
Mean pore size	nm	340–390
Maximum Load	N	333
Tensile Strength	N/mm^2	87.7
Elongation	%	25.8
Modulus (Automatic Young's)	MPa	612.5
Load at 1%	N	10.4
Tensile stress at 0.2%	N/mm^2	87.4
Membrane area	m^2	0.02212

Table 2. Properties of MD modules used in the study.

Properties	Side-Stream MD Module	Submerged DCMD Module
Membrane material	PVDF	PVDF
Mean pore size (μm)	0.2	0.2
Number of hollow fibers	15	13
Nominal inner diameter of the fiber (mm)	0.80	0.80
Nominal outer diameter of the fiber (mm)	1.2	1.2
Effective membrane area (m^2)	0.0113	0.00139
Effective module length (m)	0.30	0.37
Effective module's membrane area (m^2)	0.01696	0.01813

2.2. Experimental Setup Description

A bench scale integrated MBBMR-MD experimental setup was assembled to investigate the efficiency of integration of membrane bioreactor combined with UF membrane separation followed by membrane distillation. The experimental unit consists of standard process components and instrumentations of research quality mounted on a movable bench. A schematic diagram of the experimental setup as configuration-1 used in experiments 1 and 2 is shown in Figure 3, and configuration-2 used in experiment-3 is shown in Figure 4. An acrylic tank of thickness 10 mm with a removable cover (MBBMR Tank) of capacity 23.6 L has been used as a feed tank. Two ceramic air diffusers were provided at the bottom of the reactor to produce fine bubbles. A circular tank made of acrylic material with a volume of 8.5 L was used as an MD feed tank in configuration 1. The same tank was used as a membrane distillation bioreactor in configuration 2.

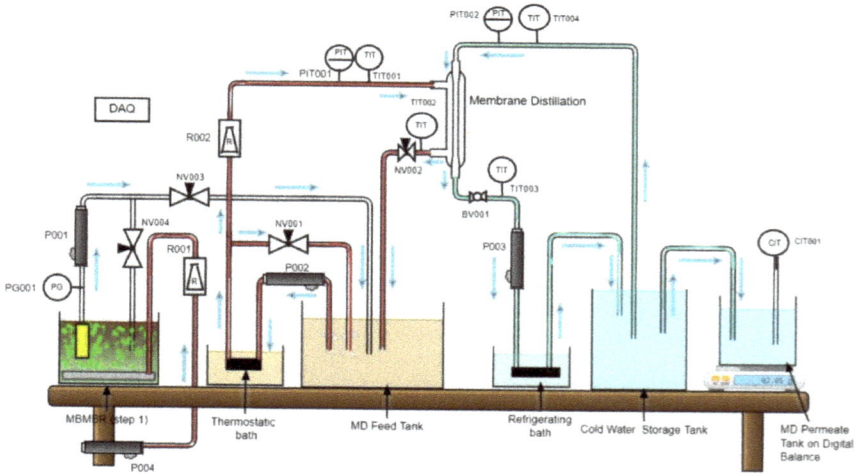

Figure 3. Schematic of the hybrid MBBMR and side stream DCMD configuration used in Experiments 1 and 2. (PG00x: pressure gauge; P00x: pump; NV00x: needle valve; R001: air flow meter, R002: water flow meter; PIT00x: pressure transducer; TIT00x:. thermocouples; CIT001: conductivity transmitter. DAQ: data acquisition).

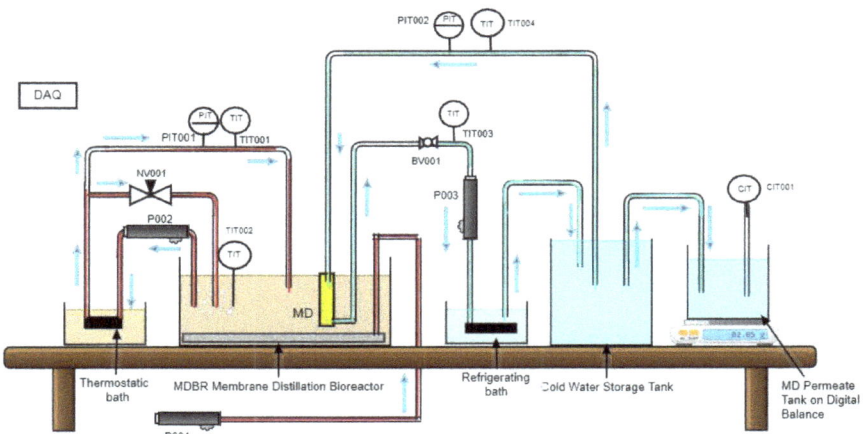

Figure 4. Schematic of the hybrid submerged MDBR configuration 2 used in Experiment 3. (PG00x: pressure gauge; P00x: pump; NV00x: needle valve; R001: air flow meter, R002: water flow meter; PIT00x: pressure transducer; TIT00x:. thermocouples; CIT001: conductivity transmitter. DAQ: data acquisition).

Polyurethane (PU) tubes were used to connect all tanks. A refrigerated bath has been used (Huber, model no MPC-K25, Germany) for cooling the MD permeate. Diaphragm pumps were used to transfer the influent from one tank to another. An air pump was used to maintain sufficient dissolved oxygen (DO) levels in the reactor. The transmembrane pressure of MBR was measured by using a pressure gauge. Needle valves were used at specific points to control the flow in the model reactor. A pressure transducer for the hot side membrane inlet is provided; similarly, a pressure transducer for the cold side membrane inlet is provided. Thermocouples measure the temperature at the hot side of the membrane (both inlet and outlet) as well as for the cold side (both inlet and outlet). A ball valve has been provided for flow control. A conductivity transmitter (Signet 9900 Transmitter) was provided to record the conductivity of treated water. A precision scale (GF-1200 precision scale from AandD Weighing) was used to measure the final treated water from the MD. All fittings and valves were made of stainless steel 316 L fittings. Finally, a data acquisition system (DAQ) was used to record the pressure, temperature, and mass of permeate on a timely basis.

3. Results and Discussion

The daily variations of the main quality parameters of water streams in terms of temperature, pH, TDS, BOD$_5$, and TSS for each configuration are presented in this section. The readings of the parameters were taken at 4 locations in the setup, i.e., MBBMR tank (pH$_0$, TDS$_0$, BOD$_{5,0}$, TSS$_0$), MBBMR filtrate (pH$_{T1}$, TDS$_{T1}$, BOD$_{5,T1}$, TSS$_{T1}$), MD feed tank (pH$_{T2}$, TDS$_{T2}$, BOD$_{5,T2}$, TSS$_{T2}$), and MD permeate (pH$_p$, TDS$_p$, BOD$_{5,p}$, TSS$_p$). Similarly, the MBBMR filtrate flux (F$_0$) and MD permeate flux (J$_p$) were recorded throughout the time of the tests. The operational duration of experiments is reported in the summary of results table (Table 3) which also shows the averages of performance parameters of the tested configurations.

Table 3. Summary of experimental results indicating averages of performance parameters of the proposed configurations for industrial wastewater treatment.

Duration of Experiment (Days)	Parameters (Pi)	Feed Wastewater (i = f)	MBBMR Tank (i = 0)	MBBMR Filtrate (i = 1)	MD Tank (i = 2)	MD Permeate (i = p)
Experiment 1 27	TDS_i (mg/L)	818.0	945.6	770.5	1947.8	1.9
	pHi	9.4	7.9	8.8	8.0	6.4
	TURi (NTU)	1.6	0.6	0.6	1.6	-
	BOD_5,i (mg/L)	6.0	2.5	1.0	-	-
	TSSi (mg/L)	89.8	97.5	82.9	-	-
	TOCi (mg/L)	11.7	11.5	12.0	-	-
	Ti, °C	-	22.1	20.2	47	19
	Jf (L/m².h)	-	44.23	-	-	-
	TMP (psi)	-	−6.1	-	-	-
	J_p (L/(m².h)	-	-	-	-	3.3
Experiment 2 16	TDSi (mg/L)	962.0	1037.4	922.7	2291.3	32.0
	pHi	4.5	6.8	7.2	7.6	6.7
	TURi (NTU)	40.0	1575.4	1.5	4.5	-
	BOD_5,i (mg/L)	-	5.1	3.3	-	-
	TSSi (mg/L)	524.1	1643.4	151.6	-	-
	TOCi (mg/L)	286.3	44.8	28.5	-	-
	Ti, °C	-	19.8	19.8	47	18
	Jf (L/m².h)	-	31.6	-	-	-
	TMP (psi)	-	−7.0	-	-	-
	J_p (L/(m².h)	-	-	-	-	2.6
Experiment 3 74	TDSi (mg/L)	818.0	-	-	1729.4	75.0
	pHi	9.4	-	-	8.3	7.1
	TURi (NTU)	1.6	-	-	0.9	0.3
	BOD_5,i (mg/L)	6.0	-	-	-	2.0
	TSSi (mg/L)	89.8	-	-	577.5	-
	TOCi (mg/L)	11.7	-	-	21.7	-
	Ti, °C	-	-	-	46	19-
	J_p (L/(m².h)	-	-	-	-	2.182

3.1. Assessment of Hybrid MBBMR and DCMD Configuration Performance for Secondary Wastewater Treatment (Experiment 1) and Primary Wastewater Treatment (Experiment 2)

3.1.1. TDS of MBBMR and MDBR Effluents

The treatment of the feed secondary wastewater was accomplished in two successive steps. The first step is the MBBMR treatment, and the second step is MD purification. As shown in Table 3, the average pH of the mixed liquor-suspended solids (MLSS) in the MBBMR tank was 7.9, and was observed to vary in the range of 7.7–8.2. This level of pH values is characteristic of industrial wastewater types. Although it was advised to maintain the pH of the activated sludge system at about 7 for the best growth of the microorganisms [25]. In this study, no pH control was adopted since the MBBR system can tolerate changes in temperature and pH. The pH of the MD permeate water was in the range of 6.0–6.57 range with an average value of 6.4, which is similar to the pH of the deionized water produced in the laboratory from tap water by a pure water RO unit. TDS readings were recorded and plotted for TDS_{T2} and TDS_P in Figure 5. For Experiment 1, the TDS_f values of the feed water ranged from 835–859 mg/L, however, an increase in TDS_{T2} values of the MLSS liquor of the MDBR tank in the range of 1214–3205 mg/L was observed due to the recirculation of MD reject back to the MD feed tank, and therefore accumulation of dissolved solids in the tank (Figure 5). The TDS values of the MD permeate were in the range of 0–3 mg/L which gives removal efficiencies of TDS in the range of 99.75–99.96%.

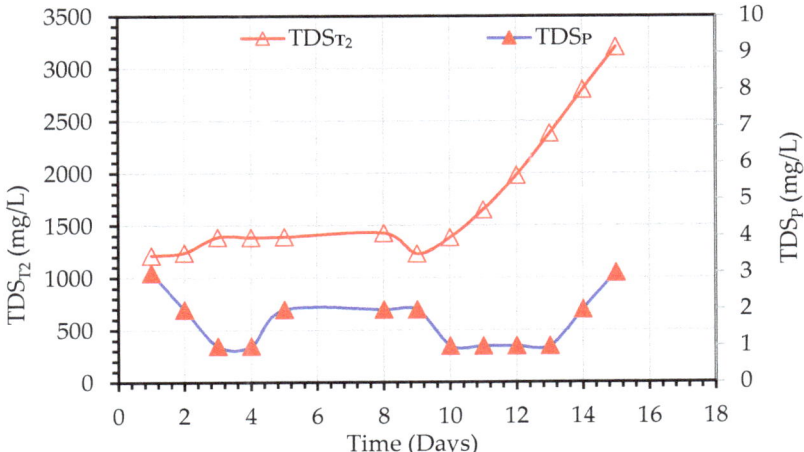

Figure 5. Variations of TDS values observed during the treatment of secondary wastewater using a hybrid MBBMR and DCMD system (Experiment 1).

For Experiment 2, the TDS_{T1} of the MBBMR effluent ranged from 931–1079 mg/L. However, a gradual increase in TDS_{T2} values in the MDBR tank was observed and ranged from 981–3033 mg/L due to recirculation of MD reject back to the MD feed as mentioned above, as shown in Figure 6. The TDS_P were in the range of 7–64 mg/L, the removal efficiency of TDS ranged between 93.7–99.3%. The TDS values of the MBBMR tank were in the higher range due to.

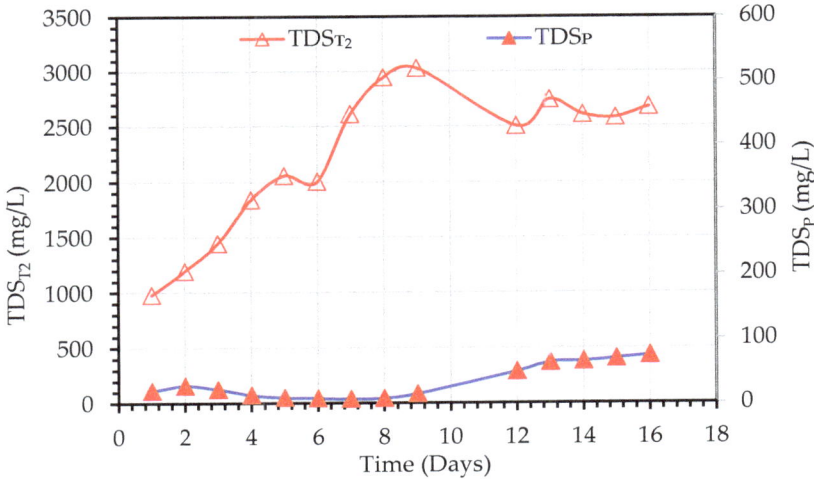

Figure 6. Variations of TDS values observed during the treatment of primary wastewater using a hybrid MBBMR and DCMD system (Experiment 2).

3.1.2. TSS of MBBMR and MDBR Effluents

The values of TSS_0 represent the mixed liquor suspended solids (MLSS) concentrations of the MBBMR tank, while TSS_{T1} values refer to the effluents of the MBBMR tank. Similarly, the $BOD_{5,0}$ is meant for the mixed liquor of the MBBMR tank, and $BOD_{5,T1}$ refers to the effluents of the MBBMR tank. The variations of these parameters are depicted in Figure 7 (Experiment 1) and Figure 8 (Experiment 2). During the experiment period, the

TSS_0 increased from 72 to 119.5 mg/L and the TSS_{T1} increased from 50 to 98.5 mg/L. The removal efficiency of MBBMR for TSS varied from around 45% at the beginning of the testing period to negative values at the end which indicated insufficient acclimatization of microorganisms in the tank. While under the normal and steady operation of MBBMR, it is expected to achieve stable TSS_{T1} values and high TSS removal efficiency [26], the TSS_{T1} values increased with the increase in TSS_0 values. This indicates that the testing period of 3 weeks is not adequate to achieve the required biodegradation rate of biomass. Arabgol et al. [27] reported that five weeks of operation are required for full inoculation of MBBR, followed by another three weeks for reaching steady-state operation. In the case of BOD_5, the removal efficiency varied from 89% to 95%.

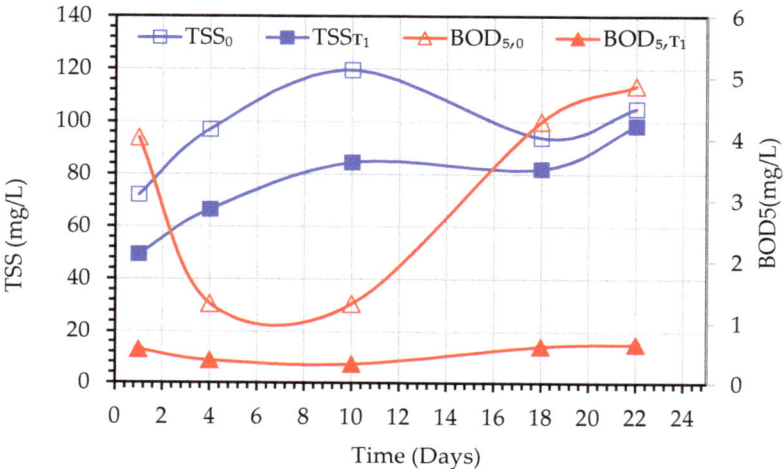

Figure 7. Variations of BOD5 and TSS values observed during the treatment of secondary wastewater using a hybrid MBBMR and DCMD system (Experiment 1).

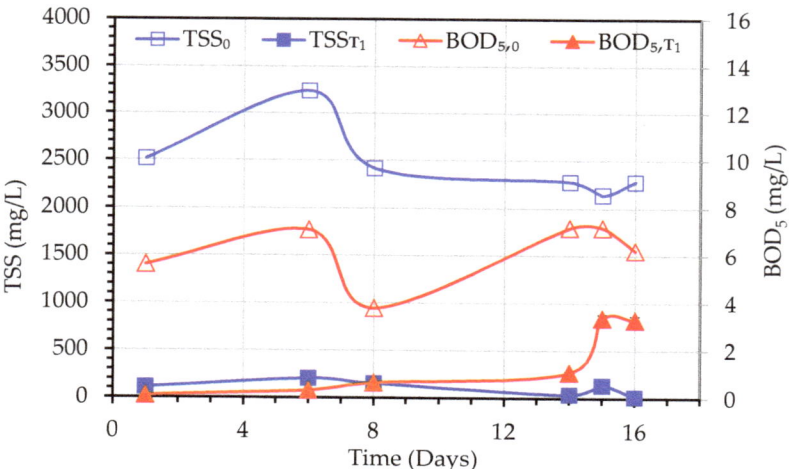

Figure 8. Variations of BOD_5 and TSS values observed during the treatment of primary wastewater using a hybrid MBBMR and DCMD system (Experiment 2).

As already reported for experiment 1, TSS and BOD_5 values were measured similarly for experiment 2 (Figure 8). In this case, the TSS of the MLSS liquor ranged from

2142–3237 mg/L and the TSS removal efficiency in experiment-2 ranged between 61–98%, while initial BOD_5 ranged from 3.76–7.18 mg/L and BOD_5 removal efficiency was around 47–98%.

3.1.3. MDBR Permeate Flux

The MDBR is a combination of membrane distillation separation and wastewater biological treatment in one process unit operation. The temperature of the bioreactor was maintained around 49 °C by recirculating the MLSS liquor in a heat exchanger placed in a heating bath operating at 50 °C. In Experiment 1 and Experiment 2, the hot MLSS liquor is recirculated through the shell side of a side-stream direct contact MD module. On the cold side of the MD module (lumen side), the temperature of the permeate water was maintained at 19 °C by recirculating the permeate water in a heat exchanger placed in a cooling bath operating at around 15 °C. The permeate water flux rate is the key performance parameter of the hybrid system and depends mainly on the temperature gradient across the MD membrane which is the driving force of the process. The MD module inlet feed temperature ($T_{H,0}$) varied between 46.1–47.7 °C, and the MD module outlet temperature ($T_{H,1}$) varied between 45.8–47.1 °C. The inlet MD condensate temperature ($T_{C,0}$) varied between 16.9–19.1 °C, and the outlet condensate (permeate) temperature ($T_{C,1}$) varied from 20.29–22.6 °C.

A comparison of permeate flux (J_p) obtained during Experiment 1 and Experiment 2 has been illustrated in Figure 9. The MD permeate flux was observed to reduce linearly, due to fouling of the membranes or pore blockage, and the TDS elevation in the MLSS liquor. The average values of permeate flux for Experiment 1 and Experiment 2 were 3.3 and 2.6 L/(m²h), respectively, while the average TSD of MLSS liquor for Experiment 1 and Experiment 2 were 1948 and 2291 mg/L, respectively (Table 3). Therefore, it can be calculated that the percentage reduction in J_p (21%) is almost correlated with the percentage increase in TDS (18%). Hence, it can be concluded that the TDS plays a crucial role in the MD permeate flux [24]. When TDS of the permeate water was increased from 1.9 mg/L in Experiment 1, when the TSD of MLSS liquor was 1948 mg/L, to 32 mg/L in Experiment 2, when the TDS of MLSS liquor was 2291 mg/L (Table 3). Membrane wetting is the only possible reason for the increase of permeate water TDS. Experiment 2 was carried out after Experiment 1 without changing the MD module, hence loss of membrane hydrophobicity, and subsequently, the occurrence of membrane wetting may be attributed to the number of days of use as well as the increase in MLSS liquor salinity [14,24].

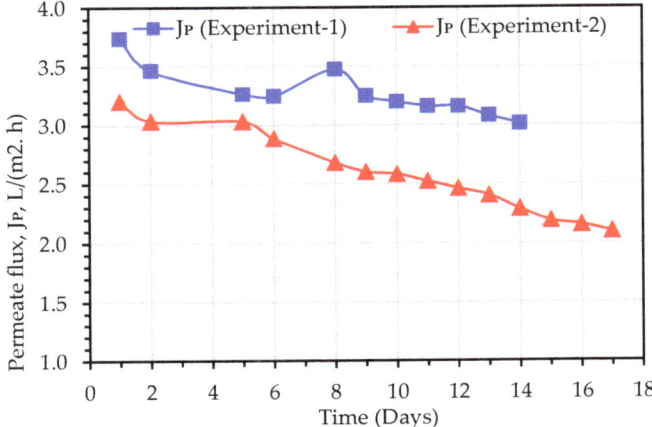

Figure 9. Comparison of MD permeate flux values observed during purification of secondary wastewater (Experiment 1) and primary wastewater (Experiment 2) using hybrid MBBMR and DCMD system.

3.2. Assessment of Hybrid MDBR Configuration Performance for Secondary Wastewater Purification (Experiment 3)

In experiment 3, a hydrophobic hollow fiber membrane module was submerged in the MDBR tank, where the shell side comes into direct contact with the hot MLSS liquor, while the cold permeate is recirculated in the lumen side of the MD module. The hydrodynamic regime in the outer boundary of the membranes is affected only by the movement of air bubbles, while the permeate water velocity inside the lumen is related to the permeate recirculation flow rate which was fixed at 2 L/min. As illustrated in Figure 10, the permeate flux, J_p, value varied in the range of 1.47–2.91 L/(m².h). The experiment was carried out for a period of 60 days. The flux was observed to decrease in a more rapid rate in the first 7 days of the experiment, where the normalized flux declined from 1 to 0.69, then the rate of decrease remains moderate for the remainder of the experiment, where the normalized flux declined from 0.69 to 0.56. As explained earlier, the reduction in flux is due to the synergy effect of two factors i.e., the effect of membrane fouling and the effect of increased concentration of the MLSS liquor.

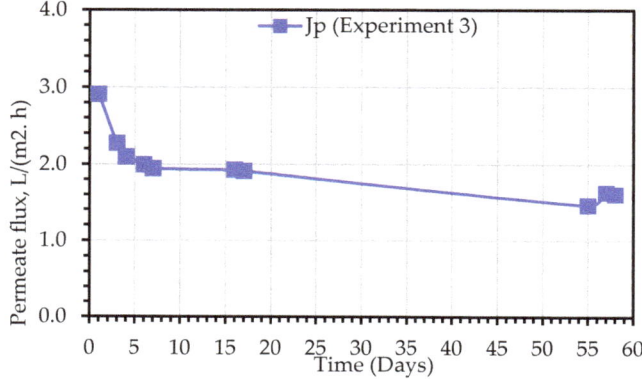

Figure 10. Permeate flux observed during treatment of primary wastewater using a hybrid DCMD system (Experiment 3).

Figure 11 shows the variations of TDS_{T2} and TDS_P observed during the purification of secondary wastewater using the MDBR system (Experiment 3). TDS_{T2} values ranged from 890–1914 mg/L, and TDS_P values were in the range of 19–167 mg/L. This increase in TDST2 is due to the rejection of the dissolved ions by the MD module which results in the accumulation of dissolved ions in the bioreactor tank. It should be noted that at Day 17 when the TDS_{T2} concentration reached 1705 mg/L, the MDBR tank was emptied and refilled again with feed secondary treated wastewater. In contradiction to the observed trend of correlation between TDS_{T2} and TDS_P (Figure 6), the TDS_P trend in Experiment 3 initially increased in the first six days of operation, then decreased to an almost stable level at Day 17 till the end of the experiment. Although this can indicate the stable hydrophobic characteristics of membranes, it can be also related to the bubbling effect on the hydrodynamic regime in the outer boundary of the membranes and the fouling mitigation effect of bubblers.

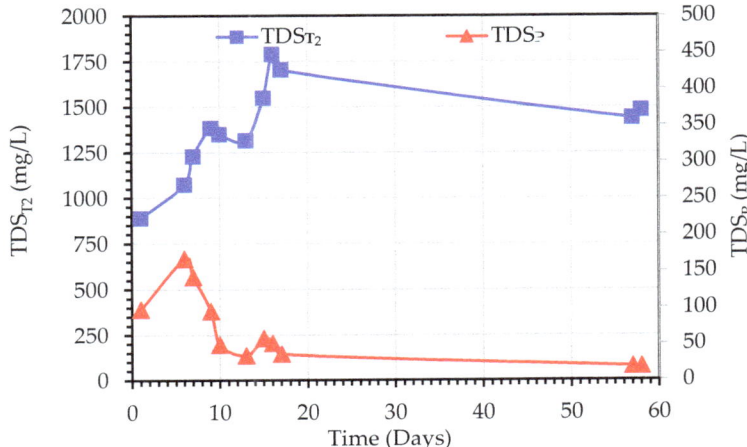

Figure 11. Variations of TDS values observed during the treatment of secondary wastewater using a hybrid DCMD system (Experiment 3).

3.3. Contact Angle Measurement

Contact angle (θ) is an essential parameter for evaluating a membrane's hydrophilicity and wetting behavior [28]. The contact angle of a surface is measured by placing a drop of liquid and measuring the angle formed between the surface and the line tangent to the edge of the drop of liquid. A low contact angle indicates high surface energy and, therefore, a high hydrophilic character of the membrane. Vice versa in the case of high contact angle. Contact angle provides quantitative data about the wettability of a surface at a molecular level [28]. The measure of the contact angle allows for assessing a material surface's quality before an adhesion process. The same sidestream MD membrane module was used for Experiment 1 and Experiment 2. After completion of Experiment 2, the MD module was disassembled, and three fibers were cut and used for the contact angle measurement. The measured values of the contact angle are shown in Table 4, where it can be seen that the average values were ranged from 60.1° to 63.2°. When compared to the pristine fiber, the contact angle of the used fiber worsened by 44.7%, which indicates the possibility of hydrophobicity loss. However, the rate of hydrophobicity loss with time is low compared to other MD membranes used for wastewater treatment [17]. Conventional hydrophobic MD membranes (i.e., membranes that display apparent contact angle θ* > 90° with high surface tension liquids such as water) suffer from membrane wetting in desalination of feedwater containing low surface energy contaminants (e.g., shale gas-produced water [29,30] and coal seam gas produced water [31]).

Table 4. Contact angle results of membranes used in Experiment 1 and 2.

Membrane	Pristine Fiber	Fiber 1 (Used)	Fiber 2 (Used)	Fiber 3 (Used)
Location 1	109.4	64	66	58.9
Location 2	112.5	62.4	55.4	61.3
Average	111	63.2	60.7	60.1

The loss of hydrophobicity and the concomitant decrease in the contact angle of the membrane progresses with the time of operation [32]. The accumulation of foulants on the membrane surface, and in the membrane pores is a time-dependent process and is the main factor for hydrophobicity degradation [33]. The contact angle of the membrane can be measured only if the membrane module is disassembled. Therefore, the practical method for assessing the changes in contact angle is by observing the trend of change of salinity and flux rate of MD permeate i.e., TDS_p and J_p values. In Experiment 1, where a new MD membrane module was used, the TDS_p values were stable around an average value of 1.7 mg/L with no significant increase with time (Figure 5). However, in Experiment 2, where the same MD membrane module was used after cleaning with deionized water, the TDS_p values maintained approximately constant level in the nine days of operation at an average of 14 mg/L, then start to increase progressively up to 75 mg/L after 16 days of operation. The manner of J_p change with time of operation is another indicator of contact angle change. In experiment 1, the permeate flux reduction of 20% was observed after 14 days of operation, while in Experiment 2, the permeate flux reduction of 35% was observed after 16 days of operation.

After the completion of Experiment 3, the contact angle of the submerged MD membrane was measured (Table 5), and the average value was found as 84.75–85.4°.

Table 5. Contact angle results of membranes used in Experiment 3.

Membrane	Pristine Fiber	Fiber 1 (Used)	Fiber 2 (Used)	Fiber 3 (Used)
Location 1	109.4	84.4	85.2	84.9
Location 2	112.5	85.3	85.6	84.6
Average	111	84.85	85.4	84.75

By comparing the average contact angle values for the side stream module and submerged module, it can be deduced that the hydrophobicity loss of the submerged module is lower. As noted by Morrow et al. [34] in sidestream configuration, fouling is mitigated with hydraulic crossflow; however, in submerged configuration, the fouling is mitigated via air scour.

3.4. Comparison of Performance Parameters of Different System Configurations

In experiment 1, the TDS_P values of the MD permeate ranged from 1–3 mg/L. Whereas in experiment 2, the TDS_P ranged from 8–64 mg/L, and in experiment 3, the TDS_P ranged from 25–167 mg/L. A higher degree of treatment was obtained in experiment 1 which had secondary treated water as the feed water. Whereas in experiment 2, when primary treated water was used as feed water, the worse values for TDS_P and J_P were obtained. Table 3 provides a summary of experimental results indicating averages of performance parameters of the proposed configurations for industrial wastewater treatment.

Membranes used in Experiment 1 were cleaned with distilled water for a period of 30 min before being used in Experiment 2. Since secondary treated water was used in Experiment 1, and the membrane was new, the permeate showed the best results. In Experiment 2, primary treated water was used together with the membrane used in experiment 1 and showed worse results comparatively.

The flux decreased in all three experiments, with a higher flux in Experiment 1. In experiment 2, the TDS decreased along with the permeate flux until day 8. After this, an increase in TDS was noticed due to the clogging of pores compared to initial conditions. In Experiment 3, the TDS and flux increased during day 2 and then decreased; during this time, there was no change in pressure or temperature of the influent on the system.

3.5. Comparison of the Quality of the Water Produced by the Three Different Configurations with the Quality of the Water Produced by Jeddah Industrial Wastewater Treatment Plant

To compare the treatment efficiency achieved with the proposed configurations in this study, water quality analysis data reports were collected from Jeddah Industrial City Wastewater Treatment Plant, operated by Modon (JICWTP). In this plant, around 40% of the secondary effluents are purified by the advanced treatment process consisting of a sand filter, UF, and RO. The comparison points were chosen at the UF product outlet and the RO product outlet. The daily water analysis reports of the advanced treatment process were collected for fifteen days, and the average values of TDS of the RO product water, and TSS of the UF product water were calculated for comparison.

The TDS values for the RO effluent at the JICWTP and MD permeate for all three experiments are shown in Figure 12a. As expected, the average TDS value of the MD product for the Exp1 configuration in which secondary treated wastewater was used as a feed for the MBBMR step was less than 2 mg/L. However, the average TDS of the advanced RO treatment product was 106 mg/L. When primary wastewater was used as feed for the MBBMR step, the TDS of the MD product increased to around 24 mg/L probably due to the wetting of the MD membrane by surfactants and oil residues present in the primary feed water. Further, the TDS of the MD permeate was worsened in the case of the Experiment 3 configuration in which the MBBMR step was excluded. Further, The TSS values of UF effluent from the JICWTP are compared to the MBBMR setup in Experiment 1 and Experiment 2, as shown in Figure 12b. The comparison of the TSS data showed higher performance of the advanced RO treatment at JICWTP compared to the configurations tested in this work. The use of a sand filter prior to the UF step eliminates oil and nanoparticles present in the secondary wastewater by absorption mechanism, resulting in a TSS value of 1 mg/L. However, in the case of Experiment 1 and Experiment 2 configurations, the MBBMR treatment is not efficient for the removal of these contaminants. Therefore, the TSS values of 76 mg/L and 110 mg/L were found.

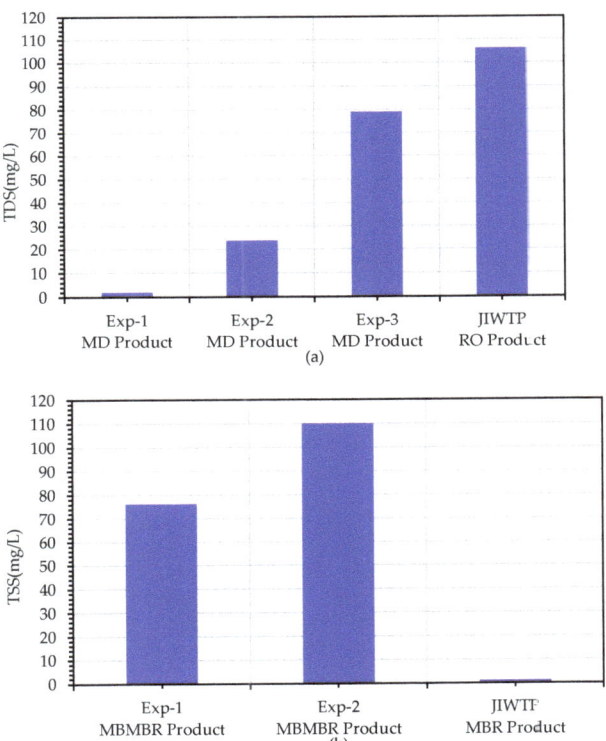

Figure 12. Comparison of TDS and TSS values obtained during treatment of secondary treated wastewater using hybrid MBBMR and DCMD system (Experiment 1 and Experiment 2) with respective values reported by Jeddah Industrial Wastewater Treatment Plant: (**a**) TDS (**b**) TSS.

4. Conclusions

Membrane distillation is a potential high rejection, the terminal process for onsite industrial wastewater treatment when waste heat is available. Membrane wettability threat is the main challenge for MD application in industrial wastewater which contains high concentrations of dissolved organic matter and surfactants. This work compares the role of pretreatment steps prior to MD on the MD performance efficiency. A new integrated membrane system configuration consisting of MBBR integrated with UF, followed by MBR integrated with MD, was tested for treatment and purification of primary and secondary wastewater treatment. This configuration was compared with a simple integrated membrane system configuration consisting of MBR integrated with MD only. The MD performance efficiency achieved by the first configuration was better than that achieved by the second configuration. Additionally, the results proved that the quality of MD permeate remains stable when applying the 3-step process. A correlation has been found between TSS in the MD feed and the TDS of MD permeate. Hence, TSS should be maintained low by introducing a sand filter for absorbing contaminants that are still present after the MBBMR step.

Supplementary Materials: The following supporting information can be downloaded at: https://www.mdpi.com/article/10.3390/membranes13010016/s1, Table S1: Characteristics of the industrial wastewater influents and effluents of the Industrial City Wastewater Treatment Plant (Modon JICWTP, Jeddah).

Author Contributions: Conceptualization, M.A., O.B., E.D., Z.C. and H.A.; methodology, O.B., H.A. and A.C.; investigation, M.S.A. and Z.W.; resources, H.O. and A.S.; data curation, M.S.A. and H.A.; writing—original draft preparation, M.S.A.; writing—review and editing, H.A., O.B., F.M. and Z.C.; visualization, M.S.A.; supervision, H.A. and O.B.; project administration, M.A., E.D. and W.J.; funding acquisition, M.A. All authors have read and agreed to the published version of the manuscript.

Funding: This work was supported by the Deputyship for Research and Innovation, Ministry of Education in Saudi Arabia through the project number 632.

Institutional Review Board Statement: Not applicable.

Informed Consent Statement: Not applicable.

Data Availability Statement: Not applicable.

Acknowledgments: The authors would like to express their appreciation for Econity, South Korea for the supply of PVDF hydrophobic membranes complementary and for the Jeddah Industrial Wastewater Treatment Plant for collaboration.

Conflicts of Interest: The authors declare no conflict of interest. The funders had no role in the design of the study; in the collection, analyses, or interpretation of data; in the writing of the manuscript; or in the decision to publish the results.

References

1. Wang, X.J.; Xia, S.Q.; Chen, L.; Zhao, J.F.; Renault, N.J.; Chovelon, J.M. Nutrients Removal from Municipal Wastewater by Chemical Precipitation in a Moving Bed Biofilm Reactor. *Process Biochem.* **2006**, *41*, 824–828. [CrossRef]
2. Chave, P. *The EU Water Framework Directive: An Introduction*; IWA Publishing: London, UK, 2001.
3. Tong, T.; Elimelech, M. The Global Rise of Zero Liquid Discharge for Wastewater Management: Drivers, Technologies, and Future Directions. *Environ. Sci. Technol.* **2016**, *50*, 6846–6855. [CrossRef] [PubMed]
4. Alkhudhiri, A.; Darwish, N.B.; Hilal, N. Analytical and Forecasting Study for Wastewater Treatment and Water Resources in Saudi Arabia. *J. Water Process Eng.* **2019**, *32*, 100915. [CrossRef]
5. Al-Jasser, A. Saudi Wastewater Reuse Standards for Agricultural Irrigation: Riyadh Treatment Plants Effluent Compliance. *J. King Saud Univ. Eng. Sci.* **2011**, *23*, 1–8. [CrossRef]
6. Boavida-Dias, R.; Silva, J.R.; Santos, A.D.; Martins, R.C.; Castro, L.M.; Quinta-Ferreira, R.M. A Comparison of Biosolids Production and System Efficiency between Activated Sludge, Moving Bed Biofilm Reactor, and Sequencing Batch Moving Bed Biofilm Reactor in the Dairy Wastewater Treatment. *Sustainability* **2022**, *14*, 2702. [CrossRef]
7. Muhaba, S.; Mulubrhan, F.; Darun, M.R. *Application of Petrochemical Wastewater Treatment Processes*; AIP Publishing LLC: New York, NY, USA, 2022; Volume 2454, p. 050034.
8. Luis, P. *Fundamental Modeling of Membrane Systems: Membrane and Process Performance*; Elsevier: Amsterdam, The Netherlands, 2018.
9. Radjenović, J.; Matošić, M.; Mijatović, I.; Petrović, M.; Barceló, D. Membrane Bioreactor (MBR) as an Advanced Wastewater Treatment Technology. In *Emerging Contaminants from Industrial and Municipal Waste*; Barceló, D., Petrovic, M., Eds.; Springer: Berlin/Heidelberg, Germany, 2008; Volume 5S/2, pp. 37–101, ISBN 978-3-540-79209-3.

10. Englande, A.J.; Krenkel, P.; Shamas, J. Wastewater Treatment &Water Reclamation ☆. In *Reference Module in Earth Systems and Environmental Sciences*; Elsevier: Amsterdam, The Netherlands, 2015; p. B9780124095489096000, ISBN 978-0-12-409548-9.
11. Ødegaard, H.; Rusten, B.; Westrum, T. A New Moving Bed Biofilm Reactor-Applications and Results. *Water Sci. Technol.* **1994**, *29*, 157. [CrossRef]
12. Rusten, B.; Eikebrokk, B.; Ulgenes, Y.; Lygren, E. Design and Operations of the Kaldnes Moving Bed Biofilm Reactors. *Aquac. Eng.* **2006**, *34*, 322–331. [CrossRef]
13. Sohail, N.; Ahmed, S.; Chung, S.; Nawaz, M.S. Performance Comparison of Three Different Reactors (MBBR, MBR and MBBMR) for Municipal Wastewater Treatment. *Desalin. Water Treat.* **2020**, *174*, 71–78. [CrossRef]
14. Warsinger, D.M.; Servi, A.; Connors, G.B.; Mavukkandy, M.O.; Arafat, H.A.; Gleason, K.K.; Lienhard V, J.H. Reversing Membrane Wetting in Membrane Distillation: Comparing Dryout to Backwashing with Pressurized Air. *Environ. Sci. Water Res. Technol.* **2017**, *3*, 930–939. [CrossRef]
15. Warsinger, D.E.M. Thermodynamic Design and Fouling of Membrane Distillation Systems. *arXiv* **2017**, arXiv:1711.07348.
16. Jeong, S.; Song, K.G.; Kim, J.; Shin, J.; Maeng, S.K.; Park, J. Feasibility of Membrane Distillation Process for Potable Water Reuse: A Barrier for Dissolved Organic Matters and Pharmaceuticals. *J. Hazard. Mater.* **2021**, *409*, 124499. [CrossRef] [PubMed]
17. Goh, S.; Zhang, J.; Liu, Y.; Fane, A.G. Membrane Distillation Bioreactor (MDBR)—A Lower Green-House-Gas (GHG) Option for Industrial Wastewater Reclamation. *Chemosphere* **2015**, *140*, 129–142. [CrossRef] [PubMed]
18. Yan, Z.; Jiang, Y.; Liu, L.; Li, Z.; Chen, X.; Xia, M.; Fan, G.; Ding, A. Membrane Distillation for Wastewater Treatment: A Mini Review. *Water* **2021**, *13*, 3480. [CrossRef]
19. Martin, A.; Dahl, O. Process Design of Industrial-Scale Membrane Distillation System for Wastewater Treatment in Nano-Electronics Fabrication Facilities. *MethodsX* **2020**, *7*, 101066.
20. Silva, M.; Reis, B.; Grossi, L.; Amaral, M. Improving the Energetic Efficiency of Direct-Contact Membrane Distillation in Mining Effluent by Using the Waste-Heat-and-Water Process as the Cooling Fluid. *J. Clean. Prod.* **2020**, *260*, 121035. [CrossRef]
21. Fane, T. Membrane Distillation Bioreactor (MDBR). In *Encyclopedia of Membranes*; Drioli, E., Giorno, L., Eds.; Springer: Berlin/Heidelberg, Germany, 2015; pp. 1–2, ISBN 978-3-642-40872-4.
22. Khaing, T.-H.; Li, J.; Li, Y.; Wai, N.; Wong, F. Feasibility Study on Petrochemical Wastewater Treatment and Reuse Using a Novel Submerged Membrane Distillation Bioreactor. *Sep. Purif. Technol.* **2010**, *74*, 138–143. [CrossRef]
23. Leyva-Díaz, J.C.; Martín-Pascual, J.; González-López, J.; Hontoria, E.; Poyatos, J.M. Effects of Scale-up on a Hybrid Moving Bed Biofilm Reactor—Membrane Bioreactor for Treating Urban Wastewater. *Chem. Eng. Sci.* **2013**, *104*, 808–816. [CrossRef]
24. Di Trapani, D.; Di Bella, G.; Mannina, G.; Torregrossa, M.; Viviani, G. Comparison between Moving Bed-Membrane Bioreactor (MB-MBR) and Membrane Bioreactor (MBR) Systems: Influence of Wastewater Salinity Variation. *Bioresour. Technol.* **2014**, *162*, 60–69. [CrossRef]
25. Majid, A.; Mahna, M. Application of Lab-Scale MBBR to Treat Industrial Wastewater Using K3 Carriers: Effects of HRT, High COD Influent, and Temperature. *Int. J. Environ. Sci. Nat. Resour.* **2019**, *20*, 35–42. [CrossRef]
26. Yang, X.; López-Grimau, V. Reduction of Cost and Environmental Impact in the Treatment of Textile Wastewater Using a Combined MBBR-MBR System. *Membranes* **2021**, *11*, 892. [CrossRef]
27. Arabgol, R.; Vanrolleghem, P.A.; Piculell, M.; Delatolla, R. The Impact of Biofilm Thickness-Restraint and Carrier Type on Attached Growth System Performance, Solids Characteristics and Settleability. *Environ. Sci. Water Res. Technol.* **2020**, *6*, 2843–2855. [CrossRef]
28. Decker, E.; Frank, B.; Suo, Y.; Garoff, S. Physics of Contact Angle Measurement. *Colloids Surf. A Physicochem. Eng. Asp.* **1999**, *156*, 177–189. [CrossRef]
29. Lu, K.J.; Zuo, J.; Chang, J.; Kuan, H.N.; Chung, T.-S. Omniphobic Hollow-Fiber Membranes for Vacuum Membrane Distillation. *Environ. Sci. Technol.* **2018**, *52*, 4472–4480. [CrossRef] [PubMed]
30. Shaffer, D.L.; Arias Chavez, L.H.; Ben-Sasson, M.; Romero-Vargas Castrillón, S.; Yip, N.Y.; Elimelech, M. Desalination and Reuse of High-Salinity Shale Gas Produced Water: Drivers, Technologies, and Future Directions. *Environ. Sci. Technol.* **2013**, *47*, 9569–9583. [CrossRef] [PubMed]
31. Woo, Y.C.; Chen, Y.; Tijing, L.D.; Phuntsho, S.; He, T.; Choi, J.-S.; Kim, S.-H.; Shon, H.K. CF4 Plasma-Modified Omniphobic Electrospun Nanofiber Membrane for Produced Water Brine Treatment by Membrane Distillation. *J. Membr. Sci.* **2017**, *529*, 234–242.
32. Pan, J.; Chen, K.; Cui, Z.; Bamaga, O.; Albeirutty, M.; Alsaiari, A.O.; Macedonio, F.; Drioli, E. Preparation of ECTFE Porous Membrane for Dehumidification of Gaseous Streams through Membrane Condenser. *Membranes* **2022**, *12*, 65. [CrossRef]
33. Rezaei, M.; Warsinger, D.M.; Duke, M.C.; Matsuura, T.; Samhaber, W.M. Wetting Phenomena in Membrane Distillation: Mechanisms, Reversal, and Prevention. *Water Res.* **2018**, *139*, 329–352. [CrossRef]
34. Morrow, C.P.; McGaughey, A.L.; Hiibel, S.R.; Childress, A.E. Submerged or Sidestream? The Influence of Module Configuration on Fouling and Salinity in Osmotic Membrane Bioreactors. *J. Membr. Sci.* **2018**, *548*, 583–592. [CrossRef]

Disclaimer/Publisher's Note: The statements, opinions and data contained in all publications are solely those of the individual author(s) and contributor(s) and not of MDPI and/or the editor(s). MDPI and/or the editor(s) disclaim responsibility for any injury to people or property resulting from any ideas, methods, instructions or products referred to in the content.

Article

Development of Lignin-Containing Cellulose Nanofibrils Coated Paper-Based Filters for Effective Oil-Water Separation

Anna Mittag [1], Md Musfiqur Rahman [2], Islam Hafez [2,*] and Mehdi Tajvidi [2]

[1] Department of Chemical and Biomolecular Engineering, University of Notre Dame, Notre Dame, IN 46556, USA
[2] Laboratory of Renewable Nanomaterials, School of Forest Resources, University of Maine, 5755 Nutting Hall, Orono, ME 04469, USA
* Correspondence: islam.hafez@maine.edu

Abstract: New methods of oil-water separation are needed as industrialization has increased the prevalence of oil-water mixtures on Earth. As an abundant and renewable resource with high oxygen and grease barrier properties, mechanically refined cellulose nanofibrils (CNFs) may have promising applications for oil-water separations. The unbleached form of these nanofibrils, lignin-containing CNFs (LCNFs), have also been found to display extraordinary barrier properties and are more environmentally friendly and cost-effective than CNFs. Herein, both wet and dry LCNF-modified filter papers have been developed by coating commercial filter paper with an LCNF suspension utilizing vacuum filtration. The LCNF-modified filters were tested for effectiveness in separating oil-water emulsions, and a positive relationship was discovered between a filter's LCNF coat weight and its oil collection capabilities. The filtration time was also analyzed for various coat weights, revealing a trend of high flux for low LCNF coat weights giving-way-to predictions of a coat weight upper limit. Additionally, it was found that wet filters tend to have higher flux values and oil separation efficiency values than dry filters of the same LCNF coat weight. Results confirm that the addition of LCNF to commercial filter papers has the potential to be used in oil-water separation.

Keywords: lignin-containing cellulose nanofibrils; oil-water separation; water filtration; surface modification

Citation: Mittag, A.; Rahman, M.M.; Hafez, I.; Tajvidi, M. Development of Lignin-Containing Cellulose Nanofibrils Coated Paper-Based Filters for Effective Oil-Water Separation. *Membranes* **2023**, *13*, 1. https://doi.org/10.3390/membranes13010001

Academic Editors: Hongjun Lin and Meijia Zhang

Received: 4 October 2022
Revised: 12 December 2022
Accepted: 13 December 2022
Published: 20 December 2022

Copyright: © 2022 by the authors. Licensee MDPI, Basel, Switzerland. This article is an open access article distributed under the terms and conditions of the Creative Commons Attribution (CC BY) license (https://creativecommons.org/licenses/by/4.0/).

1. Introduction

Due to rapid industrial and economic development, the need for oil and hence the prevalence of oil-water mixtures has increased dramatically in recent years. These mixtures have many damaging effects, however, threatening human health, disrupting ecosystems and the environment, and wasting valuable resources [1]. Specifically, oily wastewater pollution affects groundwater and drinking water, endangers human health, affects crop production, destructs the natural landscape, and contributes to atmospheric pollution [2]. With the international need for oil consistently increasing, it is likely that these issues of oily wastewater pollution will only compound in future years. This is incredibly problematic as it limits the amount of usable, clean water—an essential resource that is already scarce in many parts of the world. By the 2000s, 58% of the global population lived under some level of water scarcity and that number is only projected to increase [3]; hence methods of oil-water separation are crucial for the health of humans and the environment. Specifically, techniques of oil-water emulsion separation need to be improved as these mixtures are thermodynamically stable and therefore difficult to manipulate [1]. Current methods of separation include gravity separation, chemical dispersants, centrifugation, and flotation—all high energy consuming, costly, complex, and possibly polluting processes [4] and thus an energy-conscious, eco-friendly, and low-cost solution is needed.

One promising solution may be to incorporate cellulose nanofibrils (CNFs) into paper-based filters. Being nontoxic and renewable, paper is a desirable material to work with [5]. Traditional filtration approaches, however, have limited separation efficiency for stable emulsified oil-water mixtures [6]; due to their large pore size and limited wettability, commercial filter papers are not successful in effectively separating oil-water emulsions [4]. Recent research on CNFs has identified impressive oxygen and grease barrier properties that may allow us to functionalize the filter paper for use in these applications. Cellulose is advantageous due to its renewable nature, biodegradability, low cost, and nontoxicity [7], and it can be extracted from wood, plants, algae, bacteria, and even tunicates, a family of sea animals [8]. Cellulose is a linear homopolysaccharide linked by β 1–4 glucosidic bonds, with the molecular formula of $(C_6H_{10}O_5)_n$ [9]. By applying mechanical treatments such as grinding, cryocrushing, or microfluidization to both the amorphous and crystalline regions of cellulose, CNFs can be generated [8].

Recent work has found that CNFs can help produce films with increased barrier properties against oxygen and grease [10] due largely to the hydrogen bonding between hydroxyl groups [11] that forms a tight impermeable layered structure. While most literature on cellulose nanomaterials has focused on bleached nanofibrils, it is also possible to produce nanofibrils from unbleached fibers as well as recycled cardboard. These resulting nanofibrils are known as lignin-containing cellulose nanofibrils (LCNFs) and are produced from unbleached chemical pulps, thermo-mechanical pulps, or old corrugated containers (OCC) [12]. LCNFs offer many of the same advantages as CNFs but additionally have a lower production cost and environmental impact for a higher yield [13]. In fact, a previous study reported that producing 1 kg of LCNFs is 100 times cheaper than producing 1 kg of TEMPO-oxidized CNFs, a cellulose nanomaterial widely utilized in the literature [12]. There is also promising evidence that LCNF-based materials may even be superior to CNF-based materials in some applications. In recent experiments, LCNF-modified packaging has displayed excellent oil barrier properties, outperforming CNF-modified packaging, likely because of LCNF's lower polarity and surface energy [5]. Keeping all this in mind, it would be of great value to utilize LCNFs in oil-water separation as this implementation would have multi-faceted benefits.

Previous work that investigated self-standing CNF and LCNF films revealed their production is expected to be slow and energy-consuming [14]; hence an approach that modifies currently available commercial filter paper was taken in this study. One previously used method for applying CNFs and LCNFs to materials was to coat a surface and then thermally dry it [15,16]. Another study at Wuhan University [4] utilized tunicate cellulose nanocrystals to coat filter papers through physical and chemical (i.e., using a crosslinking agent) methods. Filters prepared via chemical crosslinking exhibited better oil separation than those prepared via physical modification. More recent studies have shown cellulose nanofibrils to be sturdier and easier to handle than nanocrystals [17], so this may be beneficial in regard to the reusability of our developed filters. Recently, other studies have found promising water-filtration results utilizing membranes made from natural resources. These include membranes formed out of polylactic acid and gelatin [18], chitosan-cellulose nanocrystals [19], date seed biomass [20], graphene oxide, sodium alginate, and lignin [21]. Other attempts involved incorporating hydrophobic polymers such as poly(perfluorooctylethyl methacrylate) or poly(methylhydrosiloxane) with cellulosic materials [22,23]. However, these treatments are often non-sustainable and may raise health concerns depending on the chemicals used in these polymeric materials. Despite the successful attempts, there is still a need to explore and provide a proof of concept of low-cost alternatives for oil/water separation.

The overarching goal of this study was to contribute knowledge to advance the development of an eco-friendly and low-cost filter using LCNFs for effective and efficient oil-water separation for use in oil-spill accidents and oily wastewater environments. In this work, a vacuum filtration technique was utilized to apply uniform layers of LCNFs to the filter paper without crosslinking agents or thermal drying, helping to minimize energy consumption and provide a low-cost and biodegradable option for oil filters. This approach was modeled after a previous study at the University of Maine [5] but differs from it by using filter paper as a starting material and by producing it for oil-water separation rather than packaging. Our objectives were to improve separation performance and optimize the time efficiency of paper filters through physical modification using LCNFs.

2. Materials and Methods

2.1. Materials

Whatman grade 5 filter paper (10 cm diameter and 2.5 μm pore size) was purchased and utilized throughout experiments, both as a control sample and as the base for LCNF-modified filters. LCNFs were obtained from the University of Maine's Process Development Center (PDC) and were made by mechanically refining old corrugated containers (OCC). The constituents of the OCC LCNFs were 61.86% cellulose, 18.05% hemicelluloses, and 16.67% lignin [24]. The as-received LCNF contained 2 wt% solids but was diluted to 0.1 wt% solids prior to use. For the oil/water emulsion, vegetable oil was purchased from the local grocery store, analytical grade Tween80 surfactant was obtained from MilliporeSigma (Burlington, MA, USA), and analytical grade red oil O ($C_{26}H_{24}N_4O$) dye was sourced from Alfa Aesar (Haverhill, MA, USA).

2.2. Preparation of LCNF-Modified Filter Papers

The first step in preparing wet and dry LCNF-modified filter papers was to produce an LCNF suspension. The LCNF suspension was prepared by diluting the as-received LCNFs to 0.1 wt% solids, sonicating the 0.1 wt% slurry for 3 min at 90% duty cycle and output control value 3 (Branson 450 Sonifier, Ultrasonics Corporation, Danbury, CT, USA), and then agitating the mixture using a planetary centrifugal mixer (Thinky 310, Thinky Corporation, Tokyo, Japan) by mixing for 1 min at 2200 rpm and then defoaming for 30 s at 2000 rpm. The next step was to deposit the LCNF suspension onto the filter's surface. After testing a multitude of coating methods, it was determined that the most effective way to evenly coat the filters with LCNF was to use vacuum filtration. To do so, a commercial filter paper was placed into a Buchner funnel (10.5 cm diameter) and coated with water to adhere to the funnel, then the LCNF suspension was poured on top of the filter using a glass stirring rod to ensure even distribution, the vacuum filtration was run at 20 inHg until the water in the suspension had successfully passed through the filter and the LCNFs were left on top. Multiple amounts of the 0.1 wt% LCNF suspension were utilized to create different coat weights on the filters. After removing the LCNF-modified filters from the funnel, they were either used immediately in oil-water separation testing (for the wet filters) or air-dried (for the dry filters). To make sure the dry filters remained flat while drying, they were restrained and weighed down by PVC rings and metal weights. Figure 1 displays a schematic of the filter modification process. Both wet and dry filters with LCNF coat weights spanning from 0 g per square meter (gm^{-2}) to 9 gm^{-2} were created and utilized in testing.

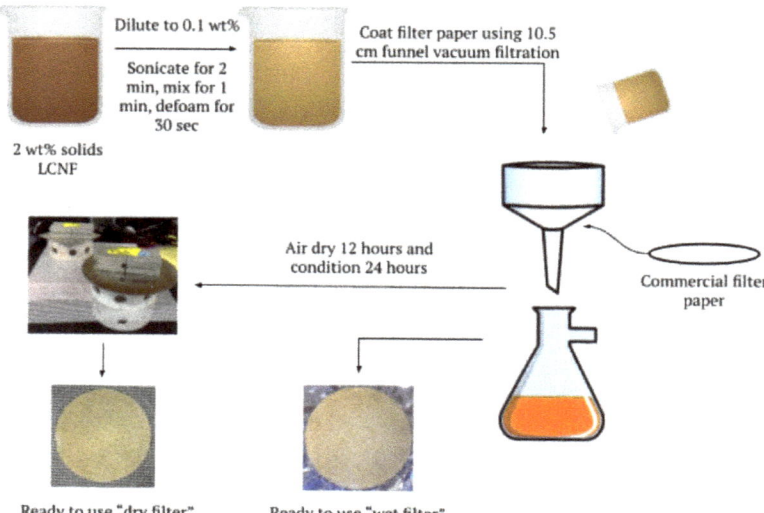

Figure 1. Summary of filter coating process of both wet and dry LCNF-modified filter papers.

2.3. Filter Characterization

The dry LCNF-modified filter paper coat weights were calculated using the mass differences of the filters before alterations and after the LCNF coating had dried. The wet LCNF-modified filter paper coat weights were calculated using the mass of LCNFs suspension that was applied to the commercial filters. In both cases, results were reported in grams of LCNF per square meter of filter paper (gm^{-2}).

The morphology of the LCNF-modified filter papers was evaluated through scanning electron microscopy (SEM) imaging. A Zeiss Nvision 40 scanning electron microscope (SEM; Oberkochen, Germany) machine was utilized to perform SEM imaging on an unmodified control filter as well as several dry LCNF-modified filters with varying coat weights. To prepare filters for SEM imaging, samples were cut using a sharp blade so that they could fit on the SEM sample stub using a double-sided carbon tape followed by a 4 nm of sputter coating of Au/Pd. SEM images were obtained at multiple magnifications. An electron high tension (EHT) voltage of 3 kV was maintained at the time of scanning. Only dry LCNF-modified filters could be visualized by SEM imaging as this SEM could not handle the moist nature of the wet LCNF-modified filters.

To determine the wettability of modified filter papers, a Krüss mobile surface analyzer (Krüss GmbH, Hamburg, Germany) was used to measure the contact angle of two drops of liquids—polar water and non-polar diiodomethane. Each drop was approximately 1 µL in volume and the contact angles were measured after 1 s of the drop being on the surface. After measuring the contact angles for each LCNF coat weight, the surface free energy (SFE) and its polar and disperse components were calculated using the Owens, Wendt, Rabel, and Kaelble (OWRK) model [25].

2.4. Preparation of Oil-Water Emulsions

Oil-in-water emulsions were prepared using a 1:99 (oil:water) mass ratio. To create a stable oil-water emulsion, a surfactant was needed. Surfactants are classified by hydrophilic-lipophilic balance (HLB), which is described as a numeric value that conveys the balance of the size and strength of two opposite groups, hydrophobic and lipophilic groups, in an emulsifier [26]. To create a stable oil-in-water emulsion, an HLB between 8 and 18 is needed. In our experiments, Polysorbate 80, a non-ionic surfactant commonly known as Tween80, was utilized. Tween80 has an HLB of 15 and thus was an ideal emulsifier for our experiment. For an 80 mL emulsion, 0.16 g of Tween80 was used. Additionally, red oil dye was utilized to help visualize results. The oil, water, surfactant, and dye were mixed for 1 min at 2200 rpm and then defoamed for 30 s at 2000 rpm in the planetary centrifugal mixer (Thinky 310, Thinky Corporation, Tokyo, Japan) to effectively prepare a homogenous emulsion. Using optical microscopy, the emulsion particle size was measured—on average oil particles were 6.8 microns in diameter—emulsions were determined to be stable if they did not separate after 15 min.

2.5. Oil-Water Separation Process

The oil-water separation tests were carried out on a vacuum filtration setup, using a 9.5 cm diameter funnel so that the filters formed a cup-like shape, ultimately preventing liquid from bypassing the filter. A total of 80 mL of the oil-water emulsion was poured over the LCNF-modified filter papers using a glass stir rod to ensure even distribution, and the vacuum filtration was performed under a pressure of 27 inHg. A container was used to collect the filtrate, as ideally the water passed through the filter and the oil was collected by the filter. The amount of filtrate collected was measured to compare with the initial quantity of liquid deposited onto the filter. Additionally, the time of filtration was recorded, with the filtration being considered complete when the time between two consecutive drops surpassed 10 s. After filtration, the used filter was placed in an oven at 80 °C for 2 h to allow the excess water to evaporate while still retaining the oil collected. These conditions proved to be an ample amount of time and a high enough temperature to evaporate the water in our experiments. After drying, the collected oil mass was calculated gravimetrically by subtracting the initial dry weight of the filter from its dry weight after filtration. A schematic representation of this oil-water separation procedure is displayed in Figure 2.

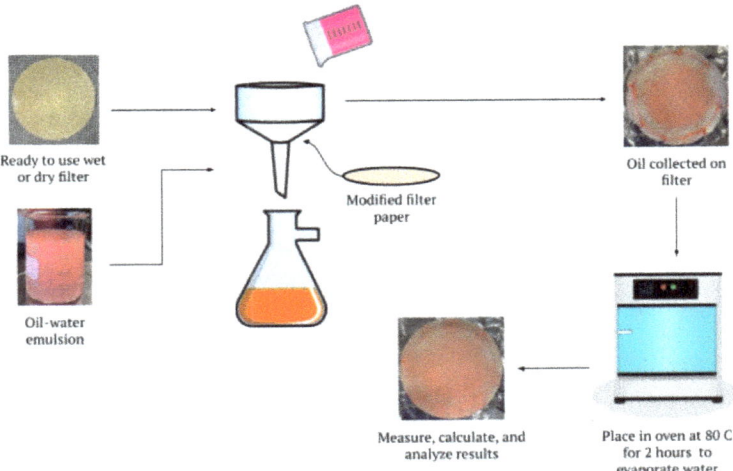

Figure 2. Summary of oil-water separation process using vacuum filtration.

Various equations were utilized in order to compare useful variables between the filters tested. The water flux, J (L m^{-2} h^{-1} bar^{-1}), was calculated using Equation (1):

$$J = V/(At\Delta P), \tag{1}$$

where V (L) is the permeated water volume, A (m^2) is the surface area of the funnel, t (h) is the drain time, and ΔP (bar) is the pressure across the filter paper.

The oil separation efficiency, R_1 (%), was calculated with Equation (2):

$$R_1 = (m_{oil\ f}/m_{oil\ i}) \times 100\% \tag{2}$$

in which $m_{oil\,f}$ (g) and $m_{oil\,i}$ (g) represented the mass of oil in the filtrate and the initial mass of oil used in the emulsion, respectively.

The water separation efficiency, R_2 (%), was calculated using Equation (3):

$$R_2 = (m_{water\,f}/m_{water\,i}) \times 100\% \quad (3)$$

where $m_{water\,f}$ (g) was the mass of water after the separation process and $m_{water\,i}$ (g) was the mass of water before the separation process.

Optical microscopy was also utilized in order to visualize oil droplets in the emulsions before and after filtration through the various filters. Image J software (U.S. National Institutes of Health, Bethesda, ML, USA) was used to estimate the size of the droplets.

3. Results and Discussion

3.1. Filtration Outcomes

As the LCNF coat weight increased on filter papers, so did the oil separation efficiencies, R_1. In other words, the more LCNF that was deposited on a filter, up to a coat weight of around 6.6 gm^{-2} gsm, the more oil was collected by it in the filtration process. At approximately 6.6 gm^{-2}, however, the increase in oil separation efficiencies seems to level off a bit. Figure 3 displays an oil collection graph in which the x-axis represents a filter's LCNF coat weight, and the y-axis represents the oil collection efficiency as calculated in Equation (2). As seen in the graph, an unmodified commercial filter could collect approximately 5% of the oil contained within an emulsion, while the LCNF-modified filters produced in this study collected up to 61% of the oil. This is due to the superior oil barrier properties of LCNFs that previous studies have identified [5]. While the mechanism for these properties is not well known, it is possible that lignin adds water resistance and crack-fold resistance. To further investigate the role of lignin, a control sample of dry CNF-modified filter paper was tested (coat weight: 7.5 gm^{-2}). At 7.5 gm^{-2}, the collected oil percent of the CNF-based filter was 75%, whereas that of the LCNF-modified filter was 51% at 6.6 gm^{-2}. The key difference between LCNF and CNF layers used in modifying the filter paper is the greater extent of hydrogen bonding within the CNF film as opposed to LCNF film, possibly resulting in a less porous coating layer. This result indicates the need for a tight network of micro- and nano-sized fibers to achieve a favorable oil separation. Based on this comparison, the role of lignin as a factor in the separation process is not clear. However, using LCNF-based materials instead of CNF-based materials is still favorable as they require less energy to produce.

As explained in the Methods section, the filters were prepared via dry and wet approaches. It was hypothesized that a wet filter could result in a better oil separation if nano-sized fibrils were collected on the fibers of filter paper, hence resulting in better separation efficiency. However, from Figure 3, wet LCNF-modified filters had comparable oil separation efficiencies, R_1, to dry LCNF-modified filters. Furthermore, it is worth noting that a re-wetted dry filter is not the same as a wet filter as during the drying process, shrinkage occurs and layers of LCNFs dry together, resulting in decreased swelling abilities when re-wetted (also known as hornification) [27]. Based on these findings, it is evident that the application of a uniform and tightly packed LCNF layer on the filter paper enabled the rejection of oil from the oil-water emulsion. However, further experiments are needed to verify whether or not adsorption contributes to the separation mechanism.

The visual results of the separation experiments also confirmed these findings. Figure 4 shows the oil-water emulsions before and after filtration through a variety of wet and dry LCNF-modified filters, as well as through an unmodified control filter paper. Both photos of the emulsions as well as optical microscopy images of the particles are included for the unmodified filter and three LCNF-modified filters (wet 2.44 gm^{-2}, wet 8.34 gm^{-2}, and dry 8.22 gm^{-2}). The oil in the emulsions was dyed red so that its presence could be clearly seen in the filtration process. It is apparent that the unmodified filter paper (Figure 4a) did a poor job removing oil from the emulsion, as seen in the high prevalence of red dye in the filtrate image and in the high number of large oil droplets in the post-filtration microscopy imaging. The low coat weight wet LCNF-modified filter (Figure 4b) also performed rather poorly, displaying visual results comparable to the unmodified filter—with the filtrate color and the post-filtration oil particle sizes being relatively similar. The two high coat weight LCNF-modified filters (Figure 4c,d) show the large effect of LCNFs on oil collection ability. The post-filtration photos in both these cases are much clearer than the previous examples, showing very little red dye, and the post-filtration microscopy images show much smaller oil droplets. Between these two filters, which have relatively the same coat weight, it is apparent that the wet LCNF-modified filter (Figure 4c) is more effective than the dry LCNF-modified filter (Figure 4d) in removing oil, as seen by the clearer filtrate and smaller oil particles of Figure 4c. The use of microscopy imaging to analyze the size and morphology of oil droplets was applied to a number of other LCNF-modified filters in order to further explore the effects of both coat weight and wet versus dry filter conditions. Table 1 displays the post-filtration average oil particle size (calculated using Image J) and coefficient of variation of these values for several wet and dry LCNF-modified filters, as well as an unmodified control filter. Few clear conclusions could be made from the values collected through microscopy images as a significant pattern did not emerge. For instance, the average oil particle size post-filtration using an unmodified filter was 8.4 µm, and using a dry LCNF-modified filter of coat weight 7.2 gm^{-2} it was nearly identical at 8.1 µm. Additionally, it may be possible that smaller droplets coalesce after passing through the filter to make larger droplets, hence increasing the average particle size seen through microscopy. One trend that was evident through microscopy images, however, was that dry LCNF-modified filters had a much lower coefficient of variation (ranging from 12.9–94.9%) than the wet LCNF-modified filters (ranging from 73.1–660.9%). The lower variation of the oil particles' sizes after passing through the dry LCNF-modified filters may be due to the LCNF coat layer being more uniform after drying. When re-wetting dry filters during filtration, the fibers have a decreased ability to swell [27], rendering them more uniform than never-dried filters.

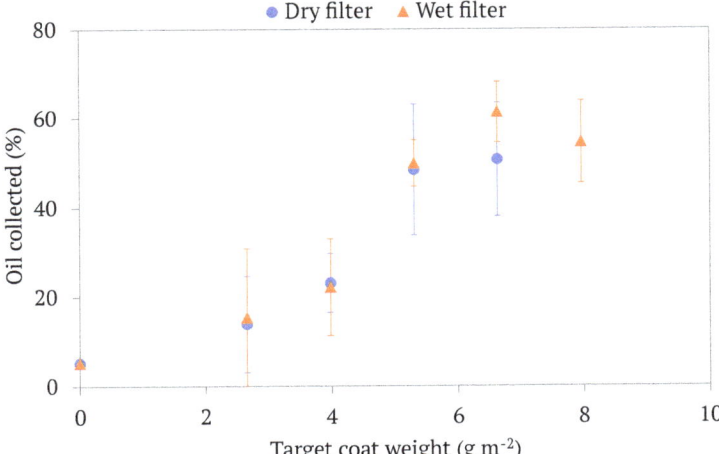

Figure 3. The oil separation efficiencies, calculated by Equation (2), of different LCNF-modified filters by coat weight. Both dry and wet filters are displayed. Each point represents the average oil separation efficiency of multiple trials of modified filters with the same target coat weight, with the standard deviations of said values being displayed in the error bars.

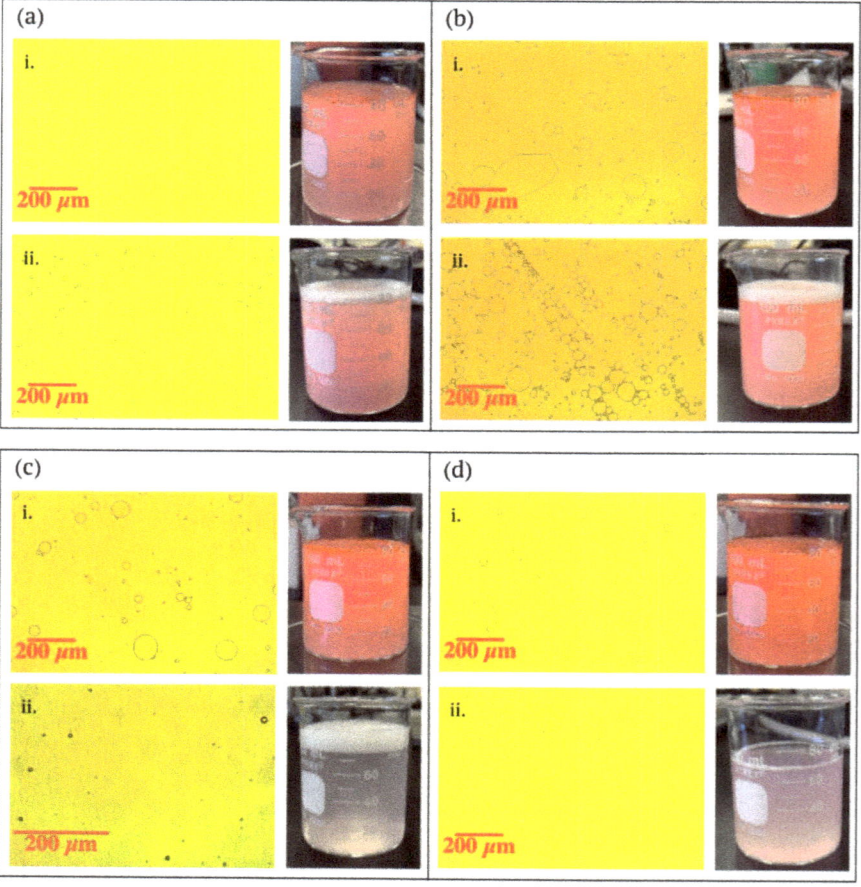

Figure 4. Photographs and microscopy imaging of oil-water emulsions before (i) and after (ii) filtration through an (**a**) unmodified control filter, (**b**) wet 2.44 gm^{-2} filter, (**c**) wet 8.34 gm^{-2} filter, and (**d**) dry 8.22 gm^{-2} filter.

Table 1. The average size of oil particles in filtrates after running an oil-water emulsion through wet and dry filters at various LNCF coat weights.

Wet Filters			Dry Filters		
Coat Weight (gm^{-2})	Average Oil Particle Size (μm)	Coefficient of Variation (%)	Coat Weight (gm^{-2})	Average Oil Particle Size (μm)	Coefficient of Variation (%)
0	8.4	73.1	0	8.4	73.1
2.44	7.0	158.0	7.16	8.1	94.9
3.79	2.3	660.9	8.22	4.7	74.8
7.49	5.5	181.6	8.76	2.8	60.0
8.34	1.0	170.3	10.62	2.7	12.9

While oil collection is of utmost importance in the development of our filters for oil-water separation, it is crucial to balance collection effectiveness with time efficiency so that the filters are usable in the real world. As expected, filters with a higher coat weight of LCNFs had a longer filtration time, t. Figure 5 shows this trend, displaying filtration time curves in which the x-axis represents a filter's LCNF coat weight, and the y-axis represents the time to filter 80 mL of an emulsion through said filter. Trendlines for both the wet and dry LCNF-modified filters exhibit positive slopes, illustrating that filtration time increases greatly with increased LCNF coat weights. Another finding from our study was that for filters with similar coat weights, wet LCNF-modified filters had a shorter filtration time, t, than dry LCNF-modified filters. One possible explanation for this finding is that as the filter dries, the space between layers of LCNF decreases, and thus there is less free space for water to travel, so there is simply more liquid trying to go through a tighter space.

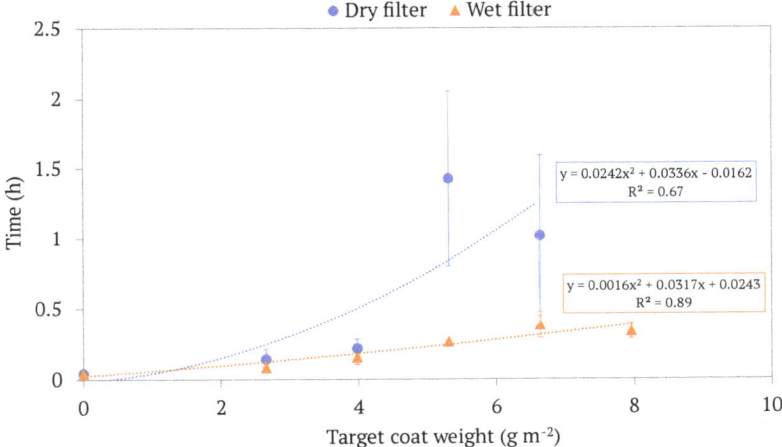

Figure 5. The filtration time of different LCNF-modified filters by coat weight. Both dry and wet filters are displayed, each set with its own respective trendline. Each point represents the average filtration time of multiple trials of modified filters with the same target coat weight, with the standard deviations of said values being displayed in the error bars.

With the efficiency of time and energy in mind, it is critical to determine an upper limit coat weight, i.e., the maximum amount of LCNF that can be applied to a filter without massively inhibiting its ability to allow water to pass through. To aid in this analysis, we calculated flux values, J, of each filter tested using Equation (1). Figure 6 displays a graph of the filters' flux values in relation to their coat weight. Filters with a higher coat weight displayed a lower flux than filters with a low coat weight, and dry LCNF-modified filters displayed a lower flux (on average) than wet LCNF-modified filters. By incorporating trendlines into the flux graphs, we are able to visualize the coat weight at which water flow levels off, helping to predict the maximum coat weight of LCNF that one could apply to a filter paper in both wet and dry conditions. This is an important aspect of the study as it is the balance between oil collection efficiency (which increases with coat weight) and flux (which decreases with coat weight) that will enable a filter to be both effective and realistically usable. The flux of both dry and wet LCNF-modified filters appeared to level off around a coat weight of 5.3 gm^{-2}. While previous works have not investigated the flux of filter papers coated in OCC LCNFs, there have been studies that utilize tunicate cellulose nanocrystals (TCNCs) and bamboo-based LCNFs in similar applications [4,28]. Flux rates are much higher in each of these studies (reaching up to 317.7 L m^{-2} h^{-1} bar^{-1})

than the values found in this experiment, but this is likely because both the TCNCs and the bamboo-based LCNFs are able to coat the inside of the filter paper pores rather than just coating the surface of the paper as we are. While using OCC LCNF to coat commercial filter paper means our fluxes are much lower than these reported values, it also means the process of modifying filters is simpler and more cost-effective.

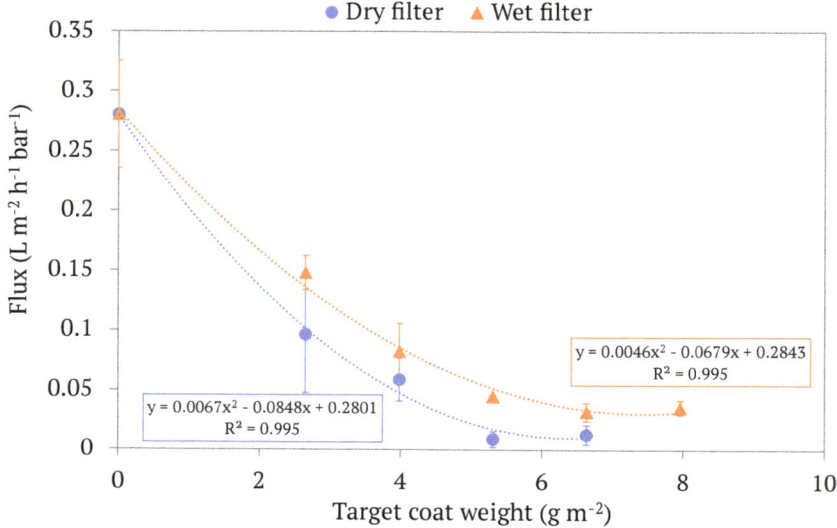

Figure 6. The flux values, calculated by Equation (1), of different LCNF-modified filters by coat weight. Both dry and wet filters are displayed, each set with its own respective trendline. Each point represents the average flux of multiple trials of modified filters with the same target coat weight, with the standard deviations of said values being displayed in the error bars.

Another important factor to consider regarding our filters' usability and efficiency is what we call the water separation efficiency, R_2, as noted in Equation (3). Water separation efficiency is defined as the percent of water originally in the emulsion that is recovered after the filtration process. A high water separation efficiency means that most of the water was able to pass through the filter, while a low value would signify that some water was collected with the oil (ultimately wasting it). The water separation efficiencies, R_2, for various filters are displayed in Figure 7. The unmodified control filters had water separation efficiencies around 98%, meaning that even in uncoated filters approximately 2% of the water contained in the emulsion was lost. The water separation efficiencies of wet LCNF-modified filters remained relatively constant even as coat weight increased (R_2 values ranged from 95.98–98.29%), while those of dry LCNF-modified filters decreased as coat weight increased (with R_2 values below 90% for all of the higher coat weights). This means that the wet LCNF-modified filters waste less water in the separation process, a finding once again is contributed to the smaller pores and more densely packed layers characteristic of dry LCNF-modified filters.

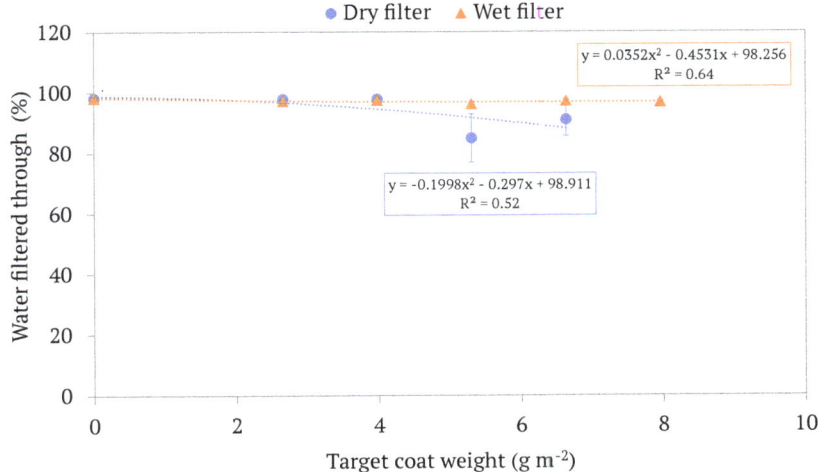

Figure 7. The water separation efficiency values, calculated by Equation (3), of different LCNF-modified filters by coat weight. Both dry and wet filters are displayed, each set with its own respective trendline. Each point represents the average water separation efficiency of multiple trials of modified filters with the same target coat weight, with the standard deviations of said values being displayed in the error bars.

3.2. Filter Properties

SEM images of the surface of an unmodified filter and of dry LCNF-modified filters of various coat weights were captured and are displayed in Figure 8. The commercially available filter paper was composed of heterogeneous microfibers with distinct borders. SEM imaging and analysis clearly showed the entangled network of individual fibers within the unmodified filter paper (Figure 8a). The LCNF-modified filter papers, on the other hand, displayed a more dense and uniform morphology due to the formation of a tight LCNF layer on the surface. SEM images of the LCNF-modified filters show a much smoother filter surface (Figure 8b–e) than the control filter (Figure 8a). As the LCNF coat weight of filters increased, the size of the pores in the filters visibly decreased and the presence and entanglement of LCNFs surrounding filter paper fibers increased. While the filter lightly coated in LCNF (3.11 gm^{-2}) still had a number of voids visible in the 85× magnified images (Figure 8b), the filter most heavily coated in LCNF (9.54 gm^{-2}) had no visible voids (Figure 8e), even at the 500× magnification we utilized. The tight network created by high coat weights of LCNF is believed to be one of the reasons responsible for creating barrier properties against a number of substances [29]. While SEM imaging could only be performed on the dry LCNF-modified filters, it is assumed that the wet LCNF-modified filters had similar trends in decreasing pore size with increasing LCNF coat weights. One potential difference, however, between the morphologies of the dry and wet LCNF-modified filters could be that wet filters are packed comparatively less tightly. This assumption is due to the fact that LCNF experiences shrinkage after drying, and therefore it is likely that the pores of dry LCNF-modified filters would be slightly smaller than the pores of wet LCNF-modified filters of relatively similar coat weights.

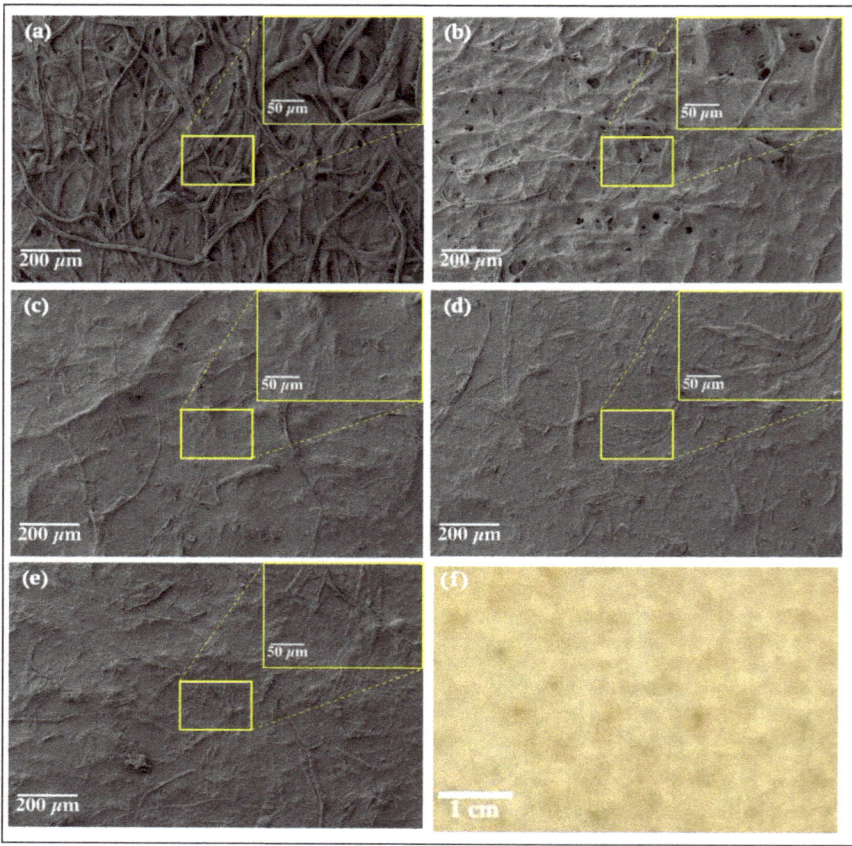

Figure 8. Surface SEM images of dry LCNF-modified filters at various coat weights: (**a**) uncoated control filter, (**b**) 3.11 gm^{-2}, (**c**) 5.49 gm^{-2}, (**d**) 6.02 gm^{-2}, and (**e**) 9.54 gm^{-2} at different magnifications (85× and 500×). (**f**) Magnified image of dry filter paper.

Surface free energies (SFE) were calculated in order to evaluate the barrier properties of the various filters, as SFE has a great influence on a material's wetting and adsorption of water, oil, and grease [13]. Previous literature has shown that LCNF promotes a larger water contact angle and lower surface energy than CNF due to lignin creating greater water repellency [5]. Table 2 summarizes the water contact angle, diiodomethane contact angle, surface free energy, dispersive free energy, and polar surface free energy of filters coated in various weights of LCNF. Once again, only dry LCNF-modified filters were characterized due to the limitations of the Krüss mobile surface analyzer. Values for the unmodified commercial filters could not be measured either due to these filters' extreme porosity or high hydrophilicity. Our findings show that filters with higher LCNF coat weights tend to have lower SFE values. These SFE values (which range from 39.5 mN m^{-1} to 62.8 mN m^{-1}) are comparable with those reported in the literature, with a prior study finding SFEs of LCNF-coated materials ranging from (43.6–62.48 mN m^{-1}) [30]. The SFE of our filter with a low LCNF coat weight was much higher than our other filters' SFE values. This is likely due to the fact that our lightly coated filter had an LCNF coat weight of 2.87 gm^{-2}, which as seen in previous sections, leaves many free pores in the filter surface, while our other filters have an LCNF coat weight between 5.49–9.23 gm^{-2}.

Table 2. Water contact angles, diiodomethane contact angles, and surface free energy and its components for dry LCNF-modified filters at various coat weights.

Coat Weight (gm^{-2})	2.87	5.49	6.02	9.23
Water contact angle (°)	49.9 (±11.4)	78.5 (±6.1)	81.1 (±5.7)	84.9 (±4.4)
Diiodomethane contact angle (°)	19.2 (±5.4)	46.1 (±3.1)	43.0 (±4.5)	45.0 (±6.2)
Surface free energy (mN m^{-1})	62.8 (±7.5)	41.2 (±4.0)	41.5 (±4.3)	39.5 (±4.7)
Dispersive surface energy (mN m^{-1})	48.0 (±1.5)	36.4 (±1.7)	38.1 (±2.4)	37.0 (±3.3)
Polar surface energy (mN m^{-1})	14.8 (±6.0)	4.7 (±2.4)	3.4 (±1.9)	2.5 (±1.4)

4. Conclusions

Coating commercial filter papers with LCNFs can improve oil-water separation capabilities. Wet LCNF-modified filters collected up to 61% of oil, while dry LCNF-modified filters collected up to 51% of oil in experiments. Both of these modification techniques resulted in oil collection improvement, however, as unmodified filters only collected up to 5% of the oil. Wet LCNF-modified filters exhibited a higher flux than dry LCNF-modified filters, allowing for more time- and energy-efficient processes. Water waste was also lower when wet LCNF-modified filters were used compared to dry LCNF-modified filters, with water separation efficiency values above 95% for wet LCNF-modified filters but some water separation efficiency values falling below 90% for the dry LCNF-modified filters. Increasing the LCNF coat weight increased the oil collection in both the wet LCNF-modified and dry LCNF-modified filters. At the same time, however, flux decreased as LCNF coat weight increased. From SEM images, we can see both these trends are caused by the smaller pores created by densely packed and entangled lignin-containing cellulose nanofibrils in highly coated films. A surface analysis of the modified filters showed that filters more densely coated with LCNF displayed lower surface free energies than lightly coated filters, 9.23 gm^{-2} filters had SFEs of 39.5 mN m^{-1}, while 2.87 gm^{-2} filters had SFEs of 62.8 mN m^{-1}, which also helps to explain the findings of this study. The modification techniques described in this work are low-cost, readily available, easily replicable, and energy-efficient, thus showing promise for a broader impact. Given well-established platforms for coating with CNF-based materials, there is potential for scale-up applications, but that is not within the scope of this project. Additionally, the filters and LCNFs are biodegradable and thus can decompose naturally without leaving a larger footprint. Ultimately, this method of modifying commercial filter papers with LCNFs can help produce a filter that is more economical, environmentally friendly, and attainable than many other oil-water filtration technologies. Future work will involve collecting real-time data for an extended period of time to gain further insights into the mechanism of separation.

Author Contributions: Conceptualization, M.T. and I.H.; methodology, A.M., M.M.R., I.H. and M.T.; validation, A.M. and M.M.R.; formal analysis, A.M. and M.M.R.; investigation, A.M. and M.M.R.; resources, M.T. and I.H.; writing—original draft preparation, A.M.; writing—review and editing, A.M., M.M.R., I.H. and M.T.; supervision, M.T. and I.H. All authors have read and agreed to the published version of the manuscript.

Funding: This research was funded by National Science Foundation (NSF) REU Award EEC 1757529 "Explore It! Building the Next Generation of Sustainable Forest Bioproduct Researchers".

Institutional Review Board Statement: Not applicable.

Informed Consent Statement: Not applicable.

Data Availability Statement: Not applicable.

Acknowledgments: The authors would like to thank the National Science Foundation (NSF) REU Award EEC 1757529 "Explore It! Building the Next Generation of Sustainable Forest Bioproduct Researchers." for funding this research.

Conflicts of Interest: The authors declare no conflict of interest.

References

1. Zhu, W.; Huang, W.; Zhou, W.; Qiu, Z.; Wang, Z.; Li, H.; Wang, Y.; Li, J.; Xie, Y. Sustainable and antibacterial sandwich-like Ag-Pulp/CNF composite paper for oil/water separation. *Carbohydr. Polym.* **2020**, *245*, 116587. [CrossRef] [PubMed]
2. Yu, L.; Han, M.; He, F. A review of treating oily wastewater. *Arab. J. Chem.* **2017**, *10*, S1913–S1922. [CrossRef]
3. Kummu, M.; Guillaume, J.H.; de Moel, H.; Eisner, S.; Flörke, M.; Porkka, M.; Siebert, S.; Veldkamp, T.I.; Ward, P.J. The world's road to water scarcity: Shortage and stress in the 20th century and pathways towards sustainability. *Sci. Rep.* **2016**, *6*, 38495. [CrossRef] [PubMed]
4. Huang, Y.; Zhan, H.; Li, D.; Tian, H.; Chang, C. Tunicate cellulose nanocrystals modified commercial filter paper for efficient oil/water separation. *J. Membr. Sci.* **2019**, *591*, 117362. [CrossRef]
5. Tayeb, A.H.; Tajvidi, M.; Bousfield, D. Paper-based oil barrier packaging using lignin-containing cellulose nanofibrils. *Molecules* **2020**, *25*, 1344. [CrossRef]
6. Chen, C.; Weng, D.; Mahmood, A.; Chen, S.; Wang, J. Separation mechanism and construction of surfaces with special wettability for oil/water separation. *ACS Appl. Mater. Interfaces* **2019**, *11*, 11006–11027. [CrossRef] [PubMed]
7. Li, Z.; Zhong, L.; Zhang, T.; Qiu, F.; Yue, X.; Yang, D. Sustainable, flexible, and superhydrophobic functionalized cellulose aerogel for selective and versatile oil/water separation. *ACS Sustain. Chem. Eng.* **2019**, *7*, 9984–9994. [CrossRef]
8. Moon, R.J.; Martini, A.; Nairn, J.; Simonsen, J.; Youngblood, J. Cellulose nanomaterials review: Structure, properties and nanocomposites. *Chem. Soc. Rev.* **2011**, *40*, 3941–3994. [CrossRef]
9. Habibi, Y.; Lucia, L.A.; Rojas, O.J. Cellulose nanocrystals: Chemistry, self-assembly, and applications. *Chem. Rev.* **2010**, *110*, 3479–3500. [CrossRef]
10. Hubbe, M.A.; Pruszynski, P. Greaseproof paper products: A review emphasizing ecofriendly approaches. *BioResources* **2020**, *15*, 1978–2005. [CrossRef]
11. Tayeb, A.H.; Amini, E.; Ghasemi, S.; Tajvidi, M. Cellulose nanomaterials—Binding properties and applications: A review. *Molecules* **2018**, *23*, 2684. [CrossRef] [PubMed]
12. Delgado-Aguilar, M.; González, I.; Tarrés, Q.; Pèlach, M.À.; Alcalà, M.; Mutjé, P. The key role of lignin in the production of low-cost lignocellulosic nanofibres for papermaking applications. *Ind. Crops Prod.* **2016**, *86*, 295–300. [CrossRef]
13. Rojo, E.; Peresin, M.S.; Sampson, W.W.; Hoeger, I.C.; Vartiainen, J.; Laine, J.; Rojas, O.J. Comprehensive elucidation of the effect of residual lignin on the physical, barrier, mechanical and surface properties of nanocellulose films. *Green Chem.* **2015**, *17*, 1853–1866. [CrossRef]
14. Horseman, T.; Tajvidi, M.; Diop, C.I.; Gardner, D.J. Preparation and property assessment of neat lignocellulose nanofibrils (LCNF) and their composite films. *Cellulose* **2017**, *24*, 2455–2468. [CrossRef]
15. Mazhari Mousavi, S.M.; Afra, E.; Tajvidi, M.; Bousfield, D.W.; Dehghani-Firouzabadi, M. Cellulose nanofiber/carboxymethyl cellulose blends as an efficient coating to improve the structure and barrier properties of paperboard. *Cellulose* **2017**, *24*, 3001–3014. [CrossRef]
16. Kumar, V.; Elfving, A.; Koivula, H.; Bousfield, D.; Toivakka, M. Roll-to-Roll Processed Cellulose Nanofiber Coatings. *Ind. Eng. Chem. Resour.* **2016**, *55*, 3603–3613. [CrossRef]
17. Wang, L.; Chen, C.; Wang, J.; Gardner, D.J.; Tajvidi, M. Cellulose nanofibrils versus cellulose nanocrystals: Comparison of performance in flexible multilayer films for packaging applications. *Food Packag. Shelf Life* **2020**, *23*, 100464. [CrossRef]
18. Yang, C.; Topuz, F.; Sang-Hee, P.; Szekely, G. Biobased thin-film composite membranes comprising priamine–genipin selective layer on nanofibrous biodegradable polylactic acid support for oil and solvent-resistant nanofiltration. *Green Chem.* **2022**, *13*, 5291–5303. [CrossRef]
19. Wang, Y.; He, Y.; Li, H.; Yu, J.; Zhang, L.; Chen, L.; Bai, Y. Layer-by-layer construction of CS-CNCs multilayer modified mesh with robust anti-crude-oil-fouling performance for efficient oil/water separation. *J. Membr. Sci.* **2021**, *639*, 119776. [CrossRef]
20. Alammar, A.; Hardian, R.; Szekely, G. Upcycling agricultural waste into membranes: From date seed biomass to oil and solvent-resistant nanofiltration. *Green Chem.* **2021**, *1*, 365–374. [CrossRef]
21. Jiang, Y.; Zhang, Y.; Gao, C.; An, Q.; Xiao, Z.; Zhai, S. Superhydrophobic aerogel membrane with integrated functions of biopolymers for efficient oil/water separation. *Sep. Purif. Technol.* **2022**, *282*, 120138. [CrossRef]
22. Liu, Z.; Yu, J.; Lin, W.; Yang, W.; Li, R.; Chen, H.; Zhang, X. Facile method for the hydrophobic modification of filter paper for applications in water-oil separation. *Surf. Coat. Technol.* **2018**, *352*, 313–319. [CrossRef]
23. Wu, H.; Wu, L.; Lu, S.; Lin, X.; Xiao, H.; Ouyang, X.; Cao, S.; Chen, L.; Huang, L. Robust superhydrophobic and superoleophilic filter paper via atom transfer radical polymerization for oil/water separation. *Carbohydr. Polym.* **2018**, *181*, 419–425. [CrossRef] [PubMed]
24. Amini, E.; Hafez, I.; Tajvidi, M.; Bousfield, D. Cellulose and lignocellulose nanofibril suspensions and films: A comparison. *Carbohydr. Polym.* **2020**, *250*, 117011. [CrossRef] [PubMed]
25. Kaelble, D.H. Dispersion-Polar Surface Tension Properties of Organic Solids. *J. Adhes.* **1970**, *2*, 66–81. [CrossRef]
26. Griffin, W.C. Classification of surface-active agents by HLB. *Off. J. Soc. Cosmet. Chem.* **1949**, *1*, 311–326.
27. Ding, Q.; Zeng, J.; Wang, B.; Tang, D.; Chen, K.; Gao, W. Effect of nanocellulose fiber hornification on water fraction characteristics and hydroxyl accessibility during dehydration. *Carbohydr. Polym.* **2019**, *207*, 44–51. [CrossRef]

28. Yang, S.; Wang, T.; Tang, R.; Yan, Q.; Tian, W.; Zhang, L. Enhanced permeability, mechanical and antibacterial properties of cellulose acetate ultrafiltration membranes incorporated with lignocellulose nanofibrils. *Int. J. Biol. Macromol.* **2020**, *151*, 159–167. [CrossRef]
29. Yook, S.; Park, H.; Park, H.; Lee, S.Y.; Kwon, J.; Youn, H.J. Barrier coatings with various types of cellulose nanofibrils and their barrier properties. *Cellulose* **2020**, *27*, 4509–4523. [CrossRef]
30. Hossain, R.; Tajvidi, M.; Bousfield, D.; Gardner, D.J. Multi-layer oil-resistant food serving containers made using cellulose nanofiber coated wood flour composites. *Carbohydr. Polym.* **2021**, *267*, 118221. [CrossRef]

Disclaimer/Publisher's Note: The statements, opinions and data contained in all publications are solely those of the individual author(s) and contributor(s) and not of MDPI and/or the editor(s). MDPI and/or the editor(s) disclaim responsibility for any injury to people or property resulting from any ideas, methods, instructions or products referred to in the content.

Article

Enhancing the Performance of PVDF/GO Ultrafiltration Membrane via Improving the Dispersion of GO with Homogeniser

Xin Sun [1], Hana Shiraz [2], Riccardo Wong [2], Jingtong Zhang [1], Jinxin Liu [1], Jun Lu [1] and Na Meng [1,2,*]

[1] School of Environmental Engineering, Xuzhou University of Technology, Xuzhou 221018, China
[2] Department of Chemical Engineering, Monash University, Melbourne, VIC 3800, Australia
* Correspondence: mengna309@163.com

Abstract: In this study, PVDF/GO-h composite membranes were synthesised using a homogeniser to improve the dispersion of GO nanosheets within the composite membrane's structure, and then characterised and contrasted to PVDF/GO-s control samples, which were synthesised via traditional blending method-implementing a magnetic stirrer. By characterizing membrane via X-ray diffraction (XRD), Fourier transform infrared spectroscopy (FTIR), scanning electron microscopy (SEM), water contact angle (WCA) and membrane performance. SEM results showed that the number of the finger-like structure channels and pores in the sponge like structure of PVDF/GO-h composite membranes become more compared with PVDF/GO-s membranes. Water contact angle tests showed that the PVDF/GO-h composite membranes have lower contact angle than PVDF/GO-s control, which indicated the PVDF/GO-h composite membranes are more hydrophilic. Results also showed that composite membranes blended using homogeniser exhibited both improved water flux and rejection of target pollutants. In summary, it was shown that the performance of composite membranes could be improved significantly via homogenisation during synthesis, thus outlining the importance of further research into proper mixing.

Keywords: homogeniser; PVDF; GO nanosheets; ultrafiltration membrane

Citation: Sun, X.; Shiraz, H.; Wong, R.; Zhang, J.; Liu, J.; Lu, J.; Meng, N. Enhancing the Performance of PVDF/GO Ultrafiltration Membrane via Improving the Dispersion of GO with Homogeniser. *Membranes* **2022**, *12*, 1268. https://doi.org/10.3390/membranes12121268

Academic Editor: Hongjun Lin

Received: 20 November 2022
Accepted: 9 December 2022
Published: 15 December 2022

Publisher's Note: MDPI stays neutral with regard to jurisdictional claims in published maps and institutional affiliations.

Copyright: © 2022 by the authors. Licensee MDPI, Basel, Switzerland. This article is an open access article distributed under the terms and conditions of the Creative Commons Attribution (CC BY) license (https://creativecommons.org/licenses/by/4.0/).

1. Introduction

Ultrafiltration (UF) processes have received increased attention in liquid separation in the past several decades, especially in wastewater treatment, medical, food, chemical and biochemical fields [1]. Compared to micro filtration membrane, ultrafiltration membranes prepared by ultrafiltration technology have a smaller surface pore size, between 1 and 100 nm [2], and can remove macromolecular organic matter (protein, bacteria), colloids, suspended solids [3], which makes ultrafiltration membranes play a key role in protein purification and separation.

Based on its excellent chemical resistance, antioxidation activity, thermal stability and membrane forming properties, polyvinylidene fluoride (PVDF), a semi-crystalline material, is used as an UF membrane in wastewater treatment [4–7]. However, due to the inherent hydrophobicity of PVDF material, the membrane prepared by PVDF often has serious membrane contamination, which is caused by the physical or chemical interaction between the membrane surface and the macromolecules or microorganisms in the separation solution during the membrane separation process [8]. Based on the hydrophobicity of PVDF, PVDF films tend to have a higher scaling tendency than hydrophilic films with similar separation characteristics and pore size [9]. An effective approach to solving this problem is to integrate nanomaterials into the PVDF membrane. The PVDF ultrafiltration membrane prepared by H. Younas et al. [10] by adding inorganic TiO_2 nanoparticles (NPs) has good hydrophilicity and flux, and also has a high rejection rate of humic acid (HA). In addition, the study showed that the PVDF hybrid membrane containing vermiculite nanoparticles

(Verm NPs) was prepared by the opposite method, which had higher anti-pollution performance [11]. These results indicate that the hydrophilic, permeable and antifouling properties can be improved by incorporating organic materials into PVDF polymers [12].

Graphene oxide (GO)surface contains rich groups, such as carboxyl group, hydroxyl group and epoxy group [13], which makes GO have good hydrophilicity. In addition, GO also has good mechanical strength, electrical conductivity, alkaline resistance and other excellent physical and chemical properties. Due to its excellent properties, graphene oxide is widely used in membrane separation. Therefore, graphene oxide and its derivates, as a nanofiller, may be preferred over other nanofillers owing to high aspect ratio, hydrophilicity, tensile strength, thermal conductivity and electrical conductivity [9,14–16]. The superior properties of graphene compared to polymers are also reflected in polymer/graphene nanocomposites membranes [4,12,17]. It was reported that the performance of polymer membrane was enhanced after GO was embedded [1,3,18–25]. Most polymer/GO hybrid membranes are prepared via an electronic stirring [1,3,18–28]. However, GO sheets can only be dispersed in aqueous media, which is incompatible with most organic polymers; this kind of polymer/GO membranes face the problem of poor distribution of graphene oxide into polymer [4,20,21,29]. The improvement in the properties of the nanocomposites depends on the distributions of graphene oxide layers in the polymer matrix as well as interfacial bonding between the graphene oxide layers and polymer matrix [4]. Therefore, the good distribution of GO nanosheets in the polymer matrix is of great significance to improve the performance of hybrid membranes.

Homogeniser is a commonly used mechanical method to reduce the particle in material field [30–35]. For example, Long et al. [34] used a high-speed homogeniser to treat the microcrystalline cellulose (MCC). Scanning electron microscope (SEM) results showed that the particle size of the MCC was reduced from micrometre scale down to nanoscale. Sun-Young et al. [32] prepared cellulose nanofibrils by employing a high-pressure homogenizer, and SEM results showed that the complete fibrosis of the bulk cellulose fibrils to nanoscale with high aspect ratio was accomplished by homogenization process. T. J. Nacken et al. [36] used a high pressure homogeniser to produce graphene and few layers of graphene (FLG) in a mixture of methyl pyrrolidone and water-surfactant. It was found that the high pressure homogeniser could obtain a high enough concentration of FLG suspension with low defect concentration. To the best of our knowledge, no previous study has been conducted to prepare GO/PVDF hybrid membranes using a homogeniser to disperse GO nanosheets with PVDF. Therefore, this study aims to fill this research gap.

This study aimed to fabricate a high performance PVDF/GO membrane with better GO distribution via using a homogeniser. For comparison, PVDF/GO hybrid membranes were fabricated by both conventional magnetic stirring method (PVDF/GO-s) and homogeniser dispersing method (PVDF/GO-h) and compared with PVDF membrane. XRD and FTIR analysis were conducted to ensure GO nanosheets were successfully incorporated into PVDF membranes. The WCA, water flux and rejection of hybrid membranes were tested to study the effect of GO distribution on the performance of hybrid membranes. The results of this study could shed light on the synthesis of nanomaterials incorporated membranes, which have promising application in liquid separation.

2. Experimental Section
2.1. Materials

Natural graphite power, sodium nitrate ($NaNO_3$), sulfuric acid (H_2SO_4), potassium permanganate ($KMnO_4$), hydrogen peroxide (H_2O_2) and hydrochloric acid (HCl; 32%) were purchased from Sigma Aldrich (St. Louis, MI, USA). PVDF (FR-904) was obtained from Shanghai 3 F new materials Co., Ltd. (Shanghai, China). The molecular weight (Mw) of PVDF is approximately 1.02×10^6 g/mol, measured by GPC (waters, 515). N,N-Dimethylacetamide (DMAc) and polyethylene glycol (PEG; with MW of 3,535,000 g/mol) were purchased from Sigma Aldrich (St. Louis, MI, USA).

2.2. Preparation of GO Nanosheets, PVDF, PVDF/GO-s and PVDF/GO-h Homogenised Membranes

2.2.1. Preparation of GO Nanosheets

Graphene oxide (GO) was prepared using the modified Hummers' method from graphite powder. The synthesis procedure was reported in our previous research [37]. Briefly, NaNO$_3$ (1.25 g) and natural graphite (2.5 g) were first mixed in an ice water bath. Then 60 mL of sulfuric acid was added. After 30 min, 7.5 g of KMnO$_4$ was added into the mixture. After that, the ice water bath was removed, and the mixture was further stirred overnight at room temperature. Subsequently, 135 mL of deionised (DI) water and 25 mL of H$_2$O$_2$ were added in sequence. A bright yellow mixture was obtained after the solution cooled down. GO nanosheets were obtained after the mixture was washed several times. The obtained GO nanosheets are consistent with the results of previous experiments [38].

2.2.2. Preparation of PVDF Membranes

PVDF membranes were synthesised via a phase inversion method following the these steps: 3 g PVDF powder and 17 g of DMAc solution were added into a 25 mL glass vial. Afterwards, the mixture was stirred for approximately 24 h (overnight) on a magnetic stirrer in a 50 °C oil bath. When the casting solution was fully dissolved, the vial was taken out from the oil bath and allowed to rest at room temperature for another 12 h to remove the bubbles within the solution. Then, the PVDF casting solution was cast on a clean and oven-dried glass plate by using a casting knife (Elcometer 3580) with a gap of 200 μm. The whole composite was then immediately immersed in a coagulation bath of water and allowed to sit for 15 min to enable phase inversion to occur. Subsequently, the support membrane was transferred into deionised (DI) water before further use.

2.2.3. Preparation of PVDF/GO-s and PVDF/GO-h Membranes

The synthetic procedure of PVDF/GO hybrid membrane was the same as that of PVDF membrane, except that 0.03 g GO was added into 3 g PVDF powder and 17 g DMAc solution to form 0.15 wt% GO casting solution. The GO-containing solution was stirred for 24 h on a magnetic stirrer in a 50 °C oil bath to completely disperse GO into the casting solution and then was allowed to rest at room temperature for another 12 h. For PVDF/GO-h membranes, after stirring for 24 h, the solution was homogenised by a homogeniser (AD500S-H) for another 5 min at 2000 rpm to further disperse GO nanosheets. And then the PVDF/GO-h solution was allowed to rest at room temperature for at least two days or until all bubbles within the solution have disappeared. Finally, both solutions were casted on glass plates to allow phase inversion to occur.

2.3. Characterization of PVDF and PVDF/GO Membranes

The functional groups and structure of PVDF and PVDF/GO membranes were characterised by powder X-ray diffraction (PXRD; Rigaku Mini Flex, Cu Kα radiation, Tokyo, Japan) and Fourier transform infrared spectrometer (FTIR spectrometer, PerkinElmer, Waltham, MA, USA). The surface and cross section morphologies of membranes were examined by a field emission scanning electron microscopy (FESEM; Magellan 400, Nova Nano SEM 450, FEI, New York, NY, USA). Membrane hydrophilicity was analysed via contact angle measurements (OCA-15EC, Dataphysics, Stuttgart, Germany). The static contact angle of different polymerised films was tested by the suspension drop method. After drying the film to be tested, it was flatly pasted on the slide, and then placed on the test table at room temperature. One microliter of deionised water was dropped onto the membrane surface with a microinjector. The contact angle was measured after the water drop stabilised. At least 10 contact angles at different places for each membrane were averaged to obtain a reliable value.

2.4. Membrane Performance Evaluation
Membrane Permeability and Salt Rejection

Membrane performance testing was conducted using a dead-end filtration (DEF) system (effective area is 14.2 cm^2). The detailed filtration process was as following: (1) The membrane was first compacted at 2 bars for 3 h to achieve a steady flux; (2) the trans-membrane pressure was reduced to 1 bar and the pure water flux was recorded every 1 min. At least 60 measurements were collected to obtain an average flux value; (3) The DI water was replaced by a PEG feed solution, filtration cells were stirred at 400 rpm using a stir to minimise concentration polarization and the trans-membrane pressure was returned to 1 bar. After 1 h of filtration, a sample of the permeation solution was collected. For each membrane performance evaluations, at least three samples were tested. A total organic carbon (TOC) analyser was used to determine the concentration of PEG in the feed and permeation solution; the analyser uses combustion catalytic oxidation method at 680 °C. The rejection was calculated by the following Equation (1):

$$R = (1 - \text{TOC}_{filtrate} / \text{TOC}_{feed}) \times 100\% \tag{1}$$

where $\text{TOC}_{filtrate}$ is the TOC concentration of PEG in the filtrate and TOC_{feed} is the TOC concentration in the PEG feed solution.

3. Results and Discussion
3.1. Characterisation of PVDF, PVDF/GO-s and PVDF/GO-h Membranes
3.1.1. FTIR

Figure 1 shows the FT-IR spectra of PVDF, PVDF/GO-s and PVDF/GO-h membranes and GO. As can be seen in Figure 1, pristine PVDF membrane shows peaks at 1396 cm^{-1} and 1175 cm^{-1}, attributing to C-H and C-F stretching and deformation [39], which are also prominent in all other composite membranes. The prominent features of the GO spectrum is the adsorption peaks at ~3340 cm^{-1} and ~1734 cm^{-1}, which are corresponding to O-H and C=O stretching vibrations, respectively [4]. Due to the nucleation effect of nano-filler in the PVDF matrix [40], it can be noticed that the intensity of α phase (at 760 cm^{-1}) decreases when GO is embedded in the PVDF matrix, while the intensity of β phase (at 840 cm^{-1}) increases. This indicates that GO nanosheets contain sufficient carbonyl groups to nucleate most of PVDF chains into β-phase.

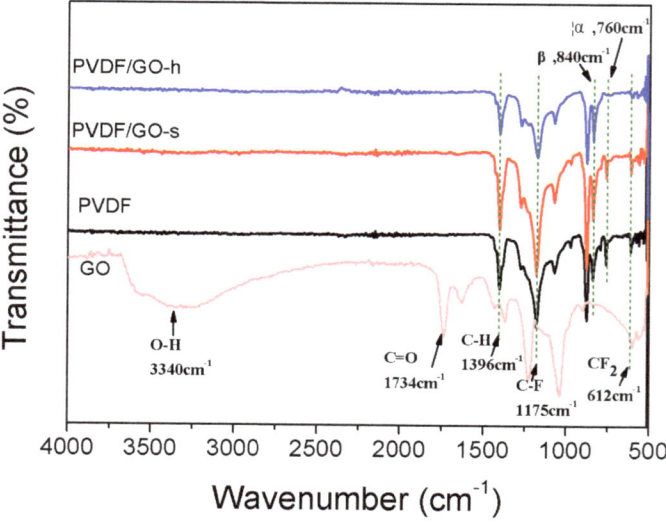

Figure 1. FTIR spectra of PVDF, PVDF/GO-s and PVDF/GO-h membranes.

Yu et al. calculated the absorption energy of α-and ß-polyform [41], and found that the significant difference in the adsorption energy of α and β phase made the energy barrier between trans-gauche-trans-gauche0 (TGTG0) and trans-trans (TT) structure increase, and it became difficult to convert TGTG0 to TT structure in the process of polymer crystallization. It can be seen from Figure 1 that compared with PVDF/GO-s, α phase of PVDF/GO-h membrane after homogenisation process decreases or even disappears. This is because the use of homogeniser can not only improve the dispersibility of GO in solution, but also effectively reduce the adsorption energy difference between α phase and β phase, overcome the energy barrier [42], help PVDF chain adsorption to the GO surface, promote the interaction between the oxygen-containing groups on GO and the hydrogen atoms on the PVDF chain. However, the peak value of β phase of PVDF-h was lower than that of PVDF-s phase, which may be because the dispersibility of GO in solution was improved by the use of homogeniser, and thus the concentration of solution was increased. This will lead to GO as a filler particle agglomeration under high concentration, resulting in reduced PVDF chain constraint and resulting in decreased β phase content [43]. L. He et al. [44] changed the crystal distribution in PVDF by adding hyperbranched chain copolymer (HBCs) modified multi-walled carbon nanotubes, and improved the β phase and thermal stability of the membrane. We used a homogeniser to enhance the β phase of the PVDF membrane. After homogenisation, the α phase of PVDF hybrid membrane was almost completely transformed into β phase. Compared with the former, the conversion rate of α phase to β phase of PVDF hybrid membrane was greatly improved by using the homogeniser. Therefore, homogeniser can change PVDF α-phase to β-phase to a greater extent. These results are in good agreement with previous studies using other carbon materials as fillers in PVDF membranes [39,45,46].

The composite membranes show no absorption peak at ~3340 cm^{-1}, which would be indicative of O-H stretching of carboxylic acid. One possibility is that because of the strong compatibility between the carbonyl group in GO and the fluorine in PVDF [47]. Another explanation is that the casting solution concentration using 0.15 wt% GO is too low for the functional groups to present at any significant level detected by FT-IR.

3.1.2. XRD

To compare the molecular structure of PVDF/GO-s membrane and PVDF/GO-h membrane and confirm that GO components were successfully integrated into the PVDF polymer matrix, X-ray diffraction was performed. Figure 2 shows the results of XRD analysis of PVDF, PVDF/GO-s, and PVDF/GO-h membranes in the range of 15° to 40°, in terms of arbitrary scale of intensity. For pristine PVDF membrane, the characteristic peaks at 18.4°, 19.9° and 26.5° can be observed, which are attributed to α-phase. Both PVDF/GO-s and PVDF/GO-h membranes display a new diffraction peak at 20.6°, which corresponds to the β-phase. This is most likely due to the crystal transformation of PVDF [48]. The formation of β-polymorph is attributed to the interaction between the CF_2 segments in PVDF polymer and the carbonyl groups (-C=O) present in GO nanosheets [39,49]. In addition, for PVDF/GO-s membranes, the intensity of α-phase peak at 18.4° dropped and the α-phase peaks at 26.5° disappeared, however, for the PVDF/GO-h membrane, both the α-phase peaks at 18.4° and 26.5° disappeared. Thus, it can be concluded that the disappearance of α-phase in PVDF/GO-h membrane indicate the enhanced crystal transformation of PVDF membrane, which is caused by the better dispersion of GO by homogeniser.

3.2. Membrane Morphology
3.2.1. SEM Image of the Membrane

Figure 3 displays the surface and cross-section SEM images of the pristine PVDF, PVDF/GO-s and PVDF/GO-h membranes. More SEM images obtained at the surface for the different membranes are provided in Figure 3a–c featuring the increase in the number of pores in the membrane through the addition of GO to the PVDF membrane structure.

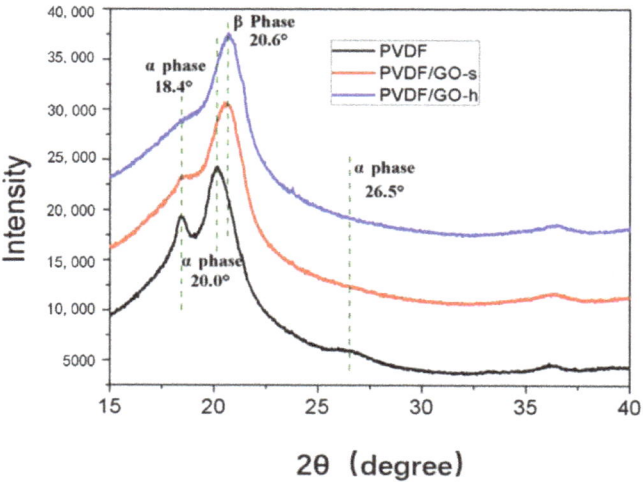

Figure 2. XRD spectra of PVDF, PVDF/GO-s and PVDF/GO-h membranes.

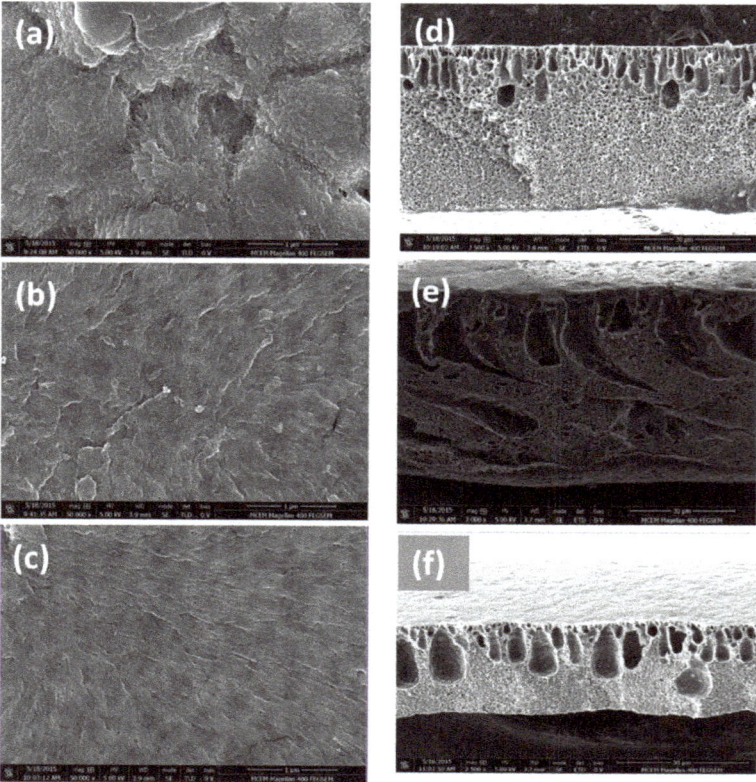

Figure 3. SEM images of top surface for (**a**) pure PVDF, (**b**) PVDF/GO-s, and (**c**) PVDF/GO-h membranes and images (**d**–**f**) are SEM cross sections for the same membranes, respectively.

As visible from the images in Figure 3, PVDF membrane contains clusters of pores which are large and apparent. The pore distribution is uneven and concentrated on certain areas of the membrane. The size of the pores is decreased with the addition of GO whilst the number of pores significantly increased, especially in the case of the membranes which were homogenised with GO. The reason for the GO embedded membranes showed an increase in the number of pores on the surface may be that the presence of the rich oxygen-containing functional groups which increases the rate of diffusion and thereby increases pore formation. Homogenised solutions contained more GO elements which suggests proper mixing has taken place between PVDF and GO compared to PVDF/GO prepared using a magnetic stirrer. This could explain the greater number of pores on the surface of the membrane for Figure 3c compared to Figure 3b.

As can be seen from cross-section SEM images in Figure 3d–f, the morphological changes between the GO embedded membranes and the pristine PVDF membrane were compared. All the membranes displayed a thin dense top-layer, along with a porous finger-like sublayer [50]. The skin layer is brought about as a result of the polymer concentration gradient that takes place when the membrane is immersed in the water bath immediately after preparation. The outer surface solidifies creating a dense skin layer. The fingers-like pores are created due to the phase inversion method. As this process occurs, demixing takes place between the water and the solvent, which slows down eventually due to the presence of the solid membrane. As a result of this time lag, caused by the delay, a dense yet porous sponge layer forms towards the bottom of the membrane [50,51].

For the membranes with GO added, the sub layer was visibly different. The finger-like pores in the PVDF/GO membranes were much wider than that of the pure PVDF membrane. The longer pore channels result from an increased rate of diffusion brought about by the hydrophilicity of GO. Rapid solidification from this diffusion creates wider pore channels [45]. The images further show the formation of a sponge-like cross-section for the PVDF membrane. This was not the case for the PVDF/GO membranes as the addition of GO into the structure of the membrane mostly prevented and strongly controlled the formation of this type of cross-section. The addition of GO also creates a floppy inner cross section because of the increased mass transformation that occurs between the solvent and the non-solvent during the process of phase inversion.

Compared to the cross-section of the PVDF/GO-s membranes, the finger-like structure channels in PVDF/GO-h membranes become thinner and shorter, also the pores in the sponge-like structure become smaller; however, the number of the finger-like structure channels and pores in the sponge-like structure increases. One reason may be that in the homogenisation process, the casting solution experienced extremely strong shear and thrust forces [34], the turbulence occurred in the shear gap between the rotor and stator also provided strong mixing power to the suspension, which improved the phase inversion. M. Hmamm et al. [52] found that when the crystallinity of polymer increased, the free volume would decrease correspondingly. Another possibility is that the synergistic effect of GO and PVDF is enhanced during the homogenization process, which eliminates the unique GO peak, improves the crystallinity of the PVDF hybrid film and decreases the free volume size of the hybrid film.

3.2.2. Surface Hydrophilicity

The hydrophilicity of PVDF, PVDF/GO-s and PVDF/GO-h membranes were characterised by the water contact angle. As can be seen in Figure 4, for GO embedded membranes, the water contact angle is reduced compared with pure PVDF membranes, indicating the improved hydrophilicity after GO incorporation. This could be because hydrophilic GO migrates spontaneously to the membrane/water interface to reduce the interface energy during the phase inversion process [21,50,53]. This also can be verified by the different colour between the surface and bottom, the colour of the surface is darker than the bottom. Previous research also found the same phenomenon [21,24]. The contact angle of the PVDF/GO-h membrane is slightly lower than that of PVDF/GO-s membrane, suggesting the surface of the PVDF/GO-h membrane is more hydrophilic than PVDF/GO-s membrane. The reason may be that the surface of PVDF/GO-h membrane becomes smooth, which is due to the large peaks and valleys on the surface are replaced by many smaller ones. It is also possible that GO is well dispersed in the polymer matrix after the action of the homogeniser, and the abundant oxygen-containing groups on the surface of GO can be evenly distributed on the membrane surface, thus effectively improving the surface hydrophilicity of the PVDF hybrid membrane and making the surface of the PVDF/GO-h membrane more hydrophilic than that of the PVDF/GO-s membrane.

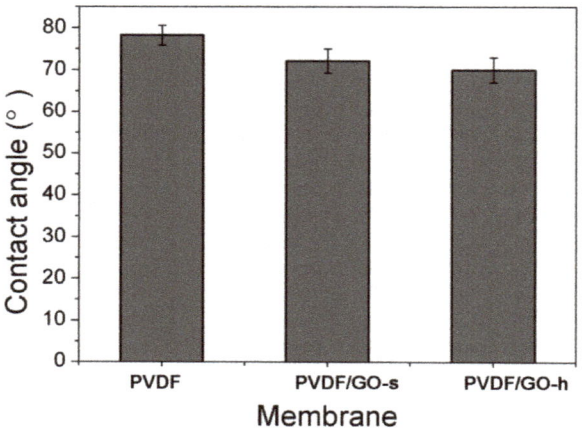

Figure 4. Water contact angles of PVDF, PVDF/GO-s and PVDF/GO-h membranes.

3.3. Membrane Evaluation

3.3.1. Membrane Flux

The permeability of the PVDF, PVDF/GO and PVDF/GO-h membranes were evaluated by measuring water flux. Figure 5 shows the water flux of these membranes. Both PVDF/GO-s membranes and PVDF/GO-h membranes exhibit higher water flux compared with the pure PVDF membrane. One reason could be due to the enhanced surface hydrophilicity after GO incorporation, as shown in the contact angle test (Figure 4). Another reason could be the enhanced phase inversion of solvent and non-solvent due to the presence of hydrophilic GO [20,24]. As shown in Figure 4, the 'finger-like' structure pores of PVDF/GO-s and PVDF/GO-h membranes become wider and longer.

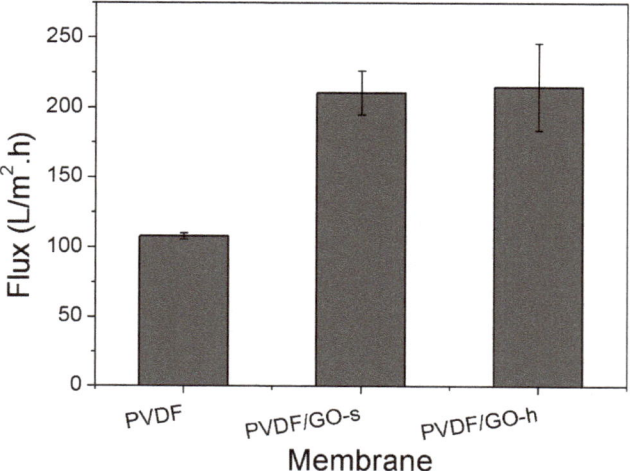

Figure 5. Water fluxes of the prepared PVDF, PVDF/GO-s and PVDF/GO-h membranes (200 MPa).

Comparing PVDF/GO-s and PVDF/GO-h membranes, it is found that PVDF/GO-h membrane has a slightly higher flux. As shown by SEM results (Figure 3), compared with the internal structure of PVDF/GO-h membrane, the free volume of PVDF/GO-h membrane decreased, while the number of free volumes increased correspondingly. H.F.M. Mohamed et al. [54] studied the relationship between the free volume in Nafion films and the permeability of O_2 and H_2 and found that the larger the free volume, the better the gas permeability. This is consistent with the conclusion of another study by H.F. Mohamed et al. [55]. The decrease in free volume in the PVDF/GO-h membrane reduces the aqueous permeability of the PVDF/GO-h membrane, while the higher porosity increases the aqueous permeability, which makes the PVDF/GO-h membrane flux slightly higher than that of the PVDF/GO-s membrane.

3.3.2. Membrane Rejection

To evaluate the effect of homogeniser on the membrane performance, the PEG (35 K) rejection of the hybrid membrane was measured. From Figure 6, it can be seen that the PEG rejections of pristine PVDF, PVDF/GO-s and PVDF/GO-h membranes are 26.99%, 25.02% and 52.81%, respectively. There are no significant difference of the rejection between PVDF and PVDF/GO-s membranes, it is similar as our previous research [37]. However, after using homogeniser, the PEG rejection is increased significantly from 26.99% to 52.81%. This is caused by the change of membrane morphology using the homogeniser. GO can be used as a pore-making agent to improve the number of pores in the PVDF membrane. According to the sieving principle of pore size, larger pores, such as PVDF/GO-S membrane, allow PEG molecules to pass through the membrane more easily than membranes with smaller pores on the surface (such as PVDF/GO-H membrane). After the PVDF/GO solution is homogenised by the homogeniser, the dispersibility of GO in the mixed solution is greatly improved, and the porosity of PVDF will be improved. Second, after the solution is homogenised, the oxygen-containing groups on GO surface will also be fully embedded into the PVDF membrane, the water molecules around PEG will be replaced by the hydroxyl groups on the GO surface to form a hydration layer, making the diameter of PEG larger than the diameter of the channel gap, thus improving the retention rate of PEG. It is not difficult to explain that PVDF/GO-h membrane has better interception effect on PEG than PVDF/GO-s membrane.

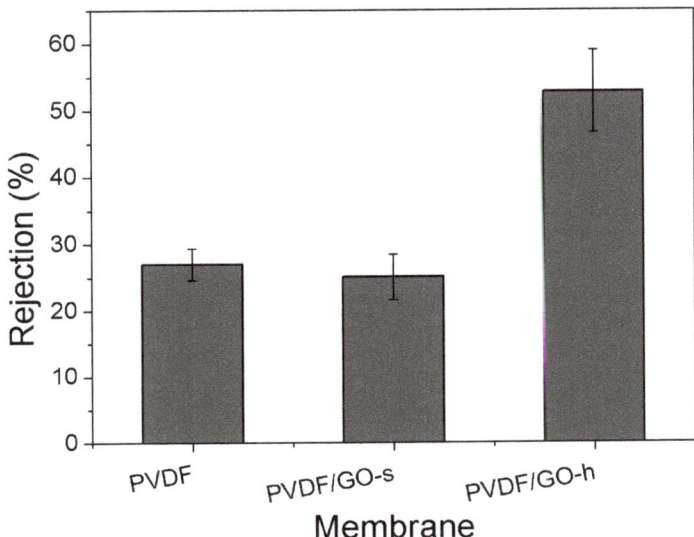

Figure 6. Water rejections of the prepared PVDF, PVDF/GO-s and PVDF/GO-h membranes (PEG 35 K).

4. Conclusions

In this study, PVDF/GO-h composite membranes were synthesised using a homogeniser (AD500S-H), and results from dead-end filtration (DEF) showed that these membranes exhibited higher water flux and rejection of target pollutants compared to control samples of PVDF/GO-s prepared solely from mixing with a magnetic stirrer. This suggests that using a homogeniser disperser to mix GO into a PVDF membrane yields significantly better characteristics and should therefore be utilised as a future method for membrane preparation. Whilst these qualities are attractive in a mem-brane for use in ultrafiltration methods, results of the water contact angle test indicate that the composite membranes exhibited increased hydrophobicity as opposed to the control membranes. This implies that there is still potential for improvement of the experiment where additional parameters should be tested for or if other additives should be incorporated in the methodology to increase membrane performance.

Author Contributions: Conceptualization, N.M. and H.S.; methodology, H.S. and R.W.; software, J.Z.; validation, H.S., R.W. and J.Z.; formal analysis, J.L. (Jinxin Liu) and J.L. (Jun Lu); investigation, X.S.; data curation, J.L. (Jinxin Liu) and J.L. (Jun Lu); writing—original draft preparation, X.S.; writing—review and editing, H.S. and R.W.; supervision, N.M.; project administration, N.M.; funding acquisition, J.L. (Jinxin Liu) and J.L. (Jun Lu). All authors have read and agreed to the published version of the manuscript.

Funding: This research was funded by the financial support of Science and Technology Plan Projects of Xuzhou City (Grant No. KC19209) and the Natural Science Foundation of the Jiangsu Higher Education Institutions of China (Grant No. 19KJA580002).

Institutional Review Board Statement: Not applicable.

Data Availability Statement: Data is contained within this article.

Conflicts of Interest: The authors declare no conflict of interest.

References

1. Zhao, H.; Wu, L.; Zhou, Z.; Zhang, L.; Chen, H. Improving the antifouling property of polysulfone ultrafiltration membrane by incorporation of isocyanate-treated graphene oxide. *Phys. Chem. Chem. Phys.* **2013**, *15*, 9084–9092. [CrossRef] [PubMed]
2. Huang, L.; Zhao, S.; Wang, Z.; Wu, J.; Wang, J.; Wang, S. In situ immobilization of silver nanoparticles for improving permeability, antifouling and anti-bacterial properties of ultrafiltration membrane. *J. Membr. Sci.* **2016**, *499*, 269–281. [CrossRef]
3. Zhang, J.; Xu, Z.; Mai, W.; Min, C.; Zhou, B.; Shan, M.; Li, Y.; Yang, C.; Wang, Z.; Qian, X. Improved hydrophilicity, permeability, antifouling and mechanical performance of PVDF composite ultrafiltration membranes tailored by oxidized low-dimensional carbon nanomaterials. *J. Mater. Chem. A* **2013**, *1*, 3101–3111. [CrossRef]
4. Kuilla, T.; Bhadra, S.; Yao, D.; Kim, N.H.; Bose, S.; Lee, J.H. Recent advances in graphene based polymer composites. *Prog. Polym. Sci.* **2010**, *35*, 1350–1375. [CrossRef]
5. Abdel-hameed, M.; Isawi, H.; El-Noss, M.; El-Kholy, R.A.; Said, M.M.; Shawky, H.A. Design and fabrication of continuous flow photoreactor using semiconductor oxides for degradation of organic pollutants. *J. Water Process Eng.* **2019**, *32*, 100922.
6. Isawi, H. Using zeolite/polyvinyl alcohol/sodium alginate nanocomposite beads for removal of some heavy metals from wastewater. *Arab. J. Chem.* **2020**, *13*, 5691–5716. [CrossRef]
7. Moustafa, H.; Isawi, H.; Abd El Wahab, S.M. Utilization of PVA nano-membrane based synthesized magnetic GO-Ni-Fe$_2$O$_4$ nanoparticles for removal of heavy metals from water resources. *Environ. Nanotechnol. Monit. Manag.* **2022**, *18*, 100696. [CrossRef]
8. Liu, F.; Hashim, N.A.; Liu, Y.; Abed, M.M.; Li, K. Progress in the production and modification of PVDF membranes. *J. Membr. Sci.* **2011**, *375*, 1–27. [CrossRef]
9. Li, H.; Song, Z.; Zhang, X.; Huang, Y.; Li, S.; Mao, Y.; Ploehn, H.J.; Bao, Y.; Yu, M. Ultrathin, molecular-sieving graphene oxide membranes for selective hydrogen separation. *Science* **2013**, *342*, 95–98. [CrossRef]
10. Younas, H.; Bai, H.; Shao, J.; Han, Q.; Ling, Y.; He, Y. Super-hydrophilic and fouling resistant PVDF ultrafiltration membranes based on a facile prefabricated surface. *J. Membr. Sci.* **2017**, *541*, 529–540. [CrossRef]
11. Isawi, H. Evaluating the performance of different nano-enhanced ultrafiltration membranes for the removal of organic pollutants from wastewater. *J. Water Process Eng.* **2019**, *31*, 100833. [CrossRef]
12. Stankovich, S.; Dikin, D.A.; Dommett, G.H.; Kohlhaas, K.M.; Zimney, E.J.; Stach, E.A.; Piner, R.D.; Nguyen, S.T.; Ruoff, R.S. Graphene-based composite materials. *Nature* **2006**, *442*, 282–286. [CrossRef] [PubMed]
13. Dreyer, D.R.; Park, S.; Bielawski, C.W.; Ruoff, R.S. The chemistry of graphene oxide. *Chem. Soc. Rev.* **2010**, *39*, 228–240. [CrossRef] [PubMed]

14. Joshi, R.; Carbone, P.; Wang, F.-C.; Kravets, V.G.; Su, Y.; Grigorieva, I.V.; Wu, H.; Geim, A.K.; Nair, R.R. Precise and ultrafast molecular sieving through graphene oxide membranes. *Science* **2014**, *343*, 752–754. [CrossRef]
15. Kim, H.W.; Yoon, H.W.; Yoon, S.-M.; Yoo, B.M.; Ahn, B.K.; Cho, Y.H.; Shin, H.J.; Yang, H.; Paik, U.; Kwon, S. Selective gas transport through few-layered graphene and graphene oxide membranes. *Science* **2013**, *342*, 91–95. [CrossRef]
16. Sun, P.; Zhu, M.; Wang, K.; Zhong, M.; Wei, J.; Wu, D.; Xu, Z.; Zhu, H. Selective ion penetration of graphene oxide membranes. *ACS Nano* **2013**, *7*, 428–437. [CrossRef]
17. Young, R.J.; Kinloch, I.A.; Gong, L.; Novoselov, K.S. The mechanics of graphene nanocomposites: A review. *Compos. Sci. Technol.* **2012**, *72*, 1459–1476. [CrossRef]
18. Wang, Z.; Yu, H.; Xia, J.; Zhang, F.; Li, F.; Xia, Y.; Li, Y. Novel GO-blended PVDF ultrafiltration membranes. *Desalination* **2012**, *299*, 50–54. [CrossRef]
19. Ganesh, B.; Isloor, A.M.; Ismail, A.F. Enhanced hydrophilicity and salt rejection study of graphene oxide-polysulfone mixed matrix membrane. *Desalination* **2013**, *313*, 199–207. [CrossRef]
20. Yu, L.; Zhang, Y.; Zhang, B.; Liu, J.; Zhang, H.; Song, C. Preparation and characterization of HPEI-GO/PES ultrafiltration membrane with antifouling and antibacterial properties. *J. Membr. Sci.* **2013**, *447*, 452–462. [CrossRef]
21. Ionita, M.; Pandele, A.M.; Crica, L.; Pilan, L. Improving the thermal and mechanical properties of polysulfone by incorporation of graphene oxide. *Compos. Part B Eng.* **2014**, *59*, 133–139. [CrossRef]
22. Xu, Z.; Zhang, J.; Shan, M.; Li, Y.; Li, B.; Niu, J.; Zhou, B.; Qian, X. Organosilane-functionalized graphene oxide for enhanced antifouling and mechanical properties of polyvinylidene fluoride ultrafiltration membranes. *J. Membr. Sci.* **2014**, *458*, 1–13. [CrossRef]
23. Zhao, C.; Xu, X.; Chen, J.; Wang, G.; Yang, F. Highly effective antifouling performance of PVDF/graphene oxide composite membrane in membrane bioreactor (MBR) system. *Desalination* **2014**, *340*, 59–66. [CrossRef]
24. Zinadini, S.; Zinatizadeh, A.A.; Rahimi, M.; Vatanpour, V.; Zangeneh, H. Preparation of a novel antifouling mixed matrix PES membrane by embedding graphene oxide nanoplates. *J. Membr. Sci.* **2014**, *453*, 292–301. [CrossRef]
25. Kumar, M.; McGlade, D.; Ulbricht, M.; Lawler, J. Quaternized polysulfone and graphene oxide nanosheet derived low fouling novel positively charged hybrid ultrafiltration membranes for protein separation. *RSC Adv.* **2015**, *5*, 51208–51219. [CrossRef]
26. Zhao, Y.; Xu, Z.; Shan, M.; Min, C.; Zhou, B.; Li, Y.; Liu, L.; Qian, X. Effect of graphite oxide and multi-walled carbon nanotubes on the microstructure and performance of PVDF membranes. *Sep. Purif. Technol.* **2013**, *103*, 78–83. [CrossRef]
27. Zhao, C.; Xu, X.; Chen, J.; Yang, F. Effect of graphene oxide concentration on the morphologies and antifouling properties of PVDF ultrafiltration membranes. *J. Environ. Chem. Eng.* **2013**, *1*, 349–354. [CrossRef]
28. Zhang, J.; Xu, Z.; Shan, M.; Zhou, B.; Li, Y.; Li, B.; Niu, J.; Qian, X. Synergetic effects of oxidized carbon nanotubes and graphene oxide on fouling control and anti-fouling mechanism of polyvinylidene fluoride ultrafiltration membranes. *J. Membr. Sci.* **2013**, *448*, 81–92. [CrossRef]
29. Stankovich, S.; Piner, R.D.; Nguyen, S.T.; Ruoff, R.S. Synthesis and exfoliation of isocyanate-treated graphene oxide nanoplatelets. *Carbon* **2006**, *44*, 3342–3347. [CrossRef]
30. Qian, C.; McClements, D.J. Formation of nanoemulsions stabilized by model food-grade emulsifiers using high-pressure homogenization: Factors affecting particle size. *Food Hydrocoll.* **2011**, *25*, 1000–1008. [CrossRef]
31. Fu, Z.-Q.; Wang, L.-J.; Li, D.; Wei, Q.; Adhikari, B. Effects of high-pressure homogenization on the properties of starch-plasticizer dispersions and their films. *Carbohydr. Polym.* **2011**, *86*, 202–207. [CrossRef]
32. Lee, S.-Y.; Chun, S.-J.; Kang, I.-A.; Park, J.-Y. Preparation of cellulose nanofibrils by high-pressure homogenizer and cellulose-based composite films. *J. Ind. Eng. Chem.* **2009**, *15*, 50–55. [CrossRef]
33. Bae, H.J.; Park, H.J.; Hong, S.I.; Byun, Y.J.; Darby, D.O.; Kimmel, R.M.; Whiteside, W.S. Effect of clay content, homogenization RPM, pH, and ultrasonication on mechanical and barrier properties of fish gelatin/montmorillonite nanocomposite films. *LWT-Food Sci. Technol.* **2009**, *42*, 1179–1186. [CrossRef]
34. Jiang, L.; Morelius, E.; Zhang, J.; Wolcott, M.; Holbery, J. Study of the poly (3-hydroxybutyrate-co-3-hydroxyvalerate)/cellulose nanowhisker composites prepared by solution casting and melt processing. *J. Compos. Mater.* **2008**, *42*, 2629–2645. [CrossRef]
35. Vladisavljević, G.T.; Shimizu, M.; Nakashima, T. Production of multiple emulsions for drug delivery systems by repeated SPG membrane homogenization: Influence of mean pore size, interfacial tension and continuous phase viscosity. *J. Membr. Sci.* **2006**, *284*, 373–383. [CrossRef]
36. Nacken, T.; Damm, C.; Walter, J.; Rüger, A.; Peukert, W. Delamination of graphite in a high pressure homogenizer. *RSC Adv.* **2015**, *5*, 57328–57338. [CrossRef]
37. Meng, N.; Priestley, R.C.E.; Zhang, Y.; Wang, H.; Zhang, X. The effect of reduction degree of GO nanosheets on microstructure and performance of PVDF/GO hybrid membranes. *J. Membr. Sci.* **2016**, *501*, 169–178. [CrossRef]
38. Meng, N.; Zhao, W.; Shamsaei, E.; Wang, G.; Zeng, X.; Lin, X.; Xu, T.; Wang, H.; Zhang, X. A low-pressure GO nanofiltration membrane crosslinked via ethylenediamine. *J. Membr. Sci.* **2018**, *548*, 363–371. [CrossRef]
39. el Achaby, M.; Arrakhiz, F.; Vaudreuil, S.; Essassi, E.; Qaiss, A. Piezoelectric β-polymorph formation and properties enhancement in graphene oxide—PVDF nanocomposite films. *Appl. Surf. Sci.* **2012**, *258*, 7668–7677. [CrossRef]
40. Hasegawa, R.; Takahashi, Y.; Chatani, Y.; Tadokoro, H. Crystal structures of three crystalline forms of poly (vinylidene fluoride). *Polym. J.* **1972**, *3*, 600–610. [CrossRef]

41. Yu, S.; Zheng, W.; Yu, W.; Zhang, Y.; Jiang, Q.; Zhao, Z. Formation mechanism of β-phase in PVDF/CNT composite prepared by the sonication method. *Macromolecules* **2009**, *42*, 8870–8874. [CrossRef]
42. Mishra, S.; Sahoo, R.; Unnikrishnan, L.; Ramadoss, A.; Mohanty, S.; Nayak, S.K. Investigation of the electroactive phase content and dielectric behaviour of mechanically stretched PVDF-GO and PVDF-rGO composites. *Mater. Res. Bull.* **2020**, *124*, 110732. [CrossRef]
43. Rahman, M.A.; Chung, G.-S. Synthesis of PVDF-graphene nanocomposites and their properties. *J. Alloys Compd.* **2013**, *581*, 724–730. [CrossRef]
44. He, L.; Sun, J.; Wang, X.; Yao, L.; Li, J.; Song, R.; Hao, Y.; He, Y.; Huang, W. Enhancement of β-crystalline phase of poly (vinylidene fluoride) in the presence of hyperbranched copolymer wrapped multiwalled carbon nanotubes. *J. Colloid Interface Sci.* **2011**, *363*, 122–128. [CrossRef] [PubMed]
45. Layek, R.K.; Samanta, S.; Chatterjee, D.P.; Nandi, A.K. Physical and mechanical properties of poly (methyl methacrylate)-functionalized graphene/poly (vinylidine fluoride) nanocomposites: Piezoelectric β polymorph formation. *Polymer* **2010**, *51*, 5846–5856. [CrossRef]
46. Manna, S.; Nandi, A.K. Piezoelectric β polymorph in poly (vinylidene fluoride)-functionalized multiwalled carbon nanotube nanocomposite films. *J. Phys. Chem. C* **2007**, *111*, 14670–14680. [CrossRef]
47. Islam, A.; Khan, A.N.; Shakir, M.F.; Islam, K. Strengthening of β polymorph in PVDF/FLG and PVDF/GO nanocomposites. *Mater. Res. Express* **2019**, *7*, 015017. [CrossRef]
48. Yu, J.; Jiang, P.; Wu, C.; Wang, L.; Wu, X. Graphene nanocomposites based on poly (vinylidene fluoride): Structure and properties. *Polym. Compos.* **2011**, *32*, 1483–1491. [CrossRef]
49. Jaleh, B.; Jabbari, A. Evaluation of reduced graphene oxide/ZnO effect on properties of PVDF nanocomposite films. *Appl. Surf. Sci.* **2014**, *320*, 339–347. [CrossRef]
50. Burns, D.B.; Zydney, A.L. Buffer effects on the zeta potential of ultrafiltration membranes. *J. Membr. Sci.* **2000**, *172*, 39–48. [CrossRef]
51. Meng, N.; Wang, Z.; Low, Z.-X.; Zhang, Y.; Wang, H.; Zhang, X. Impact of trace graphene oxide in coagulation bath on morphology and performance of polysulfone ultrafiltration membrane. *Sep. Purif. Technol.* **2015**, *147*, 364–371. [CrossRef]
52. Hmamm, M.; Zedan, I.; Mohamed, H.F.; Hanafy, T.; Bekheet, A. Study of the nanostructure of free volume and ionic conductivity of polyvinyl alcohol doped with NaI. *Polym. Adv. Technol.* **2021**, *32*, 173–182. [CrossRef]
53. Hagmeyer, G.; Gimbel, R. Modelling the rejection of nanofiltration membranes using zeta potential measurements. *Sep. Purif. Technol.* **1999**, *15*, 19–30. [CrossRef]
54. Mohamed, H.F.; Ito, K.; Kobayashi, Y.; Takimoto, N.; Takeoka, Y.; Ohira, A. Free volume and permeabilies of O_2 and H_2 in Nafion membranes for polymer electrolyte fuel cells. *Polymer* **2008**, *49*, 3091–3097. [CrossRef]
55. Mohamed, H.F.; Kobayashi, Y.; Kuroda, C.; Takimoto, N.; Ohira, A. Free volume, oxygen permeability, and uniaxial compression storage modulus of hydrated biphenol-based sulfonated poly (arylene ether sulfone). *J. Membr. Sci.* **2010**, *360*, 84–89. [CrossRef]

Article

Impact of Sidestream Pre-Treatment on Ammonia Recovery by Membrane Contactors: Experimental and Economic Evaluation

Miguel Aguilar-Moreno [1,2,*], Sergi Vinardell [1,2], Mònica Reig [1,2], Xanel Vecino [1,2], César Valderrama [1,2] and José Luis Cortina [1,2,3]

1 Chemical Engineering Department, Escola d'Enginyeria de Barcelona Est (EEBE), Universitat Politècnica de Catalunya (UPC)-BarcelonaTECH, C/Eduard Maristany 10-14, Campus Diagonal-Besòs, 08930 Barcelona, Spain
2 Barcelona Research Center for Multiscale Science and Engineering, Campus Diagonal-Besòs, 08930 Barcelona, Spain
3 CETaqua, Carretera d'Esplugues, 75, 08940 Cornellà de Llobregat, Spain
* Correspondence: miguel.aguilar.moreno@upc.edu; Tel.: +34-93-4016997

Citation: Aguilar-Moreno, M.; Vinardell, S.; Reig, M.; Vecino, X.; Valderrama, C.; Cortina, J.L. Impact of Sidestream Pre-Treatment on Ammonia Recovery by Membrane Contactors: Experimental and Economic Evaluation. *Membranes* **2022**, *12*, 1251. https://doi.org/10.3390/membranes12121251

Academic Editors: Hongjun Lin and Meijia Zhang

Received: 18 November 2022
Accepted: 8 December 2022
Published: 10 December 2022

Publisher's Note: MDPI stays neutral with regard to jurisdictional claims in published maps and institutional affiliations.

Copyright: © 2022 by the authors. Licensee MDPI, Basel, Switzerland. This article is an open access article distributed under the terms and conditions of the Creative Commons Attribution (CC BY) license (https://creativecommons.org/licenses/by/4.0/).

Abstract: Membrane contactor is a promising technology for ammonia recovery from the anaerobic digestion centrate. However, high suspended solids and dissolved organic matter concentrations can reduce the effectiveness of the technology. In this study, coagulation–flocculation (C/F) and aeration pre-treatments were evaluated to reduce chemical oxygen demand (COD), turbidity, suspended solids and alkalinity before the ammonia recovery stage using a membrane contactor. The mass transfer coefficient (K_m) and total ammonia (TAN) recovery efficiency of the membrane contactor increased from 7.80×10^{-7} to 1.04×10^{-5} m/s and from 8 to 67%, respectively, after pre-treating the real sidestream centrate. The pre-treatment results showed that dosing aluminium sulphate ($Al_2(SO_4)_3$) at 30 mg Al/L was the best strategy for the C/F process, providing COD, turbidity and TSS removal efficiencies of 50 ± 5, 95 ± 3 and $90 \pm 4\%$, respectively. The aeration step reduced $51 \pm 6\%$ the HCO_3^- content and allowed reducing alkaline consumption by increasing the pH before the membrane contactor. The techno-economic evaluation showed that the combination of C/F, aeration and membrane contactor can be economically feasible for ammonia recovery. Overall, the results of this study demonstrate that C/F and aeration are simple and effective techniques to improve membrane contactor performance for nitrogen recovery from the anaerobic digestion centrate.

Keywords: gas permeable membrane; coagulation–flocculation; resource recovery; circular economy; techno-economic evaluation

1. Introduction

Nutrient pollution is one of the major environmental problems due to excessive discharge of nitrogen and phosphorus into the environment. Anthropogenic activities and population growth have increased the amount of nitrogen contained in wastewater. The recovery of this nitrogen is particularly important considering that ammonia is the second most produced chemical in the world [1–3]. Ammoniacal nitrogen recovery has the potential (i) to reduce the dependency of the Haber–Bosch process to obtain nitrogen-based fertilizers, (ii) to produce a fertilizer (e.g., NH_4NO_3, $(NH_4)_2HPO_4$, $(NH_4)_2SO_4$) suitable for commercialization and (iii) to reintroduce nitrogen into its cycle contributing to the circular economy [4,5]. For this reason, it is important to develop efficient technologies for nitrogen recovery to support the transition of wastewater treatment plants (WWTPs) towards water resource recovery facilities (WRRF) [6].

Several technologies have been proposed to recover nitrogen from wastewater treatment plants (WWTPs), such as ion exchange (IX) technologies [7], membrane contactors (MC) [8–10] or ultrafiltration (UF) [11]. For instance, Wan et al. [12] effectively recovered nutrients from the sludge fermentation liquor in a WWTP (N-NH_4^+ and P-PO_4^{3-}) using natural zeolites and proposed a model to predict that a maximum recovery of 94% ammonium and 98% phosphate could be achieved. Among them, ammoniacal nitrogen recovery through membrane contactors has been reported as a suitable technology to achieve high nitrogen recovery efficiencies with relatively low energy inputs [5]. By this technology, ammonia in gas form diffuses through a porous hydrophobic membrane from the feed solution to the acidic stripping solution. Subsequently, it can be recovered in ammonium form as a nitrogen-rich fertilizer. [13]. Vecino et al. [14] used a membrane contactor for ammonium recovery as a nutrient-based fertilizer product and achieved a maximum ammonium recovery of 94% using a regenerated stream with ion exchange from an initial sidestream wastewater. Sheikh et al. [15] also achieved similar values (>95%) of recovery using synthetic water and liquid–liquid hollow fibre MC (LL-HFMC). Additionally, both membrane contactors and ion exchange technologies can be combined as proposed by Sancho et al. [16]. In that study, a concentrated ammonium stream was generated by means of liquid–liquid membrane contactors, by previously passing it through zeolites, achieving a recovery of 95% [16]. Thus, these publications highlight that membrane contactors have potential to achieve high recovery efficiencies and to obtain ammonium-free streams.

However, membrane contactors still need to overcome some challenges when using streams with high concentration of organic matter. Membrane fouling, caused by organic matter and/or suspended solids, can lead to the deposition of solids as a thin cake layer and increase pore clogging [17]. This phenomenon generates a reduction in the flux during long-term operation. Thus, to maintain adequate flux levels, it is necessary to increase energy and chemical consumption with a direct impact on the membrane lifetime and economic feasibility [18]. In this regard, some pre-treatment strategies have been proposed to reduce fouling of membrane contactors, such as UF [19], coagulation–flocculation (C/F) processes [17] or ion exchange [20]. For example, Rivadeneyra et al. [20] used ion exchange technology and observed a maximum chemical oxygen demand (COD) removal efficiency of 70% with an initial COD load of 4500 mg O_2/L. Raghu et al. [21] combined ion exchange with coagulation–flocculation and achieved a COD removal of 80% from an industrial wastewater effluent.

C/F consists of destabilization of colloids by surface modification. This reduces the electrostatic repulsive forces between the particles and leads to the formation of larger flocs with improved settling properties [22]. The most common coagulants and flocculants used are iron and aluminium salts because these chemicals have demonstrated their effectiveness to reduce the chemical oxygen demand (COD) of liquid streams [23,24]. C/F has been widely applied in wastewater treatment applications as it allows removing organic and inorganic matter with relatively low costs [19,25]. For instance, Al-Juboori et al. [26] evaluated the use of PAX/polymer or starch as a coagulant to pre-treat the centrate before a membrane contactor.

Besides C/F, aeration could also be a useful pre-treatment to reduce the amount of chemicals needed to increase the pH before the membrane contactor stage. Garcia-Gonzalez et al. [27] applied low flow-rate aeration and increased the pH above 8.5 before the membrane contactor, which allowed reducing the operating costs of ammonia recovery by 57%. However, to the best of the authors' knowledge, the combination of C/F technology with aeration has not yet been used to pre-treat anaerobic digester centrate prior to a membrane contactor. Therefore, an experimental and economic study is needed to understand how C/F pre-treatment impacts the technical and economic competitiveness of implementing a membrane contactor system for nitrogen recovery.

The aim of this work is to evaluate the combination of C/F, aeration and membrane contactor processes to recover ammoniacal nitrogen from the effluent of an anaerobic digester (centrate). To this end, different operating conditions and chemical reagents were evaluated for the C/F process. After the C/F process, an aeration stage was used to reduce the amount of bicarbonates in the centrate with a direct impact on the amount of chemicals needed for pH adjustment. Subsequently, the pre-treated centrate was fed to a membrane contactor system to understand how pre-treatment conditions impacted the performance of the membrane contactor and ammonium recovery efficiency. Finally, the economic potential of implementing these pre-treatment technologies before the membrane contactor was analyzed.

2. Materials and Methods

2.1. Chemical Reagent and Wastewater Source

Three types of coagulants were used for the coagulation–flocculation tests: (i) aluminium sulphate ($Al_2(SO_4)_3 \cdot 18 \cdot H_2O$) from Panreac® with a 96% of purity, (ii) iron chloride ($FeCl_3$) from Acros Organics® with a 98% of purity and (iii) a commercial coagulant HT20 from Derypol®. On the other hand, a mixture of Magnetite (Fe_3O_4) from Aldrich® with a 98% purity and silicon oxide (SiO_2) from Merck® with a purity of 98% (relation of 30:70%) was used as flocculant.

Different reagents were used for the chromatographic analysis: Methanesulfonic acid (CH_3SO_3H, 99%), sodium hydrogen carbonate ($NaHCO_3$, 99%), anhydrous sodium carbonate (Na_2CO_3, 99%), nitric acid (HNO_3 69%) and sodium hydroxide (NaOH 1 M). All these chemicals were analytical grade reagents and were supplied by Sigma-Aldrich.

The wastewater used in this study was the anaerobic digester centrate from a municipal WWTP located in the region of Barcelona (Spain). The centrate was decanted before the tests for 24 h to reduce its concentration of COD, total suspended solids (TSS) and turbidity. The centrate used for the C/F tests contained COD and total ammonia nitrogen (TAN) concentrations of 786 mg COD/L and 650 mg N/L, respectively, which were within the range reported in the literature [28,29]. It is worth mentioning that the water used for the flocculant tests came from the same location and had a similar ion concentration to that used in the other tests, although it contained a higher COD concentration (1650 mg COD/L).

2.2. Experimental Design

The study was divided into 2 distinct stages (Figure 1). The first stage corresponded to the pre-treatment stage, selection of the optimum coagulant reagent and setting the optimum operating conditions with a specialized experimental design program. The specialized software allowed optimization of the mixing speed, mixing time and sedimentation time to maximize COD, TSS and turbidity removal efficiencies. Besides C/F, an aeration column for the removal of carbonate and the consequent increase in the pH was also considered. In the second stage, the performance of the membrane contactor (pH, concentration factor, ammonium recovery percentage) was tested with the untreated sidestream water and with the pre-treated water to evaluate the effectiveness of the pre-treatment on membrane contactor performance. Finally, an economic analysis was conducted to evaluate the feasibility of the application of this process train.

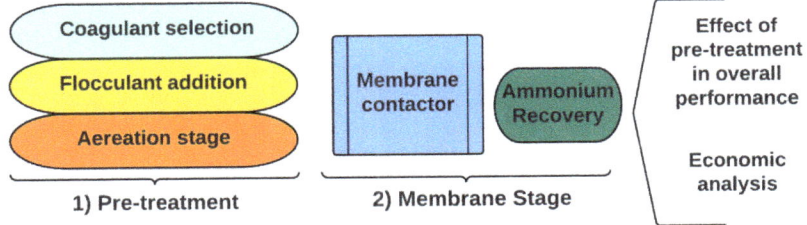

Figure 1. General scheme of the different anaerobic sidestream treatment stages used in the present study.

2.3. Experimental Set-Up

2.3.1. Coagulant Selection

The selection of the best coagulant reagent and dosage was based on combining literature screening and lab-scale tests. Initial bibliographic research was carried out to determine the most common coagulants (Table 1) and it was observed that the most widely used coagulants were based on metals, such as aluminium or iron. After this initial screening, aluminium sulphate ($Al_2(SO_4)_3$), iron chloride ($FeCl_3$) and a commercial coagulant Derypol® HT20 (which is in the category of vegetable coagulants) were chosen.

Table 1. Most frequently used coagulants in water treatment according to bibliography.

N.º	Coagulant Used	Author
1	Tanfloc POP	[30]
2	$Al_2(SO_4)_3$	[31]
3	$FeCl_3$	[32]
4	$FeCl_3$ + Clay Minerals	[33]
5	Lactic Acid	[34]
6	$AlCl_3$	[34]

The lab-scale tests were conducted in a Jar-test set-up (Jar-test *OVAN*® *JT60 E*), which consists of (i) six rotating stirring rods with adjustable speed and height and (ii) six beakers filled with 500 mL of the centrate under study. Two set of experiments were conducted to determine the best coagulant and the dosage strategy for the C/F process.

The first set of experiments was designed to determine the two most favourable coagulants. In these tests, the type of coagulant was changed, while keeping the operating conditions constant. The dosage was set at 50 mg/L and the mixing time was 5 min at a mixing speed of 200 rpm (see Table S1 of the Supplementary Information), which was based on available literature [31,34–36]. The experiments were conducted in triplicate. The second set of experiments was designed (i) to determine the optimum dosage for the two flocculants selected in the previous experiments and (ii) to obtain the most favourable coagulant at this optimum dosage. All the coagulant dosages referred to the quantity of metal added.

The impact of dosage on the efficiency of the C/F process was evaluated for the best coagulant. To this end, the dosage was varied from 10 to 800 mg/L with the Jar-test conditions mentioned above. Table 2 lists the experimental conditions for these tests. The experiments were conducted in triplicate.

Table 2. Experimental conditions for optimal dosage determination.

Coagulant	Dosage (mg/L)	Mixing Time (min)	Mixing Speed (rpm)	Settling Time (min)
Optimal coagulant	10 30 50 100 200 400 800	5	200	30

2.3.2. Determination of the Optimal Operational Conditions for the C/F Process

Once the optimum coagulant chemical and dosage were selected, the most favourable operational parameters (i.e., mixing time, mixing speed and settling time) were determined by using the Jar-test equipment. For this purpose, a design program was used to optimize the number of tests required and to determine the best operational conditions for the C/F process.

The Design Expert® 11 software was used following the factorial design of Box–Behnken, which is based on dependent and independent variables [37]. The dependent variables were those investigated and measured in the study, whereas the independent variables were modified to study their effect on the dependent variables [38]. Table 3 summarizes the dependent variables studied in this work. The coded variables were assigned values of +1 (maximum), 0 (central) and −1 (minimum) depending on the variation of each variable.

Table 3. Individual dependent variables and their range of values.

Variable	Units	Studied Range
Mixing time (MT)	min	5; 15; 25
Mixing velocity (MV)	Rpm	100; 175; 250
Resting time (RT)	min	15; 30; 45

The Box–Behnken design is a rotating or quasi-rotating second-order experimental design based on incomplete three-level factorial designs. The number of experiments (N) needed according to the Box–Behnken design can be obtained from Equation (1).

$$N = 2 \cdot k(k-1) + C_0 \qquad (1)$$

where k is the number of variables, and C_0 is the number of central points [8,39]. In this case, three variables (MT, MV and RT) and five central points were studied resulting in seventeen experiments. The Box–Behnken experimental designs were applied by means of Equations (3) and (4) [8].

$$y = \beta_0 \sum_{i=1}^{k} \beta_i X_i \sum_{i=1}^{k}\sum_{j\geq 1}^{k} \beta_{ij} X_i X_j + \varepsilon \qquad (2)$$

$$y = \beta_0 \sum_{i=1}^{k} \beta_i X_i + \beta_0 \sum_{i=1}^{k} \beta_{ii} X_i^2 + \sum_{i=1}^{k}\sum_{j\geq 1}^{k} \beta_{ij} X_i X_j + \varepsilon \qquad (3)$$

where β_o is the constant factor, β_i represents the coefficients of the linear parameters, k is the number of variables, X_i and X_j represent the independent variables, ε is the residual factor associated with the experiments, y is the dependent variable, β_{ij} represents the coefficients of the interaction parameters and β_{ii} represents coefficients of the quadratic values.

Finally, the software allows for analysis of the obtained results to provide the optimal conditions (e.g., removal of each of COD, TSS and turbidity) through the analysis of graphics and data.

2.3.3. Coagulation Test for the Optimal Coagulant Conditions and Dosage

The optimal coagulant and dosage obtained from stage 1 and 2 were tested to determine the experimental COD, TSS and turbidity removal efficiencies under the most favourable conditions. In this assay, the optimal conditions determined by the two previous tests were applied in the Jar-test equipment and it was verified if the theoretical results provided by the experimental design software were experimentally fulfilled. The experiments were conducted in triplicate.

2.3.4. Flocculation Test

Flocculation tests were conducted to evaluate if combining coagulant and flocculant addition improves solids removal efficiency when compared with stand-alone coagulant addition. The flocculation experiments were carried out with the optimal conditions obtained from the previous experiments and adding different dosages (0–50 mg/L) of a clay-based flocculant (Fe_3O_4(s) and SiO_2(s)) that works effectively with metal-based coagulants for COD reduction [33]. The flocculant was prepared by pulverizing and mixing Fe_3O_4

and SiO$_2$ with a relation of 30% Fe$_3$O$_4$ and 70% of SiO$_2$. Table S2 of the Supplementary Information shows the operational parameters used for the flocculation tests.

2.3.5. Aeration Tests

The possibility of adding an aeration stage [27,40,41] was evaluated: (i) to increase the pH of the centrate and (ii) to reduce the concentration of carbonates present in the sample. The aeration tests were carried out in an open aeration column of 3.5 m height and 30 cm diameter with a capacity of 25 L. The air was introduced at the bottom of the tank through an electric compressor at a flow rate of 2 Nm^3/h. The column was filled with the centrate and a constant air flow rate (364 L/h) was applied for a period of time adequate to cause reactions described by Equations (4)–(6).

$$HCO_{3(aq)}^- + H_{(aq)}^+ \leftrightarrow CO_{2(aq)} + H_2O_{(l)} \qquad (4)$$

$$CO_{2(aq)} \leftrightarrow CO_{2(g)} \qquad (5)$$

$$NH_{4(aq)}^+ \leftrightarrow NH_{3(g)} + H_{(g)}^+ \qquad (6)$$

Thus, these experiments allowed bicarbonate conversion to CO$_2$(g) (aq) /Equation (4) due to the aeration process promoting the removal of dissolved CO$_2$(g) (aq) as CO$_2$(g) (Equation (5)) and consequently increasing the pH. Subsequently, the pH increased allowed the conversion of NH$_4^+$ into ammonia. (Equation (6)). The aeration experiments were performed in duplicate.

2.3.6. Flat-Sheet Membrane Contactor

The different pre-treatment processes were aimed at conditioning the centrate to reduce fouling and clogging in the membrane contactor. A flat-sheet membrane contactor similar to the one used by Hasanoğlu et al. [10] was used in this study. The polytetrafluoroethylene (PTFE) membrane had a surface area of 90 cm^2 and a pore size of 0.2 µm. The pH of the feed solution was increased up to 10.2 with NaOH 1 M, to displace the equilibrium towards NH$_3$. The feed solution was stored in a 5 L tank, whereas the acid stripping solution (0.4 M nitric acid) was stored in a 1.5 L tank. Both tanks were continuously agitated, while nitric acid was continuously added to maintain the pH of the stripping solution in the acidic regime (pH < 2). The feed and stripping solutions were circulated at 450 mL/min in counter current mode towards both sides of the membrane. Further details of the membrane contactor set-up can be found elsewhere [9].

The ammonia flux through the membrane is driven by the difference between the partial pressure on both sides of the membrane, ($p_{NH_3,f} - p_{NH_3,s}$) and the mass transfer coefficient $\left(K_{m(NH_3)}\right)$ (Equation (7)).

$$J_{NH_3} = \frac{K_{m(NH_3)}\left(p_{NH_3,f} - p_{NH_3,s}\right)}{RT} \qquad (7)$$

where $p_{NH_3,s}$ is the partial pressure of ammonia in the shell side (atm), $p_{NH_3,f}$ is the partial pressure in the feed side (atm), $K_{m(NH_3)}$ is the ammonia mass transfer coefficient (m/s), R is the universal gas constant coefficient (0.082 atm·m^3/k mol·K) and T is the temperature of the system (K).

Subsequently, Equation (7) can be expressed as Equation (8) considering that: (i) the partial pressure of ammonia on both sides of the membrane can be assumed as the concentration of ammonia on either side, (ii) the pH does not vary during the experimental procedure, meaning that the concentration of ammonia is proportional to the TAN concentration in the feed solution and (iii) the ammonia partial pressure in the stripping side is negligible [8,14].

$$\ln \frac{C_{0(NH_3)f}}{C_{t(NH_3)f}} = \frac{K_{m(NH_3)} A_m}{V_f} t \tag{8}$$

where A_m is the membrane area (m^2), $C_{0(NH_3)f}$ and $C_{t(NH_3)f}$ are the feed ammonia concentration (mg/L) at the initial time and at the experimental time, respectively, and V_f is the feed volume (m^3).

The tests were conducted for both untreated and pre-treated centrate to evaluate and compare the membrane contactor performance before and after pre-treatment implementation.

2.4. Analytical Methods

The anions and cations were analyzed by an ion chromatography system (Dionex ICS-1000 and ICS-1100 Thermo-Fisher Scientific, USA) equipped with a cationic detector (ICS-1000) and an anionic detector (ICS-1100) and controlled by Chromeleon® chromatographic software. A CS16 column (4 × 250 mm) and an AS23 column (4 × 250 mm) (Phenomenex, Barcelona, Spain) were used for cation and anion determination and quantification, respectively. The mobile phase was a 0.03 mol/L CH$_3$SO$_3$H solution for the cation system, and a mixture of 0.8 mmol/L NaHCO$_3$ and 4.5 mmol/L Na$_2$CO$_3$ for the anion system.

The COD was analyzed through the Standard Method 5220C using a multiparametric photometer HI83224 (Hanna Instruments, Padua, Italy), whereas TSS were analyzed through the Standard Method 2540D [42]. A turbidimeter HI 93703 (Hanna instruments, Padua, Italy) was used to measure the turbidity. Total alkalinity was measured by titration following the Standard Method 2320B and using a T70 titrator (Mettler Toledo, Columbus, OH, United States).

2.5. Economic Analysis

An economic analysis was conducted to evaluate the techno-economic implications of implementing a membrane contactor system for ammonia recovery from the anaerobic digester centrate. Figure S1 of the Supplementary Information shows the configuration evaluated in the economic analysis, which included four different stages: (i) C/F with Al$_2$(SO$_4$)$_3$ to enhance solids sedimentation, (ii) precipitation for suspended solids removal, (iii) aeration to desorb part of the solubilized CO$_2$ and reduce the alkalinity and (iv) membrane contactor system for nitrogen recovery. The membrane contactor system was operated by using an HNO$_3$ trapping solution and considering a relation between the feed and trapping solution flow rate of 1:1. The pH of the feed solution was adjusted to 10.2 with NaOH to displace the NH$_4^+$/NH$_3$ equilibrium towards NH$_3$. The trapping solution was continuously recirculated from the acid tank to the membrane contactor and replaced when the pH increased by up to 6 [28]. The mass balance was obtained considering that the WWTP generated 150 m^3/day of centrate, containing TAN and TSS concentrations of 0.71 g N/L and 0.24 g TSS/L, respectively. Detailed information on the mass balance can be found in Table S3 of the Supplementary Information.

The capital costs, operating costs and revenues were calculated using both lab-scale data and literature average values. The capital costs accounted for membrane contactor, tanks, stirrers, blowers and pumps, whereas the operating costs accounted for energy consumption, sludge disposal, equipment replacement and the purchase of chemicals (i.e., Al$_2$(SO$_4$)$_3$, NaOH and HNO$_3$). Finally, the revenues were obtained considering (i) the commercialization of the produced NH$_4$NO$_3$ and (ii) the lower nitrogen load to be treated in the mainstream of the WWTP. Tables S4 and S5 of the Supplementary Information summarize the main design and cost parameters used for the economic analysis.

The present value (PV) of the gross cost and revenues was calculated for the nitrogen recovery configuration by using Equations (9) and (10), respectively. Subsequently, Equation (11) was used to calculate the net present value (NPV):

$$PV_{GC} = CAPEX + \sum_{t=1}^{T} \frac{OPEX_t}{(1+i)^t} \quad (9)$$

$$PV_R = \sum_{t=1}^{T} \frac{R_t}{(1+i)^t} \quad (10)$$

$$NPV = \sum_{t=1}^{T} \frac{R_t - OPEX_t}{(1+i)^t} - CAPEX \quad (11)$$

where CAPEX is the capital expenditure (EUR), $OPEX_t$ is the operating expenditure at year t (EUR), R_t is the revenue at year t (EUR), PV_{GC} is the PV of the gross cost (EUR), PV_R is the PV of the revenues (EUR), NPV is the net present value (EUR), i is the discount rate (5%) and T is the plant lifetime (20 years).

3. Results and Discussion

The following sections discuss the results concerning the application of C/F and aeration pre-treatments before a membrane contactor. Table 4 shows the COD, TSS, turbidity and ion concentrations of the centrate wastewater used for these tests.

Table 4. Initial centrate characterization.

Parameter	Value	Unity
Sodium	474.4 ± 18.4	mg/L
TAN	650 ± 64.5	mg/L
Potassium	146.6 ± 7.6	mg/L
Magnesium	33.6 ± 13.4	mg/L
Calcium	90.5 ± 26.8	mg/L
Chlorine	348.0 ± 15.4	mg/L
Nitrate	30.7 ± 8.8	mg/L
Phosphate	138.1 ± 30.2	mg/L
Sulphate	37.5 ± 10.8	mg/L
Carbonates	3366.7 ± 792.5	mg/L
Turbidity	275.1 ± 106.2	NTU
COD	786.0 ± 126.7	mg O_2/L
TSS	235.0 ± 104.7	mg/L
pH	8.2 ± 0.1	–

3.1. Coagulant and Dosage Selection for the C/F Process

Table 5 collects the COD and turbidity removal efficiencies for the three coagulants ($FeCl_3$, $Al_2(SO_4)_3$ and Derypol® HT20) analyzed in this study. $Al_2(SO_4)_3$ reported the best COD removal efficiencies (50.2 ± 1.1%), followed by $FeCl_3$ (38.9 ± 0.3%) and Derypol HT20 (36.0 ± 0.3%). Thus, $Al_2(SO_4)_3$ and $FeCl_3$ were selected for the next set of experiments. The turbidity removal efficiencies ranged from 74.2 to 84.7%. The lowest turbidity values were obtained by using $FeCl_3$ (74.2 mg/L) and they were similar to those achieved by Abdessemed et al. [43], which achieved turbidity removal values of 66.1% using $FeCl_3$.

Table 5. Results obtained on COD removal (%) and turbidity reduction for the coagulation assay coagulant test.

Coagulant	COD Removal (%)	Turbidity Reduction (%)
$Al_2(SO_4)_3$	50.2 ± 1.1	82.3 ± 1.1
Derypol HT20	36.0 ± 0.3	84.7 ± 0.4
$FeCl_3$	38.9 ± 0.3	74.2 ± 1.7

Table 6 lists the COD and turbidity removal efficiencies of $Al_2(SO_4)_3$ and $FeCl_3$ for concentrations ranging from 10–800 mg/L. The results showed that $Al_2(SO_4)_3$ provided better COD removal performance in comparison to $FeCl_3$, which reinforces the idea that $Al_2(SO_4)_3$ is the most favourable coagulant–flocculant to be used as a membrane contactor pre-treatment. On the one hand, the COD removal efficiency increased from 42.5 to 51.8% as the $FeCl_3$ concentration increased from 10 to 800 mg/L, respectively. On the other hand, the COD removal efficiency increased from 51.5 to 62.1% as the $Al_2(SO_4)_3$ concentration increased from 10 to 200 mg/L, respectively. However, in the case of $Al_2(SO_4)_3$, dosages above 200 mg/L only led to minimal improvements in the COD removal efficiency. This behaviour is due to the fact that applying coagulant dosages above the optimal level does not lead to considerable improvements [44].

Table 6. Results of water quality improvement for the coagulation experiments (COD removal (%), turbidity reduction (%)) as a function of coagulant type and coagulant dose.

Dosage (mg/L)	$Al_2(SO_4)_3$			$FeCl_3$		
	COD Removal	Turbidity Reduction	pH	COD	Turbidity Reduction	pH
10	51.5 ± 1.2	80.4 ± 2.8	8.0	42.5 ± 0.7	60.3 ± 1.2	8.0
30	56.2 ± 1.0	85.5 ± 4.4	7.7	48.0 ± 0.9	71.2 ± 1.2	7.9
50	50.1 ± 1.7	82.3 ± 3.5	7.4	38.9 ± 1.6	74.2 ± 2.4	7.7
100	41.1 ± 0.9	76.7 ± 1.2	7.1	41.5 ± 1.9	80.6 ± 3.4	7.4
200	62.1 ± 1.2	86.6 ± 4.0	6.9	45.1 ± 2.1	87.9 ± 3.3	7.1
400	66.7 ± 2.5	82.2 ± 1.7	6.1	50.0 ± 1.9	90.4 ± 4.1	6.7
600	64.7 ± 2.1	55.5 ± 2.4	4.3	52.5 ± 1.8	95.7 ± 3.4	6.4
800	66.9 ± 1.0	27.3 ± 3.3	4.1	51.8 ± 1.7	97.0 ± 3.0	5.8

The results also showed that the pH progressively decreased as the coagulant dosage increased. In the case of $Al_2(SO_4)_3$, when the metal ion (Al^{+3}) hydrolyzes in water, it reacts to form complex $(Al(OH))_n^{+(n-3)}$ compounds. This leads to the formation of $CO_2(g)$, which increases the acidity of the solution [23]. From the results of Table 6, it can be concluded that dosing 30 mg/L of Al ($Al_2(SO_4)_3$) can be considered as the optimum strategy because this dosage achieved similar COD removal efficiencies than those achieved above 200 mg/L, while reducing the coagulant dosage more than seven times.

3.2. Optimization of the Operating Conditions for the C/F Process

After selecting the optimum coagulant and dosage ($Al_2(SO_4)_3$, 30 mg Al/L), the impact of the operational conditions (i.e., mixing time, mixing speed and settling time) on the C/F efficiency was evaluated. Seventeen experiments were tested based on the outputs provided by the Design Expert 11 software (see Table S6 for further details on the experimental conditions tested). These experiments were conducted changing the mixing time, the mixing speed and the settling time. Figure 2 shows the theoretical TSS, turbidity and COD removal values obtained from the Design Expert 11 software for the different mixing time and mixing speed conditions at a fixed settling time of 30 min. It is worth mentioning that only the results of 30 min settling time are illustrated because this condition provided the best results when compared with the other settling times. The results highlighted that reducing the mixing time to 5 min and the mixing speed to 100 rpm, would theoretically increase removal values up to 100% in turbidity and suspended solids and up to 70% in COD. Accordingly, the software revealed that there was better removal when mixing time and speed were reduced to the minimum tested values. This behaviour was in agreement with Kan et al. [45], who reported that higher mixing speed did not give a better coagulation performance.

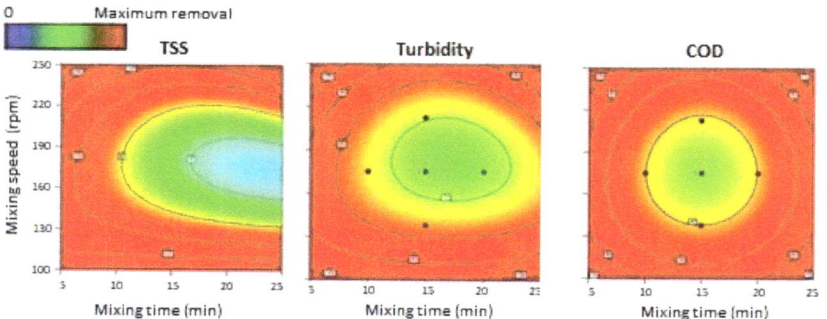

Figure 2. Theoretical TSS, turbidity and COD removal values for different mixing times and mixing speeds, at a fixed settling time of 30 min (graphics obtained from the Design Expert 11 software).

Subsequently, coagulation tests were carried out with the optimum conditions obtained from the software. Table 7 illustrates the results of these tests in terms of TSS, turbidity and COD removal values.

Table 7. Experimental removal using optimal conditions extracted from Design Expert 11. The errors represent standard deviation (n = 3).

Variables	Studied Conditions	Parameters	Experimental Removal (%)
Mixing time	5 min	COD	58.1 ± 0.3
Mixing velocity	100 rpm	TSS	94.9 ± 0.2
Settling time	30 min	Turbidity	89.8 ± 0.8

The removal values showed an improvement compared with the previous test (58.1 ± 0.3 COD, 94.9 ± 0.2 TSS and 89.8 ± 0.8 turbidity), although the values predicted by the design software were not achieved. Guimarães et al. [46] tested several coagulants (including aluminium sulphate at 40 mg/L Al) and reached COD removal efficiencies (38%) below those achieved in this study (58%). On the other hand, Salem et al. [47] reported turbidity removal efficiencies of 86%, which were similar than those achieved in this study (90%).

3.3. Flocculation Stage

Figure 3 shows the obtained values of COD and turbidity removal for the different dosages of flocculant Fe_3O_4/SiO_2 (30–70% (w/w)) added. A test without flocculant was also conducted, which consisted of applying the optimum dosages and parameters obtained from the coagulant stage tests (Section 3.1). The results illustrated maximum COD removal (89.7%) when the flocculant dosage was 10 mg/L and maximum turbidity removal (83.6%) when the dosage was increased up to 30 mg/L. In all the tests, the TSS removal values remained practically constant around 95%. Sultana et al. [48] treated wastewater with an organic concentration (745 mg O_2/L) similar to the present study water (786 mg O_2/L) using aluminium sulphate coagulant and clay-based flocculant. The authors obtained COD removal efficiencies of 46.7%, which are below those achieved in this study. On the other hand, Preston et al. [49] worked with wastewater with a similar turbidity (300 NTU) than that of the present study (275 NTU), using aluminium sulphate as coagulant and Moringa as natural flocculant, and reached a similar turbidity removal of 96.2%. Overall, Figure 3 results revealed that the addition of $Fe_3O_4(s)/SiO_2(s)$ only led to small improvements concerning removal values.

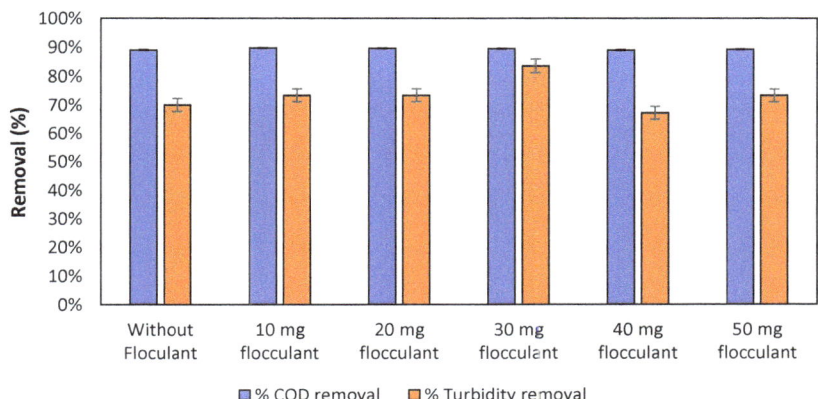

Figure 3. Removal of COD (%) and turbidity (%) from anaerobic centrate after $Fe_3O_4(s)/SiO_2(s)$ addition.

According to the results obtained, it could be concluded that the addition of coagulant + flocculant did not provide a consistent positive improvement compared to the addition of only coagulant.

3.4. Aeration Stage

An aeration step was added after coagulation–flocculation to promote $CO_2(g)$ stripping to reduce alkalinity and increase the pH before the membrane contactor system [41]. Figure 4 shows the evolution of HCO_3^- removal and pH over the aeration time. The HCO_3^- present in the centrate was reduced by about 50% after 240 min of constant aeration, although almost 30% of elimination was reached after 15 min. The results showed that after 1 h of operation time, a compromise between carbonates removal (34%) and pH increase (8.83) was achieved, although higher removal values could be reached at expenses of higher times of operation. This agrees with the pH results, where a sudden increase was observed after 15 min of aeration, reaching a constant value after 240 min. It is worth mentioning that the application of aeration could also lead to NH_3 losses due to volatilization, although they did not account for more than 2% in our study (data not shown).

García-González et al. [27] also used an aeration system as a membrane contactor pre-treatment stage. The aeration system increased the pH above 8.5, which allowed the partial displacement of NH_4^+/NH_3 equilibrium towards NH_3 without the addition of external chemicals. Besides technical aspects, aeration implementation has the potential to reduce the total cost of the process by 70% due to the reduction in alkaline purchasing cost (Dube et al., 2016). It is also relevant to mention that it is possible to use recycled chemicals to further reduce the operating cost of the system.

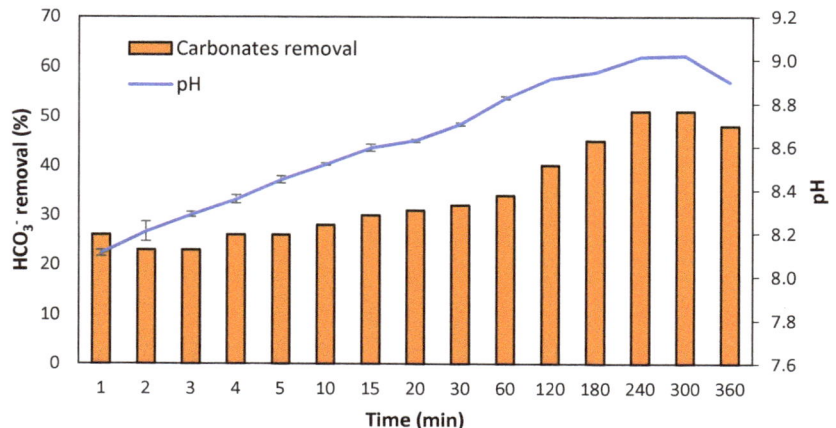

Figure 4. Variation of pH and the efficiency of HCO_3^- removal with time in the aeration stage.

3.5. Flat-Sheet Membrane Contactor Stage

Figure 5 shows the membrane contactor results for the treated and untreated centrate during the experimental time. The results illustrated that the TAN recovery efficiency increased from 7.5 to 66.6% after implementing the pre-treatment train (Figure 5B). This highlighted that C/F and aeration pre-treatments are crucial to improve the TAN recovery efficiency from the anaerobic digester centrate using membrane contactors. In the case of the pre-treated centrate, the TAN concentration in the feed solution decreased from 0.9 g/L to 0.3 g/L (Figure 5A), whereas the TAN concentration in the acid solution increased from 0 to 2.7 g/L (Figure 5C). This agrees with the outputs of other studies recovering TAN using membrane contactors [10,14]. Similarly, the results obtained in terms of concentration factor are in line with the results of TAN in the acid tank. The concentration factor corresponded to 3.8 and was obtained from the relationship between the ammonium concentration in the acid tank (3.5 g/L) and the initial ammonium concentration in the feed tank (0.9 g/L).

Besides the TAN recovery efficiency, the ammonia mass transfer coefficient (K_m) was also calculated. The K_m of the pre-treated centrate (1.04×10^{-5} m/s) was almost two orders of magnitude higher than that achieved with the non-treated centrate (7.80×10^{-7} m/s). These results corroborate that the implementation of C/F and aeration before the membrane contactor is needed to achieve efficient TAN recoveries from the anaerobic digester centrate. Interestingly, the K_m achieved in the present study with the pre-treated centrate and flat-sheet membrane contactors was higher in comparison with K_m values reported in the literature using hollow fibre contactors (Table 8). The highest Km achieved in this study could be attributed to the high efficiency of the pre-treatment process since COD, TSS and turbidity were substantially reduced. This led to almost negligible fouling, no clogging and no reduction in ammonia transfer during the operation of the membrane contactor for the pre-treated centrate.

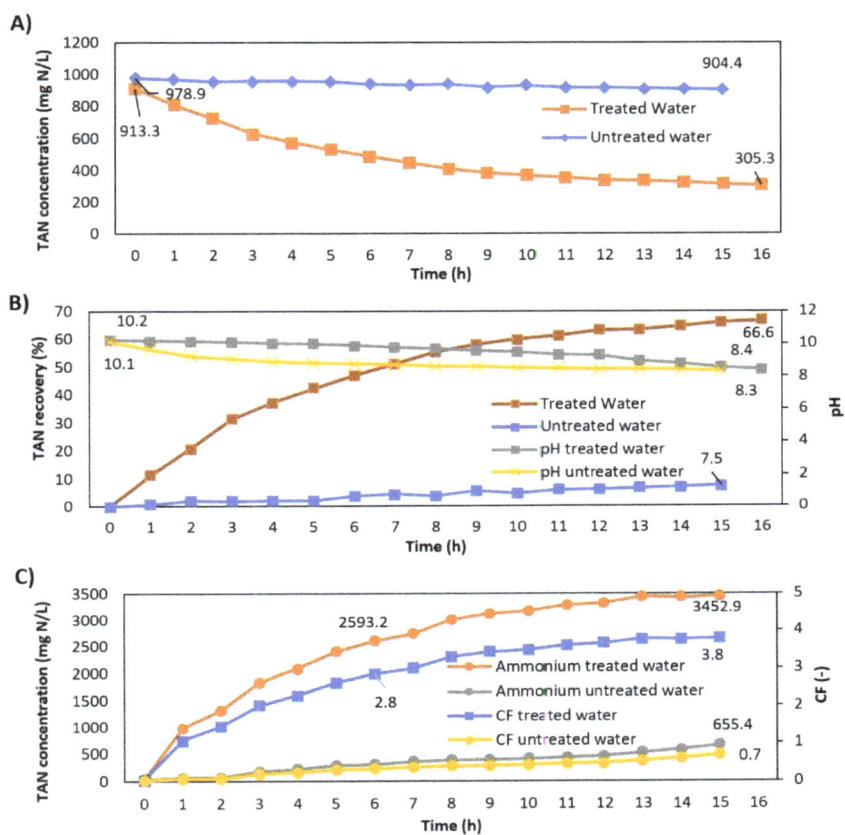

Figure 5. Membrane contactor results during operation: (**A**) TAN concentration evolution in the feed tank for pre-treated and untreated centrate, (**B**) TAN recovery and pH variation and (**C**) TAN concentration evolution and concentration factor in the acid tank.

Table 8. Km values obtained in different studies with hollow fibre liquid–liquid membrane contactors.

Study	Mass Transfer (m/s)	Flow Rate (mL/min)	Type of Contactor	Initial [NH_3] g/L	% Removal	Pre-Treatment	Water
This study	1.0×10^{-5}	450 mL/min	FS-LLMC (PTFE)	0.9	66.6	C/F and Aeration	Sidestream
[14]	8.8×10^{-7}	450 mL/min	HF-LLMC (PP)	3.9	76.1	Ion-exchange	Sidestream
[50]	8.9×10^{-6}	920 mL/min	HF-LLMC (PVDF)	2–10	90.0	-	Synthetic
[15]	2.9×10^{-7}	770 mL/min	HF-LLMC (PMP)	5.0	93.1	-	Synthetic
[51]	1.89×10^{-6}	450 mL/min	HF-LLMC (PP)	1.7	85	Sorption	Sidestream

The results of this study clearly confirmed that, in the case of a centrate with a high concentration of organic matter and suspended solids, pre-treatment using C/F and aeration can improve the performance of the membrane contactor. The pre-treatment application allows avoiding operating problems, such as loss of hydrophobicity due to biofouling and clogging of the membrane, improving the membrane recovery performance and making it technically feasible.

3.6. Economic Analysis

3.6.1. Economic Feasibility of Membrane Contactor Implementation

Figure 6 illustrates the economic balance of implementing a membrane contactor system to recover ammonia from the anaerobic digester centrate. The results show that membrane contactor implementation in a WWTP led to a negative NPV. Ammoniacal nitrogen recovery from the anaerobic digester centrate allows (i) achieving revenues from the ammonium nitrate fertilizer produced and (ii) reducing the nitrogen load to the mainstream of the WWTP with a direct impact on energy consumption. However, these revenues did not offset the additional costs associated with the construction and operation of the different process units. From these results, it is conceivable to state that further improvements are still necessary to make nitrogen recovery through membrane contactors economically attractive. Besides economic considerations, ammoniacal nitrogen recovery from the anaerobic digester centrate has the potential to reduce disturbances in the mainstream nitrification–denitrification process and improve the WWTP effluent quality [52,53].

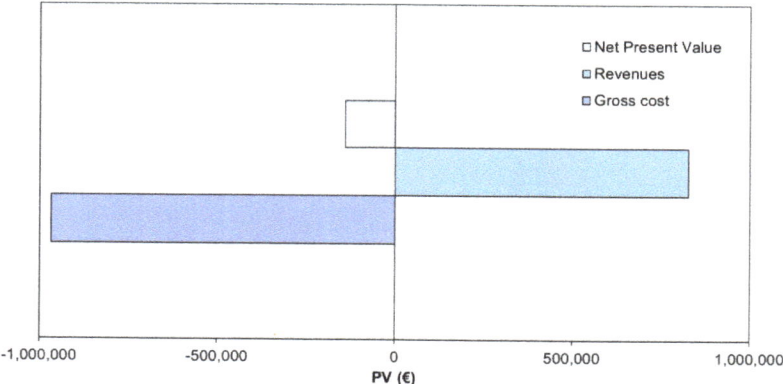

Figure 6. Gross cost, revenues and net present value for the nitrogen recovery scenario under study.

The membrane contactor system was the costliest process (55%), followed by aeration (36%) and coagulation–flocculation (9%) (see Figure S2A of the Supplementary Information). The high cost of the membrane contactor system is mainly associated with the intensive consumption of HNO_3 and, to a lesser extent, NaOH. In this regard, chemical consumption features the highest cost contribution, representing 57% of the gross cost (Figure S2B of the Supplementary Information). Energy consumption also represents an important fraction of the gross cost (34.1%), which can be attributed to the high energy requirements of the air blower system. These results highlight that chemical consumption and aeration requirements are two important operational factors influencing the economic competitiveness of the system.

3.6.2. Sensitivity Analysis

Figure 7 shows the sensitivity analysis for a ± 30% variation of the main economic parameters. The results illustrate that the NH_4NO_3 price featured the highest impact on the NPV. This is particularly important considering that the cost of fertilizers is expected to increase in the future due to the progressive increase in fuel and electricity costs [54]. To better understand how NH_4NO_3 price impacts the economic balance of the system, a sensitivity analysis was conducted for NH_4NO_3 prices between 0.30 and 0.70 EUR/kg (Figure 8). The results show that the NPV of ammoniacal nitrogen recovery increased from EUR −350,000 to 300,000 as the NH_4NO_3 price increased from 0.30 to 0.70 EUR/kg, respectively. This implies that a positive NPV was achieved at NH_4NO_3 prices above 0.52 EUR/kg. Overall, these results highlight that the commercialization of the produced NH_4NO_3 fertilizer has the potential to make membrane contactor configuration economically feasible.

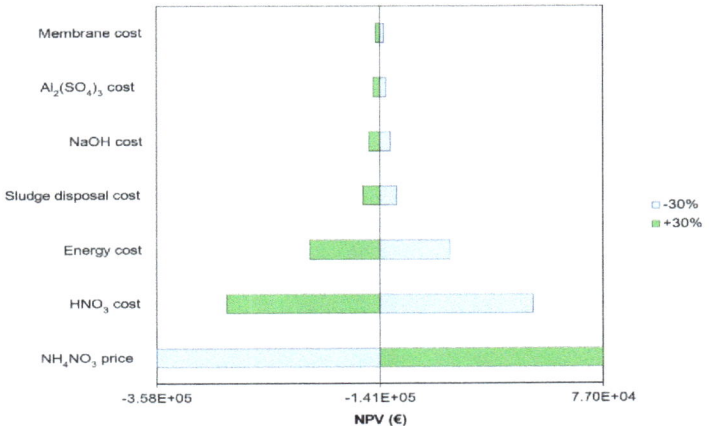

Figure 7. Sensitivity analysis for a ±30% variation of the main economic parameters.

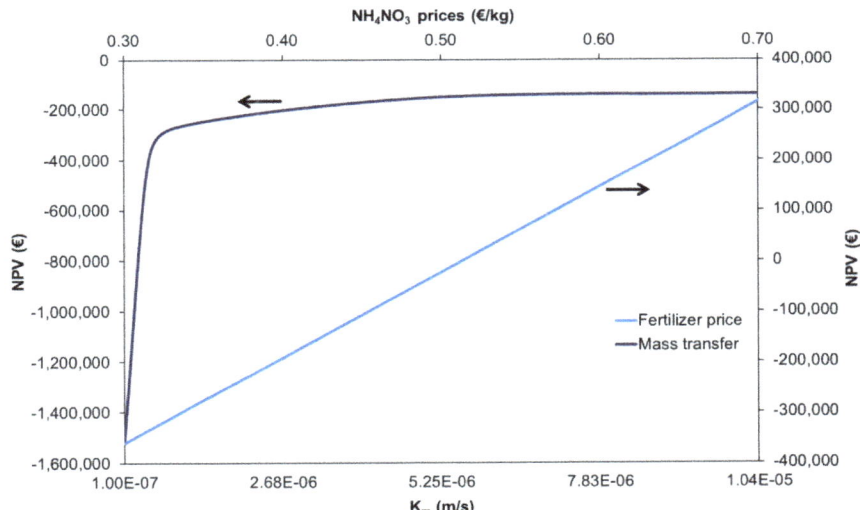

Figure 8. Sensitivity analysis for the NH_4NO_3 prices and mass transfer coefficient (K_m). *The arrows indicated the y-axis corresponding to each line of the graph.*

Nitric acid and electricity costs also feature a noticeable impact on the NPV of the system (Figure 7). This reinforces the idea that chemical consumption and aeration requirements are two important aspects influencing the economics of this configuration. Conversely, membrane purchase cost variation did not lead to important changes in the NPV. The low impact of the membrane purchase cost on NPV can be attributed to the high K_m coefficient (1.04×10^{-5} m/s) achieved in this study, which is substantially higher than in other studies [5,55]. However, it is worth mentioning that the K_m could be substantially lower during long-term membrane contactor operation due to organic and inorganic membrane fouling development on the membrane surface. For this reason, a sensitivity analysis was conducted to evaluate the impact of K_m on the economic balance of the nitrogen recovery scheme under study (Figure 8).

The results show that the NPV slightly decreased from EUR $-140,000$ to $-260,000$ as the K_m decreased from 1×10^{-5} to 1×10^{-6} m/s, respectively (Figure 8). However, a sharp decrease in the NPV was observed at K_m values below 1×10^{-6} m/s. These results highlight that K_m could have a large influence on the economic balance due to its impact on the membrane requirements of the system. For this reason, it is important to look for suitable physical and chemical cleaning strategies able to achieve effective control of long-term membrane fouling without excessive consumption of chemicals and energy.

4. Conclusions

This study evaluated the implementation of C/F and aeration pre-treatments prior to a membrane contactor stage to recover nitrogen from the anaerobic digester centrate. The results revealed that dosing aluminium sulphate at 30 mg Al/L was the best strategy for the coagulation process. The maximum COD, turbidity and TSS removals (58 and 95 and 90%, respectively) were achieved with a mixing speed of 100 rpm, a mixing time of 5 min and a settling time of 30 min. The flocculation stage using $Fe_3O_4(s)/SiO_2(s)$ (30–70% (w/w)) did not lead to noticeable improvements in the removal efficiencies. The aeration stage reduced HCO_3^- content up to 51% and increased the pH up to 9, without the addition of external chemicals. Subsequently, the effluent from the C/F and aeration stages was fed to the membrane contactor for nitrogen recovery. The membrane contactor recovered 67% of TAN and achieved a concentration factor in the acid solution of 3.8. Finally, the techno-economic evaluation showed that the combination of C/F, aeration and membrane contactor has the potential to be an economically competitive alternative for nitrogen recovery.

Supplementary Materials: The following supporting information can be downloaded at: https://www.mdpi.com/article/10.3390/membranes12121251/s1; Table S1: Initial conditions for coagulant selection; Table S2: Flocculation parameters; Figure S1: Schematic representation of the nitrogen recovery scheme; Table S3: Main flow data for the nitrogen recovery scenario under study; Table S4: Main design parameters used for the economic evaluation; Table S5: Main economic parameters used for the economic evaluation; Table S6: Experiment sets of Design Expert 11 software; Figure S2: Gross cost contribution of the nitrogen recovery scenario under study for: (A) the different processes and (B) for the different capital and operating costs. References [56–65] are cited in Supplementary Materials.

Author Contributions: Conceptualization, M.A.-M., S.V., M.R. and J.L.C.; methodology, M.A.-M., M.R. and X.V.; validation, S.V., M.R., X.V. and C.V.; formal analysis, M.A.-M. and M.R.; investigation, M.A.-M., S.V., M.R., X.V., C.V. and J.L.C.; resources, J.L.C.; data curation M.A.-M., M.R. and X.V.; writing—original draft preparation, M.A.-M. and S.V.; writing—review and editing, M.A.-M., S.V., M.R., X.V., C.V. and J.L.C.; visualization, C.V.; supervision, J.L.C.; project administration, J.L.C. and funding acquisition, J.L.C. All authors have read and agreed to the published version of the manuscript.

Funding: This research was supported by the resources recycling from agri-food urban and industrial wastes by integration of hybrid separation processes (W4V) project (PID2020-114401RB-C21) financed by the Agencia Española de Investigación, by the R^2MIT project (CTM2017-85346-R) financed by the Spanish Ministry of Economy and Competitiveness, and by the Catalan Government (ref. 2017-SGR-312), Spain.

Institutional Review Board Statement: Not applicable.

Informed Consent Statement: Not applicable.

Data Availability Statement: Not applicable.

Acknowledgments: Miguel Aguilar Moreno gratefully acknowledges the Agencia Española de Investigación (PRE2018-086214), and Xanel Vecino acknowledges the Spanish Ministry of Science and Innovation for her financial support under project PID2019-103873RJ-I00. Additionally, the authors acknowledge the Open Innovation—Research Translation and Applied Knowledge Exchange in Practice through University–Industry Cooperation (OpenInnoTrain), grant agreement number

(GAN): 823971, H2020-MSCA-RISE-2018-823971. Finally, the author would like to thank C. Mesa from Aigües de Barcelona for the supply of samples of municipal sewage sludge anaerobic sidestream as well as the information provided on the analytical characterization techniques and the inputs provided for the development of the experimental work.

Conflicts of Interest: The authors declare no conflict of interest.

References

1. Beckinghausen, A.; Odlare, M.; Thorin, E.; Schwede, S. From Removal to Recovery: An Evaluation of Nitrogen Recovery Techniques from Wastewater. *Appl. Energy* **2020**, *263*, 114616. [CrossRef]
2. Lee, W.; An, S.; Choi, Y. Ammonia Harvesting via Membrane Gas Extraction at Moderately Alkaline PH: A Step toward Net-Profitable Nitrogen Recovery from Domestic Wastewater. *Chem. Eng. J.* **2021**, *405*, 126662. [CrossRef]
3. Razon, L.F. Reactive Nitrogen: A Perspective on Its Global Impact and Prospects for Its Sustainable Production. *Sustain. Prod. Consum.* **2018**, *15*, 35–48. [CrossRef]
4. González Montiel, J.M. Acetato de Etilo En La Industria. *J. Chem. Inf. Model.* **2008**, *53*, 287.
5. Darestani, M.; Haigh, V.; Couperthwaite, S.J.; Millar, G.J.; Nghiem, L.D. Hollow Fibre Membrane Contactors for Ammonia Recovery: Current Status and Future Developments. *J. Environ. Chem. Eng.* **2017**, *5*, 1349–1359. [CrossRef]
6. Puyol, D.; Batstone, D.J.; Hülsen, T.; Astals, S.; Peces, M.; Krömer, J.O. Resource Recovery from Wastewater by Biological Technologies: Opportunities, Challenges, and Prospects. *Front. Microbiol.* **2017**, *7*, 1–23. [CrossRef]
7. Kurniawan, T.A.; Lo, W.H.; Chan, G.Y.S. Physico-Chemical Treatments for Removal of Recalcitrant Contaminants from Landfill Leachate. *J. Hazard. Mater.* **2006**, *129*, 80–100. [CrossRef] [PubMed]
8. Licon Bernal, E.E.; Maya, C.; Valderrama, C.; Cortina, J.L. Valorization of Ammonia Concentrates from Treated Urban Wastewater Using Liquid-Liquid Membrane Contactors. *Chem. Eng. J.* **2016**, *302*, 641–649. [CrossRef]
9. Reig, M.; Vecino, X.; Gibert, O.; Valderrama, C.; Cortina, J.L. Study of the Operational Parameters in the Hollow Fibre Liquid-Liquid Membrane Contactors Process for Ammonia Valorisation as Liquid Fertiliser. *Sep. Purif. Technol.* **2021**, *255*, 117768. [CrossRef]
10. Hasanoğlu, A.; Romero, J.; Pérez, B.; Plaza, A. Ammonia Removal from Wastewater Streams through Membrane Contactors: Experimental and Theoretical Analysis of Operation Parameters and Configuration. *Chem. Eng. J.* **2010**, *160*, 530–537. [CrossRef]
11. Hermassi, M.; Valderrama, C.; Gibert, O.; Moreno, N.; Querol, X.; Batis, N.H.; Cortina, J.L. Recovery of Nutrients (N-P-K) from Potassium-Rich Sludge Anaerobic Digestion Side-Streams by Integration of a Hybrid Sorption-Membrane Ultrafiltration Process: Use of Powder Reactive Sorbents as Nutrient Carriers. *Sci. Total Environ.* **2017**, *599–600*, 422–430. [CrossRef] [PubMed]
12. Wan, C.; Ding, S.; Zhang, C.; Tan, X.; Zou, W.; Liu, X.; Yang, X. Simultaneous Recovery of Nitrogen and Phosphorus from Sludge Fermentation Liquid by Zeolite Adsorption: Mechanism and Application. *Sep. Purif. Technol.* **2017**, *180*, 1–12. [CrossRef]
13. Serra-Toro, A.; Vinardell, S.; Astals, S.; Madurga, S.; Llorens, J.; Mata-Álvarez, J.; Mas, F.; Dosta, J. Ammonia Recovery from Acidogenic Fermentation Effluents Using a Gas-Permeable Membrane Contactor. *Bioresour. Technol.* **2022**, *356*, 127273. [CrossRef] [PubMed]
14. Vecino, X.; Reig, M.; Bhushan, B.; Gibert, O.; Valderrama, C.; Cortina, J.L. Liquid Fertilizer Production by Ammonia Recovery from Treated Ammonia-Rich Regenerated Streams Using Liquid-Liquid Membrane Contactors. *Chem. Eng. J.* **2019**, *360*, 890–899. [CrossRef]
15. Sheikh, M.; Reig, M.; Vecino, X.; Lopez, J.; Rezakazemi, M.; Valderrama, C.A.; Cortina, J.L. Liquid–Liquid Membrane Contactors Incorporating Surface Skin Asymmetric Hollow Fibres of Poly(4-Methyl-1-Pentene) for Ammonium Recovery as Liquid Fertilisers. *Sep. Purif. Technol.* **2022**, *283*, 120212. [CrossRef]
16. Sancho, I.; Licon, E.; Valderrama, C.; de Arespacochaga, N.; López-Palau, S.; Cortina, J.L. Recovery of Ammonia from Domestic Wastewater Effluents as Liquid Fertilizers by Integration of Natural Zeolites and Hollow Fibre Membrane Contactors. *Sci. Total Environ.* **2017**, *584–585*, 244–251. [CrossRef]
17. Leiknes, T.O. The Effect of Coupling Coagulation and Flocculation with Membrane Filtration in Water Treatment: A Review. *J. Environ. Sci.* **2009**, *21*, 8–12. [CrossRef]
18. Zarebska, A.; Nieto, D.R.; Christensen, K.V.; Norddahl, B. Ammonia Recovery from Agricultural Wastes by Membrane Distillation: Fouling Characterization and Mechanism. *Water Res.* **2014**, *56*, 1–10. [CrossRef] [PubMed]
19. Jiang, J.Q. The Role of Coagulation in Water Treatment. *Curr. Opin. Chem. Eng.* **2015**, *8*, 36–44. [CrossRef]
20. Rivadeneyra, G.; Flores, M.C.; Alvarado, A.; Norma, A.; Cantú, V. Utilización de Una Resina de Intercambio Iónico Para El Desarrollo de Biopelícula Aerobia Para El Tratamiento de Agua Residual Industrial Combinada. *Conciencia Tecnológica.* **2007**, *34*, 41–42.
21. Raghu, S.; Ahmed Basha, C. Chemical or Electrochemical Techniques, Followed by Ion Exchange, for Recycle of Textile Dye Wastewater. *J. Hazard. Mater.* **2007**, *149*, 324–330. [CrossRef] [PubMed]
22. Dosta, J.; Rovira, J.; Galí, A.; Macé, S.; Mata-Álvarez, J. Integration of a Coagulation/Flocculation Step in a Biological Sequencing Batch Reactor for COD and Nitrogen Removal of Supernatant of Anaerobically Digested Piggery Wastewater. *Bioresour. Technol.* **2008**, *99*, 5722–5730. [CrossRef] [PubMed]

23. Krupińska, I. The Effect of the Type of Hydrolysis of Aluminum Coagulants on the Effectiveness of Organic Substances Removal from Water. *Desalin. Water Treat.* **2020**, *186*, 171–180. [CrossRef]
24. Postolachi, L.; Rusu, V.; Lupascu, T. Effect of Aluminium Sulphate Aging on Coagulation Process for the Prut River Water Treatment. *Chem. J. Mold.* **2016**, *11*, 27–32. [CrossRef] [PubMed]
25. Taboada-Santos, A.; Rivadulla, E.; Paredes, L.; Carballa, M.; Romalde, J.; Lema, J.M. Comprehensive Comparison of Chemically Enhanced Primary Treatment and High-Rate Activated Sludge in Novel Wastewater Treatment Plant Configurations. *Water Res.* **2020**, *169*, 115258. [CrossRef]
26. Al-Juboori, R.A.; Uzkurt Kaljunen, J.; Righetto, I.; Mikola, A. Membrane Contactor Onsite Piloting for Nutrient Recovery from Mesophilic Digester Reject Water: The Effect of Process Conditions and Pre-Treatment Options. *Sep. Purif. Technol.* **2022**, *303*, 122250. [CrossRef]
27. García-González, M.C.; Vanotti, M.B.; Szogi, A.A. Recovery of Ammonia from Swine Manure Using Gas-Permeable Membranes: Effect of Aeration. *J. Environ. Manage.* **2015**, *152*, 19–26. [CrossRef]
28. Richter, L.; Wichern, M.; Grömping, M.; Robecke, U.; Haberkamp, J. Ammonium Recovery from Process Water of Digested Sludge Dewatering by Membrane Contactors. *Water Pract. Technol.* **2020**, *15*, 84–91. [CrossRef]
29. Eskicioglu, C.; Galvagno, G.; Cimon, C. Approaches and Processes for Ammonia Removal from Side-Streams of Municipal Effluent Treatment Plants. *Bioresour. Technol.* **2018**, *268*, 797–810. [CrossRef]
30. Huang, A.K.; Veit, M.T.; Juchen, P.T.; Gonçalves, G.D.C.; Palácio, S.M.; Cardoso, C.D.O. Sequential Process of Coagulation/Flocculation/Sedimentation—Adsorption—Microfiltration for Laundry Effluent Treatment. *J. Environ. Chem. Eng.* **2019**, *7*, 103226. [CrossRef]
31. Fragoso, R.A.; Duarte, E.A.; Paiva, J. Contribution of Coagulation-Flocculation Process for a More Sustainable Pig Slurry Management. *Water. Air. Soil Pollut.* **2015**, *226*, 4–9. [CrossRef]
32. Verma, S.; Prasad, B.; Mishra, I.M. Pretreatment of Petrochemical Wastewater by Coagulation and Flocculation and the Sludge Characteristics. *J. Hazard. Mater.* **2010**, *178*, 1055–1064. [CrossRef] [PubMed]
33. Aygun, A.; Yilmaz, T. Improvement of Coagulation-Flocculation Process for Treatment of Detergent Wastewaters Using Coagulant Aids. *Int. J. Chem. Environ. Eng.* **2010**, *1*, 97–101.
34. Devesa-Rey, R.; Bustos, G.; Cruz, J.M.; Moldes, A.B. Evaluation of Non-Conventional Coagulants to Remove Turbidity from Water. *Water. Air. Soil Pollut.* **2012**, *223*, 591–598. [CrossRef]
35. Hu, H.; Ding, L.; Geng, J.; Huang, H.; Xu, K.; Ren, H. Effect of Coagulation on Dissolved Organic Nitrogen (DON) Bioavailability in Municipal Wastewater Effluents. *J. Environ. Chem. Eng.* **2016**, *4*, 2536–2544. [CrossRef]
36. Wongcharee, S.; Aravinthan, V.; Erdei, L. Removal of Natural Organic Matter and Ammonia from Dam Water by Enhanced Coagulation Combined with Adsorption on Powdered Composite Nano-Adsorbent. *Environ. Technol. Innov.* **2020**, *17*, 100557. [CrossRef]
37. Ferreira, S.L.C.; Bruns, R.E.; Ferreira, H.S.; Matos, G.D.; David, J.M.; Brandão, G.C.; da Silva, E.G.P.; Portugal, L.A.; dos Reis, P.S.; Souza, A.S.; et al. Box-Behnken Design: An Alternative for the Optimization of Analytical Methods. *Anal. Chim. Acta* **2007**, *597*, 179–186. [CrossRef]
38. Vecino, X.; Reig, M.; Valderrama, C.; Cortina, J.L. Ion-Exchange Technology for Lactic Acid Recovery in Downstream Processing: Equilibrium and Kinetic Parameters. *Water* **2021**, *13*, 1572. [CrossRef]
39. Apostol, G.; Kouachi, R.; Constantinescu, I. Optimization of Coagulation-Flocculation Process with Aluminum Sulfate Based on Response Surface Methodology. *UPB Sci. Bull. Ser. B Chem. Mater. Sci.* **2011**, *73*, 77–84.
40. Vanotti, M.B.; Dube, P.J.; Szogi, A.A.; García-González, M.C. Recovery of Ammonia and Phosphate Minerals from Swine Wastewater Using Gas-Permeable Membranes. *Water Res.* **2017**, *112*, 137–146. [CrossRef]
41. Dube, P.J.; Vanotti, M.B.; Szogi, A.A.; García-González, M.C. Enhancing Recovery of Ammonia from Swine Manure Anaerobic Digester Effluent Using Gas-Permeable Membrane Technology. *Waste Manag.* **2016**, *49*, 372–377. [CrossRef] [PubMed]
42. Lenore, S.; Clesceri, A.E.; Greenberg, A.D.E. *APHA Standard Methods for the Examination of Water and Wastewater*, 20th ed.; American Public Health Association: Washington, DC, USA, 1999; p. 1496.
43. Abdessemed, D.; Nezzal, G.; Ben Aim, R. Coagulation-Adsorption-Ultrafiltration for Wastewater Treatment and Reuse. *Desalination* **2000**, *131*, 307–314. [CrossRef]
44. Duan, J.; Gregory, J. Coagulation by Hydrolysing Metal Salts. *Adv. Colloid Interface Sci.* **2003**, *100–102*, 475–502. [CrossRef]
45. Kan, C.; Huang, C.; Pan, J.R. Time Requirement for Rapid-Mixing in Coagulation. *Colloids Surfaces A Physicochem. Eng. Asp.* **2002**, *203*, 1–9. [CrossRef]
46. Guimarães, N.R.; Dörr, F.; Marques, R.d.O.; Pinto, E.; Ferreira Filho, S.S. Removal Efficiency of Dissolved Organic Matter from Secondary Effluent by Coagulation-Flocculation Processes. *J. Environ. Sci. Health Part A Toxic/Hazardous Subst. Environ. Eng.* **2020**, *56*, 161–170. [CrossRef] [PubMed]
47. Salem, A.T.; AL-Musawi, N.O. Water Treatment With Conventional and Alternative Coagulants: A Review. *J. Eng.* **2021**, *27*, 20–28. [CrossRef]
48. Sultana, S.; Karmaker, B.; Saifullah, A.S.M.; Galal Uddin, M.; Moniruzzaman, M. Environment-Friendly Clay Coagulant Aid for Wastewater Treatment. *Appl. Water Sci.* **2022**, *12*, 6. [CrossRef]
49. Preston, K.; Lantagne, D.; Kotlarz, N.; Jellison, K. Turbidity and Chlorine Demand Reduction Using Alum and Moringa Flocculation before Household Chlorination in Developing Countries. *J. Water Health* **2010**, *8*, 60–70. [CrossRef] [PubMed]

50. Liu, H.; Wang, J. Separation of Ammonia from Radioactive Wastewater by Hydrophobic Membrane Contactor. *Prog. Nucl. Energy* **2016**, *86*, 97–102. [CrossRef]
51. Vecino, X.; Reig, M.; Gibert, O.; Valderrama, C.; Cortina, J.L. Integration of Liquid-Liquid Membrane Contactors and Electrodialysis for Ammonium Recovery and Concentration as a Liquid Fertilizer. *Chemosphere* **2020**, *245*, 125606. [CrossRef]
52. Rodriguez-Garcia, G.; Frison, N.; Vázquez-Padín, J.R.; Hospido, A.; Garrido, J.M.; Fatone, F.; Bolzonella, D.; Moreira, M.T.; Feijoo, G. Life Cycle Assessment of Nutrient Removal Technologies for the Treatment of Anaerobic Digestion Supernatant and Its Integration in a Wastewater Treatment Plant. *Sci. Total Environ.* **2014**, *490*, 871–879. [CrossRef] [PubMed]
53. Vinardell, S.; Astals, S.; Koch, K.; Mata-Alvarez, J.; Dosta, J. Co-Digestion of Sewage Sludge and Food Waste in a Wastewater Treatment Plant Based on Mainstream Anaerobic Membrane Bioreactor Technology: A Techno-Economic Evaluation. *Bioresour. Technol.* **2021**, *330*, 124978. [CrossRef] [PubMed]
54. Panos, E.; Densing, M. The Future Developments of the Electricity Prices in View of the Implementation of the Paris Agreements: Will the Current Trends Prevail, or a Reversal Is Ahead? *Energy Econ.* **2019**, *84*, 104476. [CrossRef]
55. Noriega-Hevia, G.; Serralta, J.; Borrás, L.; Seco, A.; Ferrer, J. Nitrogen Recovery Using a Membrane Contactor: Modelling Nitrogen and PH Evolution. *J. Environ. Chem. Eng.* **2020**, *8*, 103880. [CrossRef]
56. Verrecht, B.; Maere, T.; Nopens, I.; Brepols, C.; Judd, S. The Cost of a Large-Scale Hollow Fibre MBR. *Water Res.* **2010**, *44*, 5274–5283. [CrossRef]
57. Noriega-Hevia, G.; Serralta, J.; Seco, A.; Ferrer, J. Economic Analysis of the Scale-up and Implantation of a Hollow Fibre Membrane Contactor Plant for Nitrogen Recovery in a Full-Scale Wastewater Treatment Plant. *Sep. Purif. Technol.* **2021**, *275*, 119128. [CrossRef]
58. Vinardell, S.; Astals, S.; Mata-Alvarez, J.; Dosta, J. Techno-Economic Analysis of Combining Forward Osmosis-Reverse Osmosis and Anaerobic Membrane Bioreactor Technologies for Municipal Wastewater Treatment and Water Production. *Bioresour. Technol.* **2020**, *297*, 122395. [CrossRef]
59. Vu, H.P.; Nguyen, L.N.; Lesage, G.; Nghiem, L.D. Synergistic Effect of Dual Flocculation between Inorganic Salts and Chitosan on Harvesting Microalgae Chlorella Vulgaris. *Environ. Technol. Innov.* **2020**, *17*, 100622. [CrossRef]
60. Bouzas, A.; Martí, N.; Grau, S.; Barat, R.; Mangin, D.; Pastor, L. Implementation of a Global P-Recovery System in Urban Wastewater Treatment Plants. *J. Clean. Prod.* **2019**, *227*, 130–140. [CrossRef]
61. Das, S.; Gaustad, G.; Sekar, A.; Williams, E. Techno-Economic Analysis of Supercritical Extraction of Rare Earth Elements from Coal Ash. *J. Clean. Prod.* **2018**, *189*, 539–551. [CrossRef]
62. Eurostat Electricity Price Statistics. 2021. Available online: https://ec.europa.eu/eurostat/statistics-explained/index.php?title=Electricity_price_statistics (accessed on 7 December 2022).
63. Foladori, P.; Andreottola, G.; Ziglio, G. *Sludge Reduction Technologies in Wastewater Treatment Plants*; IWA Publishing: London, UK, 2015; ISBN 9781780401706.
64. Ministerio de Agricultura Pesca y Alimentación Índices y Precios Pagados Agrarios 2022. Available online: https://www.mapa.gob.es/es/estadistica/temas/estadisticas-agrarias/economia/precios-percibidos-pagados-salarios/precios-pagados-por-los-agricultores-y-ganaderos/default.aspx (accessed on 7 December 2022).
65. Horstmeyer, N.; Weißbach, M.; Koch, K.; Drewes, J.E. A Novel Concept to Integrate Energy Recovery into Potable Water Reuse Treatment Schemes. *J. Water Reuse Desalin.* **2018**, *8*, 455–467. [CrossRef]

Article

Polystyrene Sulfonate Particles as Building Blocks for Nanofiltration Membranes

Philipp Jahn [1], Michael Zelner [2], Viatcheslav Freger [2] and Mathias Ulbricht [1,*]

1. Institute of Technical Chemistry II and Center for Water and Environmental Research, University of Duisburg-Essen, 45117 Essen, Germany
2. Wolfson Department of Chemical Engineering, Technion-Israel Institute of Technology, 3200003 Haifa, Israel
* Correspondence: mathias.ulbricht@uni-essen.de

Abstract: Today the standard treatment for wastewater is secondary treatment. This procedure cannot remove salinity or some organic micropollutants from water. In the future, a tertiary cleaning step may be required. An attractive solution is membrane processes, especially nanofiltration (NF). However, currently available NF membranes strongly reject multivalent ions, mainly due to the dielectric effect. In this work, we present a new method for preparing NF membranes, which contain negatively and positively charged domains, obtained by the combination of two polyelectrolytes with opposite charge. The negatively charged polyelectrolyte is provided in the form of particles (polystyrene sulfonate (PSSA), d ~300 nm). As a positively charged polyelectrolyte, polyethyleneimine (PEI) is used. Both buildings blocks and glycerol diglycidyl ether as crosslinker for PEI are applied to an UF membrane support in a simple one-step coating process. The membrane charge (zeta potential) and salt rejection can be adjusted using the particle concentration in the coating solution/dispersion that determine the selective layer composition. The approach reported here leads to NF membranes with a selectivity that may be controlled by a different mechanism compared to state-of-the-art membranes.

Keywords: nanofiltration; polyelectrolyte complex membrane; polystyrene sulfonate particles; charged mosaic membranes

1. Introduction

Nanofiltration (NF) is gaining increasing importance because it offers new possibilities for more effective water purification and it has also great potential for the recovery of valuable resources from water [1]. In many cases, a tailored selectivity, for instance, between different ions, is of large interest, but the permeance of the membrane should also be competitive. Most frequently used commercially available NF membranes are thin-film composite membranes, most of which are fabricated using the interfacial polymerization of polyamides as a separation layer [2]. One of the promising emerging alternatives is NF membranes with polyelectrolytes as building blocks for their separation layer, with the layer-by-layer (LBL) technology as one effective fabrication method [3,4]. The combination of polymers with complementary charged groups (polyelectrolytes) on a suited ultrafiltration (UF) membrane forms a selective thin film with controllable properties. To obtain such thin films, different LBL methods can be used, e.g., dip coating, spray coating, and spin coating [3–5]. The formation and structure of these films are strongly influenced by pH, ionic strength, and temperature. The LBL process in general is not limited to polymeric materials; for instance, it can be used to prepare layers from charged particles [6]. The most significant drawbacks of LBL-enabled processes are the cumbersome multi-step coating process and the fact that the polyelectrolyte-based membranes may exhibit a lack of stability to high ionic strengths and extreme pH values. Membranes prepared using the LBL method often exhibit very similar separation properties compared to simply charged NF membranes because the separation performance is often largely determined by the last applied layer [7]. LBL-prepared polyelectrolyte membranes are stable in organic solvents; therefore, they are suitable for solvent-resistant nanofiltration (SRNF) [8].

The use of a combination of polyelectrolytes of opposite charge in membrane fabrication, e.g., via the LBL process, results in polyelectrolyte complex (PEC) membranes. Polyelectrolyte membranes have been known for many decades and were first described by Meyer and Sievers in the 1930s [9,10]. The transport through these membranes is described by the Donnan Steric Pore Model with Dielectric Exclusion (DSPM-DE) theory [11]. Different contributions to selectivity can be discussed, based on three mechanisms, i.e., size exclusion, Donnan exclusion, and dielectric exclusion. The selectivity of different types of NF membranes is affected differently by individual contributions. Membranes whose selectivity is based to a large extent on dielectric exclusion are characterized by the fact that they often have a higher rejection of multivalent ions compared to monovalent ions. For example, the well-known polyamide membrane NF270 from DuPont exhibits a strong dependence on dielectric exclusion due to its dense structure, low dielectric constant, and limited swelling in water. By this membrane, both kinds of divalent ions, cations and anions, are more rejected than monovalent ions; e.g., single salt rejection of both $MgCl_2$ ($CaCl_2$) and Na_2SO_4 is higher than that of NaCl ($MgCl_2$ ($CaCl_2$) = Na_2SO_4 > NaCl), although the membrane has a negative surface charge [12]. This also results in a higher scaling tendency for typical scalants, such as hydroxylapatite ($Ca_5[OH(PO_4)]_3$) or calcium sulfate ($CaSO_4$). Conversely, when the selectivity of the membrane is strongly dependent on Donnan exclusion, which is the case when relatively loosely bound swellable polyelectrolytes are used to build the selective layer, the scaling tendency can decrease due to the depletion of one of the scaling forming species. An example of a negatively charged NF membrane, with high dependence on Donnan exclusion, was presented by Bernstein et al. [13]. This membrane was synthesized by grafting cross-linked poly(vinyl sulfonic acid) onto an UF membrane, leading to a strongly negatively charged selective layer of the NF membrane. This also led to a much higher rejection of negatively charged ions than positively charged ions (Na_2SO_4 > NaCl > $CaCl_2$). These membranes showed a significantly lower scaling tendency compared to commercial polyamide (PA) membranes [14]. Previous work of Levchenko and Freger [12] demonstrated that cross-linked polyethyleneimine (PEI) generates a positively charged NF selective layer. This membrane was prepared by crosslinking of PEI on a suitable UF membrane support. Due to the strong positive charge, the single-salt rejection sequence was $MgCl_2$ > NaCl > Na_2SO_4. In addition, this membrane also showed a significantly lower scaling tendency for phosphates and sulfates, based on the depletion of these species from the retentate.

The combination of loosely bound, swellable polyelectrolytes of different charge can lead to a charged mosaic (CM) membrane. The concept of CM membrane was developed by Sollner in 1932 [15]. The selective layer of CM membranes is characterized by differentially charged domains, and their separation mechanism is also strongly influenced by Donnan exclusion [16,17]. This leads to a depletion of both kinds of charged species of higher charge density via interaction with the complementary domains and results in a unique rejection pattern (NaCl > Na_2SO_4 ~ $CaCl_2$). This could also be an advantage for scaling prevention, but it of interest as well for tertiary treatment of saline wastewater, where there is interest in removal of NaCl, but ions such as Ca^{2+}, Mg^{2+}, or HPO_4^{2-} should remain in the treated water. CM membranes have been under development for a long time. A CM membrane was prepared, for example, via demixing of a charged and an uncharged polymer during membrane casting and subsequent functionalization of the uncharged polymer with oppositely charged groups [16]. However, no CM membrane with the competitive selectivity and permeability can yet be fabricated with a well-scalable method [3].

In a recent perspective article on new materials and approaches to membrane fabrication [18], the utilization of nano- and microparticles as building blocks for membranes was also emphasized as one promising route. Very much research is devoted to nanocomposite membranes with porous inorganic or organic/inorganic particles as part of the selective layer [19]. However, the focus of this work is on purely organic particles that can act as permeable domains in the selective layer. This approach is much less explored. Among the few examples in the literature are zwitterionic polymeric nanoparticles that have been integrated via interfacial polymerization into PA layers [20].

In this work, we present a new method for preparing NF membranes, which contain negatively and positively charged domains, obtained by the combination of two polyelectrolytes with opposite charge. The negatively charged polyelectrolyte is provided in the form of particles with a diameter of about 300 nm. Particles are synthesized by batch emulsion polymerization of 4-styrene sulfonic acid ethyl ester with divinylbenzene as a crosslinker monomer and subsequent conversion to polystyrene sulfonate (PSSA). As a positively charged polyelectrolyte, PEI is used to act as the matrix for incorporation/immobilization of the PSSA particles. Both buildings blocks and glycerol diglycidyl ether (GDE) as the cross-linker for PEI are applied to a UF membrane support in a simple one-step coating process, followed by thermal curing. The fraction of PSSA particles in the coating solution/dispersion was varied, and this yielded tunable composition and net charge of the selective layer, as shown by IR spectroscopy and zeta potential analyses. NF characterization revealed that the salt rejection could also be tuned from the typical behavior of a cationic membrane (without PSSA) to that of an anionic membrane (at a high PSSA content). For medium values of PSSA concentration used for the coating, it was possible to obtain net-charge-balanced NF membranes that had equal rejections of Na_2SO_4 and $CaCl_2$ and lower rejection of NaCl. Hence, the feasibility of integration of polyanionic particles as building blocks in PEC NF membranes was demonstrated, but no CM behavior could be obtained.

2. Materials and Methods

2.1. Materials

Polyethersulfone (PES) flat sheet ultrafiltration membranes with a molecular weight cut-off (MWCO) of 30 kDa, provided by Sartorius (type: 14659, batch number: 2050123), were used as a support membrane. Divinylbenzene (DVB) from Fluka was used as a crosslinker monomer. The functional monomer styrene sulfonate sodium salt (SSA-Na), bromoethane (EtBr), potassium persulfate (KPS), polyethyleneimine (PEI), 270 kDa), crosslinker glycerol diglycidyl ether (GDE), and the surfactant sodium dodecyl sulfate (SDS) were purchased from Sigma Aldrich. The solvents acetonitrile and dichloromethane were obtained from VWR. The salts sodium chloride (NaCl), sodium sulfate (Na_2SO_4), calcium chloride ($CaCl_2$), and sodium phosphate ($Na_3(PO_4)_2$) were received from Fluka. Silica gel (for chromatography) with a particle size of 60–200 µm from Acros Organics was used for the purification of the monomer. All chemicals were used as received. Ultrapure water was provided by the water purification system Arium from Sartorius (Göttingen, Germany).

2.2. Preparation of Negatively Charged Polyelectrolyte Particles (Polystyrene Sulfonate)

The polyelectrolyte particle synthesis via emulsion polymerization of a hydrophobic precursor, the protected polystyrene sulfonic acid, and subsequent deprotection was based on the works of Tiwari and Walther [21] and Woeste et al. [22]. Since the monomer styrene sulfonic acid ethyl ester (SSE) was not readily available, it was synthesized (Figure 1). The silver method was used to convert the sodium salt of the monomer into the corresponding sulfonic acid ester [23].

First, SSA-Na was dissolved in water, and silver nitrate in a molar ratio 1:1 was added as solid under cooling at 4 °C and protection against light. The precipitated grey solid was separated via suction filtration and washed several times with ice-cold water and diethyl ether. Then, the grey product was dissolved in acetonitrile and filtered again to remove impurities. For the second step, the double molar amount of EtBr relative to SSE-Ag was added, and the reaction was carried out for six hours at 70 °C. After cooldown, the solution was filtrated via suction filtration to remove the co-product silver bromide.

The solvent was then removed by rotary distillation. Afterwards, the white residue was dissolved in dichloromethane, and the solution was purified by passing it through a column containing silica gel. Finally, the solvent was removed using rotary distillation. The final product was a slightly yellowish viscous liquid and was stored in a freezer at $<-21\,°C$ to prevent auto-polymerization. The purity of the product SSE was confirmed by ^{1}H-NMR spectroscopy. The particles were prepared by emulsion polymerization of SSE with DVB as a crosslinker, using SDS as a surfactant and KPS as an initiator (Figure 2). A different reactor from Tiwari and Walter [20] was used, and some reaction conditions were adjusted. First, 200 mL of a solution of SDS in water with a concentration of 0.5 mmol/L was filled into the small lab-scale glass reactor with a mechanical stirrer (instead of using snap-on glass vials with magnetic stirring bar). After degassing of the SDS solution in a vacuum chamber at 200 mbar, the monomer mixture with 1 wt% SSE relative to continuous phase and 4 mol% DVB (relative to total monomer) was added. After heating to 70 °C, the mixture was stirred at 800 rpm for 30 min with a mechanical anchor stirrer. Then, KPS dissolved in a small amount of water was added to the reactor; the concentration of KPS in the mixture was 4 mmol/L. After a few seconds to minutes, the emulsion changed from turbid to a white dispersion. To ensure complete monomer conversion, the reaction was continued for 24 h. The mixture was then filtrated with an MN615 $\frac{1}{4}$ pleated filter paper (corresponding retention range > 4 µm) from Macherey-Nagel to remove big structures and thereafter filled in a dialysis bag with a nominal MWCO of 12 kDa and dialyzed against DI water. After reaching a conductivity of < 5 µS/cm in the dialysate, purification was considered complete. Then, the particles were freeze-dried (Martin Christ Alpha 1-4 100400 ISCEON, Osterode, Germany); a cotton wool-like solid was obtained. For deprotection, the particles were dispersed in 1 mol/L aqueous sodium hydroxide solution and heated to 110 °C under reflux for 12 h (Figure 2). These harsh conditions ensure a complete conversion of the sulfonic acid ester. The purification was carried out again by dialysis (MWCO 12 kDa) until a conductivity < 5 µS/cm was reached. Obtained particles were again freeze-dried.

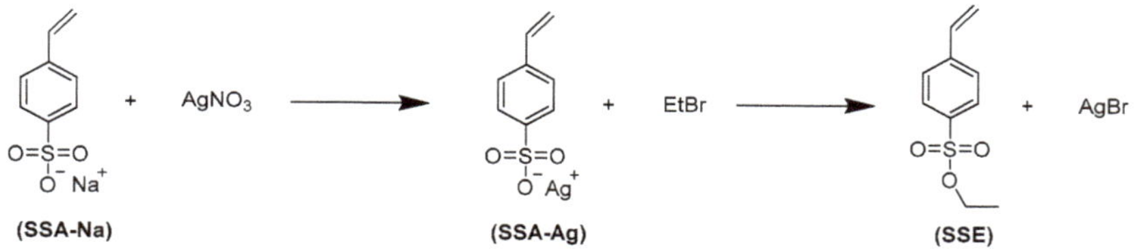

Figure 1. Reaction scheme for synthesis of styrene sulfonic acid ethyl ester.

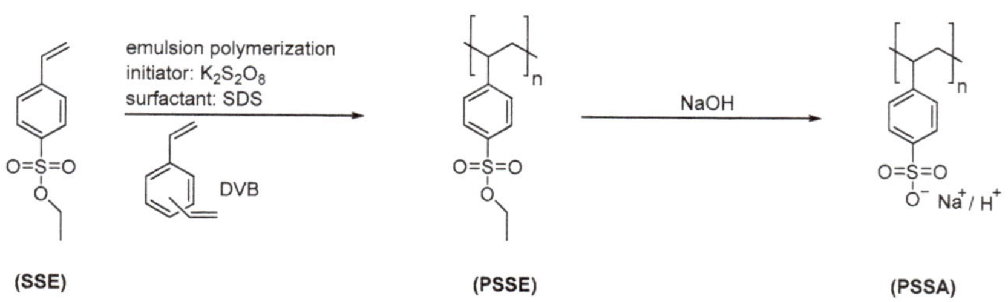

Figure 2. Reaction scheme for synthesis of the particles via emulsion polymerization of SSE and DVB initiated by KPS in an aqueous SDS solution, followed by saponification of the sulfonic acid ester ("deprotection" of ion exchange groups).

2.3. Particle Characterization

2.3.1. Calculation of Charged Group Density

The charged group density (CGD) was calculated following Equation (1),

$$CGD = \frac{z}{x_1 \cdot M_1 + x_2 \cdot M_2} \quad (1)$$

where z is the charge per repeat unit and M_1 is the molar mass of functional monomer, M_2 is the molar mass of crosslinker monomer, and x_1 and x_2 are the molar fractions of functional monomer and crosslinker monomer, respectively, in the copolymer.

2.3.2. Zeta Potential and Particle Size

The particles were re-dispersed in ultrapure water at a concentration of 1 mg/mL. Dynamic light scattering (DLS) measurements were performed on a Zetasizer UltraPro from Malvern Panalytical (Worcestershire, UK) equipped with a DTS1070 flow cell. After determining particle size, the zeta potential was analyzed in the same cell. The PDI for an individual peak of the particle size distribution was calculated with Equation (2).

$$PDI = \left(\frac{\sigma}{d}\right)^2 \quad (2)$$

where σ is the standard deviation and d is the mean particle size. For measuring the pH dependency, an automatic titration unit (MPT-2) was connected to the Zetasizer instrument. The measurements were performed with a pH increment of 0.5. The pH was adjusted by using HCl or NaOH, respectively.

2.3.3. Scanning Electron Microscopy (SEM)

For SEM image acquisition, the particles were first dispersed in water. In parallel, a single crystalline silicon wafer was immersed in a 10 g/L solution of PEI (270 kDa, branched) in water and cleaned with water after 10 min. Subsequently, the wafer was dried with compressed air and then immersed in the particle dispersion for another 10 min, followed by rinsing with water. Due to the electrostatic interactions between the PEI on the wafer surface and the particles, single particles could be imaged. To ensure sufficient conductivity of the sample, the samples were sputtered with an Au/Pd layer. The image acquisition was performed with the instrument Apreo S LoVac from Thermo Fisher Scientific (Waltham, MA, USA).

2.4. Membrane Fabrication

The support membrane was cut into rectangular shape (130 mm × 210 mm). It was first washed with a mixture of water/ethanol (50:50) for two hours to remove soluble components. Then, the membrane was soaked in a solution of 50 g/L glycerol in ethanol for 24 h. Afterwards, it was mounted in a glass frame (120 mm × 200 mm) that allows it to cover the membrane with a solution. The coating solution/dispersion with the desired concentrations was prepared by adding PSSA particles to a solution of PEI in ethanol solution, followed by sonication for 20 min to ensure that the particles were also well dispersed. Finally, the crosslinker (GDE) was added, and the solution/dispersion was stirred for 20 min. In the meantime, the surface of the mounted membrane was washed with ethanol a few times to remove the excess of glycerol from the surface, followed by a quick drying of the surface with compressed air. Next, the modification solution was spread on the membrane surface, limited by the glass frame; the volume/area ratio was always ~0.28 mL/cm². This value was chosen to ensure a complete coverage of the membrane surface with liquid. After 5 min, the liquid was discarded, and the wet membrane was transferred to an oven where is was kept in horizontal orientation at 60 °C for two hours to ensure a complete cross-linking of PEI by GDE.

2.5. Membrane Testing and Characterization

2.5.1. Water Permeance and Single Salt Rejection

The performance of the membrane was determined in a laboratory dead-end nanofiltration set-up equipped with a stirrer. The feed container had a volume of 100 mL. The active membrane area was 9.62 cm^2 and the stirring rate was set to 600 rpm. Each membrane sample was fully compacted by pure water filtration at 8 bar until constant flux was reached, before testing separation performance. Water permeance was calculated by Equation (3).

$$P = \frac{V}{p \cdot t \cdot A} \quad (3)$$

where V is the filtered volume, p is the transmembrane pressure, t is the sampling time, and A is the active membrane area. The rejection of NaCl, Na$_2$SO$_4$, and CaCl$_2$ was determined with single salt feed solutions containing 1 g/L of the individual salts in water. Conductivity was measured to determine salt concentrations. Rejection was calculated by Equation (4).

$$R = \left(1 - \frac{C_P}{C_F}\right) \cdot 100\% \quad (4)$$

where C_P and C_F are salt concentrations in initial feed and in collected permeate, respectively. The mixture of the three salts (0.25 g/L Na$_2$SO$_4$, 0,25 g/L CaCl$_2$, and 0.5 g/L NaCl) at a total concentration of 1 g/L was also used. The cation and anion concentrations in mixed salt solution of feed and permeate were determined separately. The cations were analyzed via atomic absorption spectroscopy (AAS) using M-Series FS95 from Thermo Fisher Scientific (Waltham, MA, USA), and for the anions, an ion chromatograph from Metrohm (IC 883 with Autosampler; Herisau, Switzerland) was used. For all filtrations, a maximum of 20 mL of permeate was filtered through the membrane to avoid a too strong concentration of the feed (maximum concentration factor of 1.25).

Three samples for each membrane type have been tested and mean values and standard deviations are reported.

2.5.2. Zeta Potential

Zeta potential of the membrane surface was determined by using a SurPASS1 electrokinetic analyzer from Anton Paar (Graz, Austria) equipped with an adjustable gap cell. The gap width was adjusted to 100 μm with a tolerance of 5 μm. The measurement was performed with 1 mmol/L KCl solution as electrolyte. At the beginning of each measurement, 550 mL of that KCl solution was added to a container, and the pH value was adjusted to a value of ~2.5. After 10 min of circulating the solution through the measurement cell, the measurement was started. During measurement, the pH value was automatically adjusted with an increment of 0.5 by using 0.1 mol/L KOH solutions. At every pH increment, a triple determination was performed. The zeta potential was calculated by using the Helmholtz–Smoluchowski Equation (5).

$$\zeta = \frac{dU_{str}}{dp} \cdot \frac{\eta}{\epsilon \cdot \epsilon_0} \cdot K_B \quad (5)$$

where $\frac{dU_{str}}{dp}$ is the slope of the plot streaming potential vs. differential pressure, K_B is electrolyte conductivity, η is electrolyte viscosity, ϵ is dielectric constant of electrolyte, and ϵ_0 is permittivity of vacuum.

2.5.3. ATR-IR Spectroscopy

The surface chemistry was characterized using FTIR spectroscopy in the attenuated total reflectance (ATR) mode (Bruker Alpha I). The membrane sample was measured at three different locations in the range 400–4000 cm^{-1}.

3. Results and Discussion

3.1. Poly(Styrene Sulfonic Acid) Particles

PSSA particles were synthesized as described in Section 2.2. The mechanical stirring system was used instead of simple magnetic stirring to have more control over the stirring speed. The emulsion polymerization was performed using a surfactant (SDS) concentration (0.5 mmol/L), which was well below its critical micelle concentration (CMC ~8.2 mmol/L) because it is well-known that emulsion polymerization below the CMC also lead to well-defined particles [24,25]. Lower SDS concentrations produce fewer nuclei during nucleation and lead to growth of larger particles and vice versa [21]. The specific particles selected for this work had been obtained with a crosslinker content of 4 mol% DVB in the dispersed organic phase consisting of the monomer SSE (Figure 2). The resulting moderate cross-linking degree should on the one hand provide sufficient swelling in water to allow ion transport through the particles and on the other hand yield sufficient particle stability. Table 1 shows the most important properties of the obtained particles. The target size of 200 nm was approximately obtained for the protected version (207 ± 12 nm) of the particles by using 0.5 mmol/L SDS and 4 mol% DVB with 1 wt% SSE (compared to continuous phase). Furthermore, the sulfur content was determined by elemental analysis of the dried particles and used to calculate the actual density of functional or charged groups. The values were only slightly lower than the theoretical values of 4.8 mmol/g for protected and 4.9 mmol/g for deprotected particles, calculated by Equation (1) for complete incorporation of both monomers in the copolymer.

Table 1. Size and related polydispersity index, determined by DLS, as well as sulfur content, analyzed by elemental analysis, of the particles after synthesis ("protected") and after subsequent deprotection. Mean values and standard deviation were calculated from results of three individual measurements.

Sample	Size (nm)	PDI	S (wt%)	Functional/Charged Group Density, Experimental (mmol/g)
Protected	207 ± 12	0.14 ± 0.01	14.9 ± 0.1	4.7
Deprotected	344 ± 40	0.14 ± 0.06	13.6 ± 0.1	4.3

Average particle size of as-synthesized particles increased from 207 nm to 344 nm after complete hydrolysis of all ester groups to yield sulfonic acid groups; this indicated significant swelling of the particles in water (Figure 3a, Table 1). The swelling is driven by the hydration of the charged groups of the polymer and counteracted by the chemical crosslinking of the network. The PDI was low and did not change after deprotection. The size distribution could be described as practically monodisperse. Moreover, the introduction of the sulfonic acid groups shifted the zeta potential after deprotection of the particles to more negative values (Figure 3b). The fact that the protected particles also showed a negative zeta potential can be explained by incorporation of the surfactant SDS, with sulfate groups, on the particle's surface.

To investigate the stability of the particles, their size and zeta potential in water were measured as function of pH value (Figure 4a). The size varied only slightly between 235 nm and 257 nm. The zeta potential decreased slightly in the acidic pH range. These results proved that the particles are negatively charged and that, consequently, their swelling degree did not change significantly over the entire pH range. Figure 4b shows an SEM image of the protected particles. The observed size was 133 ± 27 nm and therefore smaller than the values determined by DLS. This is due to the dry state of the particles, because the DLS method determines the hydrodynamic diameter, which is usually larger because of hydration effects.

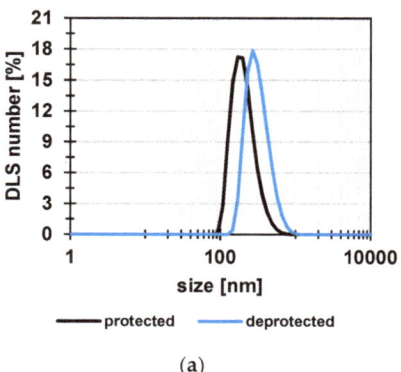 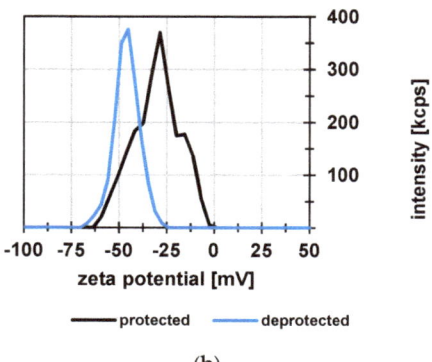

Figure 3. Comparison of protected and deprotected particles: (**a**) size; (**b**) zeta potential (kcps = kilo counts per seconds); data are shown for one of the three independent measurements (Table 1).

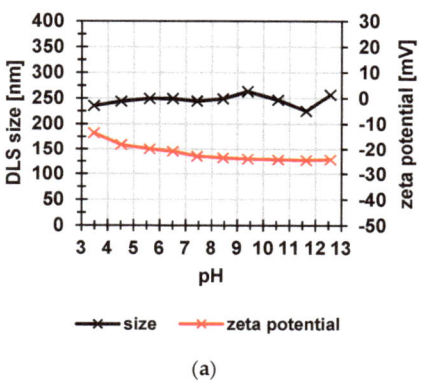 

Figure 4. (**a**) Particle size and zeta potential as function of pH value; (**b**) SEM image of protected particles deposited on a PEI-coated silicon wafer. Measured particle size 133 ± 27 nm ($n > 100$).

3.2. Membrane Performance

As the cationic matrix for incorporation of the particles, branched PEI with a molar mass of 270 kDa was used. The particles were limited in their swelling in water by crosslinking during synthesis by polymerization (Section 3.1), whereas the matrix polymer PEI should be cross-linked by GDE. The coating of the porous PES support membranes was carried out with ethanol instead of water as solvent, because the crosslinker GDE is insoluble in water. To increase pore stability during drying, the membranes were impregnated, before coating, in a glycerol/ethanol mixture. Glycerol cannot evaporate at the given conditions and thus additionally stabilizes the pores. In the first series of experiments, the PEI and the crosslinker concentrations were kept constant at 0.5 g/L and 1 g/L, respectively, while the particle concentration was varied from 0 to 1 g/L; results are shown in Figure 5. The relatively small errors of the measurements for three samples of 9.62 cm^2 from the same membrane batch indicate that the membrane fabrication is uniform on the cm length scale. The NF experiments with the single salt solutions were typically performed by first using the NaCl solution, followed by the Na$_2$SO$_4$ solution and then the CaCl$_2$ solution; finally, the NaCl solution was filtered again. The NaCl rejection values in first and last filtrations were identical within the margin of error, indicating that the membranes were stable during the series of filtrations of different salt solutions. The reference membrane without addition of particles already showed an Na$_2$SO$_4$ rejection of

24%, a similar rejection for NaCl and the highest rejection for CaCl$_2$. This can be explained by the positively charged barrier layer composed of cross-linked PEI. The amino groups were protonated and therefore positively charged and thus increased the rejection based on Donnan exclusion, especially for double-positively charged Ca^{2+} ions. The observed reduced water permeance and increased rejection for all single salts at low addition of PSSA particles (up to 0.125 g/L) indicated the promotion of crosslinking of the PEI-based barrier layer by the particles. Because of the small PSSA fraction, the effect of PEI on rejection still dominated, so that rejection of CaCl$_2$ was still highest. With a particle concentration of 0.25 g/L in the coating solution/dispersion, the resulting membrane had approximately equal rejections (~50%) for CaCl$_2$ and Na$_2$SO$_4$ and ~25% rejection for NaCl. Similar rejection of the salts with the two double-charged ions of opposite charge indicated a macroscopically neutral membrane barrier layer. For that kind of membrane, it had also been found that the rejection of the individual ions in a ternary salt mixture of the same total salt concentration was, within the range of error, identical to the data obtained for single salt feeds (Figure S1).

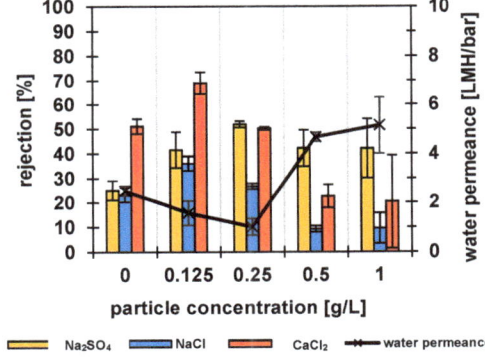

Figure 5. Influence of the variation of PSSA particle content in ethanol used for coating the porous PES support membrane onto the water permeance and rejection of single salts. The PEI and GDE contents were 0.5 g/L and 1 g/L, respectively.

By increasing the particle content further, up to 1 g/L, the water permeance increased strongly and salt rejection decreased. However, the influence of the negatively charged particles on salt rejection became larger because the rejection of Na$_2$SO$_4$ was much higher than for the other salts. Hence, the particles dominated the properties of the selective layer. However, they seemed to interfere with the chemical cross-linking of the PEI. Instead, the proportion of ionic crosslinking, due to interactions between PEI (+) and particles (−), became larger. This formation of polyelectrolyte complexes, in combination with a lower degree of chemical crosslinking of the layer, could be considered a reason for higher water permeance and overall lower salt rejection.

In a second series of experiments, both PEI and crosslinker concentrations were kept constant at 1 g/L, and the particle concentration was varied from 0 to 1 g/L; results are shown in Figure S2. Because of the higher PEI concentration, the salt rejection of the reference membrane (0 g/L PSSA) was much higher than in the first series and similar to other nanofiltration membranes with the cross-linked PEI layer reported in the literature [12]. The effects of the PSSA particles were similar to the first series, but because of the higher rejection values, the trends were less clear. Therefore, the further analysis is focused on composite membranes from the first series. Overall, the permeance of the obtained membranes was rather low compared to other NF membranes, with PEI as part of the selective layer (e.g., [12,26]). The main reason is likely that the PES UF membrane that had been used as support had not been developed for this purpose and that its structure had not been fully protected against pore collapse during the thermal curing step.

Zeta potential data are shown in Figure 6a. The results for the composite membranes reflect the influence of the particle concentration very well. The membrane prepared only with PEI (equivalent to PSSA content of 0 g/L in Figure 6) had the typical zeta potential with an isoelectric point of pH 9. As the fraction of particles within the coating solution/dispersion increased, the isoelectric point shifted to a more negative value, and the zeta potential became correspondingly more negative. This related well to the single salt rejections and the corresponding water permeabilities. Relevant parts of IR spectra are shown in Figure 6b (complete IR spectra are shown in Figure S3). The band at ~1038 cm^{-1} could be assigned to the symmetric S-O stretching vibration of the sulfonic acids groups as a signature of the PSSA particles. Data clearly reveal that with increasing particle concentration in the coating solution/dispersion, the intensity of the corresponding band also increased.

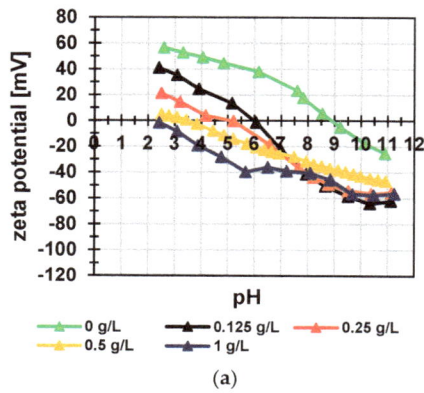 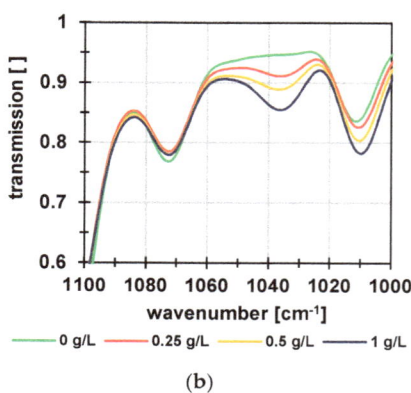

Figure 6. (**a**) Zeta potential as function of pH value; (**b**) IR spectra; membranes coated with 0.5 g/L PEI and variable PSSA content.

SEM data for selected membranes can be seen in Figure 7. The images show an added layer containing particles on the surface of the PES support membrane in both cases. The distribution of the particles appeared relatively inhomogeneous, but an increase in particle density on the surface can be observed from Figure 7a (0.25 g/L) to Figure 7b (0.5 g/L), which is consistent with the increase in particle fraction used for coating. Particles seem to be partially embedded in a thinner layer; this could be explained by the partial formation of an interpenetrating structure upon mixing both polyelectrolytes that is then cross-linked. It should be considered that images had been taken for the dry membranes; because the cross-linked polyelectrolytes forming the barrier layer will swell in water, the wet structure in operating mode will be different.

Because the barrier layer is formed by the combination of a strong (PSSA particles) and a weak (PEI) polyelectrolyte, the effective charge of the barrier layer will change with the pH value (Figure 6a). Therefore, it can be expected that the rejection pattern for different salts for each specific membrane type (with a certain PSSA:PEI ratio; Figure 5) will also depend on the pH value (analogous to, for example, previous work where Nafion and polyvinylamine had been combined in the barrier layer of a polyelectrolyte complex NF membrane [17]). Because the focus of this work was the demonstration of the feasibility of using PSSA particles as building blocks for the fabrication of tunable polyelectrolyte complex membranes in a simple one-step coating process, this aspect was not further investigated.

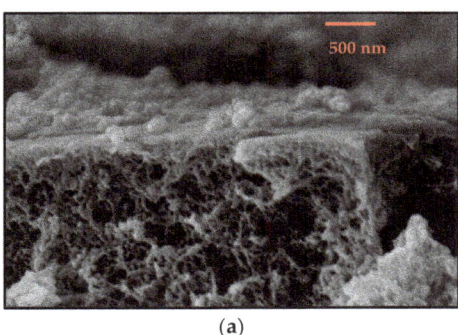 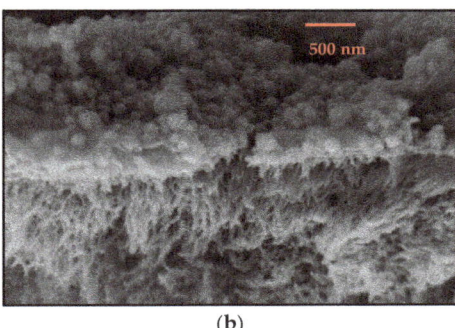

(a)   (b)

Figure 7. Cross-section SEM images of composite membranes obtained by using different solutions/dispersions for coating: (**a**) 0.25 g/L PSSA particles, 0.5 g/L PEI, and 1 g/L GDE; (**b**) 0.5 g/L PSSA particles, 0.5 g/L PEI, and 1 g/L GDE.

In summary, polyelectrolyte barrier layers with structures and properties that are tunable by the particle fraction have been obtained. Up to particle concentrations of 0.25 g/L, the covalent crosslinking between PEI molecules by GDE dominated, and ionic cross-linking between PEI and PSSA (PEC formation) might have an additional contribution. By further increasing the particle concentration up to 1 g/L, the chemical crosslinking between PEI molecules was disturbed, and ionically crosslinked structures PEI and PSSA were predominantly formed. In the apparently charge-balanced barrier layers (according to results of single salt and salt mixture nanofiltration, Figure 5 and Figure S1) obtained at 0.25 g/L, the particles may act as negatively charged domains (with sulfonic acid group) and the PEI (with ammonium groups) as a positively charged matrix. It is believed that the way the two polyelectrolytes are applied to the membrane surface results in the formation of polyelectrolyte complexes, which is also thermodynamically favored [27,28]. However, due to the cross-linking of the PSSA particles (during synthesis) and of the PEI (during membrane fabrication), a complete interpenetration of the two polyelectrolytes is impossible, so the evoked domain structure is plausible.

Unfortunately, none of the membranes prepared here, neither charge-balanced nor with an excess of one charge, showed the rejection pattern expected for CMs in conventional theory [16,29]. Specifically, no membrane showed the rejection of salt of both divalent cation and divalent anions lower than that of monovalent ions. The charge-balanced membranes could fail to yield the expected salt rejection patterns for several reasons. First, it might be because the charged domains were not continuous over the entire barrier layer (e.g., because PEI could fully engulf particles on the PES surface and partially block support membrane pores) or because the particle's size was too large. However, images in Figure 7 indicate that the PEI matrix and most particles could cross the entire top-layer thickness. It is then likely that the reason might be of a more fundamental nature, as explained below.

In a recent paper, Fan et al. [30] reported that conductivity of an ion-exchange membrane loaded with a counter-ion of different valences was largest for monovalent counterions and decreased with valency. This trend was related to the mobilities of counter-ions suppressed by the non-homogeneity of the electric field around fixed charges, predicted within the Manning counter-ion condensation theory, for media of reduced permittivity and increasing with the valency. Essentially, the same conclusion follows from a recently proposed and distinctly different physical picture based on the Bjerrum-like association of fixed charges and counter-ions [31]. Since—as a defining feature of CMs—all ions permeate an ideal CM as counter-ions via their respective domains, the conductivity of domains is directly related to the permeability of the CM. Although partitioning factors—the basis of conventional theory of CMs based on Donnan model—favors multivalent ions, their reduced mobility highlighted by Fan et al. [30] may override the effect of partitioning and preclude the expected CM performance. This scenario obviously needs further investigation and will be clarified in future studies.

An overview of literature data for polyelectrolyte complex membranes made from building blocks with similar ionic groups but using the LBL approach is provided in Table S1 [32–37]. The comparison of overall separation performance, i.e., considering the trade-off between water permeance and salt rejection, reveals that the charge-balance membranes, taken as one example, are not yet fully competitive. One option to increase performance due to higher water permeance with same rejection pattern has already been indicated above, i.e., preventing the pore collapse of the support membrane during the thermal curing step.

4. Conclusions

In this work, we have reported a novel polyelectrolyte complex NF membrane prepared by a simple one-step coating on a PES ultrafiltration membrane. The negatively charged PSSA particles, as a novel kind of building block, were synthesized by batch emulsion polymerization. Such PSSA particles and positively charged PEI were used for membrane preparation. The membrane charge (zeta potential) and salt rejection could be adjusted by the particle concentration in the coating solution/dispersion that determined the selective layer composition. In this way, membranes with low permeabilities, but balanced charge and the symmetric rejection of divalent ions, could be obtained. The membrane barrier structure and separation performance were controlled by the fraction of particles. At low particle concentrations, the crosslinking of PEI seemed to be more effective. At higher particle concentrations, an increase in permeability could be observed, possibly due to less chemical crosslinking of the PEI and more ionic interaction between PEI and PSSA particles. The membranes obtained at a specific ratio between PSSA and PEI contained domains with either an excess of positive or negative charges, resulting in approximately equal rejection of positively and negatively charged species of higher valency. Overall, the feasibility of synthesizing tailored polyanionic particles and of their integration as building blocks in PEC NF membranes was demonstrated. However, CM behavior was not obtained. The reason might lie in the membrane morphology or be of a more fundamental nature. The domain structure achieved in this work might not meet the stringent requirements that both domains in a CM span the entire thickness and have commensurate permeability controlled by the Donnan mechanism. As an alternative, in the next stage towards an improved structure, positively charged particles may be synthesized and combined with the already established negatively charged particles. On the other hand, the ideal CM performance may also be precluded by the reduced mobility of multivalent ions compared to their monovalent counterparts. Hence, the feasibility and the development of a true CM NF membrane still remain a challenge for future research.

Supplementary Materials: The following supporting information can be downloaded at https://www.mdpi.com/article/10.3390/membranes12111138/s1, Figure S1. Water permeance and rejection of individual ions in a mixture of three salts for the composite membrane obtained by a coating at a PSSA particle concentration of 0.25 g/L and PEI and GDE concentrations of 0.5 g/L and 1 g/L, respectively. Figure S2. Influence of the variation in PSSA particle content in ethanol used for coating the porous PES support membrane onto water permeance and rejection of single salts. The PEI and GDE contents were 1 g/L (in the second series of experiments). Figure S3. Complete IR spectra of NF membranes from the first series.

Author Contributions: Conceptualization, M.U. and V.F.; methodology, P.J., M.Z. and M.U.; formal analysis, P.J. and M.U.; investigation, P.J. and M.Z.; data curation, P.J.; writing—original draft preparation, P.J.; writing—review and editing, M.U. and V.F.; visualization, P.J.; supervision, M.U.; project administration, M.U.; funding acquisition, V.F. and M.U. All authors have read and agreed to the published version of the manuscript.

Funding: This work has been part of "TertNF" project (02WIL1488) funded by German Federal Ministerium of Education and Research (BMBF) within the frame of the "German-Israel Water Technology Cooperation".

Institutional Review Board Statement: Not applicable.

Informed Consent Statement: Not applicable.

Data Availability Statement: Not applicable.

Acknowledgments: The authors are grateful for the support provided by Tobias Kallweit and Pascale Wünscher at University of Duisburg-Essen.

Conflicts of Interest: The authors declare no conflict of interest.

References

1. Zhang, H.; He, Q.; Luo, J.; Wan, Y.; Darling, S.B. Sharpening Nanofiltration: Strategies for Enhanced Membrane Selectivity. *ACS Appl. Mater. Interfaces* **2020**, *12*, 39948–39966. [CrossRef]
2. Freger, V.; Ramon, G.Z. Polyamide desalination membranes: Formation, structure, and properties. *Prog. Polym. Sci.* **2021**, *122*, 101451. [CrossRef]
3. Ji, Y.-L.; Gu, B.-X.; An, Q.-F.; Gao, C.-J. Recent Advances in the Fabrication of Membranes Containing "Ion Pairs" for Nanofiltration Processes. *Polymers* **2017**, *9*, 715. [CrossRef] [PubMed]
4. Durmaz, E.N.; Sahin, S.; Virga, E.; de Beer, S.; de Smet, L.C.P.M.; de Vos, W.M. Polyelectrolytes as Building Blocks for Next-Generation Membranes with Advanced Functionalities. *ACS Appl. Polym. Mater.* **2021**, *3*, 4347–4374. [CrossRef] [PubMed]
5. Joseph, N.; Ahmadiannamini, P.; Hoogenboom, R.; Vankelecom, I.F.J. Layer-by-layer preparation of polyelectrolyte multilayer membranes for separation. *Polym. Chem.* **2014**, *5*, 1817–1831. [CrossRef]
6. Lee, D.; Rubner, M.F.; Cohen, R.E. All-Nanoparticle Thin-Film Coatings. *Nano Lett.* **2006**, *6*, 2305–2312. [CrossRef] [PubMed]
7. Ghostine, R.A.; Markarian, M.Z.; Schlenoff, J.B. Asymmetric Growth in Polyelectrolyte Multilayers. *J. Am. Chem. Soc.* **2013**, *135*, 7636–7646. [CrossRef] [PubMed]
8. Joseph, N.; Ahmadiannamini, P.; Jishna, P.S.; Volodin, A.; Vankelecom, I.F. 'Up-scaling' potential for polyelectrolyte multilayer membranes. *J. Membr. Sci.* **2015**, *492*, 271–280. [CrossRef]
9. Meyer, K.H.; Sievers, J.-F. La perméabilité des membranes I. Théorie de la perméabilité ionique. *Helv. Chim. Acta* **1936**, *19*, 649–664. [CrossRef]
10. Meyer, K.H.; Sievers, J.-F. La perméabilité des membranes. II. Essais avec des membranes sélectives artificielles. *Helv. Chim. Acta* **1936**, *19*, 665–677. [CrossRef]
11. Fievet, P. SEDE (Steric, Electric, and Dielectric Exclusion) Model: Approximated Versions. In *Encyclopedia of Membranes*; Springer: Berlin/Heidelberg, Germany, 2016; pp. 1755–1758.
12. Levchenko, S.; Freger, V. Breaking the Symmetry: Mitigating Scaling in Tertiary Treatment of Waste Effluents Using a Positively Charged Nanofiltration Membrane. *Environ. Sci. Technol. Lett.* **2016**, *3*, 339–343. [CrossRef]
13. Bernstein, R.; Antón, E.; Ulbricht, M. UV-Photo Graft Functionalization of Polyethersulfone Membrane with Strong Polyelectrolyte Hydrogel and Its Application for Nanofiltration. *ACS Appl. Mater. Interfaces* **2012**, *4*, 3438–3446. [CrossRef] [PubMed]
14. Kaganovich, M.; Zhang, W.; Freger, V.; Bernstein, R. Effect of the membrane exclusion mechanism on phosphate scaling during synthetic effluent desalination. *Water Res.* **2019**, *161*, 381–391. [CrossRef]
15. Sollner, K. Über Mosaikmembranen. *Biochem. Z.* **1932**, *244*, 370.
16. Linder, C.; Kedem, O. Asymmetric ion exchange mosaic membranes with unique selectivity. *J. Membr. Sci.* **2001**, *181*, 39–56. [CrossRef]
17. Zelner, M.; Jahn, P.; Ulbricht, M.; Freger, V. A mixed-charge polyelectrolyte complex nanofiltration membrane: Preparation, performance and stability. *J. Membr. Sci.* **2021**, *636*, 119579. [CrossRef]
18. Nunes, S.P.; Culfaz-Emecen, P.Z.; Ramon, G.Z.; Visser, T.; Koops, G.H.; Jin, W.; Ulbricht, M. Thinking the future of membranes: Perspectives for advanced and new membrane materials and manufacturing processes. *J. Membr. Sci.* **2020**, *598*, 117761. [CrossRef]
19. Bassyouni, M.; Abdel-Aziz, M.H.; Zoromba, M.S.; Abdel-Hamid, S.M.S.; Drioli, E. A review of polymeric nanocomposite membranes for water purification. *J. Ind. Eng. Chem.* **2019**, *73*, 19–46. [CrossRef]
20. Ibrahim, G.P.S.; Isloor, A.M.; Bavarian, M.; Nejati, S. Integration of Zwitterionic Polymer Nanoparticles in Interfacial Polymerization for Ion Separation. *ACS Appl. Polym. Mater.* **2020**, *2*, 1508–1517. [CrossRef]
21. Tiwari, R.; Walther, A. Strong anionic polyelectrolyte microgels. *Polym. Chem.* **2015**, *6*, 5550–5554. [CrossRef]
22. Woeste, G.; Meyer, W.H.; Wegner, G. Copolymers of ethyl p-vinylbenzenesulfonate for the preparation of polyelectrolytes of reproducible ion content. *Die Makromol. Chem.* **1993**, *194*, 1237–1248. [CrossRef]
23. Emmons, W.D.; Ferris, A.F. Metathetical Reactions of Silver Salts in Solution. II. The Synthesis of Alkyl Sulfonates1. *J. Am. Chem. Soc.* **1953**, *75*, 2257. [CrossRef]
24. Chern, C. Emulsion polymerization mechanisms and kinetics. *Prog. Polym. Sci.* **2006**, *31*, 443–486. [CrossRef]
25. Priest, W.J. Particle Growth in the Aqueous Polymerization of Vinyl Acetate. *J. Phys. Chem.* **1952**, *56*, 1077–1082. [CrossRef]

26. Ghiasi, S.; Behboudi, A.; Mohammadi, T.; Ulbricht, M. High-performance positively charged hollow fiber nanofiltration membranes fabricated via green approach towards polyethyleneimine layer assembly. *Sep. Purif. Technol.* **2020**, *251*, 117313. [CrossRef]
27. Schlenoff, J.B.; Yang, M.; Digby, Z.A.; Wang, Q. Ion Content of Polyelectrolyte Complex Coacervates and the Donnan Equilibrium. *Macromolecules* **2019**, *52*, 9149–9159. [CrossRef]
28. Yang, M.; Shi, J.; Schlenoff, J.B. Control of Dynamics in Polyelectrolyte Complexes by Temperature and Salt. *Macromolecules* **2019**, *52*, 1930–1941. [CrossRef]
29. Zelner, M.; Stolov, M.; Tendler, T.; Jahn, P.; Ulbricht, M.; Freger, V. Elucidating ion transport mechanism in polyelectrolyte-complex membranes. *J. Membr. Sci.* **2022**, *658*, 120757. [CrossRef]
30. Fan, H.; Huang, Y.; Billinge, I.H.; Bannon, S.M.; Geise, G.M.; Yip, N.Y. Counterion Mobility in Ion-Exchange Membranes: Spatial Effect and Valency-Dependent Electrostatic Interaction. *ACS ES&T Eng.* **2022**, *2*, 1274–1286. [CrossRef]
31. Freger, V. Ion partitioning and permeation in charged low-T* membranes. *Adv. Colloid Interface Sci.* **2020**, *277*, 102107. [CrossRef]
32. van der Poel, S. Parting Ways—Removal of Salts and Organic Micropollutants by Direct Nanofiltration: Pretreatment of Surface Water for the Production of Dune Infiltration Water. 2020. Available online: https://repository.tudelft.nl/islandora/object/uuid%3A6774b91c-6850-4c82-b3c0-a3110f0c40b9 (accessed on 4 August 2022).
33. Han, Y.; Jiang, Y.; Gao, C. High-Flux Graphene Oxide Nanofiltration Membrane Intercalated by Carbon Nanotubes. *ACS Appl. Mater. Interfaces* **2015**, *7*, 8147–8155. [CrossRef] [PubMed]
34. Liu, C.; Shi, L.; Wang, R. Crosslinked layer-by-layer polyelectrolyte nanofiltration hollow fiber membrane for low-pressure water softening with the presence of SO_4^{2-} in feed water. *J. Membr. Sci.* **2015**, *486*, 169–176. [CrossRef]
35. de Grooth, J.; Oborný, R.; Potreck, J.; Nijmeijer, K.; de Vos, W.M. The role of ionic strength and odd–even effects on the properties of polyelectrolyte multilayer nanofiltration membranes. *J. Membr. Sci.* **2015**, *475*, 311–319. [CrossRef]
36. Reurink, D.M.; Willott, J.D.; Roesink, H.D.W.; de Vos, W.M. Role of Polycation and Cross-Linking in Polyelectrolyte Multilayer Membranes. *ACS Appl. Polym. Mater.* **2020**, *2*, 5278–5289. [CrossRef]
37. Liu, Y.; Chen, G.Q.; Yang, X.; Deng, H. Preparation of Layer-by-Layer Nanofiltration Membranes by Dynamic Deposition and Crosslinking. *Membranes* **2019**, *9*, 20. [CrossRef]

Article

Water Molecular Dynamics in the Porous Structures of Ultrafiltration/Nanofiltration Asymmetric Cellulose Acetate–Silica Membranes

João Cunha [1,2], Miguel P. da Silva [1,3], Maria J. Beira [1,2,*], Marta C. Corvo [4], Pedro L. Almeida [4,5], Pedro J. Sebastião [1,2], João L. Figueirinhas [1,2] and Maria Norberta de Pinho [1,3]

1. Center of Physics and Engineering of Advanced Materials (CeFEMA), Laboratory for Physics of Materials and Emerging Technologies (LaPMET), Instituto Superior Técnico (IST), Universidade de Lisboa (ULisboa), Av. Rovisco Pais 1, 1049-001 Lisboa, Portugal; joaotiagocunha@gmail.com (J.C.); miguel.pereira.da.silva@tecnico.ulisboa.pt (M.P.d.S.); pedro.jose.sebastiao@tecnico.ulisboa.pt (P.J.S.); joao.figueirinhas@tecnico.ulisboa.pt (J.L.F.); marianpinho@tecnico.ulisboa.pt (M.N.d.P.)
2. Department of Physics (DF), Instituto Superior Técnico (IST), Universidade de Lisboa (ULisboa), Av. Rovisco Pais 1, 1049-001 Lisboa, Portugal
3. Department of Chemical Engineering (DEQ), Instituto Superior Técnico (IST), Universidade de Lisboa (ULisboa), Av. Rovisco Pais 1, 1049-001 Lisboa, Portugal
4. Centro de Investigação em Materiais (CENIMAT), Faculdade de Ciências e Tecnologia, Universidade Nova de Lisboa, Campus da Caparica, 2829-516 Caparica, Portugal; marta.corvo@fct.unl.pt (M.C.C.); pla@fct.unl.pt (P.L.A.)
5. Department of Physics, ISEL, R. Conselheiro Emídio Navarro 1, 1959-007 Lisboa, Portugal
* Correspondence: maria.beira@tecnico.ulisboa.pt

Abstract: This study presents the characterization of water dynamics in cellulose acetate–silica asymmetric membranes with very different pore structures that are associated with a wide range of selective transport properties of ultrafiltration (UF) and nanofiltration (NF). By combining ^{1}H NMR spectroscopy, diffusometry and relaxometry and considering that the spin–lattice relaxation rate of the studied systems is mainly determined by translational diffusion, individual rotations and rotations mediated by translational displacements, it was possible to assess the influence of the porous matrix's confinement on the degree of water ordering and dynamics and to correlate this with UF/NF permeation characteristics. In fact, the less permeable membranes, CA/SiO$_2$-22, characterized by smaller pores induce significant orientational order to the water molecules close to/interacting with the membrane matrix's interface. Conversely, the model fitting analysis of the relaxometry results obtained for the more permeable sets of membranes, CA/SiO$_2$-30 and CA/SiO$_2$-34, did not evidence surface-induced orientational order, which might be explained by the reduced surface-to-volume ratio of the pores and consequent loss of sensitivity to the signal of surface-bound water. Comparing the findings with those of previous studies, it is clear that the fraction of more confined water molecules in the CA/SiO$_2$-22-G20, CA/SiO$_2$-30-G20 and CA/SiO$_2$-34-G20 membranes of 0.83, 0.24 and 0.35, respectively, is in agreement with the obtained diffusion coefficients as well as with the pore sizes and hydraulic permeabilities of 3.5, 38 and 81 kg h^{-1} m^{-2} bar^{-1}, respectively, reported in the literature. It was also possible to conclude that the post-treatment of the membranes with Triton X-100 surfactants produced no significant structural changes but increased the hydrophobic character of the surface, leading to higher diffusion coefficients, especially for systems associated with average smaller pore dimensions. Altogether, these findings evidence the potential of combining complementary NMR techniques to indirectly study hydrated asymmetric porous media, assess the influence of drying post-treatments on hybrid CA/SiO$_2$ membrane' surface characteristics and discriminate between ultra- and nano-filtration membrane systems.

Keywords: NMR; spectroscopy; diffusometry; relaxometry; cellulose acetate; asymmetric membranes; ultrafiltration; nanofiltration

1. Introduction

It is established that the structure and dynamics properties of pore-confined molecules are greatly affected by the morphology of porous media [1–3]. In membrane separation, the state of water within a membrane's three-dimensional porous network plays a role in elucidating the mechanisms of its selective mass transfer task. Concertedly, the separation performance of a membrane can be gauged by the interplay of factors such as the pore size, electrical charge, hydrophilic/hydrophobic characteristics of the membrane polymeric or hybrid matrix and the solutes [4,5]. Therefore, the membranes' porous structure and the state of water within its porous matrix are crucial to understanding the mechanisms of membrane selective transport.

The determination of the accurate morphological features of porous media still represents a challenge as many properties depend not only on the void size's distribution but also on their connectivity and liquid–surface interactions [6]. Although there is a vast amount of scientific literature focused on microscopic and spectroscopic characterisation for elucidating the mechanisms of membrane selective transport in the active layer structures of integrally skinned cellulose acetate (CA) or cellulose esters membranes [4,7–12], this subject is more complex in the study of hybrid CA and silica, CA/SiO_2, asymmetric membranes constituting the system of this work [13,14]. Previous studies by de Pinho et al. [15,16] on the characterisation of the water order and dynamics in asymmetric CA/SiO_2 hybrid membranes, covering a wide range of ultrafiltration (UF) and nanofiltration (NF) permeation properties, pointed to an essential indication that Nuclear Magnetic Resonance (NMR) relaxometry observables, which are strongly dependent on water–surface interactions due to confinement, can be reliably correlated with the membranes' asymmetric porous structures and selective permeation performance.

Nuclear Magnetic Resonance (NMR) relaxometry is a widely used experimental technique that enables the study of a large variety of chemical compounds, such as liquid crystals, polymers, ionic liquids and complex food systems, just to name a few [17–20]. The ^1H NMR longitudinal relaxation rate dispersion (R_1 in the function of the ^1H Larmor frequency) is sensitive to molecular motions occurring at timescales ranging from milli- to picoseconds and from slower collective motions in liquid crystalline phases to fast molecular rotations. ^1H NMR relaxometry is especially sensitive to the existence of some degree of confinement, enabling an indirect study of a confining matrix by introducing a well-known liquid, usually water, into its structure. Relaxation-inducing interactions of the probing liquid with the surrounding surfaces, often referred to as rotations mediated by translational displacements, enables the characterization of a given matrix in terms of the effective mean square displacement of the liquid molecules confined in the porous system as well as the degree of order induced by these interactions [21–24].

In the present work, the main objective is to probe the water molecular dynamics within the porous structure of asymmetric CA/SiO_2 hybrid membranes over a wide range of UF and NF permeation properties by ^1H NMR relaxometry as a means to assess the effect of the drying post-treatments on the membranes' asymmetric structure modification.

2. Experimental Section

2.1. Membrane Preparation and Characterization

A series of flat asymmetric CA/SiO_2 hybrid membranes were made in a laboratory by coupling the wet phase inversion [25] with sol–gel techniques [26]. The synthesis methodology is described by de Pinho et al. [13]. Membranes were made from casting solutions containing 16.4 wt.% cellulose acetate (CA) polymer (≈30,000 average molecular weight), supplied by Sigma-Aldrich (Steinheim, Germany), a SiO_2 content equal to 5 wt.%, and three different solvent system ratios of formamide (enhancing pore-forming agent) and acetone. The acid catalysed hydrolysis of the SiO_2 alkoxide sol–gel precursor was promoted in situ by adding deionised water, tetraethyl orthosilicate (TEOS), supplied by Sigma-Aldrich (Steinheim, Germany), and nitric acid to the polymer casting solution. All chemicals were of reagent grade and 65% nitric acid was of technical grade. Membrane films were cast with the aid of a 250 μm calibrated doctor blade, followed by evaporation for 30 s before coagulation in an ice-cold deionised water bath. Table 1 shows the membranes' casting solution compositions and film-casting conditions used in the preparation of three membranes with distinct UF porous structures, labelled as CA/SiO_2-22, CA/SiO_2-30 and CA/SiO_2-34. In these membrane labels, the second field is represented by numbers 22, 30 and 34, which correspond to the formamide contents of 21.3%, 29% and 32.9% (wt.%), respectively, in the casting solutions.

Table 1. Asymmetric CA/SiO_2 hybrid membranes film casting solutions and casting conditions.

	Casting Solution Composition (wt.%)		
Membrane	CA/SiO_2-22	CA/SiO_2-30	CA/SiO_2-34
CA	16.4	16.4	16.4
Formamide	21.3	29.0	32.9
Acetone	58.8	51.1	47.2
TEOS (SiO_2 precursor)	3.0	3.0	3.0
H_2O	0.5	0.5	0.5
HNO_3	4 drops (pH ≈ 2)	4 drops (pH ≈ 2)	4 drops (pH ≈ 2)
Casting Conditions			
Temperature of casting solution (°C)	20–25		
Temperature of casting atmosphere (°C)	20–25		
Relative humidity of casting atmosphere (%)	40–50		
Solvent evaporation time (min)	0.5		
Gelation medium	Ice-cold deionised water (2 h)		

Following preparation, the asymmetric CA/SiO_2 hybrid membranes were conditioned in surfactant mixtures by a procedure adapted from Vos et al. [27]. This treatment was carried out using aqueous solutions of non-ionic surface-active agents composed of glycerol, supplied by PanReac (Darmstadt, Germany), and/or triton X-100, supplied by VWR (Briare, France). In that regard, membrane films were immersed for 15 min in one of the following solutions: (a) an aqueous solution of glycerol 20 vol.% (G20) or (b) an aqueous solution of triton X-100 4 vol.% and glycerol 20 vol.% (GT). All chemicals used in the treatments were of reagent grade and the conductivity of the deionised water was lower than 10 μS cm^{-1}. For NMR sample preparation, to access the water behavior within the membranes' porous matrices, the membrane films were immersed in deionised water for 48 h. Excess surface water was gently removed before enclosing a roll of hydrated membrane film in a sealed 5 mm outer diameter NMR tube. The membranes are identified throughout this work by a three-field code: the first code refers to the membrane hybrid matrix (CA/SiO_2), followed by a second field relative to the formamide content (in wt.%) in the casting solutions (of 22, 30 and 34), and the third corresponds to the drying membrane post-treatment of G20 or GT.

The membranes were characterised in terms of pure water hydraulic permeability (L_p) and a molecular weight cut-off (MWCO) referring to the molecular weight of the solute that is 95 % retained by the membrane. Details on the characterisation of the membranes studied are described in da Silva et al. [15].

2.2. Methods

^{1}H NMR Spectroscopy: The series of spectra obtained from the high resolution ^{1}H NMR relaxometry experiments performed at 7T was analyzed in order to extract the number of Lorentzian components and their respective longitudinal relaxation rates and signal contribution.

^{1}H NMR Diffusometry: At controlled temperatures and using a probe head with field gradient coils (Bruker Diff 30, Billerica, MA, USA) and a Bruker 7T superconductor connected to a Bruker Avance III NMR console, it was possible to measure the self-diffusion coefficient, D, of water molecules entrapped in the membrane matrix. The applied Pulsed

Gradient Stimulated Echo (PGSE) sequence produces an attenuation of the signal intensity for increasing magnetic field gradient strengths, expressed by Equation (1):

$$I = I_0 exp\left\{-\gamma_{^1H}^2 g^2 D\delta^3 \left(\frac{\Delta}{\delta} - \frac{1}{3}\right)\right\},\qquad(1)$$

where $\gamma_{^1H}$ is the proton gyromagnetic ratio, g is the gradient strength, δ is the length of the gradient pulses and Δ is the delay between pulsed gradients. Expression (1) does not take into account that water molecules are confined, which means that the obtained diffusion coefficients can be viewed as having apparent values with an order of magnitude that is well-estimated. More exact estimations of the diffusion coefficients would require the development of robust models that take into account the experimental conditions, namely magnetic field gradient pulse durations, which, as far as the authors know, were not yet achieved. In the case of the studied systems, except for pure water, multi-exponential decays were observed, which lead to the addition of the corresponding number of components to Equation (1).

^{1}H NMR Relaxometry: The longitudinal relaxation rate, R_1, was measured across a broad frequency range at controlled temperatures. For ^{1}H Larmor frequencies ranging between 10 kHz and 9 MHz, the measurements were made using a home-developed Fast Field Cycling (FFC) relaxometer [28]. For the remaining frequencies, the conventional inversion recovery technique was applied using the Bruker Avance II console paired with a variable field iron-core Bruker BE-30 electromagnet (10–100 MHz) or with a Bruker Widebore 7T superconductor magnet for the measurements at 300 MHz.

3. Results and Discussion

3.1. Membrane Characterization

Table 2 shows the hydraulic permeability, L_p, and molecular weight cut-off, MWCO, of the asymmetric CA/SiO$_2$ hybrid membranes.

Table 2. Characteristics of the asymmetric CA/SiO$_2$ hybrid membranes [15].

Membrane		Hydraulic Permeability, L_p (kg h^{-1} m^{-2} bar^{-1})	Molecular Weight Cut-Off, MWCO (kDa)
CA/SiO$_2$-22	G20	3.5 ± 0.2	4
	GT	2.2 ± 0.2	3
CA/SiO$_2$-30	G20	38 ± 2	14
	GT	40 ± 3	29
CA/SiO$_2$-34	G20	81 ± 4	35
	GT	62 ± 4	21

As it can be observed by looking at the hydraulic permeabilities previously obtained by de Pinho et al. [15] for the membrane systems studied in the present work, the CA/SiO$_2$-30 and CA/SiO$_2$-34 membranes present marked ultrafiltration characteristics, whereas the CA/SiO$_2$-22 membrane has a hydraulic permeability tjhat is one order of magnitude lower, thus standing within the border between nano- and ultrafiltration.

3.2. ^{1}H NMR Spectroscopy

Generally, the results from relaxometry experiments are obtained by integrating over the entire ^{1}H NMR spectrum and fitting the varying amplitudes, proportional to the magnetization along the fixed external magnetic field, to Equation (2). In Figure 1, the model fitting results following spectral integration are exemplified.

$$M_z = M_\infty + (M_0 - M_\infty)e^{-\tau R_1}\qquad(2)$$

In the case of the present work, the high resolution spectrum, obtained at a 7T external magnetic field, was divided into the minimum number of Lorentzian components for which it was possible to determine the longitudinal relaxation rate and the fraction of the population corresponding to each contribution. The obtained results are presented in Appendix A.1.

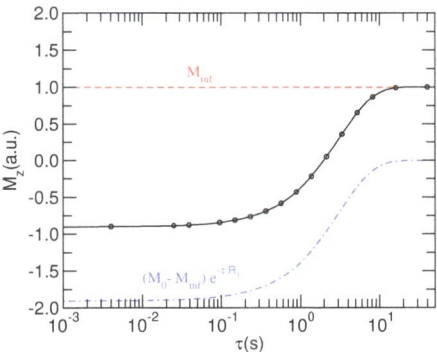

Figure 1. Sample curve showing the magnetization recovery for each inversion recovery delay, τ.

For the majority of the studied systems, two components are observed. In these cases, the fraction of more confined water molecules, q, relates with the shortest relaxation time, $T_1 = R_1^{-1}$, which is highlighted in red in Figures A1–A3. For samples CA/SiO$_2$-22 G20 and CA/SiO$_2$-34 G20, four and three contributions were, respectively, detected. In the case of sample CA/SiO$_2$-22 G20, one of the contributions was immediately disregarded in view of the extremely small T_1 (0.038 s), which would make it undetectable at lower frequencies. The other contribution having the shortest longitudinal relaxation time represented a very small percentage of the signal (3%), and it was, therefore, also not considered. For the CA/SiO$_2$-34 G20 system, we simply considered the contribution with the shortest relaxation time to be that of water molecules in a more confined environment. The list of more confined water population fraction is presented in Table 3.

Table 3. Considered fraction of more confined water molecules to apply in the ^{1}H NMR relaxometry analysis.

	CA/SiO$_2$-22		CA/SiO$_2$-30		CA/SiO$_2$-34	
	G20	GT	G20	GT	G20	GT
More confined population ratio, q	0.83	0.40	0.24	0.40	0.35	0.50

As it can be immediately concluded from the results presented in Table 3, the CA/SiO$_2$-22 G20 membrane is dramatically different from CA/SiO$_2$-30 G20 and CA/SiO$_2$-34 G20 systems in terms of more confined population fraction or, in other words, the surface-to-volume ratio is much larger for CA/SiO$_2$-22 G20. This result is consistent with the smaller pores observed for the CA/SiO$_2$-22 membranes and the consequent lower hydraulic permeability of this system (see Table 2). The post-treatment with triton X-100 (GT) appears to have uniformized the confined population ratio for the three membrane compositions.

3.3. ^{1}H NMR Diffusometry

Figure 2 shows the model fitting analyses made for each of the studied hydrated membranes, and Table 4 presents the obtained diffusion coefficients. The model fitting to the diffusometry and relaxometry data was performed using the open access online platform at fitteia.org (accessed in 1 September 2022), *fitteia*®, which applies the non-linear least squares minimization method with a global minimum target provided by the powerful MINUIT numerical routine from the CERN library [29,30].

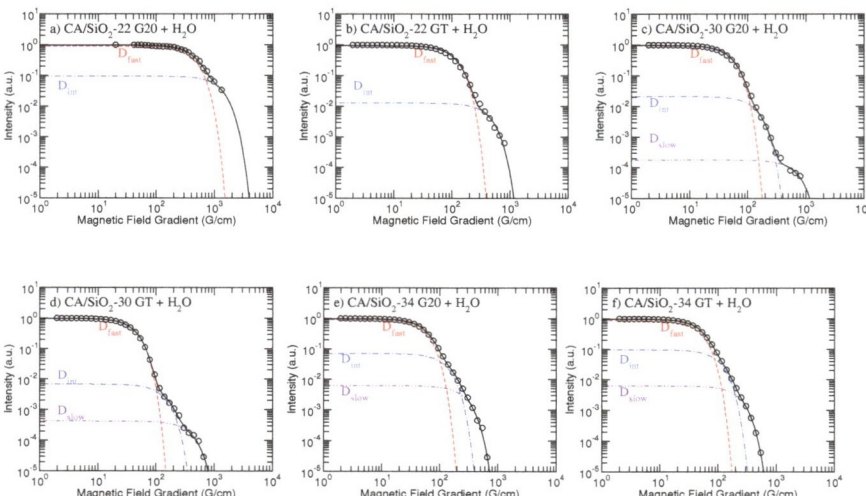

Figure 2. Diffusometry model fitting results obtained for the G20 and GT versions of membranes CA/SiO$_2$-22—(**a**,**b**); CA/SiO$_2$-30—(**c**,**d**); and CA/SiO$_2$-34—(**e**,**f**)—membranes at 22 °C. The dashed-red line represents the fast diffusion contribution, the dot-dashed-blue line represents the intermediate diffusion contribution and the dot-dot-dashed-violet line represents the slow diffusion contribution.

As it can be observed, all hydrated membranes present at least two diffusion coefficients that can be associated with the water molecules experiencing different degrees of confinement. For the CA/SiO$_2$-30 (G20 and GT) and CA/SiO$_2$-34 (G20 and GT) systems, three diffusion components were observed. In all these cases, the third residual component can only be observed in the logarithmic scale. In the case of CA/SiO$_2$-22 (G20 and GT), the slowest component is probably not observable due to its smaller value, which may fall out of the measurable range for this technique.

Table 4. Diffusion coefficients obtained from the PGSE ^{1}H NMR experiments performed at 25 °C. The model fitting was performed considering by an uncertainty equal to 5% of the signal intensity for each point.

Membrane		D_{fast} (10^{-10} m^2/s)	D_{int} (10^{-11} m^2/s)	D_{slow} (10^{-11} m^2/s)
CA/SiO$_2$-22	G20	0.09	0.11	–
	GT	1.40	0.97	
CA/SiO$_2$-30	G20	6.50	9.50	0.38
	GT	9.50	9.60	1.10
CA/SiO$_2$-34	G20	5.50	10.0	2.20
	GT	6.90	16.0	3.10

From Table 4, it is possible to conclude that the CA/SiO$_2$-22 systems present much smaller diffusion coefficients than the CA/SiO$_2$-30 and CA/SiO$_2$-34 systems, which is expected in view of their smaller pores. Membranes CA/SiO$_2$-30 and CA/SiO$_2$-34 seem to be harder to distinguish in terms of the diffusion coefficient, possibly because their higher permeability increases the relative amount of less confined water. The slower and intermediate diffusion coefficients, D_{slow} and D_{int}, respectively, seem to be smaller for CA/SiO$_2$-30 systems, which is consistent with the smaller pore sizes observed for these membranes [15]. However, the faster diffusion component is larger for membranes CA/SiO$_2$-30 than for membranes CA/SiO$_2$-34, which might be a consequence of a the pore size distribution in membranes CA/SiO$_2$-30 varying across a broader range of characteristic lengths. The fact that previous SEM studies have shown a wide distribution of pore sizes in these membranes makes it difficult to compare the ^{1}H NMR diffusometry results obtained for membranes CA/SiO$_2$-30 and CA/SiO$_2$-34 [15]. Nevertheless, the CA/SiO$_2$-22 systems are markedly less permeable and lead to much smaller diffusion coefficients, rendering the comparison between this and the CA/SiO$_2$-30 and CA/SiO$_2$-34 systems meaningful.

3.4. ^{1}H NMR Relaxometry

3.4.1. Raw Data and Theoretical Models

In Figure 3, the ^{1}H NMR relaxometry profiles obtained for the membranes studied in the present work are obtained. In order to enable a comparison between the profiles of membranes that were subject to a different drying process, the results previously obtained by de Pinho et al. [16] for membranes dried using the solvent exchange procedure were also added to the figure.

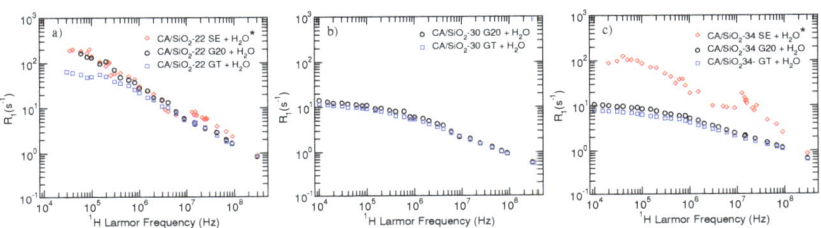

Figure 3. NMRD profiles obtained for the CA/SiO$_2$-22—(**a**); CA/SiO$_2$-30—(**b**); and CA/SiO$_2$-34—(**c**)—membranes at 22 °C. (*) Data extracted from previous works by de Pinho et al. [16] related to membranes dried using the solvent exchange procedure.

As it can be immediately concluded from the observation of the longitudinal relaxation profiles displayed in Figure 3, for the systems studied in the present work (black circles—G20—and blue squares—GT), the CA/SiO$_2$-30 and CA/SiO$_2$-34 membranes present rather similar relaxometry profiles, while CA/SiO$_2$-22 membranes presented significantly different results.

It is also possible to see that post-treatment with triton X-100 leads to very small differences for CA/SiO$_2$-30 and CA/SiO$_2$-34 membranes, while it produces a significant R_1 decrease across the lower frequency range in the case of membranes CA/SiO$_2$-22.

Furthermore, comparing the results obtained in the present work with those related to the solvent-exchange—(SE) dried membranes (red diamonds)—it is possible to observe a significant difference for the CA/SiO$_2$-34 membranes while the CA/SiO$_2$-22 systems seem almost insensitive to the drying process. This result may be explained by the fact that membranes that are more permeable, such as CA/SiO$_2$-34, are bound to be more impacted by the drying process than systems with smaller pores. In fact, the hydraulic permeabilities found for these systems, reported in previous studies, also show that the permeability of the CA/SiO$_2$-22 membrane is almost unaffected by post-treatment drying processes, while permeabilities obtained for the CA/SiO$_2$-30 and CA/SiO$_2$-34 systems vary over a wider range of values, especially when comparing the SE drying process to the G20 and GT post-treatments [15].

The curves presented in Figure 3 representing the longitudinal relaxation rate, R_1, obtained at different magnetic fields (or ^{1}H Larmor frequencies) and called NMR dispersion (NMRD) curves encode information on the molecular dynamics of the systems under analysis. In the present work, it was considered that the water entrapped in the membranes' pores may relax as a result of rotational and translational diffusions and rotations mediated by translational displacements, which are motivated by the interactions of water molecules with the porous matrix. Furthermore, assuming that these mechanisms are effective at different time scales and, thus, independent of one another, the total relaxation rate may be written as the sum of the individual rates (Equation (3)):

$$R_1 = R_1^{Rot} + (1-q)R_1^{SD} + qR_1^{RMTD}, \tag{3}$$

where q is the fraction of water molecules interacting with the pore walls, which was determined with the analysis of the spectral components of the signals obtained at a ^{1}H Larmor frequency of 300 MHz:

- Rotational diffusion (Rot):
  The model by Bloembergen, Purcell and Pound, better known as the BPP model, was applied in order to describe rotations of water molecules in the membranes [31,32]. The contribution of this mechanism to the NMR dispersion curves of water ^{1}H spins is given by Equation (4).

$$R_1^{Rot} = A_{Rot}\left[\frac{\tau_{Rot}}{1+\omega^2\tau_{Rot}^2} + \frac{4\tau_{Rot}}{1+4\omega^2\tau_{Rot}^2}\right]. \tag{4}$$

The prefactor A_{Rot} depends on the effective intramolecular distance between ^{1}H nuclear spins, r_{eff} (1.58 Å in the case of the water molecule), via Expression (5), which can easily be calculated for the water molecule:

$$A_{Rot} = \frac{3}{10}\left(\frac{\mu_0}{4\pi}\right)^2 \gamma_I^4 \hbar^2 \frac{1}{r_{eff}^6}, \tag{5}$$

with μ_0 denoting the vacuum magnetic permeability ($4\pi \times 10^{-7}$ H/m), γ_I denoting the magnetic ratio of the nucleus with spin I and $\hbar = h/(2\pi)$ denoting the reduced Planck constant ($1.0545718 \times 10^{-34}$ m^{2}Kg/s). Given that A_{rot} can be estimated and fixed, the only parameter in Equation (4) that needs to be determined via the model-fitting analysis is the rotational correlation time, τ_{Rot}.

- Translational Diffusion (SD):
  Self-diffusion of water molecules may be accounted for using the Torrey model [33,34]. Torrey assumed that molecules have equal probabilities of jumping in any direction from an initial state into another, reaching a random jump-like solution. The associated longitudinal relaxation rate frequency dependence is described by Equation (6).

$$R_1^{SD} = \frac{3}{2}\left(\frac{\mu_0}{4\pi}\right)^2 \gamma_I^4 \hbar^2 I(I+1)\left[j^{(1)}(\omega,\tau_D,d,r,n) + j^{(2)}(2\omega,\tau_D,d,r,n)\right]. \tag{6}$$

Parameter n is the ^{1}H spin density, and d is the average intermolecular interspin distance. τ_D, the translational diffusion correlation time, $<r^2>$, the mean square jump distance, and the diffusion coefficient, D, are related by the following equation.

$$<r^2> = 6\tau_D D. \tag{7}$$

The functions $j^{(i)}(\omega,\tau_D,d,r,$ and $n)$ are the spectral density functions described in references [33,34].

- Rotations mediated by translational displacements (RMTD):
  The water motion in the confined system gives rise to a relaxation mechanism associated with rotations mediated by translational displacements. This model describes the movement of water molecules near the pores' walls and, therefore, is related to the interaction of those molecules with the membranes' surfaces. The contribution of this model to the longitudinal relaxation rate is given by [35,36] the following:

$$R_1^{RMTD} = \frac{A_{RMTD}}{\nu^p} G(\nu, \nu_{max}, \nu_{min})$$
$$= A_{RMTD} \left[\frac{f\left(\frac{\nu_{max}}{\nu}\right) - f\left(\frac{\nu_{min}}{\nu}\right)}{\nu^p} - 4 \frac{f\left(\frac{\nu_{max}}{2\nu}\right) - f\left(\frac{\nu_{max}}{2\nu}\right)}{2\nu^p} \right], \quad (8)$$

where

$$f(x) = \frac{1}{\pi} \left[arctan(\sqrt{2x} + 1) + arctan(\sqrt{2x} - 1) - arctan\left(\frac{\sqrt{2x}}{x+1}\right) \right] \quad (9)$$

This contribution exhibits one high cut-off frequency, ν_{max}, and one low cut-off frequency, ν_{max}, which are, respectively, associated with the largest and smallest possible translational relaxation modes and, therefore, to the smallest and largest possible average displacements, respectively: $\nu_{max}^{-1} = l_{min}^2 \pi/2D$ and $\nu_{min}^{-1} = l_{max}^2 \pi/2D$, where D is the diffusion coefficient and l is the average displacement. Exponent p can vary between 0.5 and 1, where $p = 0.5$ corresponds to a situation where there is an isotropic distribution of coupled rotations and self-diffusion motions along the pore/channel's surfaces, while for $p = 1$, there is a preferential orientation of the rotations/translations relaxation modes along the constraining surfaces. The parameter A_{RMTD} is inversely proportional to the square root of the diffusion coefficient and to the range of wave numbers related to the motional modes induced by the surface, Δq. This parameter is proportional to the square of the fraction of molecules interacting with the surface and to the square of the order parameter, representing the long time limit residual correlation of restricted tumbling.

3.4.2. Model Fitting

In Figures 4 and 5, the model fitting results produced by the *Master* module of the online platform *fitteia*® [37] are presented. Figure 4 show the results obtained for pure water and the CA/SiO$_2$-22 G20 and GT-hydrated membranes. The model fitting analysis of CA/SiO$_2$-30 and CA/SiO$_2$-34 systems is presented in Figure 5. The model fitting parameters resulting from the NMRD curves analysis are summarized in Table 5.

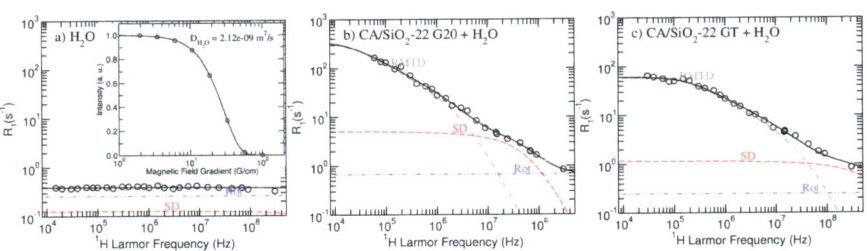

Figure 4. NMRD profiles and model fitting results obtained for the pure water—(**a**); CA/SiO$_2$-22 G20—(**b**); and CA/SiO$_2$-22 GT—(**c**)—membranes. The dot-dashed-brown lines represents the RMTD contribution, the dashed-red-line represents the self-diffusion contributions and the dot-dot-dashed-blue line represents the rotations/reorientations contribution to the longitudinal relaxation rate profiles.

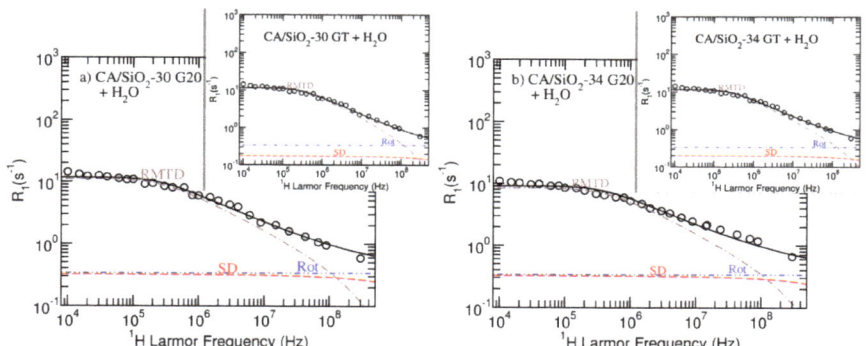

Figure 5. NMRD profiles and model fitting results obtained for systems CA/SiO$_2$-30 G20 and GT—(**a**); and CA/SiO$_2$-34 G20 and GT—(**b**). The dot-dashed-brown lines represents the RMTD contribution, the dashed-red-line represents the self-diffusion contributions and the dot-dot-dashed-blue line represents the rotations/reorientations contribution to the longitudinal relaxation rate profiles.

Table 5. Parameters obtained from the NMRD model fitting analysis made on the studied membranes. The model fitting was performed considering an uncertainty of the relaxation rate equal to 10% of its value. The ^{1}H spin density, n, needed for the Torrey model for translational self-diffusion was fixed to the calculated value of 6.69 × 10^{28} ^{1}H nuclear spins per cubic meter. Parameters D_{SD} and D_{RMTD} were fixed to the D_{fast} and D_{int} diffusion coefficients presented in Table 4, respectively. In the case of pure water, the diffusion coefficient was fixed to that presented in Figure 4. The fraction q, representing the more confined water, was also not a free parameters, and its value was set to that presented in Table 3 for each studied hydrated membrane. Parameter A_{rot} was also calculated and fixed as explained in the rotations model section.

Parameters	CA/SiO$_2$-22		CA/SiO$_2$-30		CA/SiO$_2$-34		H$_2$O
	G20	GT	G20	GT	G20	GT	
A_{Rot} (10^{10} s^{-2})	1.08	1.08	1.08	1.08	1.08	1.08	1.08
τ_{Rot} (10^{-12} s)	13	5	6	6	6	6	5
D_{SD} (10^{-10} m^2/s)	0.09	1.40	6.50	9.50	5.50	6.90	21
r (Å)	3.0	3.0	3.0	3.0	3.0	3.0	3.0
d (Å)	2.7	2.7	2.7	2.7	2.7	2.7	2.7
q	0.83	0.40	0.24	0.40	0.35	0.50	–
D_{RMTD} (10^{-11} m^2/s)	0.11	0.97	9.50	9.60	10.0	16.0	–
A_{RMTD} (10^{3} s$^{-(1+p)}$)	27	18	5.8	3.7	3.8	2.9	–
l_{max} (10^{-9} m)	5.3	5.3	13	12	10	16	–
p	0.56	0.51	0.50	0.50	0.50	0.50	–

The model proposed in this work and the combination of ^{1}H NMR relaxometry and diffusometry experimental techniques allowed for a consistent analysis of all the studied hydrated membranes, as it can be concluded by the good quality of the fits.

In Figure 4, it is possible to observe the striking NMRD profile difference when comparing pure water with confined water. Water molecules entrapped in the matrix have the additional RMTD relaxation pathway, which significantly increases the longitudinal relaxation rate. Moreover, confined water presented diffusion coefficients that are up to three orders of magnitude smaller than that measured for free water (see inserted image in Figure 4a and Table 5).

The parameter q was fixed to the value obtained from the analysis of the spectral components. D_{SD} and D_{RMTD} were set equal to the value of D_{fast} and D_{int} obtained from the diffusometry analysis and presented in Table 4, respectively. D_{SD} corresponds to a less confined fraction of water that does not interact directly with the matrix, while D_{RMTD} corresponds to a more confined fraction of water that relaxes as a result of interactions with the surface.

Despite the apparent similarities between the relaxometry profiles obtained for the G20 and GT versions of membranes CA/SiO_2-30 and CA/SiO_2-34, the model-fitting analysis evidences a decrease in the self-diffusion relaxation rate contribution for the membranes that were post-treated with triton X-100 (compare the dashed red line of the sub figures with that of the respective inserted image in Figure 5). This contribution decrease is more significant for the CA/SiO_2-22 membranes, as observed in Figure 4. This observation is consistent with triton X-100 increasing the hydrophobicity of the cellulose acetate matrix, making the water less bound to it and leading to higher diffusion coefficients (see D_{SD} and D_{RMTD} in Table 5 or, respectively, D_{fast} and D_{int} in Table 4). The fact that this increase is more significant for the CA/SiO_2-22 hydrated membranes may be explained by the fact that smaller pore sizes relate to a larger ratio of water/surface interactions. Furthermore, the increased hydrophobicity suggested from this model-fitting analysis might explain the uniformization of the bound water fraction, q, found for the GT porous membranes (see Table 3).

This analysis enabled the estimation of the characteristic pore size given by the parameter l_{max}, that, on average, induces more effective ^{1}H NMR relaxation through rotations mediated by translational displacements. As it can be observed, the additional treatment with triton X-100 does not significantly affect this dimension, except in the case of CA/SiO_2-34 systems. Combining the previously described increased hydrophobicity with the fact that this membrane is the most permeable, it is possible that the signal from more bound water molecules is masked by the signal of unbound water, leading to an apparently larger characteristic dimension.

Regarding the fact that A_{RMTD} is inversely proportional to the square root of the diffusion coefficient, the values obtained for this parameter seem to be consistent for all the samples and further support the increased hydrophobicity conferred upon treating the matrix with triton X-100 (GT).

Parameter p shows that there is an isotropic distribution of coupled rotations and self-diffusion motions along the matrix' pores for all systems, except for the CA/SiO_2-22 pair, where some degree of anisotropy is detected. The fact that CA/SiO_2-22 membranes have smaller pores is expected to increase the degree of confinement, thus evidencing water-ordering induced by the surface.

4. Conclusions

In this study, ^{1}H NMR spectroscopy, diffusometry and relaxometry were successfully combined in order to consistently analyze three pairs of hydrated ultrafiltration/nanofiltration asymmetric cellulose acetate–silica membranes. Each CA/SiO_2-22, CA/SiO_2-30 and CA/SiO_2-34 pair of membranes was composed of one membrane in which the post-treatment involved an aqueous solution of glycerol with 4 vol.% of triton X-100 (GT) and another where triton was not involved in the post-treatment (G20).

The results seem to be consistent with the post-treatment with triton X-100 rendering the matrix surfaces more hydrophobic and increasing the self-diffusion coefficients obtained for water molecules in different confinement environments. This impact is more significant when the characteristic pore sizes are smaller given the increased probability of water/matrix interactions.

Comparing the results obtained in the present work with those related to membranes dried using the solvent-exchange procedure, presented in previous studies, it becomes clear that the drying process has a much less pronounced impact on the cellulose acetate–silica matrix when the pores are characterized by smaller dimensions.

The surface-bound water population variation observed between the CA/SiO_2-22 G20 and the two analogous membrane systems is in line not only with the diffusion coefficients obtained in the present work but also with the hydraulic permeabilities reported in previous studies.

On the whole, this work evidences the advantage of combining complementary experimental techniques with a relatively simple relaxation model to study and differentiate between ultrafiltration/nanofiltration porous media and track their sensitivity to different post-treatments/drying processes.

Author Contributions: Conceptualization, all authors; software, P.J.S. and J.L.F.; formal analysis, J.C., J.L.F., M.J.B. and P.J.S.; investigation, J.C., M.P.d.S., J.L.F., M.J.B., M.C.C. and P.L.A.; resources, M.C.C., P.L.A., M.P.d.S. and M.N.d.P.; writing—original draft preparation, J.C., J.L.F., M.N.d.P., M.P.d.S., M.J.B. and P.J.S.; writing—review and editing, all authors; supervision, P.J.S., J.L.F. and M.N.d.P. All authors have read and agreed to the published version of the manuscript.

Funding: This research was funded by the Portuguese Fundação para a Ciência e a Tecnologia (FCT) grant number PTDC/CTM-BIO/6178/2014, UID/CTM/04540/2019, UIDB/04540/2020 and UIDB/50025/2020-2023; and MCC researcher's contract (2021.03255.CEECIND). M. J. Beira was funded by FCT grant number PD/BD/142858/2018.

Institutional Review Board Statement: Not applicable.

Data Availability Statement: Not applicable.

Acknowledgments: The NMR spectrometers at IST and FCT NOVA are part of the National NMR network (PTNMR), supported by FCT (ROTEIRO/0031/2013-PINFRA/22161/2016). The authors would like to thank European COST Action EURELAX CA15029 (2016–2020) and acknowledge the article processing charge's full waiver granted by the *Membranes* journal.

Conflicts of Interest: The authors declare no conflict of interest. The funders had no role in the design of the study; in the collection, analyses, or interpretation of data; in the writing of the manuscript; or in the decision to publish the results.

Abbreviations

The following abbreviations are used in this manuscript:

NMR	Nuclear magnetic resonance;
^{1}H	Proton, hydrogen-1;
PGSE	Pulse Gradient Stimulated Echo;
Rot	Rotations;
SD	Self-diffusion;
RMTD	Rotations Mediated by Translational Displacements;
CA	Cellulose acetate;
NF	Nanofiltration;
UF	Ultrafiltration;
CA/SiO$_2$	Cellulose acetate/silica;
G20	Surfactant conditioning with an aq. sol. of glycerol 20 vol.%;
GT	Surfactant conditioning with an aq. sol. of glycerol 20 vol.% and triton x-100 4 vol.%.

Appendix A

Appendix A.1. ^{1}H NMR Spectra Obtained from the Relaxometry Experiments Performed at 7T

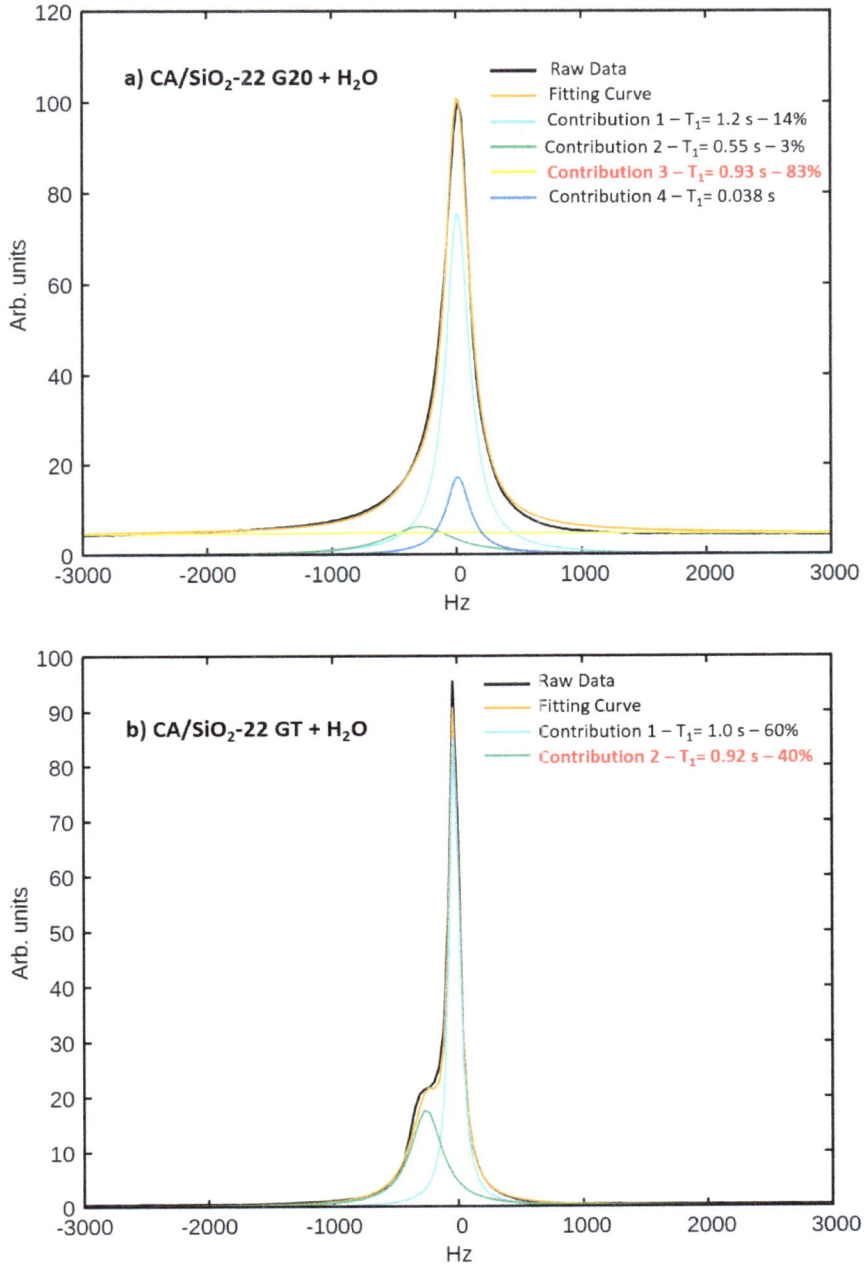

Figure A1. ^{1}H NMR spectroscopy analysis of the inversion recovery experimental spectra for membranes CA/SiO$_2$-22 G20 (**a**) and GT (**b**).

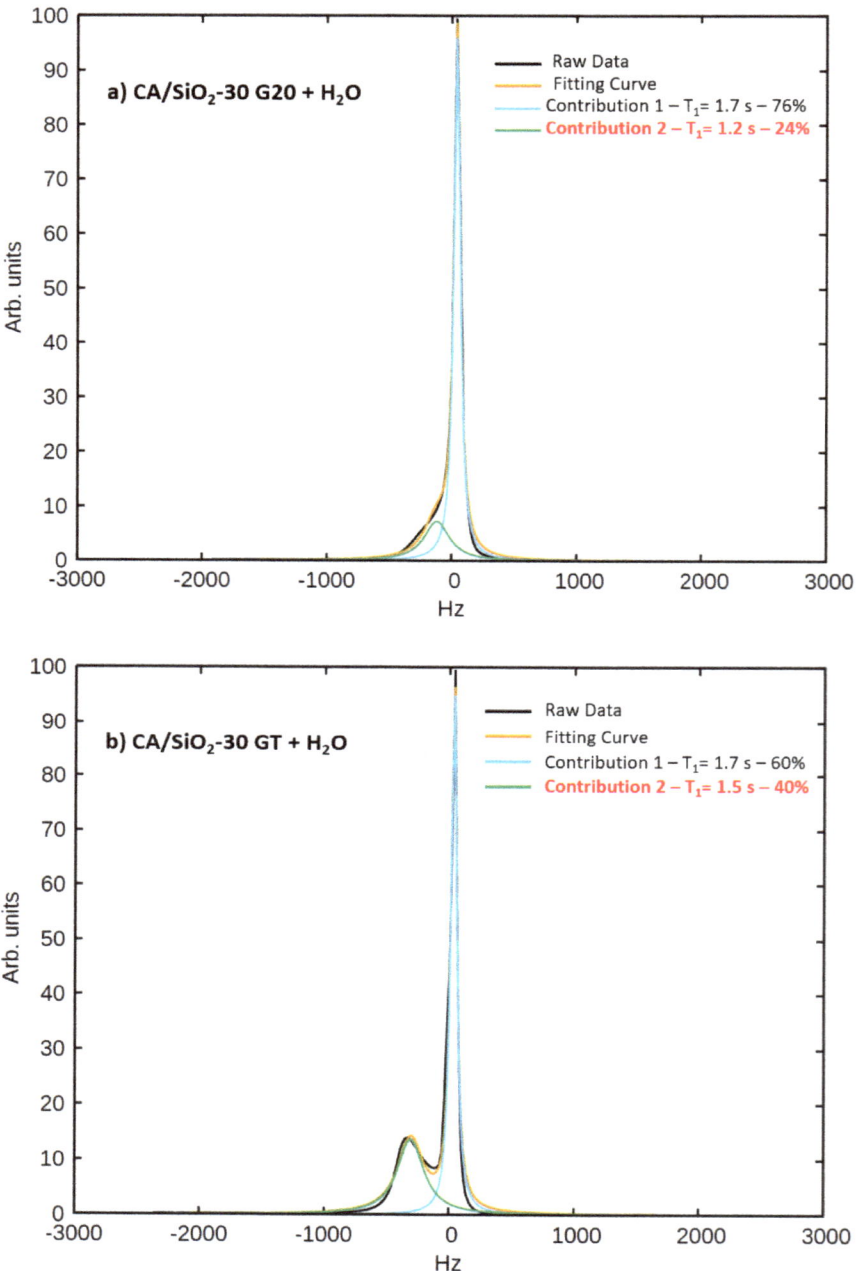

Figure A2. ^{1}H NMR spectroscopy analysis of the inversion recovery experimental spectra for membranes CA/SiO$_2$-30 G20 (**a**) and GT (**b**).

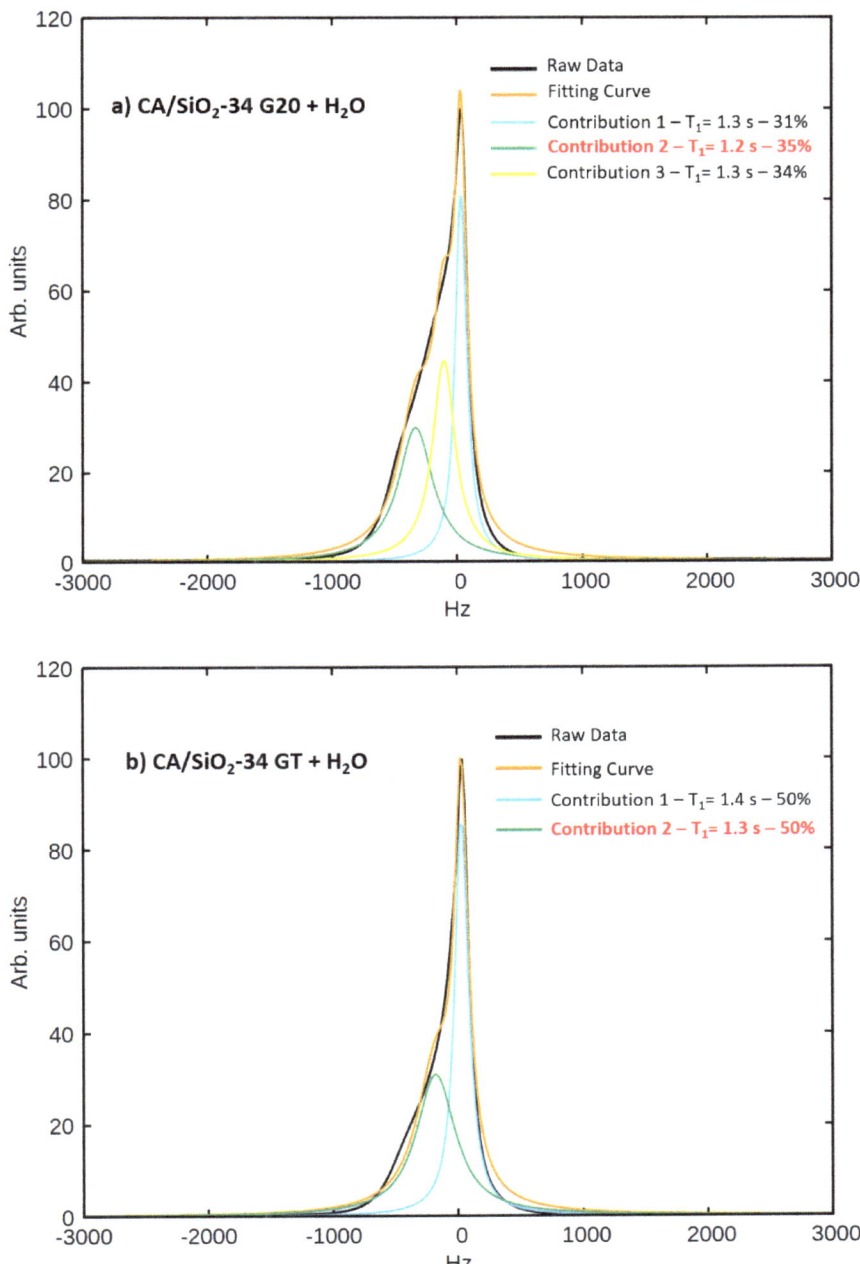

Figure A3. ^{1}H NMR spectroscopy analysis of the inversion recovery experimental spectra for membranes CA/SiO$_2$-34 G20 (**a**) and GT (**b**).

References

1. Hansen, E.W.; Fonnum, G.; Weng, E. Pore Morphology of Porous Polymer Particles Probed by NMR Relaxometry and NMR Cryoporometry. *J. Phys. Chem. B* **2005**, *109*, 24295–24303. [CrossRef] [PubMed]
2. Chakraborty, S.; Kumar, H.; Dasgupta, C.; Maiti, P.K. Confined Water: Structure, Dynamics, and Thermodynamics. *Accounts Chem. Res.* **2017**, *50*, 2139–2146. [CrossRef] [PubMed]
3. Mietner, J.B.; Brieler, F.J.; Lee, Y.J.; Fröba, M. Properties of Water Confined in Periodic Mesoporous Organosilicas: Nanoimprinting the Local Structure. *Angew. Chem. Int. Ed.* **2017**, *56*, 12348–12351. [CrossRef] [PubMed]
4. Stamatialis, D.F.; Dias, C.R.; de Pinho, M.N. Structure and Permeation Properties of Cellulose Esters Asymmetric Membranes. *Biomacromolecules* **2000**, *1*, 564–570. [CrossRef] [PubMed]
5. Xu, F.; Wei, M.; Zhang, X.; Song, Y.; Zhou, W.; Wang, Y. How Pore Hydrophilicity Influences Water Permeability? *Research* **2019**, *2019*, 2581241. [CrossRef]
6. Silletta, E.V.; Velasco, M.I.; Gomez, C.G.; Strumia, M.C.; Stapf, S.; Mattea, C.; Monti, G.A.; Acosta, R.H. Enhanced Surface Interaction of Water Confined in Hierarchical Porous Polymers Induced by Hydrogen Bonding. *Langmuir* **2016**, *32*, 7427–7434. [CrossRef]
7. Murphy, D.; de Pinho, M.N. An ATR-FTIR study of water in cellulose acetate membranes prepared by phase inversion. *J. Membr. Sci.* **1995**, *106*, 245–257. [CrossRef]
8. Dias, C.R.; Rosa, M.J.; de Pinho, M.N. Structure of water in asymmetric cellulose ester membranes—And ATR-FTIR study. *J. Membr. Sci.* **1998**, *138*, 259–267. [CrossRef]
9. Stamatialis, D.F.; Dias, C.R.; Norberta De Pinho, M. Atomic force microscopy of dense and asymmetric cellulose-based membranes. *J. Membr. Sci.* **1999**, *160*, 235–242. [CrossRef]
10. Ong, R.; Chung, T.; Helmer, B.; De Wit, J. Characteristics of water and salt transport, free volume and their relationship with the functional groups of novel cellulose esters. *Polymer* **2013**, *54*, 4560–4569. [CrossRef]
11. del Gaudio, I.; Hunter-Sellars, E.; Parkin, I.P.; Williams, D.; Da Ros, S.; Curran, K. Water sorption and diffusion in cellulose acetate: The effect of plasticisers. *Carbohydr. Polym.* **2021**, *267*, 118185. [CrossRef] [PubMed]
12. Ioan, S.; Necula, A.M.; Stoica, I.; Olaru, N.; Olaru, L.; Ioanid, G.E. Surface Properties of Cellulose Acetate. *High Perform. Polym.* **2010**, *22*, 598–608. [CrossRef]
13. Mendes, G.; Faria, M.; Carvalho, A.; Gonçalves, M.C.; de Pinho, M.N. Structure of water in hybrid cellulose acetate-silica ultrafiltration membranes and permeation properties. *Carbohydr. Polym.* **2018**, *189*, 342–351. [CrossRef] [PubMed]
14. Faria, M.; Moreira, C.; Eusébio, T.; Brogueira, P.; de Pinho, M.N. Hybrid flat sheet cellulose acetate/silicon dioxide ultrafiltration membranes for uremic blood purification. *Cellulose* **2020**, *27*, 3847–3869. [CrossRef]
15. da Silva, M.P.; Beira, M.J.; Nogueira, I.D.; Sebastião, P.J.; Figueirinhas, J.L.; de Pinho, M.N. Tailoring the Selective Permeation Properties of Asymmetric Cellulose Acetate/Silica Hybrid Membranes and Characterisation of Water Dynamics in Hydrated Membranes by Deuterium Nuclear Magnetic Resonance. *Membranes* **2022**, *12*, 559. [CrossRef]
16. Beira, M.J.; Silva, M.P.; Condesso, M.; Cosme, P.; Almeida, P.L.; Corvo, M.C.; Sebastião, P.J.; Figueirinhas, J.L.; de Pinho, M.N. Molecular order and dynamics of water in hybrid cellulose acetate–silica asymmetric membranes. *Mol. Phys.* **2019**, *117*, 975–982. [CrossRef]
17. de Almeida Martins, J.P.; Chávez, F.V.; Sebastião, P.J. NMR molecular dynamics study of chromonic liquid crystals Edicol Sunset Yellow doped with salts. *Magn. Reson. Chem.* **2014**, *52*, 540–545. [CrossRef]
18. Overbeck, V.; Appelhagen, A.; Rößler, R.; Niemann, T.; Ludwig, R. Rotational correlation times, diffusion coefficients and quadrupolar peaks of the protic ionic liquid ethylammonium nitrate by means of ^{1}H fast field cycling NMR relaxometry. *J. Mol. Liq.* **2021**, *322*, 114983. [CrossRef]
19. Calucci, L.; Pizzanelli, S.; Mandoli, A.; Birczyński, A.; Lalowicz, Z.T.; De Monte, C.; Ricci, L.; Bronco, S. Unravelling Main- and Side-Chain Motions in Polymers with NMR Spectroscopy and Relaxometry: The Case of Polyvinyl Butyral. *Polymers* **2021**, *13*, 2686. [CrossRef]
20. Ates, E.G.; Domenici, V.; Florek-Wojciechowska, M.; Gradišek, A.; Kruk, D.; Maltar-Strmečki, N.; Oztop, M.; Ozvural, E.B.; Rollet, A.L. Field-dependent NMR relaxometry for Food Science: Applications and perspectives. *Trends Food Sci. Technol.* **2021**, *110*, 513–524. [CrossRef]
21. Ordikhani Seyedlar, A.; Stapf, S.; Mattea, C. Nuclear magnetic relaxation and diffusion study of the ionic liquids 1-ethyl- and 1-butyl-3-methylimidazolium bis(trifluoromethylsulfonyl)imide confined in porous glass. *Magn. Reson. Chem.* **2019**, *57*, 818–828. [CrossRef] [PubMed]
22. Zavada, T.; Kimmich, R. The anomalous adsorbate dynamics at surfaces in porous media studied by nuclear magnetic resonance methods. The orientational structure factor and Lévy walks. *J. Chem. Phys.* **1998**, *109*, 6929–6939. [CrossRef]
23. Carvalho, A.; Sebastiao, P.J.; Fonseca, I.; Matos, J.; Goncalves, M.C. Silica and silica organically modified nanoparticles: Water dynamics in complex systems. *Microporous Mesoporous Mater.* **2015**, *217*, 102–108. [CrossRef]
24. Kumar, A.; Cruz, V.; Figueirinhas, J.L.; Sebastião, P.J.; Trindade, A.C.; Fernandes, S.N.; Godinho, M.H.; Fossum, J.O. Water Dynamics in Composite Aqueous Suspensions of Cellulose Nanocrystals and a Clay Mineral Studied through Magnetic Resonance Relaxometry. *J. Phys. Chem. B* **2021**, *125*, 12787–12796. [CrossRef]
25. Kunst, B.; Sourirajan, S. An approach to the development of cellulose acetate ultrafiltration membranes. *J. Appl. Polym. Sci.* **1974**, *18*, 3423–3434. [CrossRef]

26. Brinker, C.J.; Scherer, G.W. *Sol-Gel Science*; Academic Press: San Diego, CA, USA, 1990. [CrossRef]
27. Vos, K.D.; Burris, F.O. Drying cellulose acetate reverse osmosis membranes. *Ind. Eng. Chem. Prod. Res. Dev.* **1969**, *8*, 84–89. [CrossRef]
28. Sousa, D.; Domingos Marques, G.; Cascais, J.M.; Sebastião, P.J. Desktop fast-field cycling nuclear magnetic resonance relaxometer. *Solid State NMR* **2010**, *38*, 36–43. [CrossRef]
29. Sebastião, P.J. The art of model fitting to experimental results. *Eur. J. Phys.* **2014**, *35*, 015017. [CrossRef]
30. James, F. MINUIT Function Minimization and Error Analysis. In *MINUIT Function Minimization and Error Analysis. Reference Manual Version 94.1*; CERN: Geneva, Switzerland, 1994.
31. Bloembergen, N.; Purcell, E.M.; Pound, R.V. Relaxation effect in nuclear magnetic resonance absorption. *Phys. Rev.* **1948**, *73*, 679–712. [CrossRef]
32. Abragam, A. *The Principles of Nuclear Magnetism*; Clarendon Press: Oxford, UK, 1961.
33. Torrey, H.C. Nuclear Spin Relaxation By Translational Diffusion. *Phys. Rev.* **1953**, *92*, 962–969. [CrossRef]
34. Harmon, J.F.; Muller, B.H. Nuclear Spin Relaxation by Translational Diffusion in Liquid Ethane. *Phys. Rev.* **1969**, *182*, 400–410. [CrossRef]
35. Sebastião, P.J.; Sousa, D.; Ribeiro, A.C.; Vilfan, M.; Lahajnar, G.; Seliger, J.; Žumer, S. Field-cycling NMR relaxometry of a liquid crystal above T_{NI} in mesoscopic confinement. *Phys. Rev. E* **2005**, *72*, 061702. [CrossRef] [PubMed]
36. Vilfan, M.; Apih, T.; Sebastião, P.J.; Lahajnar, G.; Žumer, S. Liquid crystal 8CB in random porous glass: NMR relaxometry study of molecular diffusion and director fluctuations. *Phys. Rev. E* **2007**, *76*, 051708. [CrossRef] [PubMed]
37. Sebastiao, P.J.O.; Beira, M.J.; Cordeiro, R.M.O.; Kumar, A.; Fernandes, J.F.; Ferraz, A.M.P.; Gonçalves, L.N. The art of fitting ordinary differential equations models to experimental results. *Eur. J. Phys.* **2022**, *43*, 035807. [CrossRef]

Article

Impacts of Calcium Addition on Humic Acid Fouling and the Related Mechanism in Ultrafiltration Process for Water Treatment

Hui Zou [1,†], Ying Long [1,†], Liguo Shen [1], Yiming He [2], Meijia Zhang [1,*] and Hongjun Lin [1,*]

1 College of Geography and Environmental Sciences, Zhejiang Normal University, Jinhua 321004, China
2 Department of Materials Science and Engineering, Zhejiang Normal University, Yingbin Road 688, Jinhua 321004, China
* Correspondence: mzhang15@zjnu.edu.cn (M.Z.); hjlin@zjnu.cn (H.L.)
† These authors equally contributed to this work.

Abstract: Humic acid (HA) is a major natural organic pollutant widely coexisting with calcium ions (Ca^{2+}) in natural water and wastewater bodies, and the coagulation–ultrafiltration process is the most typical solution for surface water treatment. However, little is known about the influences of Ca^{2+} on HA fouling in the ultrafiltration process. This study explored the roles of Ca^{2+} addition in HA fouling and the potential of Ca^{2+} addition for fouling mitigation in the coagulation-ultrafiltration process. It was found that the filtration flux of HA solution rose when Ca^{2+} concentration increased from 0 to 5.0 mM, corresponding to the reduction of the hydraulic filtration resistance. However, the proportion and contribution of each resistance component in the total hydraulic filtration resistance have different variation trends with Ca^{2+} concentration. An increase in Ca^{2+} addition (0 to 5.0 mM) weakened the role of internal blocking resistance (9.02% to 4.81%) and concentration polarization resistance (50.73% to 32.17%) in the total hydraulic resistance but enhanced membrane surface deposit resistance (33.93% to 44.32%). A series of characterizations and thermodynamic analyses consistently suggest that the enlarged particle size caused by the Ca^{2+} bridging effect was the main reason for the decreased filtration resistance of the HA solution. This work revealed the impacts of Ca^{2+} on HA fouling and demonstrated the feasibility to mitigate fouling by adding Ca^{2+} in the ultrafiltration process to treat HA pollutants.

Keywords: membrane fouling; humic acid; calcium ion; hydraulic resistance; ultrafiltration process

1. Introduction

Due to its high efficiency in removing various pollutants, the ultrafiltration process has been widely applied to treat wastewater and surface water [1–3]. Nevertheless, the existence of natural organic matter (NOM) in natural water and wastewater bodies would cause serious membrane fouling and thus hinder the promotion of application of low-pressure membranes such as ultrafiltration membrane [4–9]. It is generally accepted that coagulation and flocculation can serve as a pretreatment step for ultrafiltration process as it can cluster foulant particles and absorb NOM, and therefore simultaneously reduce membrane fouling and improve NOM rejection [10–12].

NOM is a mixture of organic compounds that come from nature and composed of various substances such as humic acid (HA), protein, and polysaccharides [1,13–15]. Among them, HA is considered one of the most important categories that contribute to membrane fouling. It is ubiquitous in aquatic ecosystems, and the concentration distribution range varies from a few mg/L of dissolved organic carbon (DOC) to more than a few hundred mg/L DOC [16–19]. Extensive studies have reported the significant contribution of HA to membrane fouling [20–23]. Unlike protein and polysaccharides, HA has a relatively small molecular weight [20]. Therefore, it cannot be completely removed by the ultrafiltration

process and also would cause more severe irreversible membrane fouling [24]. Previous literature has reported that coagulation is an effective pretreatment approach to mitigate membrane fouling caused by HA [21]. Nevertheless, the external addition of flocculant apparently would increase the maintenance cost. Therefore, a promising and cost-effective strategy is to make full use of the flocculating substances coexisting with HA in natural water and wastewater.

Since HA bears lots of functional groups (such as hydroxylm, ethoxy, and carboxyl) and binding sites, calcium ions (Ca^{2+}) might be an excellent natural flocculant [25]. Ca^{2+} is a common metal ion in surface water and its concentration in municipal wastewater is reported to be in the range of 0.5–3 mM [26]. As a divalent cation, Ca^{2+} has a bridging effect and can promote biological flocculation in sewage, which is bound up with membrane fouling. Some studies have explored the roles of Ca^{2+} in the membrane fouling performance of HA in different membrane filtration processes [27–31]. A consistent result of the enhanced HA fouling caused by Ca^{2+} addition has been reported in the filtration processes of anion exchange, nanofiltration, and forward osmosis [29,32,33]. However, unlike the filtration processes mentioned above, the studies regarding the effects of Ca^{2+} on HA in the ultrafiltration process obtained contradictory results. Lin et al. [34] pointed out that the membrane fouling was improved due to the increased Ca^{2+} concentration. Wang et al. [31] pointed out that Ca^{2+} has a more effective capacity in ultrafiltration fouling intensification than Mg^{2+}. Nevertheless, Li et al. [35] found that Ca^{2+} can promote the formation of reversible fouling and thus can achieve a higher removal efficiency of HA. The inconsistent results suggest that the effects of Ca^{2+} on membrane fouling are complex and require further study.

The causes of the inconsistent results in the previous studies may lie in several aspects. First, the effects of Ca^{2+} on HA-induced membrane fouling depend on the membrane material. However, the materials of the ultrafiltration membrane applied in the literature differed in different studies. In addition, previous studies evaluated the HA–Ca^{2+} fouling through a single or whole filtration resistance variation. For example, Lin et al. [34] investigated the effects of Ca^{2+} on HA fouling for the polyvinylchloride (PVC) membrane through the interfacial interaction energy change. Chang et al. [36] mainly focused on the HA–Ca^{2+} effects in hydraulically irreversible fouling. Previous studies did not well distinguish the different filtration resistance components. Furthermore, the effects of specific Ca^{2+} concentration on HA fouling depend on the HA concentration. In fact, studies were seldom with regard to the HA–Ca^{2+} fouling of polyvinylidene fluoride (PVDF) ultrafiltration membrane [37]. However, PVDF is one of the most widely used membrane materials in wastewater treatment [38,39]. Therefore, more studies are of great significance for HA fouling control in the PVDF ultrafiltration process.

Therefore, a simplified model of the separation membrane that functions in the cross-flow filtration mode was adopted. The effects of Ca^{2+} on HA fouling were evaluated through different hydraulic resistance components including concentration polarization, deposit, internal, and membrane fouling. The properties of the HA–Ca^{2+} complexes were analyzed by using a series of characterization methods. Finally, thermodynamic interaction theory was used to analyze the possible mechanisms.

2. Materials and Methods

2.1. Sample Preparation

All the reagents and chemicals applied in the current study were purchased from Sinopharm Chemical Reagent Co., Ltd. The experimental sample was prepared according to the following steps. First, 1 g of HA was dissolved in 1000 mL of NaOH (pH = 13) and continuously stirred for 24 h to make sure the complete dissolution of HA. Next, the pH of the stock solution was adjusted to 7.0 by using 1 mol/L HCl solution and then stored at room temperature. In the current work, a HA concentration of 100 mg/L was adopted to simulate the HA content in natural water, and the working solution was prepared by diluting the stock solution with deionized water. A specific volume of the stock CaCl$_2$ solution was added during the dilution process to obtain the set Ca^{2+} concentration. It should be noted that the selected HA concentration (100 mg/L) is higher than that in the natural water body in order to facilitate the formation of membrane fouling. Similar concentration levels were typically used for lab-scale studies in the previous literature [40,41].

2.2. Filtration Resistance Tests

A lab-scale cross-flow filtration system (customized by Hangzhou Jiuling Technology Co., Ltd., Hangzhou, China) was applied for filtration resistance tests, and all the tests were conducted at room temperature with an operating pressure of 2 bar. The membrane utilized in this study was made of PVDF material (Shanghai SINAP Co., Ltd., Shanghai, China), and the effective membrane surface area was 25 cm^2. The membrane was characterized as having a 0.1 μm pore size with a 140 kDa molecular weight cutoff (MWCO).

The filtration resistance was determined according to the Darcy–Poiseuille equation described as follows [42]:

$$J_f = \frac{1}{R_m + R_e + R_p + R_i} \frac{\Delta P}{\eta_{water}} \quad (1)$$

where J_f is the filtration flux; R_m, R_e, R_p, and R_i are the membrane filtration resistance, membrane surface deposit resistance, concentration polarization resistance, and internal blocking resistance, respectively; ΔP is the transmembrane pressure; and η_{water} is the dynamic viscosity of water. In the Poiseuille equation, the viscosity of the filtrate is equivalent to that of water.

The membrane filtration resistances were tested by filtering deionized water through the virgin membranes. Before the tests, the membranes were pre-compressed under 5 bar for at least 1 h to obtain a steady pure water flux. For each membrane, at least 3 tests were conducted to obtain an average value. The values of the membrane filtration resistance were calculated according to Equation (2):

$$R_m = \frac{1}{J_f} \frac{\Delta P}{\eta_{water}} \quad (2)$$

By filtering the HA suspension, the total filtration resistances were estimated by Equation (3):

$$R_T = \frac{1}{J_f} \frac{\Delta P}{\eta_{water}} = R_m + R_e + R_p + R_i \quad (3)$$

After the filtration of the HA suspension, the membranes were rinsed with deionized water three times to eliminate all traces of the solution, especially the concentration polarization layer. Thereafter, deionized water was filtered through the rinsed membrane to obtain resistance R_1, which is the sum of R_m, R_e, and R_i:

$$R_1 = R_m + R_e + R_i \quad (4)$$

Afterward, the deposit formed on the membrane surface was removed by a sponge followed by ultrasonic wave treatment. R_2, the sum of R_m and R_i, was then obtained by filtration of deionized water through the cleaned membrane:

$$R_2 = R_m + R_i \quad (5)$$

Based on Equations (2)–(5), the values of R_p, R_e, and R_i were estimated by Equations (6)–(8):

$$R_p = R_T - R_1 \quad (6)$$

$$R_e = R_1 - R_2 \quad (7)$$

$$R_i = R_T - R_m - R_e - R_p \tag{8}$$

2.3. Analytical Methods

The functional groups of the samples were determined by a Fourier transform infrared spectrometer (FTIR, NEXUS 670, Waltham, MA, USA). The wavenumber range was 4000–500 cm^{-1}. The particle size distribution (PSD) of the HA suspensions with different Ca^{2+} concentrations was measured by a particle size analyzer (Mastersizer 3000, Malvern, UK). Triplicate measurements were conducted for each sample. The total organic carbon (TOC) content of the HA solution was determined by a TOC analyzer (Liqui TOCII, Elementar, Hanau, Germany). The contact angle of the PVDF membrane and HA samples was determined by a contact angle meter (Kino Industry Co., Ltd., Boston, MA, USA), and the operation was similar to the previous reports. Zeta potential of the HA solutions and membrane surface was measured by a Malvern Zetasizer Nano ZS and a zeta 90 Plus instrument, respectively. Details regarding the operations of the abovementioned characterization can be found in the previous publications [43–48].

2.4. Extended Derjaguin–Landau–Verwey–Overbeek (XDLVO) Theory

It has been reported that the short-ranged thermodynamic interactions between foulants and membrane surface play a key role in the adhesion of different foulants on the membrane. The thermodynamic interactions can be divided into three parts according to the XDLVO theory [49–51], which are van der Waals (LW), acid–base (AB), and electrostatic double-layer (EL) interaction energies. The strength of these energies at separation distance (h) ($\Delta G^{LW}(h)$, $\Delta G^{EL}(h)$, and $\Delta G^{AB}(h)$) (mJ·m^{-2}) can be quantified by the following equations [52,53]:

$$\Delta G^{LW}(h) = \Delta G^{LW}_{h_0} \frac{h_0^2}{h^2} \tag{9}$$

$$\Delta G^{EL}(h) = \varepsilon_r \varepsilon_0 \kappa \zeta_1 \zeta_3 \left(\frac{\zeta_1^2 \zeta_3^2}{2\zeta_1 \zeta_3}(1 - \coth\kappa h) + \frac{1}{\sinh\kappa h} \right) \tag{10}$$

$$\Delta G^{AB}(h) = \Delta G^{AB}_{h_0} \exp\left(\frac{h_0 - h}{\lambda} \right) \tag{11}$$

where h and h_0 are the separation distance (nm) and minimum separation distance (nm) between two entities, respectively; $\varepsilon_r \varepsilon_0$ is the solution dielectric constant (C·V^{-1}·m^{-1}); κ, ζ, and λ represent the reciprocal of the Debye length (nm^{-1}), surface zeta potential (mV), and the attenuation of AB interaction (usually assigned as 0.6), respectively; the subscripts 1, 2, and 3 mean the membrane, pure water, and foulant, respectively; $\Delta G^{LW}_{h_0}$, $\Delta G^{AB}_{h_0}$, and $\Delta G^{EL}_{h_0}$ are the interaction energies at a separation distance of h_0 (mJ·m^{-2}), which can be quantified by Equations (12)–(14), respectively:

$$\Delta G^{LW}_{h_0} = -2 \left(\sqrt{\gamma_1^{LW}} - \sqrt{\gamma_2^{LW}} \right) \left(\sqrt{\gamma_3^{LW}} - \sqrt{\gamma_2^{LW}} \right) \tag{12}$$

$$\Delta G^{EL}_{h_0} = \frac{\varepsilon_r \varepsilon_0 \kappa}{2} \left(\zeta_1^2 + \zeta_3^2 \right) \left(1 - \coth(\kappa h_0) + \frac{2\zeta_1 \zeta_3}{\zeta_1^2 + \zeta_3^2} \operatorname{csch}(\kappa h_0) \right) \tag{13}$$

$$\Delta G^{AB}_{h_0} = 2 \left[\sqrt{\gamma_2^+}(\sqrt{\gamma_1^-} + \sqrt{\gamma_3^-} - \sqrt{\gamma_2^-}) + \sqrt{\gamma_2^-}(\sqrt{\gamma_1^+} + \sqrt{\gamma_3^+} - \sqrt{\gamma_2^+}) - \sqrt{\gamma_3^-}\sqrt{\gamma_1^+} - \sqrt{\gamma_3^+}\sqrt{\gamma_1^-} \right] \tag{14}$$

The values of γ^{LW}, γ^+, and γ^- (mJ·m^{-2}) were determined by solving a Young's equation group [54]:

$$\frac{(1 + \cos\phi)}{2} \gamma_l^{TOL} = \sqrt{\gamma_l^{LW}} \sqrt{\gamma_s^{LW}} + \sqrt{\gamma_l^-} \sqrt{\gamma_s^+} + \sqrt{\gamma_l^+} \sqrt{\gamma_s^-} \tag{15}$$

where the subscripts *l* and *s* denote the probe liquid and solid surface, respectively.

3. Results

3.1. Impacts of Ca^{2+} Concentration on Filtration Behaviors of HA

Figure 1 shows the membrane filtration flux after different operational steps under different Ca^{2+} concentrations. As displayed in Figure 1, the permeation flux significantly decreases after the filtration of HA suspensions. The flux is only 6.3%, 9.6%, and 18.8% of the virgin membrane for the HA containing Ca^{2+} concentrations of 0, 1.5, and 5.0 mM, respectively. After the cleaning processes of rinsing, deposit removal, and ultrasonic wave treatment, the permeation flux recovers to 41.1%, 59.1%, and 79.5% of the virgin membrane for the HA containing Ca^{2+} concentrations of 0, 1.5, and 5.0 mM, respectively. All in all, the addition of Ca^{2+} leads to a less flux drop as compared with the pure HA. In addition, a higher Ca^{2+} addition corresponds to a higher permeation flux and lower internal blocking resistance (R_i). As shown in Figure 1, the flux decline results from four hydraulic resistances; the effects of Ca^{2+} on HA fouling should be further ascertained and analyzed in each hydraulic resistance.

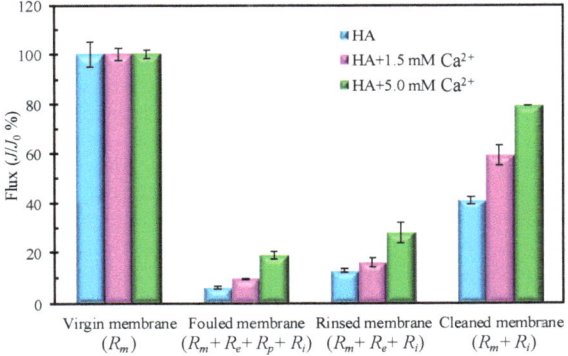

Figure 1. Comparison of membrane filtration flux after different operational steps under different Ca^{2+} concentrations (ΔP = 2 bar).

Figure 2 shows the filtration resistance distribution of HA under different Ca^{2+} concentrations. As displayed in Figure 2, the virgin membrane resistance (R_m) is comparable while the values of the other three filtration resistances (R_e, R_p, and R_i) decrease with the increased Ca^{2+} content. It indicates that the addition of Ca^{2+} can improve the anti-fouling property of HA, and the increase in the Ca^{2+} concentration can enhance this effect. However, unlike the absolute value of the hydraulic resistance, the proportion and contribution of each resistance component in the total hydraulic resistance have different variation trends. The proportion of R_m increases due to the significant reduction of total filtration resistance after the addition of Ca^{2+} into the HA solution. Among the other three resistances, the proportion of R_i is the smallest, and it decreases with the Ca^{2+} concentration (the proportion of R_i is 9.02%, 6.57%, and 4.81% for Ca^{2+} concentrations of 0, 1.5, and 5.0 mM, respectively). Similarly, the ratio of R_p to the total filtration resistance decreases with the increase in the Ca^{2+} content, which is 50.73%, 40.49%, and 32.17%, respectively. On the contrary, the proportion of R_e increases with the Ca^{2+} concentration, which is 33.93%, 43.48%, and 44.32%, respectively. The above results indicate that an increase in Ca^{2+} addition weakens the role of R_i (internal blocking resistance) and R_p (concentration polarization resistance) in the total hydraulic resistance but enhances that of R_e (membrane surface deposit resistance). This result is not completely consistent with previous studies, and further research is required to explore the underlying mechanisms.

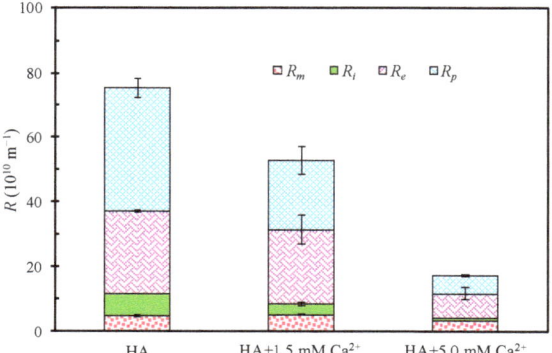

Figure 2. Influence of Ca^{2+} concentration on experimental hydraulic resistances (R_p, R_e, R_i, R_m) of HA (ΔP = 2 bar).

3.2. Characterization of HA under Different Ca^{2+} Concentrations

3.2.1. FT-IR Spectra Analysis

Figure 3 shows the FTIR spectra of HA samples with different Ca^{2+} concentrations. The broad band around 3300 cm^{-1} represents the O–H stretching vibration of phenolic compounds, and the adsorption peak at 1550 cm^{-1} can be assigned to aromatic C=C stretching and C=O stretching [55,56]. The peak around 1390 cm^{-1} represents the symmetrical stretching vibration of -COO- related to carboxylate. The vibrational frequency in the range of 650–900 cm^{-1} is usually considered aromatic C–H out of plane bending [57]. Obviously, the peak of the HA solution here is stronger than that of other cases, suggesting that the structure of HA has been changed to some extent after the addition of Ca^{2+}. However, FTIR is a qualitative characterization method, and the different peak intensities cannot strongly support the different filtration resistances shown in Figures 1 and 2. Therefore, the different filtration performances should be ascribed to other causes.

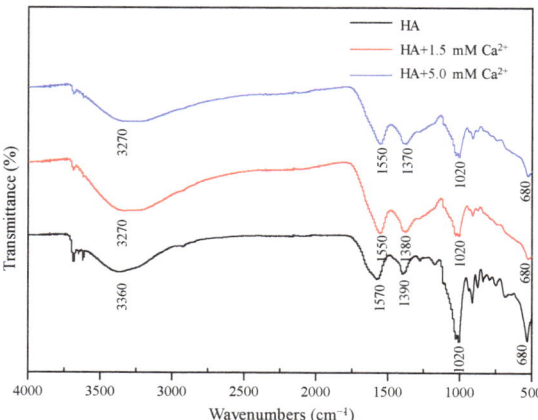

Figure 3. FTIR spectra of HA samples with different Ca^{2+} concentrations.

3.2.2. Particle Size Distribution (PSD) and TOC Removal Measurements

Figure 4 shows the PSD of HA samples containing different Ca^{2+} concentrations. As displayed in Figure 4, the pure HA solution exerts a single peak shape, and the mean size of HA flocs is about 3.31 μm. After adding 1.5 mM Ca^{2+}, the floc size of the HA suspension exhibits a double peak shape. The distribution of HA flocs in the ranges of 0–50 μm and

50–500 µm significantly decreases and increases, respectively. As a result, the mean size of HA flocs increases to 76.04 µm. After a further increase in Ca^{2+} concentration to 5.0 mM, the distribution of HA flocs in the ranges of 0–10, 70–105, and 300–500 µm increases, whereas that in the ranges of 10–70 and 105–300 µm decreases. This phenomenon suggests that the increased Ca^{2+} concentration has two different effects on HA. First, the electrostatic shielding effect leads to the compression of some negatively charged HA molecules, which leads to the reduction of some HA flocs. On the other hand, the bridging effect of Ca^{2+} results in the extension of HA molecular chains, which causes an increase in the particle size. Since the mean floc size of HA at the Ca^{2+} concentration of 5.0 mM increases to 107.28 µm, it can be concluded that the bridging effect of Ca^{2+} on HA should be much stronger than the electrostatic effect, and thus results in the formation of larger HA flocs.

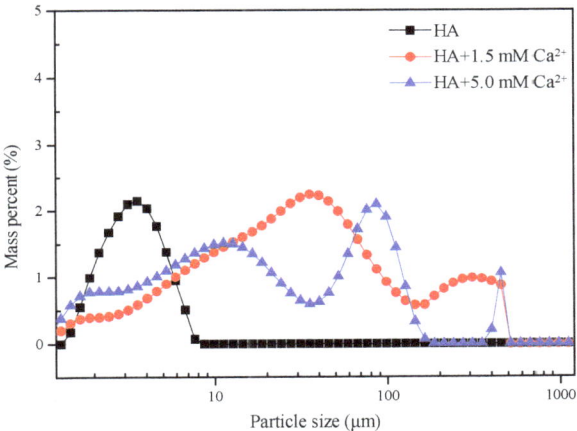

Figure 4. PSD of HA solution with different Ca^{2+} concentrations.

The enlarged particle size under the addition of Ca^{2+} is further supported by the optical image (Figure 5) and TOC removal (Figure 6). As shown in Figure 5, after 24 h of natural standing, the HA solutions with different Ca^{2+} concentrations display different sedimentation properties. When the Ca^{2+} concentration is 1.5 mM, the color of the upper layer of the mixed solution is lighter than that of the pure HA solution, although no obvious sedimentation can be seen at the bottom due to the dark color of the HA itself. Nevertheless, when the Ca^{2+} concentration increases to 5.0 mM, it can be seen that there is obvious sedimentation at the bottom of the beaker, and the upper layer solution becomes clearer. High sedimentation corresponds to a larger particle size. The observed phenomena further prove that the size of HA flocs increases with the increased Ca^{2+} concentration. Due to the enhanced sedimentary property, the TOC content in the supernatant before and after filtration significantly drops correspondingly after the addition of Ca^{2+} (Figure 6). In addition, the TOC removal efficiency after filtration also significantly increases from 44.5% (pure HA) to 78.4% (HA + 1.5 mM Ca^{2+}) and 74.1% (HA + 5.0 mM Ca^{2+}). It indicates that although the addition of Ca^{2+} is beneficial to the removal of TOC in pure HA, the addition of higher concentrations of Ca^{2+} cannot further increase the removal of TOC in the ultrafiltration membrane filtrate. In short, the above results consistently suggest that the bridging effect of Ca^{2+} can increase the particle size of HA. It is widely accepted that a larger particle size generally corresponds to a lower membrane fouling potential [58–60], which is well-consistent with the filtration resistance change. Therefore, the floc size change caused by Ca^{2+} addition is considered the main reason for the different filtration resistance of HA.

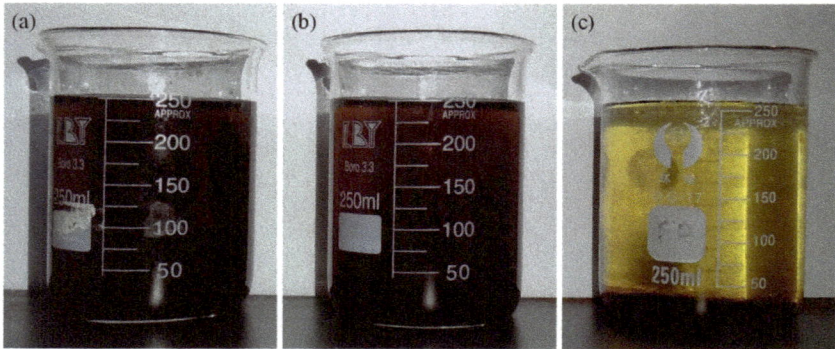

Figure 5. Optical images of HA solutions prepared after 24 h natural sedimentation with Ca^{2+} concentration of (**a**) 0 mM, (**b**) 1.5 mM, and (**c**) 5.0 mM (HA concentration = 100 mg/L).

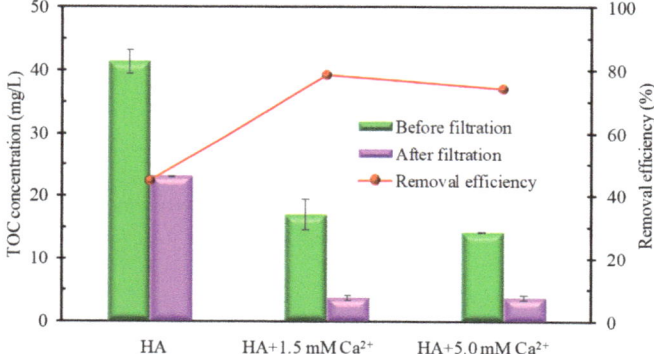

Figure 6. TOC content and removal efficiency of HA with different Ca^{2+} concentrations.

3.3. Thermodynamic Mechanism of HA Fouling Behavior

The adhesion of foulants on the membrane surface is an important process for membrane fouling formation. The XDLVO theory, which has been widely used for the quantitative calculation of the interaction energy between two surfaces, was used to evaluate the adhesion ability of HA with different Ca^{2+} concentrations. The surface contact angle and zeta potential of the PVDF membrane and HA layers with different Ca^{2+} concentrations are listed in Table 1. Based on these data, the interaction energies with separation distance were calculated, and the results are shown in Figure 7. It can be seen that the AB interaction accounts for the vast majority of the total energy for all the scenarios, and thus predominantly manipulates the fouling process. Since the AB interaction energy is positive in all three scenarios, the total interaction energy is always positive regardless of the Ca^{2+} concentration. It indicates that HA particles are difficult to adhere to the membrane surface [61], which well supports the proportion of membrane surface deposit resistance (R_e) in Figure 2 (generally more than 80% while only 33.93–44.32% in this work). In addition, the interaction energy intensity of the HA with Ca^{2+} is always higher than that of the pure HA, suggesting the improved anti-adhesion property of HA by Ca^{2+} addition.

Table 1. Surface characteristics in terms of contact angle of three probe liquids and zeta potential of the PVDF membrane and HA solutions with different Ca^{2+} concentration.

Materials	Contact Angle (°)			Zeta Potential (mV)
	Water	Glycerol	Diiodomethane	
PVDF membrane	62.16 ± 0.10	57.22 ± 1.47	23.15 ± 0.82	−25.21 ± 2.46
HA	48.93 ± 0.28	70.99 ± 0.99	34.36 ± 0.51	−26.67 ± 0.50
HA + 1.5 mM Ca^{2+}	42.36 ± 0.47	74.25 ± 0.18	41.36 ± 0.10	−22.87 ± 0.50
HA + 5.0 mM Ca^{2+}	41.75 ± 0.12	73.64 ± 0.31	38.61 ± 0.08	−19.03 ± 0.60

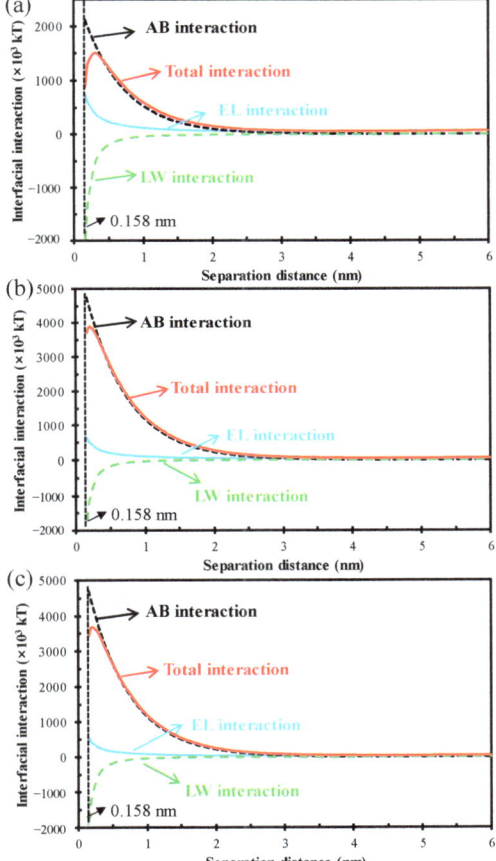

Figure 7. Profiles of interfacial interaction energies between FVDF membrane and HA with different Ca^{2+} concentrations (**a**) 0 mM, (**b**) 1.5 mM, and (**c**) 5.0 mM.

Although the interaction energy between the membrane and HA is repulsive, membrane fouling is unavoidable due to the external drag force. Figure 8 shows the schematic diagram of the hydraulic resistance variation of HA with Ca^{2+} addition in cross-flow filtration. As shown in Figure 8, small-sized HA particles are dramatically reduced due to the Ca^{2+} bridging effect, which leads to a significant reduction in internal blocking resistance (R_i). Moreover, the enlarged floc size not only can prevent the adhesion and accumulation of HA on the membrane surface but also lead to the loose structure of the foulant layer. As a result, concentration polarization resistance (R_p) decreases. Since the absolute value of R_i

and R_p decreased more with Ca^{2+}, the proportion and contribution of membrane surface deposit resistance (R_e) in the total filtration resistance correspondingly increased.

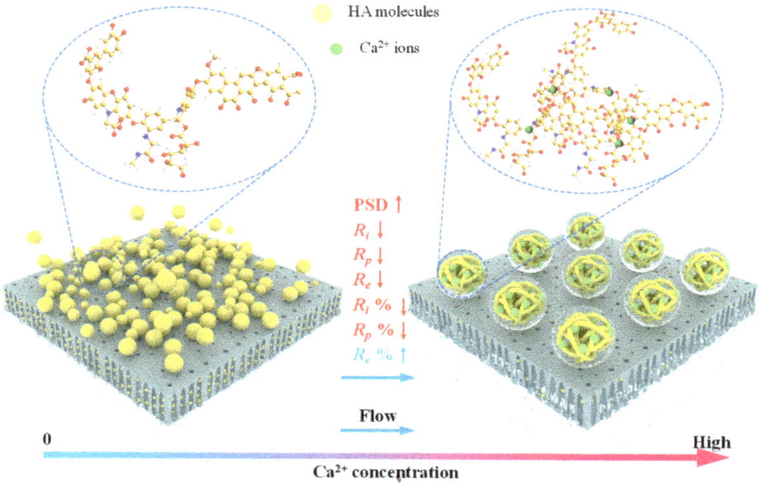

Figure 8. Schematic diagram of hydraulic resistance variation of HA with Ca^{2+} addition in cross-flow filtration.

It should be noted that the result obtained in the current work is not completely consistent with previous studies. The underlying reasons are located in several aspects. The first reason can be attributed to the membrane material. For example, the variation of membrane hydrophilicity and hydrophobicity through membrane modification may lead to dramatic changes in fouling trends. The second reason can be ascribed to the evaluation scope (a single or whole filtration resistance variation). For instance, Lin et al. [34] evaluated the effects of Ca^{2+} on HA fouling only through the interfacial interaction energy change. Chang et al. [36] mainly focused on the HA–Ca^{2+} effects in hydraulically irreversible fouling. The last reason can locate in HA concentration because the effects of specific Ca^{2+} concentration on HA fouling are dependent on the HA concentration. In this study, the SFR of HA decreased with the increase in Ca^{2+} concentration, which is inconsistent with the result (the membrane fouling firstly increased and then decreased with increasing Ca^{2+} concentration) observed by Miao et al. [37]. It is mainly attributed to the different HA concentrations applied. The HA concentration in this study was 100 mg/L, which was one-tenth that of Miao et al. (1 g/L) [37]. The Ca^{2+} concentrations selected in the current work exceeded the critical concentration, and, thus, only a monotonically decreasing variation trend was observed. As HA and Ca^{2+} concentration in the natural water bodies is in the range of 0.02–30 mg/L and 0.5–3 mM, respectively [26], the coexistence of HA and Ca^{2+} in natural water and wastewater generally can achieve a membrane mitigation effect.

4. Conclusions

In the current work, the effects of Ca^{2+} on HA fouling were evaluated with the Darcy–Poiseuille model through four different hydraulic resistance components. The results show that the increase in Ca^{2+} concentration improved the filtration flux and reduced the absolute value of each hydraulic resistance. Unlike the absolute value of the hydraulic resistance, the proportion and contribution of each resistance component in the total hydraulic resistance have different variation trends with the Ca^{2+} concentration. An increase in Ca^{2+} addition (0 to 5.0 mM) weakened the role of internal blocking resistance (R_i, 9.02% to 4.81%) and concentration polarization resistance (R_p, 50.73% to 32.17%) in the total hydraulic resistance but enhanced that of the membrane surface deposit resistance (R_e, 33.93% to 44.32%).

A series of characterizations consistently suggest that the enlarged particle size caused by Ca^{2+} addition was the main reason for the different filtration resistance of HA. The calculation results with the XDLVO theory further reveal that the anti-adhesion property of HA was improved due to the bridging effect of Ca^{2+}. This work revealed the impacts of Ca^{2+} on HA fouling in the PVDF ultrafiltration membrane and the underlying causes and demonstrated the feasibility to mitigate HA fouling in the PVDF ultrafiltration membrane by adding Ca^{2+}.

Author Contributions: Methodology, H.Z. and Y.L.; formal analysis, M.Z.; investigation, H.Z. and Y.L.; data curation, H.Z., Y.L., L.S. and Y.H.; visualization, L.S. and Y.H.; writing—original draft preparation, H.Z. and Y.L.; writing—review and editing, M.Z. and H.L.; supervision, H.L.; project administration, M.Z. and H.L.; funding acquisition, M.Z. and H.L. All authors have read and agreed to the published version of the manuscript.

Funding: This research was funded by the Natural Science Foundation of Zhejiang Province (Nos. LQ21E080010, LD21E080001), National Natural Science Foundation of China (Nos. 52000160, 51978628), and Key Research and Development Program of Zhejiang Province (No. 2022C03069).

Institutional Review Board Statement: Not applicable.

Data Availability Statement: The data presented in this study are available on request from the corresponding author.

Conflicts of Interest: The authors declare no conflict of interest.

References

1. Yamamura, H.; Okimoto, K.; Kimura, K.; Watanabe, Y. Hydrophilic fraction of natural organic matter causing irreversible fouling of microfiltration and ultrafiltration membranes. *Water Res.* **2014**, *54*, 123–136. [CrossRef] [PubMed]
2. Long, Y.; Yu, G.; Dong, L.; Xu, Y.; Lin, H.; Deng, Y.; You, X.; Yang, L.; Liao, B.-Q. Synergistic fouling behaviors and mechanisms of calcium ions and polyaluminum chloride associated with alginate solution in coagulation-ultrafiltration (UF) process. *Water Res.* **2021**, *189*, 116665. [CrossRef] [PubMed]
3. Guo, Y.; Liang, H.; Li, G.; Xu, D.; Yan, Z.; Chen, R.; Zhao, J.; Tang, X. A solar photo-thermochemical hybrid system using peroxydisulfate for organic matters removal and improving ultrafiltration membrane performance in surface water treatment. *Water Res.* **2021**, *188*, 116482. [CrossRef] [PubMed]
4. Lee, S.; Cho, J.; Elimelech, M. Combined influence of natural organic matter (NOM) and colloidal particles on nanofiltration membrane fouling. *J. Membr. Sci.* **2005**, *262*, 27–41. [CrossRef]
5. Schulz, M.; Soltani, A.; Zheng, X.; Ernst, M. Effect of inorganic colloidal water constituents on combined low-pressure membrane fouling with natural organic matter (NOM). *J. Membr. Sci.* **2016**, *507*, 154–164. [CrossRef]
6. Chen, Y.; Shen, L.; Li, R.; Xu, X.; Hong, H.; Lin, H.; Chen, J. Quantification of interfacial energies associated with membrane fouling in a membrane bioreactor by using BP and GRNN artificial neural networks. *J. Colloid Interface Sci.* **2020**, *565*, 1–10. [CrossRef]
7. You, X.; Teng, J.; Chen, Y.; Long, Y.; Yu, G.; Shen, L.; Lin, H. New insights into membrane fouling by alginate: Impacts of ionic strength in presence of calcium ions. *Chemosphere* **2020**, *246*, 125801. [CrossRef]
8. Yu, W.; Liu, T.; Crawshaw, J.; Liu, T.; Graham, N. Ultrafiltration and nanofiltration membrane fouling by natural organic matter: Mechanisms and mitigation by pre-ozonation and pH. *Water Res.* **2018**, *139*, 353–362. [CrossRef]
9. Pan, Z.; Zeng, B.; Lin, H.; Teng, J.; Zhang, H.; Hong, H.; Zhang, M. Fundamental thermodynamic mechanisms of membrane fouling caused by transparent exopolymer particles (TEP) in water treatment. *Sci. Total Environ.* **2022**, *820*, 153252. [CrossRef]
10. Yu, C.; Gao, B.; Wang, W.; Xu, X.; Yue, Q. Alleviating membrane fouling of modified polysulfone membrane via coagulation pretreatment/ultrafiltration hybrid process. *Chemosphere* **2019**, *235*, 58–69. [CrossRef]
11. You, X.; Zhang, J.; Shen, L.; Li, R.; Xu, Y.; Zhang, M.; Hong, H.; Yang, L.; Ma, Y.; Lin, H. Thermodynamic mechanisms of membrane fouling during filtration of alginate solution in coagulation-ultrafiltration (UF) process in presence of different ionic strength and iron(III) ion concentration. *J. Membr. Sci.* **2021**, *635*, 119532. [CrossRef]
12. Modarresi, S.; Benjamin, M.M. Using adsorbent mixtures to mitigate membrane fouling and remove NOM with microgranular adsorptive filtration. *J. Membr. Sci.* **2019**, *573*, 528–533. [CrossRef]
13. Luo, M.; Wang, Z.; Zhang, C.; Song, B.; Li, D.; Cao, P.; Peng, X.; Liu, S. Advanced oxidation processes and selection of industrial water source: A new sight from natural organic matter. *Chemosphere* **2022**, *303*, 135183. [CrossRef] [PubMed]
14. Zeng, B.; Pan, Z.; Shen, L.; Zhao, D.; Teng, J.; Hong, H.; Lin, H. Effects of polysaccharides' molecular structure on membrane fouling and the related mechanisms. *Sci. Total Environ.* **2022**, *836*, 155579. [CrossRef]
15. Zhang, M.; Leung, K.-T.; Lin, H.; Liao, B. Evaluation of membrane fouling in a microalgal-bacterial membrane photobioreactor: Effects of SRT. *Sci. Total Environ.* **2022**, *839*, 156414. [CrossRef]

16. Wall, N.A.; Choppin, G.R. Humic acids coagulation: Influence of divalent cations. *Appl. Geochem.* **2003**, *18*, 1573–1582. [CrossRef]
17. Tian, J.; Wu, C.; Yu, H.; Gao, S.; Li, G.; Cui, F.; Qu, F. Applying ultraviolet/persulfate (UV/PS) pre-oxidation for controlling ultrafiltration membrane fouling by natural organic matter (NOM) in surface water. *Water Res.* **2018**, *132*, 190–199. [CrossRef]
18. Ma, B.; Wu, G.; Li, W.; Miao, R.; Li, X.; Wang, P. Roles of membrane–foulant and inter/intrafoulant species interaction forces in combined fouling of an ultrafiltration membrane. *Sci. Total Environ.* **2019**, *652*, 19–26. [CrossRef] [PubMed]
19. Wu, S.; Hua, X.; Miao, R.; Ma, B.; Hu, C.; Liu, H.; Qu, J. Influence of floc charge and related distribution mechanisms of humic substances on ultrafiltration membrane behavior. *J. Membr. Sci.* **2020**, *609*, 118260. [CrossRef]
20. Li, D.; Lin, W.; Shao, R.; Shen, Y.-X.; Zhu, X.; Huang, X. Interaction between humic acid and silica in reverse osmosis membrane fouling process: A spectroscopic and molecular dynamics insight. *Water Res.* **2021**, *206*, 117773. [CrossRef]
21. Zhao, C.; Song, T.; Yu, Y.; Qu, L.; Cheng, J.; Zhu, W.; Wang, Q.; Li, P.; Tang, W. Insight into the influence of humic acid and sodium alginate fractions on membrane fouling in coagulation-ultrafiltration combined system. *Environ. Res.* **2020**, *191*, 110228. [CrossRef] [PubMed]
22. Son, M.; Kim, H.; Jung, J.; Jo, S.; Choi, H. Influence of extreme concentrations of hydrophilic pore-former on reinforced polyethersulfone ultrafiltration membranes for reduction of humic acid fouling. *Chemosphere* **2017**, *179*, 194–201. [CrossRef] [PubMed]
23. Luo, H.; Wang, Z. A new ultrasonic cleaning model for predicting the flux recovery of the UF membrane fouled with humic acid. *J. Environ. Chem. Eng.* **2022**, *10*, 107156. [CrossRef]
24. Tin, M.M.M.; Anioke, G.; Nakagoe, O.; Tanabe, S.; Kodamatani, H.; Nghiem, L.D.; Fujioka, T. Membrane fouling, chemical cleaning and separation performance assessment of a chlorine-resistant nanofiltration membrane for water recycling applications. *Sep. Purif. Technol.* **2017**, *189*, 170–175. [CrossRef]
25. Wang, L.-F.; Wang, L.-L.; Ye, X.-D.; Li, W.-W.; Ren, X.-M.; Sheng, G.-P.; Yu, H.-Q.; Wang, X.-K. Coagulation Kinetics of Humic Aggregates in Mono- and Di-Valent Electrolyte Solutions. *Environ. Sci. Technol.* **2013**, *47*, 5042–5049. [CrossRef]
26. Arabi, S.; Nakhla, G. Impact of calcium on the membrane fouling in membrane bioreactors. *J. Membr. Sci.* **2008**, *314*, 134–142. [CrossRef]
27. Khan, A.A.; Khan, I.A.; Siyal, M.I.; Lee, C.-K.; Kim, J.-O. Optimization of membrane modification using SiO_2 for robust anti-fouling performance with calcium-humic acid feed in membrane distillation. *Environ. Res.* **2019**, *170*, 374–382. [CrossRef]
28. Xu, H.; Ma, J.; Ding, M.; Xie, Z. Mechanistic insights into the removal of PFOA by 2D MXene/CNT membrane with the influence of Ca^{2+} and humic acid. *Desalination* **2022**, *529*, 115643. [CrossRef]
29. Xie, M.; Nghiem, L.D.; Price, W.E.; Elimelech, M. Impact of humic acid fouling on membrane performance and transport of pharmaceutically active compounds in forward osmosis. *Water Res.* **2013**, *47*, 4567–4575. [CrossRef]
30. Chang, E.E.; Chang, Y.-C.; Liang, C.-H.; Huang, C.-P.; Chiang, P.-C. Identifying the rejection mechanism for nanofiltration membranes fouled by humic acid and calcium ions exemplified by acetaminophen, sulfamethoxazole, and triclosan. *J. Hazard. Mater.* **2012**, *221–222*, 19–27. [CrossRef]
31. Wang, L.-F.; He, D.-Q.; Chen, W.; Yu, H.-Q. Probing the roles of Ca^{2+} and Mg^{2+} in humic acids-induced ultrafiltration membrane fouling using an integrated approach. *Water Res.* **2015**, *81*, 325–332. [CrossRef] [PubMed]
32. Hansima, M.A.C.K.; Ketharani, J.; Samarajeewa, D.R.; Nanayakkara, K.G.N.; Herath, A.C.; Makehelwala, M.; Indika, S.; Jinadasa, K.B.S.N.; Weragoda, S.K.; Wei, Y.; et al. Probing fouling mechanism of anion exchange membranes used in electrodialysis self-reversible treatment by humic acid and calcium ions. *Chem. Eng. J. Adv.* **2021**, *8*, 100173. [CrossRef]
33. Listiarini, K.; Sun, D.D.; Leckie, J.O. Organic fouling of nanofiltration membranes: Evaluating the effects of humic acid, calcium, alum coagulant and their combinations on the specific cake resistance. *J. Membr. Sci.* **2009**, *332*, 56–62. [CrossRef]
34. Lin, T.; Lu, Z.; Chen, W. Interaction mechanisms of humic acid combined with calcium ions on membrane fouling at different conditions in an ultrafiltration system. *Desalination* **2015**, *357*, 26–35. [CrossRef]
35. Li, B.; He, X.; Wang, P.; Liu, Q.; Qiu, W.; Ma, J. Opposite impacts of K^+ and Ca^{2+} on membrane fouling by humic acid and cleaning process: Evaluation and mechanism investigation. *Water Res.* **2020**, *183*, 116006. [CrossRef] [PubMed]
36. Chang, H.; Qu, F.; Liu, B.; Yu, H.; Li, K.; Shao, S.; Li, G.; Liang, H. Hydraulic irreversibility of ultrafiltration membrane fouling by humic acid: Effects of membrane properties and backwash water composition. *J. Membr. Sci.* **2015**, *493*, 723–733. [CrossRef]
37. Miao, R.; Li, X.; Wu, Y.; Wang, P.; Wang, L.; Wu, G.; Wang, J.; Lv, Y.; Liu, T. A comparison of the roles of Ca^{2+} and Mg^{2+} on membrane fouling with humic acid: Are there any differences or similarities? *J. Membr. Sci.* **2018**, *545*, 81–87. [CrossRef]
38. Abass, O.; Wu, X.; Guo, Y.; Zhang, K. Membrane Bioreactor in China: A Critical Review. *Int. Membr. Sci. Technol.* **2015**, *2*, 29–47.
39. Zhang, M.; Lee, E.; Vonghia, E.; Hong, Y.; Liao, B. 1—Introduction to aerobic membrane bioreactors: Current status and recent developments. In *Current Developments in Biotechnology and Bioengineering*; Ng, H.Y., Ng, T.C.A., Ngo, H.H., Mannina, G., Pandey, A., Eds.; Elsevier: Cambridge, MA, USA, 2020; pp. 1–23.
40. Han, L.; Xiao, T.; Tan, Y.Z.; Fane, A.G.; Chew, J.W. Contaminant rejection in the presence of humic acid by membrane distillation for surface water treatment. *J. Membr. Sci.* **2017**, *541*, 291–299. [CrossRef]
41. Xie, L.; Lu, Q.; Mao, X.; Wang, J.; Han, L.; Hu, J.; Lu, Q.; Wang, Y.; Zeng, H. Probing the intermolecular interaction mechanisms between humic acid and different substrates with implications for its adsorption and removal in water treatment. *Water Res.* **2020**, *176*, 115766. [CrossRef]
42. Ousman, M.; Bennasar, M. Determination of various hydraulic resistances during cross-flow filtration of a starch grain suspension through inorganic membranes. *J. Membr. Sci.* **1995**, *105*, 1–21. [CrossRef]

43. Li, R.; Li, J.; Rao, L.; Lin, H.; Shen, L.; Xu, Y.; Chen, J.; Liao, B.-Q. Inkjet printing of dopamine followed by UV light irradiation to modify mussel-inspired PVDF membrane for efficient oil-water separation. *J. Membr. Sci.* **2021**, *619*, 118790. [CrossRef]
44. Wu, M.; Chen, Y.; Lin, H.; Zhao, L.; Shen, L.; Li, R.; Xu, Y.; Hong, H.; He, Y. Membrane fouling caused by biological foams in a submerged membrane bioreactor: Mechanism insights. *Water Res.* **2020**, *181*, 115932. [CrossRef] [PubMed]
45. Li, Z.; Zhang, W.; Tao, M.; Shen, L.; Li, R.; Zhang, M.; Jiao, Y.; Hong, H.; Xu, Y.; Lin, H. In-situ growth of UiO-66-NH_2 in porous polymeric substrates at room temperature for fabrication of mixed matrix membranes with fast molecular separation performance. *Chem. Eng. J.* **2022**, *435*, 134804. [CrossRef]
46. Chen, B.; Hu, X.; Wang, J.; Li, R.; Shen, L.; Xu, Y.; Zhang, M.; Hong, H.; Lin, H. Novel catalytic self-cleaning membrane with peroxymonosulfate activation for dual-function wastewater purification: Performance and mechanism. *J. Clean. Prod.* **2022**, *355*, 131858. [CrossRef]
47. Zhang, R.; Xu, Y.; Shen, L.; Li, R.; Lin, H. Preparation of nickel@polyvinyl alcohol (PVA) conductive membranes to couple a novel electrocoagulation-membrane separation system for efficient oil-water separation. *J. Membr. Sci.* **2022**, *653*, 120541. [CrossRef]
48. Zhang, W.; Guo, D.; Li, Z.; Shen, L.; Li, R.; Zhang, M.; Jiao, Y.; Xu, Y.; Lin, H. A new strategy to accelerate co-deposition of plant polyphenol and amine for fabrication of antibacterial nanofiltration membranes by in-situ grown Ag nanoparticles. *Sep. Purif. Technol.* **2022**, *280*, 119866. [CrossRef]
49. van Oss, C.J. Acid—Base interfacial interactions in aqueous media. *Colloids Surfaces A Physicochem. Eng. Asp.* **1993**, *78*, 1–49. [CrossRef]
50. Liu, Y.; Shen, L.; Huang, Z.; Liu, J.; Xu, Y.; Li, R.; Zhang, M.; Hong, H.; Lin, H. A novel in-situ micro-aeration functional membrane with excellent decoloration efficiency and antifouling performance. *J. Membr. Sci.* **2022**, *641*, 119925. [CrossRef]
51. Huang, Z.; Zeng, Q.; Liu, Y.; Xu, Y.; Li, R.; Hong, H.; Shen, L.; Lin, H. Facile synthesis of 2D TiO_2@MXene composite membrane with enhanced separation and antifouling performance. *J. Membr. Sci.* **2021**, *640*, 119854. [CrossRef]
52. Hoek, E.M.V.; Agarwal, G.K. Extended DLVO interactions between spherical particles and rough surfaces. *J. Colloid Interface Sci.* **2006**, *298*, 50–58. [CrossRef] [PubMed]
53. Liu, J.; Shen, L.; Lin, H.; Huang, Z.; Hong, H.; Chen, C. Preparation of Ni@UiO-66 incorporated polyethersulfone (PES) membrane by magnetic field assisted strategy to improve permeability and photocatalytic self-cleaning ability. *J. Colloid Interface Sci.* **2022**, *618*, 483–495. [CrossRef] [PubMed]
54. Brant, J.A.; Childress, A.E. Colloidal adhesion to hydrophilic membrane surfaces. *J. Membr. Sci.* **2004**, *241*, 235–248. [CrossRef]
55. Zhao, Y.; Lu, D.; Xu, C.; Zhong, J.; Chen, M.; Xu, S.; Cao, Y.; Zhao, Q.; Yang, M.; Ma, J. Synergistic oxidation-filtration process analysis of catalytic $CuFe_2O_4$ - Tailored ceramic membrane filtration via peroxymonosulfate activation for humic acid treatment. *Water Res.* **2020**, *171*, 115387. [CrossRef] [PubMed]
56. Sundararajan, M.; Rajaraman, G.; Ghosh, S.K. Speciation of uranyl ions in fulvic acid and humic acid: A DFT exploration. *Phys. Chem. Chem. Phys.* **2011**, *13*, 18038–18046. [CrossRef]
57. Niemeyer, J.; Chen, Y.; Bollag, J.M. Characterization of Humic Acids, Composts, and Peat by Diffuse Reflectance Fourier-Transform Infrared Spectroscopy. *Soil Sci. Soc. Am. J.* **1992**, *56*, 135–140. [CrossRef]
58. Wang, J.; Guan, J.; Santiwong, S.R.; Waite, T.D. Characterization of floc size and structure under different monomer and polymer coagulants on microfiltration membrane fouling. *J. Membr. Sci.* **2008**, *321*, 132–138. [CrossRef]
59. Shen, L.-G.; Lei, Q.; Chen, J.-R.; Hong, H.-C.; He, Y.-M.; Lin, H.-J. Membrane fouling in a submerged membrane bioreactor: Impacts of floc size. *Chem. Eng. J.* **2015**, *269*, 328–334. [CrossRef]
60. Zhang, M.; Leung, K.-T.; Lin, H.; Liao, B. Membrane fouling in a microalgal-bacterial membrane photobioreactor: Effects of P-availability controlled by N:P ratio. *Chemosphere* **2021**, *282*, 131015. [CrossRef]
61. Teng, J.; Wu, M.; Chen, J.; Lin, H.; He, Y. Different fouling propensities of loosely and tightly bound extracellular polymeric substances (EPSs) and the related fouling mechanisms in a membrane bioreactor. *Chemosphere* **2020**, *255*, 126953. [CrossRef]

Communication

Recovery of Valuable Aromas from Sardine Cooking Wastewaters by Pervaporation with Fractionated Condensation: Matrix Effect and Model Validation

M. João Pereira [1], Manuela Pintado [2], Carla Brazinha [1,*] and João Crespo [1]

[1] LAQV-REQUIMTE, Department of Chemistry, NOVA School of Science and Technology, FCT NOVA, Universidade NOVA de Lisboa, 2829-516 Caparica, Portugal
[2] CBQF/Escola Superior de Biotecnologia, Universidade Católica Portuguesa, Rua Diogo Botelho, 1327, 4169-005 Porto, Portugal
* Correspondence: c.brazinha@fct.unl.pt

Abstract: Due to the lack of studies addressing the influence of real food matrices on integrated organophilic pervaporation/fractionated condensation processes, the present work analyses the impact of the real matrix of sardine cooking wastewaters on the fractionation of aromas. In a previous study, a thermodynamic/material balance model was developed to describe the integrated pervaporation—a fractionated condensation process of aroma recovery from model solutions that emulate seafood industry aqueous effluents, aiming to define the best conditions for off-flavour removal. This work assesses whether the previously developed mathematical model, validated only with model solutions, is also applicable in predicting the fractionation of aromas of different chemical families from real effluents (sardine cooking wastewaters), aiming for off-flavour removals. It was found that the food matrix does not influence substantial detrimental consequences on the model simulations, which validates and extends the applicability of the model.

Keywords: modelling of organophilic pervaporation; vacuum fractionated condensation; aroma recovery; valorisation of canning industry effluents; removal of off-flavours

1. Introduction

The large majority of studies performed for aroma recovery by pervaporation have been accomplished using model solutions [1]. The use of model systems is effective for a simple and detailed analysis of process performance and optimisation. However, model solutions cannot reproduce all the complex varieties of constituents of the feed stream, with diverse concentrations and chemical and organoleptic properties which contribute to the overall aroma profile [2]. The pervaporation of real feed mixtures should also be studied because the concentration of volatiles is usually lower than in model solutions, due to potential interferences of lipids and proteins in the aroma profile [3], which is mostly neglected when studying model solutions. However, there are still a few studies that use real feeds [3–6].

In a previous study, Pereira et al. [7] proposed a mathematical model for the pervaporation-fractionated condensation aiming at the recovery of aromas free from off-flavours using a model solution that mimicked seafood cooking wastewaters. This model allows for simulating the mass and composition of each compound in the condensers arranged in a series, mousing as input information the permeate fluxes of each aroma under study (obtained experimentally), operating conditions used in the process, and thermodynamic parameters of each aroma. The model is based on the mass balances and thermodynamic equilibrium in each condenser.

For many years, the production of commercial seafood flavourings used solid by-products. Nowadays, seafood cooking water has emerged as a promising source for producing "natural-like" aroma concentrates, valuable for the food and feed market sectors [8]. The presence of off-flavours in the agro-industrial effluents composition is one of the constraints associated with the valorisation of aromas. Even though they are frequently innocuous, off-flavours might degrade the quality of a food product, which can be quite expensive for the food and beverage sectors. The Maillard reaction or lipid oxidation, which produces numerous food smells and certain off-flavours, develops during thermal processing. The need for non-thermal processes or the use of gentler conditions is growing as a result [9]. In this work, the fractionation and separation of desirable target aromas from off-flavours are explored, benefiting from both the membrane's intrinsic selectivity and the selectivity of fractionated condensation consecutive steps.

The main objective of this work is to study the effect of the matrix on the aroma recovery from sardine cooking wastewaters by the integrated process of organophilic pervaporation/fractionated condensation, assuring off-flavour removal. Concretely, the objective is to validate the mathematical model previously developed for model solutions, extending it to apply to a real matrix, a complex sardine cooking wastewater. If the model correctly predicts the fractionation of the different aromas, despite the complexity of the real solution, the applicability of the model will increase significantly, opening opportunities for use with other real matrices.

2. Materials and Methods

The sardine cooking wastewater was kindly provided by the company A Poveira S.A. (Laúndos, Portugal). This effluent is the result of steaming the fish for 7 min at 100 °C. An acorn extract with antioxidant properties was added to the sardine cooking wastewater at the outlet of cooking chambers at a 1% (v/v) concentration to prevent lipid oxidation and suppress aroma deterioration. The effluent was collected, transported, and stored at −20 °C until needed.

The experimental setup and analytical methods of study were the same as described in previous studies [7,10].

A radial flow flat module (GKSS, Germany) was employed, presented, and discussed in detail in Schafer [11]. The membrane used was a PervapTM 4060 (DeltaMem AG, Switzerland), an organophilic dense membrane with a membrane area of 10^{-2} m^2. The active layer of polydimethylsiloxane (PDMS) was shown to have an excellent performance for the permeation of organic compounds by pervaporation, as well as a good affinity for seafood aromas [12,13].

The operation conditions applied in this study were the optimised conditions obtained previously in the studies performed with a model solution [7]. According to these, the permeate pressure applied was 1500 Pa. The temperature of the first condenser T1, condens was set at −100 °C, and the temperature of the second condenser $T_{2,condens}$ was at −196 °C.

At the end of the trials, the membrane used was rinsed with a known amount of water at room temperature, and the content of lipids, proteins, and aromas present in this solution was characterised according to the methods described in Pereira et al. [10].

3. Results

3.1. Characterisation of Sardine Cooking Wastewaters

Alcohols, aldehydes, and ketones are part of the aroma profile of the sardine cooking wastewaters, as revealed by solid-phase microextraction followed by gas chromatography mass spectrometry (SPME/GC-MS). The overall aroma profile of sardine cooking wastewaters is presented in Table 1, and it is identical to the aroma profile of sardines investigated by other researchers [14,15]. Some chemical markers were selected to study the effect of the matrix in this process, which are 1-penten-3-ol and 1-octen-3-ol, as alcohols; heptanal. (E,E)-2,4-heptadienal, (E,Z)-2,6-nonadienal, as aldehydes; and 2-nonanone as ketone. These chemical markers were selected based on the main groups of chemicals present in sardine cooking wastewaters, with diverse organoleptic properties. The main compound present in higher concentrations was 1-penten-3-ol.

Table 1. Aroma compounds identified in Sardine cooking wastewaters.

Aroma Compounds	Area *	[Ci] (ppm) *	Aroma Compounds	Area *	[Ci] (ppm) *
Aldehydes			Alcohols		
Hexanal	123246866		1-Penten-3-ol	79964041	0.100
Heptanal	28381616	0.006	1-Octen-3-ol	181778114	0.008
2-Hexenal, (E)-	58129859		(5Z)-Octa-1,5-dien-3-ol	86516933	
Octanal	16922895		2-Ethylhexanol	7501041	
Nonanal	79342642		1-Octanol	590437999	0.100
2-Octenal, (E)-	39611501		1-Penten-3-ol	79964041	0.008
2,4-Heptadienal, (E,E)-	88078708		1-Octen-3-ol	181778114	
2-Nonenal, (E)-	21644129	0.011	Ketones		
2,6-Nonadienal, (E,Z)-	92347778	0.044	2-Nonanone	44929994	0.001
2-Decenal, (E)-	12169097		3,5-Octadien-2-one	113901449	
Sulphur compounds			2-Undecanone	8019036	
			Acids		
Trans-2-(2-Pentenyl)furan	46428766		Hexanoic acid	89583410	

* Mean values of integration peak areas for all compounds identified and the concentration (ppm) of chemical markers. The aromas were identified by comparing their retention indices relative to C8–C20 n-alkanes and their mass spectra to those in the NIST Library Database. Quantification was performed with calibration curves of the pure standards, evaluated under the same circumstances. Underlined compounds are off-flavours.

3.2. Pervaporation-Fractionated Condensation Processing of Sardine Cooking Wastewaters

The permeate was generated through pervaporation experiments with sardine cooking wastewaters under upstream operating conditions described in the previous section. The total permeate fluxes obtained in the seafood model solution experiments using the same operating conditions were 889.84 g/m^2.h and 731 g/m^2.h with sardine cooking wastewaters. This lower value for the permeate flux was expected due to the total lipid (28.13 ± 2.84 g/100 g) and protein content (25.38 ± 1.93 mg/mL) of the sardine wastewater sample [10]. The presence of lipids and proteins in the feed medium might lead to interactions with aroma compounds present and also to some degree of fouling of the pervaporation membrane.

Table 2 shows the individual fluxes [mol/(m^2.s)] and the permeabilities [mol/(m.s.Pa)] to the aromas under study, as well as the separation factors obtained (calculated against water).

Table 2. Experimental parameters of pervaporation were performed with a downstream pressure of 1500 Pa, with real wastewater: aroma flowrate (J_i), permeability (L_i), and selectivity of each aroma (against water).

Compound	J_i [mol/m^2.s]	L_i [mol/(m.s.Pa)]	Separation Factor [–]
Sardine cooking wastewater			
1-Penten-3-ol	$6.58 \times 10^{-7} \pm 8.64 \times 10^{-9}$	$6.25 \times 10^{-11} \pm 4.15 \times 10^{-12}$	4.20 ± 0.28
1-Octen-3-ol	$3.91 \times 10^{-8} \pm 1.66 \times 10^{-9}$	$2.33 \times 10^{-11} \pm 7.95 \times 10^{-13}$	11.19 ± 0.38
2-Nonanone	$1.89 \times 10^{-8} \pm 9.67 \times 10^{-10}$	$1.48 \times 10^{-11} \pm 9.17 \times 10^{-13}$	51.03 ± 3.14
Heptanal	$2.80 \times 10^{-8} \pm 9.56 \times 10^{-10}$	$3.73 \times 10^{-12} \pm 3.06 \times 10^{-13}$	7.31 ± 0.60
(E2, Z6)-Nonadienal	$4.65 \times 10^{-7} \pm 2.86 \times 10^{-8}$	$1.48 \times 10^{-11} \pm 9.17 \times 10^{-13}$	171.15 ± 19.73

The data presented are the mean ± s.d. values. Underlined compounds are off-flavours.

The important and main alcohol 1-penten-3-ol, responsible for the aroma of fresh marine products, is generated from polyunsaturated fatty acids [15]. 1-Penten-3-ol presents the highest values for the individual flux (J_i) and permeability (L_i). However, the off-flavour (E2, Z6)-nonadienal shows a close permeate flux and the highest separation factor, which reinforces the importance of conjugating fractionated condensation to the pervaporation process to enable the off-flavour's removal.

At the end of the process, to better understand the effect of the matrix in the pervaporation process, the content of the total proteins and lipids that remained adsorbed to the membrane, as well as the aroma content in this adsorbed layer, were analysed by Lowry and Bligh and Dyer's methods, respectively. There was no gel formation on the membrane surface, and indeed, the protein content in the membrane was quite residual (6.15–8.46 µg/m^2), only slightly more relevant in terms of lipids showing 1.3–2 µg/m^2. Concerning the aromas, a small number of aromas remained in the membrane in a very small concentration: only 2-nonanone and 1-octen-3-ol were found in residual concentrations of 10 and 20 µg/m^2, respectively.

Model Validation for the Real Sardine Cooking Wastewater

The thermodynamic/material balance model was developed to simulate the recovery of aromas at a given permeate pressure employing fractionated condensation with two condensers in a series, supported by an efficient and optimised fractionated condensation (see the complete explanation by Pereira et al. [7]). In short, starting from simple experimental inputs such as the (i) permeate flux of each aroma present in the system, (ii) thermodynamic parameters (for each compound in the feed: Antoine constant and activity coefficient at infinite dilution), and (iii) operation conditions of downstream pressure and temperature, it is possible to simulate the composition of the condensates obtained in the sequential condensers. Through a system of equations that describe the thermodynamic equilibrium conditions and with the support of required material balances, we can select the best operating conditions to achieve the best separation of desirable flavours from off-flavours. In the end, the expressions for calculating the percentage of condensation of water and aroma(s) in the first condenser are obtained, respectively, by Equations (1) and (2).

$$\%condens_{w1} = 1 - \frac{n_{inert}}{n_{w0}} \cdot \frac{p_{vw}(T_{1,\,condens})}{p_{perm} - p_{vw}(T_{1,\,condens})} \quad (1)$$

$$\%condens_{aroma1} \cong 1 - \frac{n_{inert}}{n_{aroma0}} \cdot \frac{\varkappa_{aroma1} \cdot \gamma^{\infty}_{aroma1} \cdot p_{varoma}(T_{1,\,condens})}{p_{perm} - p_{vw}(T_{1,\,condens})} \quad (2)$$

where n_{inert} is the inert gas molar flow rate in the stream, P_v is the saturation vapour pressure of water or aroma, $pperm$ is the permeate pressure applied to the system, $n_{w\,or\,aroma0}$ is the molar flow rate before the first condenser, $\varkappa_{w\,or\,aroma}$ is the molar fraction in the feed, and $\Upsilon^{\infty}_{aroma}$ is the infinite activity coefficient of the aroma.

The model developed was applied for a permeate pressure of 1500 Pa, where the percentage of compound i that is condensed/recovered in the first condenser, $\%Condens_{i1}$, was predicted for different values of $T_{1,condens}$ [°C]. Figure 1 shows the simulations obtained for each aroma present in the sardine cooking wastewater and the experimental values acquired, in terms of $\%Condens_{i1}$ (the fraction of each chemical compound i that condenses in the first condenser) versus the temperature of the condenser, $T_{1,condens}$.

Figure 1 reveals a good adherence between the experimental and the simulated results of $\%Condens_{i1}$ as a function of the temperature of the condenser and, consequently, for the composition of condensates. This result means that, although the real medium composition is much more complex than the model solution previously studied, it is not necessary to modify the thermodynamic/material balance model used, which can be applied with success to evaluate if a given fractionation of aromas (such as the fractionation between target aromas and off-flavours) can be achieved.

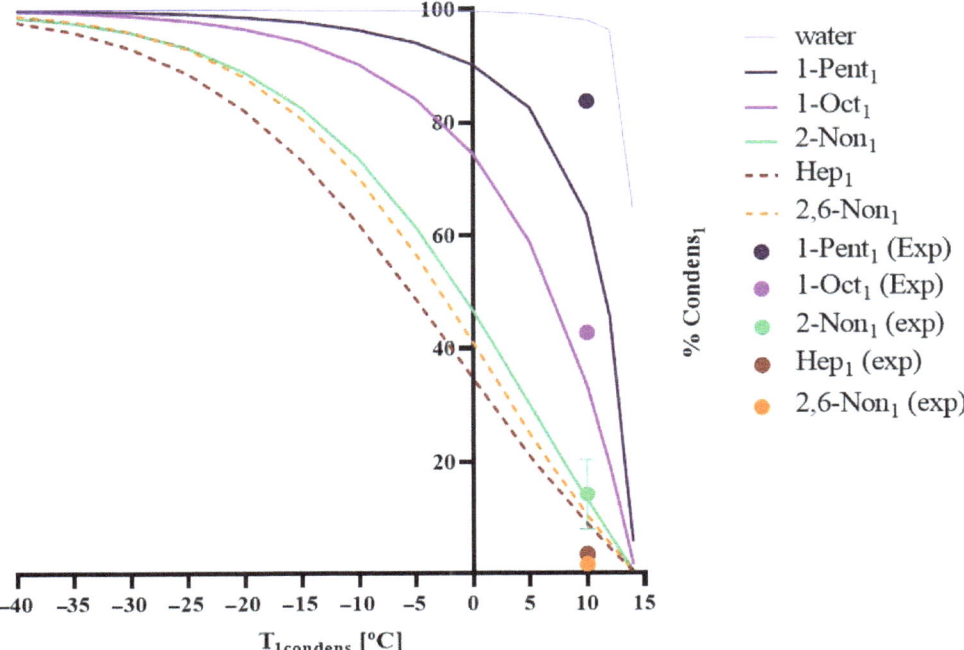

Figure 1. A model simulation was obtained for the sardine cooking wastewaters with experimental validation for five different aroma presents. Percentage of condensation of each compound (water, 1-penten-3-ol, 1-octen-3-ol, heptanal, 2,6-nonadienal, and 2-nonanone) in the 1st condenser (%$Condens_{i1}$) as a function of the temperature of the same condenser ($T_{1,condens}$). Operating conditions: Pervap 4060 membrane; T_{feed} = 60 °C; p_{perm} = 1500 Pa; lines refer to simulated values and dots to experimental data, which were analysed in triplicate.

Under the experimental conditions used in this work, 1500 Pa of permeate pressure and $T_{1,condens}$ [°C] of 10 °C (see Figure 1), a good off-flavour removal was achieved, with the partial retention of off-flavours in the 1st condenser lower than 3% for heptanal and 1.6% for (E2, Z6)-nonadienal. These retention values correspond to an off-flavour concentration of 0.02 and 0.67 mg $_{off\text{-}flavours}$/Kg$_{condensate}$ of heptanal and (E2, Z6)-nonadienal, respectively, in the 1st condensate. In a conclusion, in terms of aroma quality, the condensate recovered in the first condenser is reduced in off-flavours. Both off-flavour concentrations are below their threshold (limit of human olfactive perception) of 0.60 and 0.70 mg/L of heptanal and (E2, Z6)-nonadienal, respectively. On the other hand, it should be recognised that the recovery of desirable aromas in the first condenser is not complete, not assuring off-flavour removal: 84% of 1-penten-3-ol is recovered in the first condenser, but 43% of 1-octen-3-ol is recovered, and only 14% of the ketone 2-nonanone is recovered.

This model proves to be an excellent tool to simulate the percentage of the condensation of each aroma in each condenser at a particular downstream pressure and condenser temperature, as well as the resulting condensate composition. It enables the comparison and definition of fractionated condensation procedures based on the goal of a given industrial process.

4. Conclusions

The integrated process of pervaporation-fractionated condensation has proven to be a potential approach for the valorisation of canning industry effluents by the recovery of valuable aromas.

The model used shows a good match between experimental and predicted values, despite the high heterogeneity of the sardine cooking wastewater. The effect of the matrix did not demonstrate significant negative effects on the model simulations, making possible its use without further modification.

The model applied and validated for sardine cooking wastewaters proved to be a useful tool to predict the fractionation of aroma compounds, here illustrated by the removal of off-flavours to obtain aroma products with potential commercial value. Additionally, due to the range of chemical families evaluated, this model represents a tool that might be easily applied to other real matrices, aiming at the recovery of surplus aromas from a circular economy perspective.

Author Contributions: Conceptualisation, M.J.P. and C.B.; methodology, M.J.P. and C.B.; investigation, M.J.P.; resources, J.C.; writing—original draft preparation, M.J.P.; writing—review and editing, M.P., J.C. and C.B.; supervision, M.P., J.C. and C.B.; project administration, J.C. and C.B.; funding acquisition, J.C. and C.B. All authors have read and agreed to the published version of the manuscript.

Funding: The research leading to these results has received funding from Sistema de Apoio à Investigação Científica e Tecnológica (SAICT), Programa de Actividades Conjuntas (PAC) under Multibiore-finery Project (POCI-01-0145-FEDER-016403) and from European Regional Development Fund (ERDF), through the Incentive System to Research and Technological Development, within the Portugal2020 Competitiveness and Internationalisation Operational Program under the project "MobFood—Mobilizing scientific and technological knowledge in response to the challenges of the agri-food market" (POCI-01-0247-FEDER-024524). Maria João Pereira acknowledges Fundação para a Ciência e Tecnologia for Ph.D. fellowship SFRH/BD/119018/2016. This work was supported by the Associate Laboratory for Green Chemistry—LAQV, which is financed by national funds from FCT/MCTES (UIDB/50006/2020).

Institutional Review Board Statement: Not applicable.

Informed Consent Statement: Not applicable.

Data Availability Statement: Not applicable.

Conflicts of Interest: The authors declare no conflict of interest.

References

1. Trifunović, O.; Lipnizki, F.; Trägårdh, G. The influence of process parameters on aroma recovery by hydrophobic pervaporation. *Desalination* **2006**, *189*, 1–12. [CrossRef]
2. Schäfer, T.; Crespo, J.G. Recovery of aroma compounds from fermentation by pervaporation. *Environ. Prot. Eng.* **1999**, *25*, 73–85.
3. Martínez, R.; Sanz, M.T.; Beltrán, S. Pervaporation investigation of recovery of volatile compounds from brown crab boiling juice. *Food Sci. Technol. Int.* **2014**, *20*, 511–526. [CrossRef]
4. Schäfer, T.; Bengtson, G.; Pingel, H.; Böddeker, K.W.; Crespo, J. Recovery of aroma compounds from a wine-must fermentation by organophilic pervaporation. *Biotechnol. Bioeng.* **1999**, *62*, 412–421. [CrossRef]
5. Souchon, I.; Pierre, F.X.; Marin, M. Pervaporation as a deodorization process applied to food industry effluents: Recovery and valorisation of aroma compounds from cauliflower blanching water. *Desalination* **2002**, *148*, 79–85. [CrossRef]
6. Figoli, A.; Tagarelli, A.; Cavaliere, B.; Voci, C.; Sindona, G.; Sikdar, S.K.; Drioli, E. Evaluation of pervaporation process of kiwifruit juice by SPME-GC/Ion Trap Mass Spectrometry. *Desalination* **2010**, *250*, 1113–1117. [CrossRef]
7. Pereira, M.J.; Brazinha, C.; Crespo, J.G. Pervaporation recovery of valuable aromas from by-products of the seafood industry: Modelling of fractionated condensation for off-flavour removal. *Sep. Purif. Technol.* **2022**, *286*, 120441. [CrossRef]
8. Soares, L.S.; Vieira, A.C.F.; Fidler, F.; Nandi, L.G.; Monteiro, A.R.; Di Luccio, M. Volatile Organic Compounds Profile Obtained from Processing Steps of Pacific Oysters (Crassostrea gigas) as Perspective for Food Industry. *J. Aquat. Food Prod. Technol.* **2020**, *29*, 194–206. [CrossRef]
9. Ridgway, K.; Lalljie, S.P.D.; Smith, R.M. Analysis of food taints and off-flavours: A review. *Food Addit. Contam.* **2010**, *27*, 146–168. [CrossRef]
10. Pereira, M.J.; Grosjean, O.; Pintado, M.; Brazinha, C.; Crespo, J. Clean Technologies for Production of Valuable Fractions from Sardine Cooking Wastewaters: An Integrated Process of Flocculation and Reverse Osmosis. *Clean Technol.* **2022**, *4*, 276–295. [CrossRef]
11. Schafer, T. Recovery of Wine-Must Aroma by Pervaporation. Ph.D. Thesis, Universidade Nova de Lisboa, Caparica, Portugal, 2002.

12. Kujawska, A.; Knozowska, K.; Kujawa, J.; Kujawski, W. Influence of downstream pressure on pervaporation properties of PDMS and POMS based membranes. *Sep. Purif. Technol.* **2016**, *159*, 68–80. [CrossRef]
13. Martínez, R.; Sanz, M.; Beltrán, S.; Teresa Sanz, M.; Beltrán, S. Concentration by pervaporation of brown crab volatile compounds from dilute model solutions: Evaluation of PDMS membrane. *J. Memb. Sci.* **2013**, *428*, 371–379. [CrossRef]
14. Ganeko, N.; Shoda, M.; Hirohara, I.; Bhadra, A.; Ishida, T.; Matsuda, H.; Takamura, H.; Matoba, T. Analysis of volatile flavor compounds of sardine (Sardinops melanostica) by solid phase microextraction. *J. Food Sci.* **2008**, *73*, S83–S88. [CrossRef]
15. Mansur, M.A.; Bhadra, A.; Takamura, H.; Matoba, T. Volatile flavor compounds of some sea fish and prawn species. *Fish. Sci.* **2003**, *69*, 864–866. [CrossRef]

Article

High-Flux Ultrafiltration Membranes Combining Artificial Water Channels and Covalent Organic Frameworks

Kai Liu [1,2,†], Jinwen Guo [1,2,†], Yingdong Li [1,2], Jinguang Chen [1,2] and Pingli Li [1,2,*]

1. School of Chemical Engineering and Technology, Tianjin University, Tianjin 300072, China
2. Tianjin Key Laboratory of Membrane Science and Desalination Technology, State Key Laboratory of Chemical Engineering, Tianjin University, Tianjin 300072, China
* Correspondence: lipingli@tju.edu.cn
† These authors contributed equally to this work.

Abstract: Artificial water channels (AWCs) have been well investigated, and the imidazole-quartet water channel is one of the representative channels. In this work, covalent organic frameworks (COFs) composite membranes were fabricated through assembling COF layers and imidazole-quartet water channel. The membranes were synthesized by interfacial polymerization and self-assembly process, using polyacrylonitrile (PAN) ultrafiltration substrates with artificial water channels (HC6H) as modifiers. Effective combination of COF layers and imidazole-quartet water channels provide the membrane with excellent performance. The as-prepared membrane exhibits a water permeance above 271.7 $L \cdot m^{-2} \cdot h^{-1} \cdot bar^{-1}$, and high rejection rate (>99.5%) for CR. The results indicated that the composite structure based on AWCs and COFs may provide a new idea for the development of high-performance membranes for dye separation.

Keywords: high-flux membranes; artificial water channels; COFs; composite structure

1. Introduction

Currently, water shortage has become a serious problem due to human activities and poor water treatment [1]. Membrane-based water purification technologies have been frequently studied due to their simple process, high efficiency, and low energy consumption [2,3]. Both high selectivity and excellent permeability are required for the ideal separation membrane.

As the most important component of biofilms, the channel aquaporin (AQP) plays a crucial role in the unique exchange of particles between cells and their environment. Owing to its high water permeability and superior selectivity [4,5], AQP with proteo-liposomes has been applied in the fabrication of separation membranes [6]. However, the stability of AQP under harsh conditions and compatibility with substrate membrane remains a challenge to be solved [7].

Researchers have tried to mimic the structure of AQP in recent years. AQP has a typical hourglass structure with high enough filtration selectivity to transport water through electrostatic repulsion in the aromatic and arginine (ar/R) constriction region [8]. Owing to the reduction in the collective hydrogen bonding, the positive charges at the entrance of the AQP channel could further improve the water transport activity [9,10]. This distinctive structure of AQP provides an idea for the synthesis of AQP alternatives. As one possible substitute, artificial water channels (AWCs) can provide an effective pathway towards membrane by mimicking AQP's mechanism. AWCs, reported by Barboiu in 2011 [11], are synthetic water channels with self-assembled columnar structures. The central hydrophobic or hydrophilic pore guarantees the directional transfer of water, and the outer hydrophobic exterior matches the polymeric membrane environment.

The biomimetic hydrophobic AWCs have been investigated over the past 30 years. Ultrashort carbon porins (CNTPs) were inserted in a lipid bilayer by Noy et al. [12]. Gong et al. successfully synthesized a series of (m-phenylene-ethynylene) m-PE macrocycles, and their permeability could reach 4.9×10^7 $H_2O \cdot s^{-1} \cdot channel^{-1}$ [13]. Hydrazide-appended pillar arene was first introduced by Ogoshi et al. in 2008 [14]. Hou et al. synthesized a second generation of pillar arene PAP layer and applied it in the transmembrane transport of chiral amino-acids with a permeability of 3.5×10^8 $H_2O \cdot s^{-1} \cdot channel^{-1}$ [15]. Moreover, the imidazole-quartet (I-quartet) channels obtained by self-assembly of alkyl ureido-imidazole (HC6H) [11] was investigated to improve membrane permeability without sacrificing its selectivity. However, low compatibility with the substrate needs to be addressed.

Recently, covalent organic frameworks (COFs) have attracted a lot of attention in the fields of energy-storage, catalyst, and separation. As a new class of crystalline organic porous material, COFs are composed of H, B, C, O, N, and other light atoms, and possess unique properties, such as inherent porosity, good pore aperture, and abundant functional groups [16–18]. Compared with traditional polymer membrane materials, the separation efficiency of COF membranes has been greatly improved. However, there is still a trade-off problem between permeability and selectivity. Appropriate substrates have been proven to be a feasible way for the further enhancement of membrane performance [19]. Construction of a 2D+1D structure via inserting 1D cellulose nanofibers (CNFs) has been reported to improve both selectivity and permeability [20]. In addition to the above methods, inserting specific water channels might be an effective way to optimize the performance of COF membranes. The well-organized pores in the COFs' active layer provide more insertion sites for HC6H. The shielding effect of embedded channels could enhance the sieving ability of the membrane, and the multiple interaction between COFs and water channels could improve the membrane stability. Meanwhile, the selective water transport capacity of water channels can also ensure the enhancement of water permeability of the membrane without sacrificing rejection.

In this work, water channels derived from the self-assembly of alkyl ureido-imidazole (HC6H) were employed for the performance enhancement of COF membranes. HC6H is a typical self-assembled channel with an artificial tubular structure that relies on strong intermolecular hydrogen bonding to achieve a stable state (Figure S1). During the self-assembly process, HC6H can spontaneously insert into the hydrophobic region of the membrane matrix to form a cylindrical structure. Therefore, HC6H could be used as a molecular scaffold for the construction of I-quartet, stabilized by internal water wires (Figure S1) [11]. These I-quartets were then inserted into COF (TpPa) membranes and connected by hydrogen bonding to form a stable composite membrane structure, as shown in Figure 1. Its separation performance was investigated and optimized in term of water permeance, dye rejection, and stability. The results showed that the composite COF membranes with high permeance and selectivity exhibited great potential in the development and industrialization of high-performance membranes.

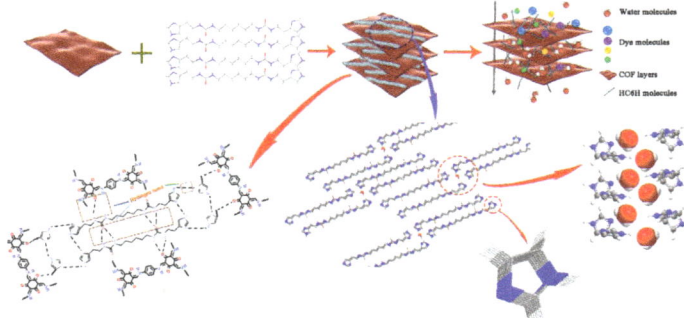

Figure 1. The composite architecture of TpPa-HC6H.

2. Experimental

2.1. Materials

Polyacrylonitrile (PAN) ultrafiltration membranes (mean pore size: 0.134 μm) were provided by Shandong Megavision Membrane Engineering and Technology Co., Ltd. (Shandong, China). P-phenylenediamine (Pa, 99.9 wt%) was obtained from Innochem (Beijing, China). 1,3,5-triformylphloroglucinol (Tp, 95 wt%) was supplied by Bide Pharmatech Ltd. (Shanghai, China). Hexamethylene diisocyanate (HDI, 97 wt%) and histamine (96 wt%) were obtained from McLean Biochemical Technology Co., Ltd. (Shanghai, China) Tetrahydrofuran (THF, 99 wt%). N, N-dimethylacetamide (99.5 wt%), ethylacetate (99.5 wt%), and acetonitrile (99.5 wt%) were purchased from Kemio Chemical Reagent Co., Ltd. (Tianjin, China). Acetic acid (99 wt%), n-hexane (99 wt%), and sodium hydroxide (NaOH, 98 wt%) were obtained from local suppliers. Dyes including methyl blue (MB), congo red (CR), acid fuchsin (AF), chrome black-T (CB-T), and rhodamine B (RB) were provided by Aladdin (Shanghai, China).

2.2. Preparation of HC6H

The synthesis of HC6H is shown in Figure S2. Hexamethylene diisocyanate and histamine (molar ratio of 1:2) were poured into a mixture of THF (5 mL), N,N-dimethylacetamide (10 mL), and ethyl acetate (5 mL). After ultrasonic treatment and shaking for 10 min, the above mixture was heated at 12 °C for 15 min and acetonitrile (5 mL) was added. The mixture was then heated at 120 °C and stirred at reflux for 1 h. Finally, the product was cooled at room temperature for 3 h. It was then washed with THF and filtrated to obtain a white powder, which was vacuum dried for 10 h.

2.3. Synthesis of Composite Matrix Membrane

Figure 2 illustrates the fabrication process of the composite matrix membranes by interfacial polymerization (IP). Firstly, HC6H powder was dissolved in anhydrous ethanol and sonicated for 30 min to obtain a HC6H solution. The solubility of HC6H is confirmed as 0.42 ± 0.01 g per 100 g ethanol at room temperature. The PAN ultrafiltration membrane was hydrolyzed in sodium hydroxide solution (1 mol/L) at 60 °C for 1.5 h, and then immersed into HC6H solution. Pa (0.2 wt%) and Tp (0.02 wt%) was dissolved in deionized water and n-hexane, respectively. The 25 mL Pa solution (including 250 μL acetic as the catalyst) was poured onto the active surface of the PAN substrate and underwent adsorption for 30 s. After that, the excess Pa solution was removed from the substrate and then a 25 mL Tp hexane solution was added onto the Pa-saturated substrate. When the n-hexane solution came into contact with the surface of the Pa-saturated substrate, the color of the membrane surface immediately changed to yellow. After 30 s of reaction, the residual solution was drained off, and the membrane was heated at 60 °C for 5 min to form a more stable hexagonal framework. Finally, the membrane was transferred to deionized water for 3 h before usage.

2.4. Membrane Characterization

The chemical structure of HC6H powder was explored using Fourier Transform Infrared Spectra (FTIR, IRAffinity-1S, Marlborough, MA, USA) in the scanning range of 500–4000 cm^{-1} and nuclear magnetic resonance spectrometry (NMR, JEOL JNM ECZ600R, Tokyo, Japan). Thermal stability of HC6H powder was determined by a thermo gravimetric analyzer (TGA, Netzsch TG 209F3, Karlsruhe, Germany).

The morphologies of the mixed-dimensional membrane were characterized by a scanning electron microscopy (SEM, Regulus 8100, Tokyo, Japan). TEM images of the composite matrix membrane were obtained using field emission transmission electron microscope (TEM, JEM-1400Flash, Tokyo, Japan). Atomic force microscopy (AFM, Dimension icon, Karlsruhe, Germany) was used to detect surface roughness.

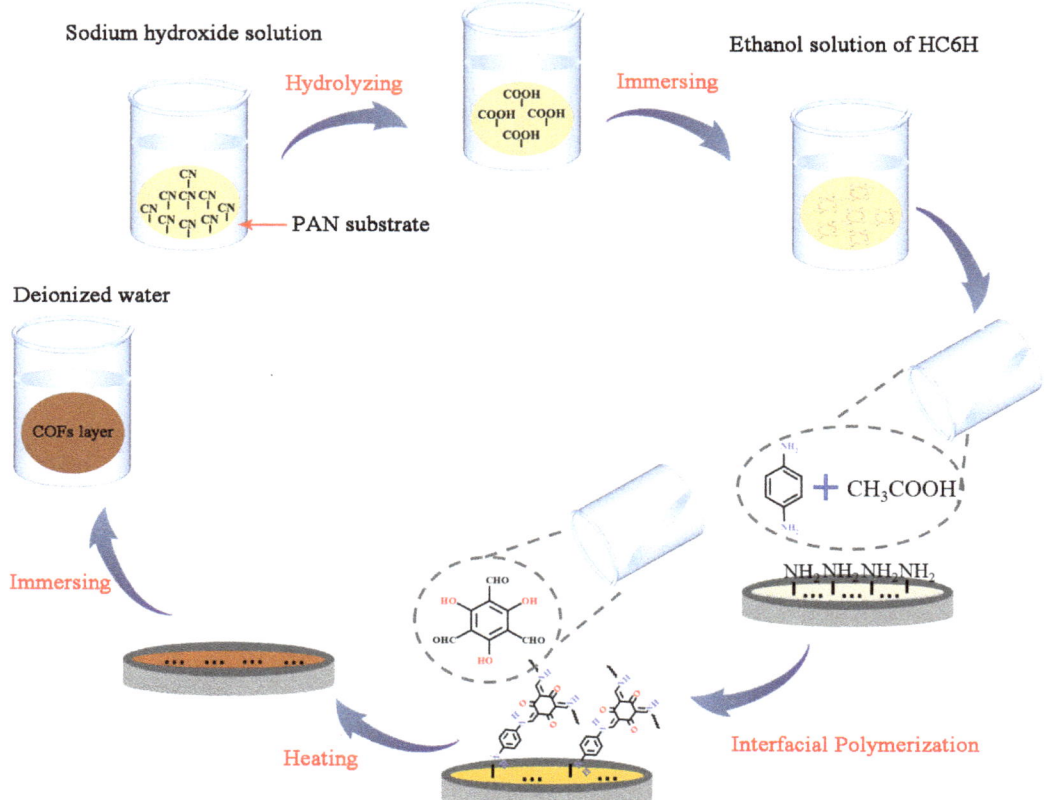

Figure 2. The synthesis process of composite matrix membranes by interfacial polymerization.

Membrane surface hydrophilicity was analyzed through water contact angle measurer (WCA, OCA15EC, Silicon Valley, CA, USA). The water contact angle measurement of each sample was repeated for five times to obtain the average value. X-ray diffractometer (XRD, D8-Focus, Karlsruhe, Germany) was used to explore the interlayer spacing of the powder and composite matrix membranes in room temperature at the range of $5° < 2θ < 30°$ with a step of $0.02°/s$. Elements and functional groups of the composite matrix membrane were obtained from the X-ray photoelectron spectroscopy (XPS, Kratos, Manchester, UK) and Fourier transform infrared (FTIR, IRAffinity-1S, Marlborough, MA, USA) spectra with the range of 400–4000 cm^{-1}. The UV/V spectrophotometer (UV3600 Shimadzu, Tokyo Japan) was used to analyze the concentrations of dyes in feed, permeated, and retentate solutions.

2.5. Membrane Separation Performance

Dye rejection and water permeability of the composite matrix membranes were determined by a cross-flow filtration system with an effective membrane area of 28.26 cm^2. Before the test, the membrane was pre-compacted under 2.5 bar for 20 min to achieve a stabilized state, and then the performance was measured at 2 bar. The dye concentrations (MB, CR, CB-T, AF, RB) in the feed solution were kept as 0.1 g/L.

Water permeance (F, L·m^{-2}·h^{-1}·bar^{-1}) was calculated by Equation (1)

$$F = V/(A * t * \Delta P) \tag{1}$$

where V, t, A, and ΔP are the volume of the permeated solution (L), filtration time (h), effective membrane area (m^2), and transmembrane pressure (bar), respectively.

The rejections for dyes were calculated by Equation (2)

$$R = (1 - C_P/C_F) * 100\% \qquad (2)$$

where C_P and C_F are the dye concentrations in the feed and permeate solutions, respectively.

3. Results and Discussion

3.1. Characterizations of HC6H

The 1H solid-state NMR and FT-IR spectra of HC6H powder were obtained to confirm its chemical structure. As shown in Figure 3a, the observed peaks in the 1H NMR spectrum w are consistent with the previously reported results [11]: δ = 1.23 (s, 2H), 1.33 (t, 2H), 2.58 (t, 2H), 2.95 (q, 2H), 3.21 (q, 2H), 5.78 (t, 1H), 5.88 (t, 1H), 6.77 (s, 1H), 7.53 (s, 1H), 11.85 (s,1H). As shown in Figure 3b, the FT-IR spectrum presents characteristic peaks at 1240 cm^{-1} (-C-C-), 1616 cm^{-1} (-NH-), 1736 cm^{-1} (-C=N-), and 2847 cm^{-1} and 2921 cm^{-1} (-CH2-), further confirming the successful synthesis of HC6H.

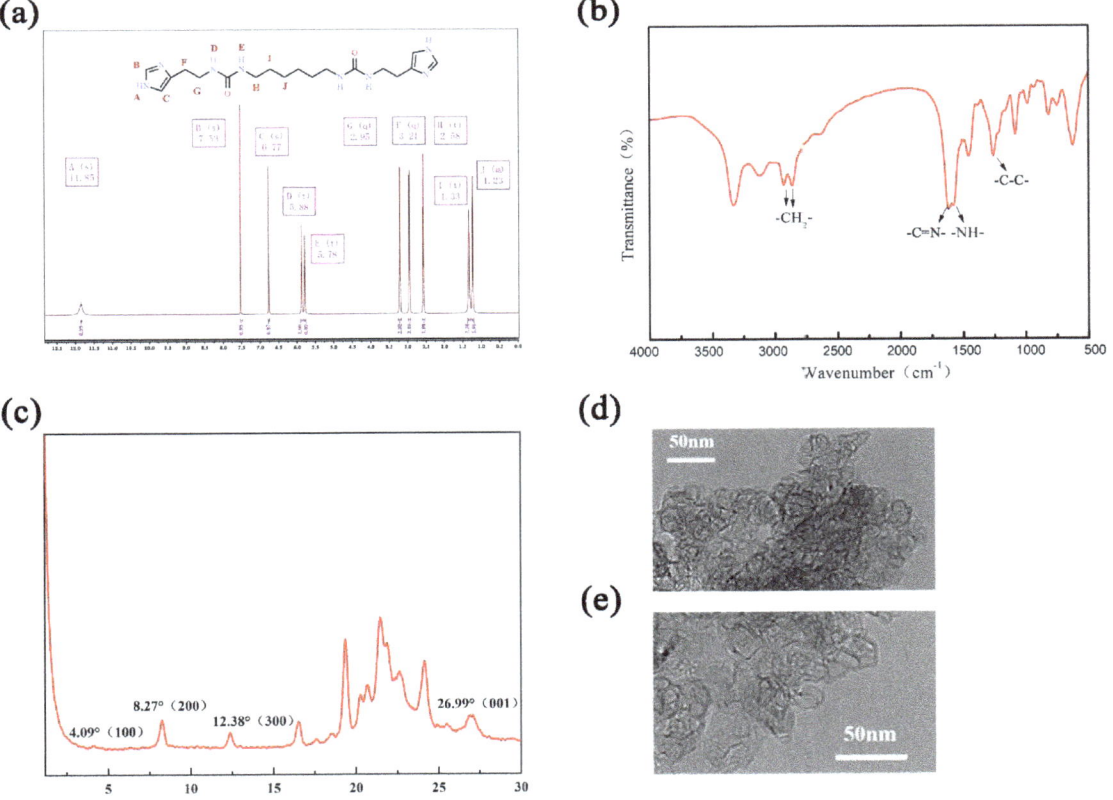

Figure 3. The characterizations of HC6H powders. (**a**) ^{1}H NMR spectroscopy of the synthesized HC6H powders. (**b**) FT−IR spectra of HC6H powders. (**c**) XRD pattern of HC6H powders. (**d**) TEM pattern of HC6H powders (60 k). (**e**) TEM pattern of HC6H powders (80 k).

The crystal structure of HC6H powder was characterized by XRD, as shown in Figure 3c. The main diffraction peaks are located at 2θ = 4.09° (100), 8.27° (200), 12.38° (300), and 26.99° (001), indicating a moderate crystallinity of the obtained HC6H powder. According to the Bragg equation [21], the minimum interlayer spacing is 3.3 Å, which could transport water and reject the dyes. The structure of HC6H powder was determined by TEM, and the results are shown in Figure 3d,e. It can be clearly seen that the HC6H exhibits a planar and ordered layered pore structure, which matches the single crystal structure in the XRD pattern. Meanwhile, the pore structure of HC6H is stacked in an ordered layered structure (Figure 3d), which matches the periodic results of the layered thin sheets in the XRD pattern, indicating the I-quartet crystal structure in the arranged layers of HC6H molecules. When the ethanol solution of HC6H was poured onto the surface of PAN substrate, the self-assembly process occurred, and the resulting pore structure provides sufficient channels for water molecules. In the next step of interfacial polymerization, the COF layer was in contact with the water channel, and a composite structure was successfully formed onto the membrane surface.

TGA was used to explore the composition and thermal stability of HC6H (Figure S3). A small amount of mass loss is exhibited in the temperature range of 100–160 °C because of the strong binding of water molecules to the channels. The powder firstly decomposed at 180 °C and the weight loss could reach 74.19% at 180–440 °C. The secondary decomposition starts at 440 °C, and the weight loss at this stage was about 7.97%, which is consistent with the previous report [11]. Meanwhile, it also indicated the HC6H powder can maintain a stable self-assembled structure at 180 °C, and its thermal stability can meet the requirements for dye separation and wastewater treatment.

3.2. Morphologies of Composite Matrix Membranes

Figures S4 and S5 display the SEM and AFM images of the top surface of composite matrix membranes fabricated with different reaction times and HC6H concentrations. The variations of surface roughness with HC6H concentrations and reaction times were shown in Figure 4a,b.

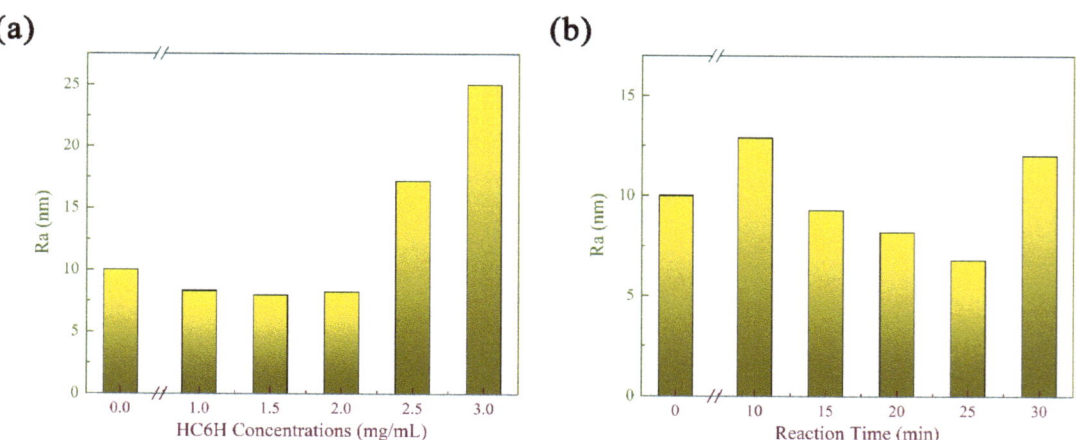

Figure 4. The effect of HC6H concentration (**a**) and reaction time (**b**) on surface roughness of HC6H–TpPa/HPAN membrane.

It can be seen that both PAN and HPAN membranes display smooth surfaces; the formation of the COF layer results in a rough and dense surface (Figure S6). When the HC6H concentration was adjusted in a small range (≤2.0 mg/mL), it had little effect on the surface morphology of the membrane (Figure S4g–i), and the surface roughness value was stable in the range smaller than the pristine membrane (Figure 4b). As the HC6H concentration increased above 2.0 mg/mL, granular substances appeared on the membrane surface (Figure S4j–k) and surface roughness increased (Figure 4b). This is because the reaction sites were saturated and a large amount of HC6H accumulated on the membrane surface. The accumulation of HC6H will inevitably lead to a decrease of permeance.

As the reaction time increased from 10 to 25 min, the surface roughness values first increased and then decreased (Figure 4a), which is consistent with the SEM images. When the reaction time was short (10 min), the membrane surface had obvious stripes and high roughness due to the rapid formation of the COF layer (similar to Figure S4a). When the time was extended to 15 min (Figure S4c), water channels formed by HC6H appeared and the surface became smooth. When the reaction time was extended to 25 min, the strip-like protrusions disappeared and the film surface was smoother. However, when the reaction time reached 30 min, the formed cross-linked structure blocked the membrane pores, so a granular structure (Figure S4f) with a surface roughness of 12 nm (Figure S5f) on the membrane surface could be clearly seen. As the reaction time was extended to 20 min, the strip-like protrusions on the membrane surface had disappeared, indicating the formation of a good planar configuration (Figure S4g–k).

3.3. Surface Hydrophilicity of Composite Matrix Membranes

Surface hydrophilicity plays a critical role in improving the permeability of the membrane. Figure 5a shows water contact angles of the membranes fabricated with different reaction times and HC6H concentrations. The water contact angles of the PAN substrate decreased from $60.0 \pm 1.5°$ to $37.2 \pm 1.5°$ after the hydrolysis in sodium hydroxide (Figure S7). After the formation of TpPa layer, the surface of TpPa/HPAN membrane displayed a water contact angle of $45.9 \pm 1.5°$. After incorporating 1.0 mg/mL of HC6H, the HC6H−TpPa/HPAN membrane surface displayed a water contact angle of $53.5 \pm 1.5°$, which might be due to the long hydrophobic chain alkanes of HC6H. When the HC6H concentration was above 1.0 mg/mL, the water contact angle of the membranes decreased continuously, probably due to the synergistic effect of hydrophilic groups and roughness. Both the number of hydrophilic groups and the surface roughness (7.94 nm to 25 nm) on the membrane surface increased with the increase of the concentration.

Figure 5b shows the effect of reaction time on water contact angle. When the reaction time was 10 min, a hydrophilic water channel has not been formed on the membrane surface, and the hydrophobic chain alkanes played a major role on the membrane surface, leading to the increase of water contact angle. When the reaction time reached 30 min, the contact angle dropped to $38.92 \pm 1.5°$. In general, due to the presence of artificial water channels, the hydrophilicity of the membrane was enhanced and the overall water contact angle was reduced.

3.4. Chemical Composition of Composite Matrix Membranes

The chemical structure of the membranes and the molecular interactions between HC6H and TpPa were determined via FT-IR and XPS spectra. Figure 5c compared the FT-IR spectra of HC6H−TpPa/HPAN membrane (red curve), TpPa/HPAN membrane (blue curve), HPAN membrane (green curve), and PAN membrane (black curve). The FT-IR spectrum HPAN exhibited adsorption bands at 1569 cm^{-1} due to the stretching vibration of –COOH compared with the spectrum of PAN; this is because, after the hydrolysis of the PAN substrate in sodium hydroxide solution, the cyano group of PAN was converted to a carboxyl group. The FT-IR spectrum of HC6H−TpPa/HPAN exhibited adsorption bands at 1230 cm^{-1} and 1650 cm^{-1} due to the stretching vibration of -C-C- and -C=N-, respectively, indicating the successful assembly of artificial water channels on the membrane surface. In addition, the adsorption bands at 3354 cm^{-1}, assigned to -OH group, indicated the existed interaction between TpPa layer and artificial water channels.

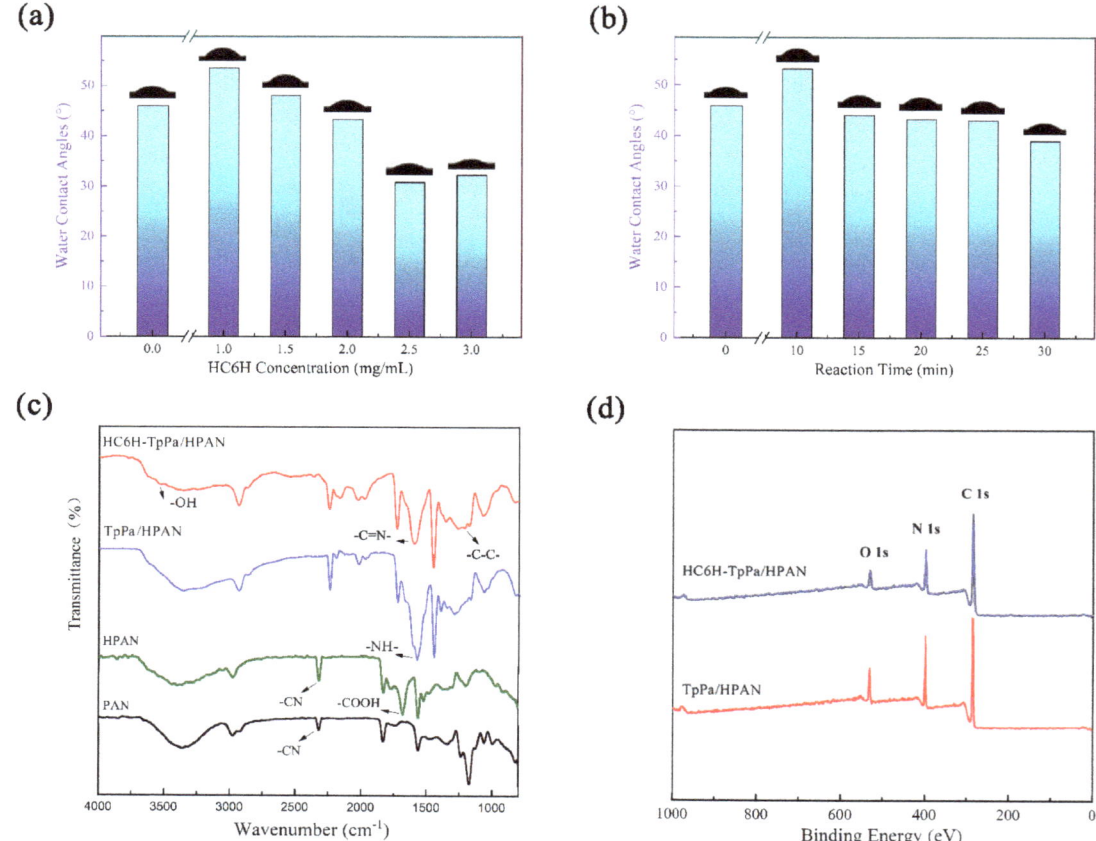

Figure 5. The influence of reaction concentration (**a**) and reaction time (**b**) on water contact angle of HC6H−TpPa/HPAN. (**c**) FT−IR spectra of HC6H−TpPa/HPAN membrane (**red curve**), TpPa/HPAN membrane (**blue curve**), HPAN membrane (**green curve**), and PAN membrane (**black curve**). (**d**) XPS spectra of TpPa/HPAN membrane and HC6H−TpPa/HPAN membrane.

As demonstrated in Figure 5d, XPS measurement was used to explore the chemical bonding and surface elements of TpPa-/HPAN and HC6H−TpPa/HPAN membranes. Both membranes showed three peaks at 285.3 eV (C 1 s), 398.68 eV (N 1 s), and 530.98 eV (O 1 s). Table S1 lists the elemental atomic percentage of carbon (C), oxygen (O), and nitrogen (N) on the membranes surface. Compared with TpPa-/HPAN membrane, the HC6H−TpPa/HPAN membrane showed a decreased O content and an increased N content, with the N/O ratio increased from 2.16 to 2.97. This might be because the HC6H exhibited a higher nitrogen element, with an N/O ratio of 3.5, than the TpPa layer. Therefore, the increase of N/O ratio also proved the successful incorporation of HC6H into the membrane. According to the change of N/O ratio in the membrane, we can calculate that the proportion of HC6H in the composite membrane was 60.4%.

3.5. Seperation Performance of Composite Matrix Membranes

The key reaction parameters in the process of HC6H self-assembly were investigated to optimize the separation performance of HC6H−TpPa/HPAN membrane, including reaction time, HC6H concentration, and reaction temperature.

In the process of membrane preparation, the HC6H concentration was a critical factor determining the generation of artificial water channels. The influence of HC6H concentration on water permeance and CR rejection was investigated, and the result is shown in Figure 6a. It can be seen that the water permeance of the pristine membrane was around 135.8 L·m^{-2}·h^{-1}·bar^{-1}. As HC6H concentration increased from 1.0 mg/mL to 2.0 mg/mL, water permeance dramatically increased from 169.8 to 271.7 L·m^{-2}·h^{-1}·bar^{-1}. These results indicated that a higher HC6H concentration is beneficial to the formation of water channels in the membrane. However, further increasing the HC6H concentration to 3.0 mg/mL caused a significant decline in water permeance to 216.5 L·m^{-2}·h^{-1}·bar^{-1}. As the assembly of HC6H in the membrane tended to be saturated, excessive HC6H stayed on the membrane surface or in the channels, leading to a decrease in water permeance. Furthermore, the rejection rate to CR of all the composite matrix membranes was above 99.9%, and the maximum water permeance was almost twice that of the pristine membrane. As a result, 2.0 mg/mL HC6H was chosen to be the optimal concentration for the following investigation.

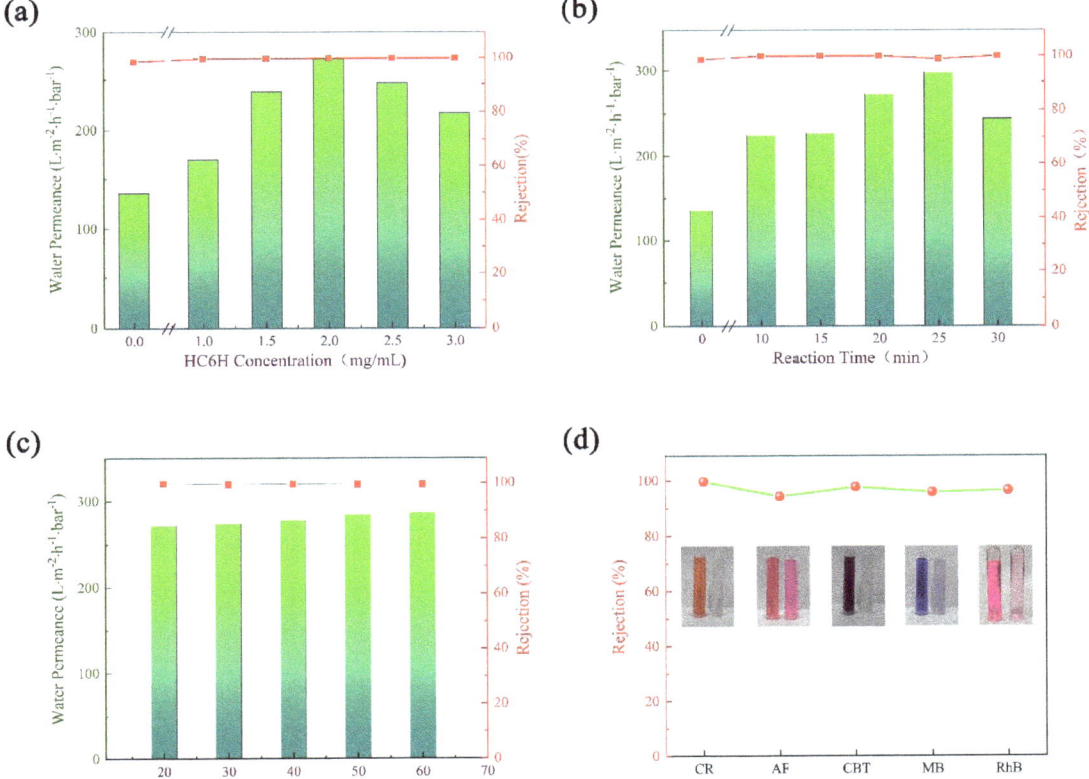

Figure 6. The influence of reaction concentration (**a**), reaction time (**b**) and reaction temperature (**c**) on the separation performance of the HC6H−TpPa/HPAN membranes. (**d**) The rejection performance of HC6H−TpPa/HPAN membranes to different dyes (CR, AF, CB−T, MB, and RhB).

Similar to the self-assembly of HC6H in other substrates [22], the reaction time with HC6H has a great influence on the membrane performance. It directly determines the number of artificial water channels formed in the membrane. Figure 6b displays the pure water permeance and CR rejection of the membranes fabricated with different reaction times. It can be seen that the water permeance of the TpPa/HPAN membrane was 135.8 L·m^{-2}·h^{-1}·bar^{-1}. As the reaction time increased from 10 to 25 min, water permeance of the HC6H−TpPa/HPAN membranes increased significantly from 225 to 297 L·m^{-2}·h^{-1}·bar^{-1}. When the reaction time was relatively short, most of the HC6H had not yet assembled into the water channel structure, but more and more water channels were formed in the membrane with the increase of self-assembly time. When the reaction time reached 30 min, water permeance reduced to 244 L·m^{-2}·h^{-1}·bar^{-1}. This might because the self-growth of HC6H, with a longer reaction time, easily led to the formation of a cross-linking structure, leading to a pore blockage. The CR rejection of all composite matrix membranes was higher than 98.7%. Based on the above results, 20 min was chosen as the optimal reaction time with HC6H.

Finally, the effect of reaction temperature on the separation performance of the membrane was also studied (Figure 6c). When the reaction temperature increased from 20 °C to 60 °C, the water permeance of the membrane increased slightly from 271.7 to 286.6 L·m^{-2}·h^{-1}·bar^{-1}, and the CR rejection rate was still maintained at a high level (>99.9%). According to the Arrhenius law [23], an appropriate increase in temperature is conducive to the formation of more water channels within a certain period of time. Since the effect of temperature on the membrane properties is not obvious, 20 °C was chosen as the self-assembly temperature.

After investigating the optimal preparation parameters of the composite matrix membranes, we further investigated their separation performance using several typical dyes (Figure 6d). It can be seen that the rejection rates of the membrane for CR (2.85 × 0.89 nm), AF (1.17 × 1.13 nm), CB-T (1.55 × 0.88 nm), MB (1.25 × 0.51 nm), and RhB (1.69 × 0.83 nm) were 99.9%, 94.3%, 97.7%, 95.8%, and 96.5%, respectively, indicating that the composite matrix membrane had excellent separation performance. Owing to the formation of artificial water channels in the membrane, they can effectively repel these five dye molecules (rejection rate > 90%) with high water permeance.

3.6. Stability of Membrane

The mechanical stability of composite matrix membranes was investigated by measuring the water permeance under different operating pressures. As shown in Figure 7a, when the pressure increased from 1.0 to 3.5 bar, the water permeance increased almost linearly from 191.08 to 360 L·m^{-2}·h^{-1}·bar^{-1}, indicating that the water channels of composite matrix membranes remained at good rigidity below this pressure.

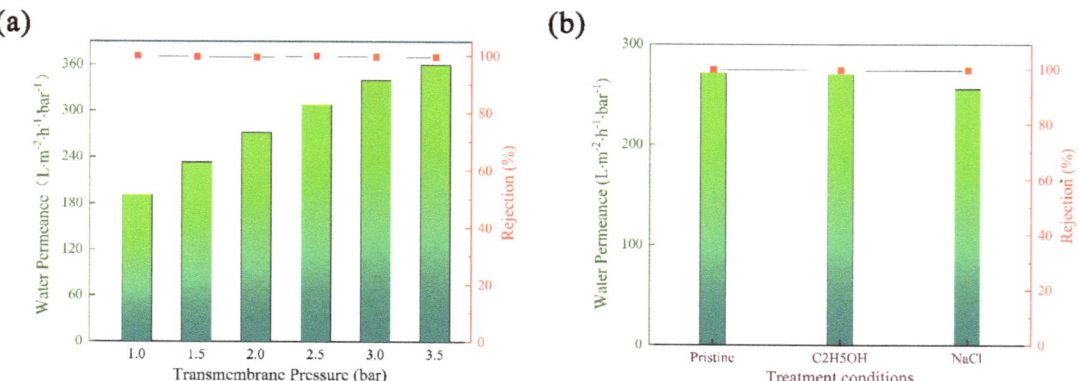

Figure 7. (**a**) Water permeance under different transmembrane pressures on composite matrix membranes. (**b**) Separation performance before and after different solution treatments for the HC6H−TpPa/HPAN membranes.

In addition, the composite matrix membranes were immersed in NaCl solution (2.0 mg/mL) or aqueous alcohol for 7 days, and then their dye rejection and water permeance were measured to reveal chemical stability. As shown in Figure 7b, the water permeance changed little, and the dye rejection remained at a high level, indicating that the NaCl solution and aqueous alcohol could hardly destroy the membrane structure. According to the above results, the composite matrix membranes have potential for use in the application of dye separation from saline and organic wastewater.

In order to highlight the superior performance of the composite matrix membranes, we compared its dye rejection and water permeance with other membranes (Figure 8). Many reported membranes exhibited a moderate water permeance (<50 L·m^{-2}·h^{-1}·bar^{-1}) with a high RhB dye rejection rate (e.g., >95%), or a high water permeance (>200 L·m^{-2}·h^{-1}·bar^{-1}) with a low RhB rejection rate (<90%) [24]. In contrast, the HC6H−TpPa/HPAN composite matrix membranes, fabricated in this work, exhibited a water permeance of 271 L·m^{-2}·h^{-1}·bar^{-1}, and rejection rates to several dyes above 94%. Such an excellent water permeance should be ascribed to the remarkable composite structure composed of COF layers and HC6H.

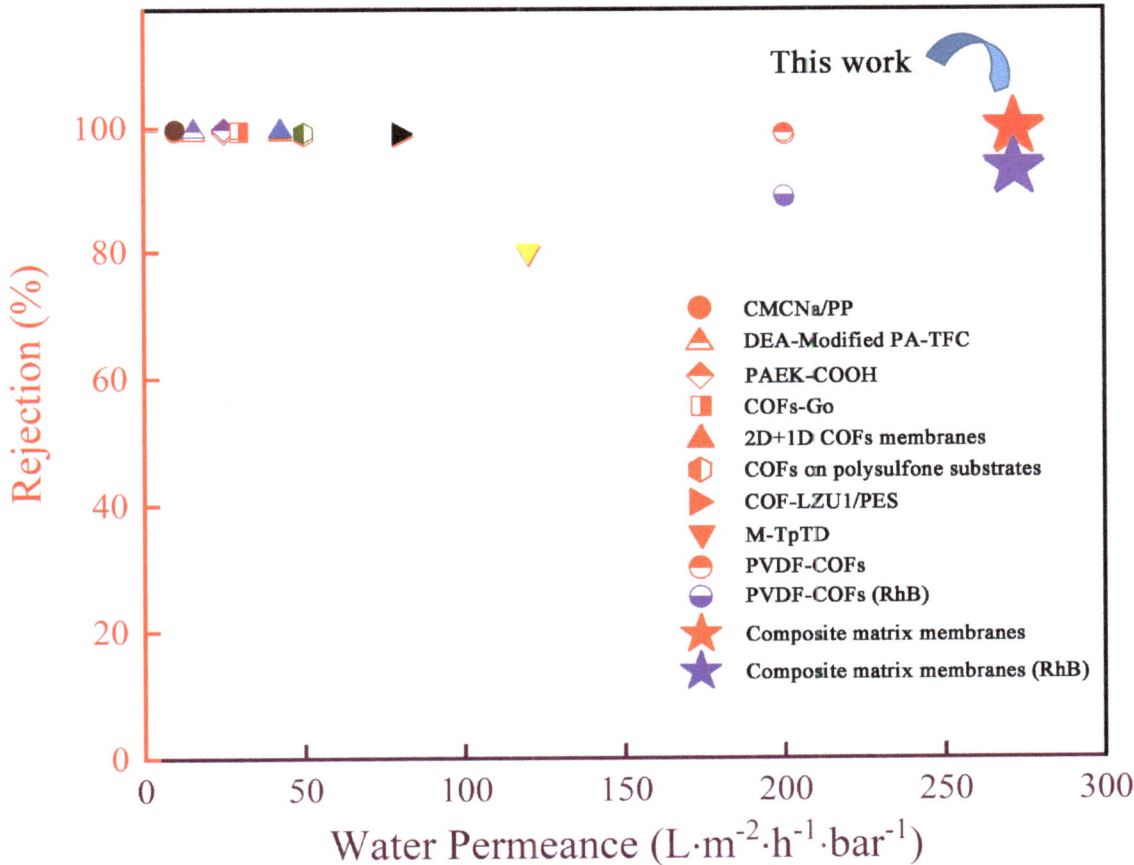

Figure 8. Comparation of dye rejection (red: CR, blue: RhB) and water permeance between HC6H−TpPa/HPAN composite matrix membranes and other membranes reported in the literature. Detailed information for these membranes is given in Table S2.

4. Conclusions

In this study, we presented a composite self-assembly strategy to prepare advanced COF membranes by introducing artificial water channels. Using the hydrogen bonding between HC6H molecules and COFs, a highly efficient artificial water channel structure in the membrane was successfully constructed. The hydrophilicity and water permeability of the membrane surface have been greatly improved by the introduction of artificial water channels, it breaks the permeability and selectivity trade-off effect, and it has a high separation performance in dye rejection. Under optimized synthesis conditions, the composite matrix membrane exhibits excellent water permeance (271.76 $L \cdot m^{-2} \cdot h^{-1} \cdot bar^{-1}$) and high rejection rates (>94%) to different dye molecules. Moreover, the membrane stability was improved due to the multiple interactions between the water channels and COF layers. Additionally, the composite matrix membranes showed stable permeation and rejection performance under different conditions; the good thermal stability of the water channel structure also enables the composite membrane to meet the higher requirements of dye separation. It follows that, using artificial water channels to design composite structures, as in this study, may prove to be an effective strategy to optimize the separation performance of the COF membrane. It shows great potential in the treatment of dyeing wastewater.

Supplementary Materials: The following supporting information can be downloaded at: https://www.mdpi.com/article/10.3390/membranes12090824/s1, Figure S1: (a) Water-free I-quartet formation via CH···N and NH···N interactions; (b) Water-assisted I-quartet formation via CH···N, N···HO, and NH···O-H interactions; (c) HC6H molecules linked via C=O···H (White: H, gray: C, blue: N, red: O, the data is atom distance (Å)), Figure S2: Synthesis of alkylureido-imidazoles (HC6H) using hexamethylene diisocyanate and histamine, Figure S3: TGA curves of the synthesized HC6H powders, Figure S4: The SEM images of the upper surfaces of TpPa/HPAN and HC6H-TpPa/HPAN composite matrix membranes prepared at different reaction times and concentrations of HC6H, Figure S5: The AFM images of the upper surfaces of TpPa/HPAN and HC6H-TpPa/HPAN mixed dimensional membranes prepared at different reaction times and concentrations of HC6H, Figure S6: SEM images of PAN and HPAN membranes. Figure S7: Water contact angles of the PAN and HPAN membranes; Table S1: The elemental atomic percentage on the membrane surface; Table S2: Performance comparison among various membranes towards different dyes rejection. References [25–30] are cited in the supplementary materials.

Author Contributions: K.L. and J.G. performed the experiment and wrote the manuscript; P.L. (corresponding author) performed the data analyses with constructive discussions; Y.L. and J.C. contributed to the conception of the study. All authors have read and agreed to the published version of the manuscript.

Funding: This research received no external funding.

Institutional Review Board Statement: Not applicable.

Informed Consent Statement: Informed consent was obtained from all subjects involved in the study.

Data Availability Statement: Not applicable.

Conflicts of Interest: The authors declare no conflict of interest.

References

1. van Loosdrecht, M.C.M.; Brdjanovic, D. Anticipating the next century of wastewater treatment. *Science* **2014**, *344*, 1452–1453. [CrossRef] [PubMed]
2. An, A.K.; Guo, J.; Jeong, S.; Lee, E.-J.; Tabatabai, S.A.A.; Leiknes, T. High flux and antifouling properties of negatively charged membrane for dyeing wastewater treatment by membrane distillation. *Water Res.* **2016**, *103*, 362–371. [CrossRef] [PubMed]
3. Marchetti, P.; Solomon, M.F.J.; Szekely, G.; Livingston, A.G. Molecular Separation with Organic Solvent Nanofiltration: A Critical Review. *Chem. Rev.* **2014**, *114*, 10735–10806. [CrossRef] [PubMed]
4. Pohl, P.; Saparov, S.M.; Borgnia, M.J.; Agre, P. Highly selective water channel activity measured by voltage clamp: Analysis of planar lipid bilayers reconstituted with purified AqpZ. *Proc. Natl. Acad. Sci. USA* **2001**, *98*, 9624–9629. [CrossRef] [PubMed]
5. Murata, K.; Mitsuoka, K.; Hirai, T.; Walz, T.; Agre, P.; Heymann, J.B.; Engel, A.; Fujiyoshi, Y. Structural determinants of water permeation through aquaporin-1. *Nature* **2000**, *407*, 599–605. [CrossRef] [PubMed]

6. Zhao, Y.; Qiu, C.Q.; Li, X.S.; Vararattanavech, A.; Shen, W.M.; Torres, J.; Helix-Nielsen, C.; Wang, R.; Hu, X.; Fane, A.G.; et al. Synthesis of robust and high-performance aquaporin-based biomimetic membranes by interfacial polymerization-membrane preparation and RO performance characterization. *J. Membr. Sci.* **2012**, *423*, 422–428. [CrossRef]
7. Huang, L.B.; Di Vincenzo, M.; Li, Y.; Barboiu, M. Artificial Water Channels: Towards Biomimetic Membranes for Desalination. *Chemistry* **2021**, *27*, 2224–2239. [CrossRef]
8. Agre, P. Aquaporin water channels (Nobel lecture). *Angew. Chem. Int. Ed.* **2004**, *43*, 4278–4290. [CrossRef]
9. Horner, A.; Siligan, C.; Cornean, A.; Pohl, P. Positively charged residues at the channel mouth boost single-file water flow. *Faraday Discuss.* **2018**, *209*, 55–65. [CrossRef]
10. Eriksson, U.K.; Fischer, G.; Friemann, R.; Enkavi, G.; Tajkhorshid, E.; Neutze, R. Subangstrom Resolution X-ray Structure Details Aquaporin-Water Interactions. *Science* **2013**, *340*, 1346–1349. [CrossRef]
11. Le Duc, Y.; Michau, M.; Gilles, A.; Gence, V.; Legrand, Y.M.; van der Lee, A.; Tingry, S.; Barboiu, M. Imidazole-quartet water and proton dipolar channels. *Angew. Chem. Int. Ed. Engl.* **2011**, *50*, 11366–11372. [CrossRef]
12. Tunuguntla, R.H.; Allen, F.I.; Kim, K.; Belliveau, A.; Noy, A. Ultrafast proton transport in sub-1-nm diameter carbon nanotube porins. *Nat. Nanotechnol.* **2016**, *11*, 639–644. [CrossRef]
13. Zhou, X.B.; Liu, G.D.; Yamato, K.; Shen, Y.; Cheng, R.X.; Wei, X.X.; Bai, W.L.; Gao, Y.; Li, H.; Liu, Y.; et al. Self-assembling subnanometer pores with unusual mass-transport properties. *Nat. Commun.* **2012**, *3*, 8. [CrossRef]
14. Ogoshi, T.; Kanai, S.; Fujinami, S.; Yamagishi, T.A.; Nakamoto, Y. Para-bridged symmetrical pillar 5 arenes: Their Lewis acid catalyzed synthesis and host-guest property. *J. Am. Chem. Soc.* **2008**, *130*, 5022–5023. [CrossRef]
15. Chen, L.; Si, W.; Zhang, L.; Tang, G.F.; Li, Z.T.; Hou, J.L. Chiral Selective Transmembrane Transport of Amino Acids through Artificial Channels. *J. Am. Chem. Soc.* **2013**, *135*, 2152–2155. [CrossRef]
16. Liang, R.R.; Jiang, S.Y.; A, R.H.; Zhao, X. Two-dimensional covalent organic frameworks with hierarchical porosity. *Chem Soc Rev.* **2015**, *7*, 905–912. [CrossRef]
17. Xu, H.; Gao, J.; Jiang, D. Stable, crystalline, porous, covalent organic frameworks as a platform for chiral organocatalysts. *Nat. Chem.* **2015**, *7*, 905–912. [CrossRef]
18. Xu, F.; Xu, H.; Chen, X.; Wu, D.; Wu, Y.; Liu, H.; Gu, C.; Fu, R.; Jiang, D. Radical Covalent Organic Frameworks: A General Strategy to Immobilize Open-Accessible Polyradicals for High-Performance Capacitive Energy Storage. *Angew. Chem. Int. Ed.* **2015**, *54*, 6814–6818. [CrossRef]
19. Su, Y.-Y.; Yan, X.; Chen, Y.; Guo, X.-J.; Chen, X.-F.; Lang, W.-Z. Facile fabrication of COF-LZU1/PES composite membrane via interfacial polymerization on microfiltration substrate for dye/salt separation. *J. Membr. Sci.* **2021**, *618*, 118706. [CrossRef]
20. Yang, H.; Yang, L.; Wang, H.; Xu, Z.; Zhao, Y.; Luo, Y.; Nasir, N.; Song, Y.; Wu, H.; Pan, F.; et al. Covalent organic framework membranes through a mixed-dimensional assembly for molecular separations. *Nat. Commun.* **2019**, *10*, 2101. [CrossRef]
21. Muniz, F.T.L.; Miranda, M.A.R.; Morilla dos Santos, C.; Sasaki, J.M. The Scherrer equation and the dynamical theory of X-ray diffraction. *Acta Crystallogr. Sect. A* **2016**, *72*, 385–390. [CrossRef] [PubMed]
22. Mao, Y.; Zhang, M.; Cheng, L.; Yuan, J.; Liu, G.; Huang, L.; Barboiu, M.; Jin, W. Bola-amphiphile-imidazole embedded GO membrane with enhanced solvent dehydration properties. *J. Membr. Sci.* **2020**, *595*, 117545 [CrossRef]
23. Liu, K.; Hu, X.; Wei, Z.; Li, Y.; Jiang, Y. Modified Gaussian Process Regression Models for Cyclic Capacity Prediction of Lithium-Ion Batteries. *IEEE Trans. Transp. Electrif.* **2019**, *5*, 1225–1236. [CrossRef]
24. Wu, C.; Wang, X.; Zhu, T.; Li, P.; Xia, S. Covalent organic frameworks embedded membrane via acetic-acid-catalyzed interfacial polymerization for dyes separation: Enhanced permeability and selectivity. *Chemosphere* **2020**, *261*, 127580. [CrossRef]
25. Yu, S.; Chen, Z.; Cheng, Q.; Lü, Z.; Liu, M.; Gao, C. Application of thin-film composite hollow fiber membrane to submerged nanofiltration of anionic dye aqueous solutions. *Sep. Purif. Technol.* **2012**, *88*, 121–129. [CrossRef]
26. Liu, M.; Chen, Q.; Lu, K.; Huang, W.; Lü, Z.; Zhou, C.; Yu, S.; Gao, C. High efficient removal of dyes from aqueous solution through nanofiltration using diethanolamine-modified polyamide thin-film composite membrane. *Sep. Purif. Technol.* **2017**, *173*, 135–143. [CrossRef]
27. Xing, L.; Guo, N.; Zhang, Y.; Zhang, H.; Liu, J. A negatively charged loose nanofiltration membrane by blending with poly (sodium 4-styrene sulfonate) grafted SiO_2 via SI-ATRP for dye purification. *Sep. Purif. Technol.* **2015**, *146*, 50–59. [CrossRef]
28. Zhang, X.; Li, H.; Wang, J.; Peng, D.; Liu, J.; Zhang, Y. In-situ grown covalent organic framework nanosheets on graphene for membrane-based dye/salt separation. *J. Membr. Sci.* **2019**, *581*, 321–330. [CrossRef]
29. Wang, R.; Shi, X.; Xiao, A.; Zhou, W.; Wang, Y. Interfacial polymerization of covalent organic frameworks (COFs) on polymeric substrates for molecular separations. *J. Membr. Sci.* **2018**, *566*, 197–204. [CrossRef]
30. Kandambeth, S.; Biswal, B.P.; Chaudhari, H.D.; Rout, K.C.; Kunjattu, H.S.; Mitra, S.; Karak, S.; Das, A.; Mukherjee, R.; Kharul, U.K.; et al. Selective Molecular Sieving in Self-Standing Porous Covalent-Organic-Framework Membranes. *Adv. Mater.* **2017**, *29*, 1603945. [CrossRef]

Article

Membrane Photobioreactor Applied for Municipal Wastewater Treatment at a High Solids Retention Time: Effects of Microalgae Decay on Treatment Performance and Biomass Properties

Hui Zou [1,†], Neema Christopher Rutta [1,†], Shilei Chen [1], Meijia Zhang [1,*], Hongjun Lin [1,*] and Baoqiang Liao [2]

1. College of Geography and Environmental Sciences, Zhejiang Normal University, Jinhua 321004, China; zouzou@zjnu.edu.cn (H.Z.); neychriss@yahoo.com (N.C.R.); slchen@zjnu.edu.cn (S.C.)
2. Department of Chemical Engineering, Lakehead University, 955 Oliver Road, Thunder Bay, ON P7B 5E1, Canada; bliao@lakeheadu.ca
* Correspondence: mzhang15@zjnu.edu.cn (M.Z.); hjlin@zjnu.cn (H.L.)
† These authors contributed equally to this work.

Abstract: Membrane photobioreactor (MPBR) technology is a microalgae-based system that can simultaneously realize nutrient recovery and microalgae cultivation in a single step. Current research is mainly focused on the operation of MPBR at a medium SRT. The operation of MPBR at a high SRT is rarely reported in MPBR studies. Therefore, this study conducted a submerged MPBR to treat synthetic municipal wastewater at a long solids retention time of 50 d. It was found that serious microalgae decay occurred on day 23. A series of characterizations, including the biomass concentration, chlorophyll-a content, nutrients removal, and physical-chemical properties of the microalgae, were conducted to evaluate how microalgae decay affects the treatment performance and biomass properties. The results showed that the biomass concentration and chlorophyll-a/MLSS dropped rapidly from 3.48 to 1.94 g/L and 34.56 to 10.71 mg/g, respectively, after the occurrence of decay. The effluent quality significantly deteriorated, corresponding to the total effluent nitrogen and total phosphorus concentration sharply rising and exceeding that of the feed. In addition, the particle became larger, the content of the extracellular polymeric substances (EPSs) decreased, and the soluble microbial products (SMPs) increased instantaneously. However, the filtration resistance had no significant increase because of the comprehensive interactions of the floc size, EPSs, and SMPs. The above results suggest that the MPBR system cannot maintain long-term operation under a high SRT for municipal wastewater treatment. In addition, the biological treatment performance of the MPBR deteriorated while the antifouling performance of the microalgae flocs improved after the occurrence of decay. The occurrence of microalgae decay was attributed to the double stresses from the light shading and intraspecific competition under high biomass concentration. Therefore, to avoid microalgae decay, periodic biomass removal is required to control the environmental stress within the tolerance range of the microalgae. Further studies are required to explore the underlying mechanism of the occurrence of decay.

Keywords: membrane photobioreactor; microalgae decay; solids retention time; treatment performance; biomass properties; municipal wastewater treatment

Citation: Zou, H.; Rutta, N.C.; Chen, S.; Zhang, M.; Lin, H.; Liao, B. Membrane Photobioreactor Applied for Municipal Wastewater Treatment at a High Solids Retention Time: Effects of Microalgae Decay on Treatment Performance and Biomass Properties. *Membranes* 2022, 12, 564. https://doi.org/10.3390/membranes12060564

Academic Editor: Joaquim Comas

Received: 13 April 2022
Accepted: 26 May 2022
Published: 28 May 2022

Publisher's Note: MDPI stays neutral with regard to jurisdictional claims in published maps and institutional affiliations.

Copyright: © 2022 by the authors. Licensee MDPI, Basel, Switzerland. This article is an open access article distributed under the terms and conditions of the Creative Commons Attribution (CC BY) license (https://creativecommons.org/licenses/by/4.0/).

1. Introduction

Wastewater reclamation and reuse have received more and more attention in the world as a result of the increasingly serious freshwater scarcity. At present, various wastewater treatment technologies, such as the activated sludge process (CAS) and membrane bioreactor (MBR), have matured and are widely used in practical applications [1–6]. However, most of these systems target organics removal using bacteria, and the treated effluent generally contains high levels of nitrogen and phosphorus. The direct disposal of such an effluent would cause eutrophication in the water body. Therefore, it generally requires additional processes targeting nutrient removal in order to meet discharge standards.

The microalgal membrane photobioreactor (MPBR), which integrates the photobioreactor (PBR) with the membrane filtration processes, is a promising technology for simultaneous microalgae cultivation and nutrient recovery [7–10]. For such a system, the use of sewage can offset the cost of the nutrients required for microalgae cultivation, and the microalgae biomass that is produced is one of the most promising precursors for biofuel production [11]. In addition, the greenhouse gas CO_2 can be fixed by the microalgae through photosynthesis during the wastewater treatment process [12]. The feasibility of using MPBR for wastewater treatment has been extensively studied in the last decade [10,11,13–19].

As a complex biological system, the performance of MPBR is highly dependent on various operating conditions, such as lighting, hydraulic retention time (HRT), and solids retention time (SRT) [10,19]. Among all of these factors, SRT is a critical factor that has a significant influence on the biomass concentration, microalgal productivity, and nutrient removal in MPBR [15,18,20,21]. It is well known that a significant advantage of MPBR over PBR is the decoupling of HRT from SRT, which can reduce the downstream microalgal harvesting and dewatering due to the higher microalgae concentration that is achieved. However, researchers currently mainly adopt a medium SRT for the operation of MPBR [10,15,18,20]. To the best of our best knowledge, only two studies have operated MPBR at a long SRT [22,23]. For instance, Xu et al. [22] conducted MPBR at a prolonged SRT of 180 d for long-term operation, and eventually achieved a high biomass concentration of 4.84 g/L. A similar result was also reported by Praveen et al. [23]. However, the successes of these two studies were mainly attributed to the utilization of low organic strength secondary wastewater and a low initial microalgae concentration.

In fact, except for secondary wastewater, MPBR has also been applied for the treatment of high organic strength wastewater such as municipal wastewater and anaerobically digested wastewater [24,25]. However, the feasibility of the long-term operation of MPBR at a high SRT to treat municipal wastewater has never been reported. Therefore, a study on MPBR in this field is expected to provide valuable insight into the application of MPBR for municipal wastewater treatment.

In this study, a lab-scale MPBR system was operated at a high SRT of 50 d to explore the feasibility of long-term operation of the MPBR system under a high SRT for municipal wastewater treatment. A serious microalgae decay phenomenon occurred on day 23. The effects of microalgae decay on the treatment performance and biomass properties were then identified by a series of characterizations, including biomass production, chlorophyll-a concentration, nutrient removal, and microalgal properties. This study could provide practical experience for the operation and management of MPBR for high organic strength wastewater treatment.

2. Materials and Methods

2.1. MPBR Setup and Operation

A lab-scale cylindrical submerged transparent MPBR system was conducted for municipal wastewater treatment. The schematic of this setup is displayed in Figure 1. Solid–liquid separation was accomplished using a flat plate membrane module. The membranes used in this work were commercial grade, and were purchased from SINAP Co. Ltd., Shanghai, China. Air was pumped into the reactor through an aeration pump to provide CO_2 for microalgae growth and to form eddy currents to scour the membrane surface for fouling control. Gentle mixing was created using a magnetic stirrer (Model 6795-61, Corning, New York, NY, USA) located at the bottom of the reactor so as to prevent microalgal precipitation. Continuous illumination was provided by four LED lamps (two on each side). Details regarding the operating conditions and membrane module properties are listed in Table 1. *Chlorella vulgaris* (CPCC 90) that was precultivated in a modified salt medium (MSM) [26] was inoculated as the seed.

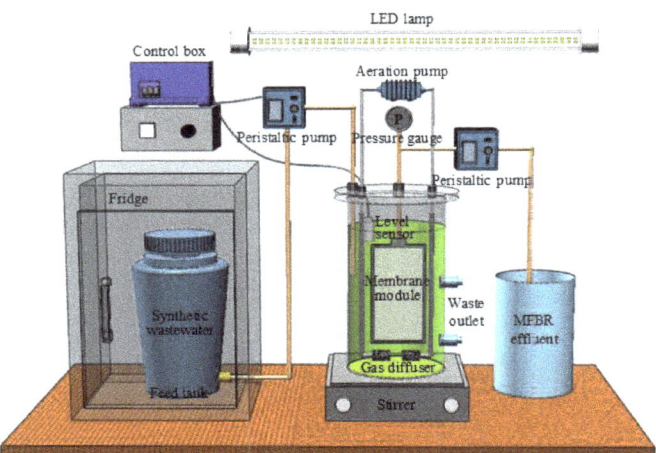

Figure 1. Experimental setup of the MPBR.

Table 1. Operating conditions and membrane module properties of the MPBR system.

Parameters	Value
Working volume	9.64 L
Aeration rate	7.5 ± 0.03 L/min
Illumination intensity	8400 lux
SRT	50 d
HRT	2.9 ± 0.1 d
Operating temperature	25.2 ± 1.0 °C
Operating pH	6.81 ± 0.66
Membrane type	Flat sheet
Membrane material	Polyvinylidene fluoride (PVDF)
Effective surface area	0.03 m^2
Pore size	0.1 µm
Membrane flux	7.30 ± 0.34 L/(h·m^2)

Simulated municipal wastewater was utilized as the feed in this study. The compositions of the synthetic influent are displayed in Table 2. The concentrations of glucose, nitrogen, and phosphorus were determined according to the corresponding concentration in the medium-strength municipal wastewater. The concentrations of trace elements were the same as those in the modified MSM medium for microalgae pre-cultivation. The feed was stored in a fridge at 5 °C and pumped by a peristaltic pump that was controlled by a level sensor (Madison Co., New York, USA). Another peristaltic pump was used to intermittently suck the permeate, using an operating mode of 3 min on and 2 min off.

Table 2. Composition of synthetic municipal wastewater.

Reagents	Element Concentration (mg/L)
Glucose	500
EDTA disodium salt dehydrate	64
NH$_4$Cl	50 (N)
K$_2$HPO$_4$	3.55 (P)
KH$_2$PO$_4$	5.9 (P)
CaCl$_2$·2H$_2$O	3.0 (Ca)
MnCl$_2$·4H$_2$O	0.4 (Mn)
CoCl$_2$·6H$_2$O	0.1 (Co)

Table 2. *Cont.*

Reagents	Element Concentration (mg/L)
$FeSO_4 \cdot 7H_2O$	1.0 (Fe)
$Na_2MoO_4 \cdot 2H_2O$	0.47 (Mo)
$ZnSO_4 \cdot 7H_2O$	2.0 (Zn)
$CuSO_4 \cdot 5H_2O$	0.4 (Cu)
H_3BO_3	2.0 (B)
$MgSO_4 \cdot 7H_2O$	6.0 (Mg)

2.2. Extraction and Analysis of Chlorophyll-a

The extraction and analysis of chlorophyll-a followed the method used by Nautiyal, Subramanian [27]. A known content of microalgae sediments was obtained by centrifugation at 8000× *g* for 10 min and was then resuspended into a certain volume of methanol. After that, the obtained microalgal suspension was immersed in a 60 °C water bath for 30 min and then cooled down to room temperature. The chlorophyll-a concentration in the solvent was spectrophotometrically determined using a visible spectrophotometer (DR2800, Hach) at three wavelengths. The content of chlorophyll-a in unit mass microalgae can be calculated using the following equation:

$$\text{Chlorophyll-a/MLSS(mg/g)} = (16.29(A^{665.2} - A^{750}) - 8.54(A^{652} - A^{750}))/\text{MLSS} \quad (1)$$

where A^{750}, $A^{665.2}$, and A^{652} represent the absorbance at 750, 665.2, and 652 nm, respectively, and MLSS is the mixed liquor suspended solids of the microalgae.

2.3. PSD Analysis and Microscopic Observation

The PSD of the microalgae suspension was measured using a Malvern Mastersizer 2000 instrument (Worcestershire, West Midlands, UK) with a detection range of 0.02–2000 μm. Each sample was automatically measured in triplicate by the machine. This measurement was conducted one to two times per week.

The micromorphology of the microalgae was observed using an inverted optical microscope (Olympus IX51). For each sample, at least 30 images were randomly taken using a digital camera connected to the microscope.

2.4. Soluble Microbial Products (SMP) and Extracellular Polymeric Substances (EPS) Measurement

The SMP sample was collected from the microalgae suspension through centrifugation at 4000× *g* for 10 min and successive filtration through a 0.45 μm membrane. The bound EPS of the microalgae was extracted through a cation exchange resin (CER) (Dowex™ Marathon™ C, Na+ form, Sigma-Aldrich, Bellefonte, PA, USA) method [28]. Details regarding the extraction process can be found in our previous publication [29]. The total content of SMP and bound EPS were normalized as the sum of protein and carbohydrates, which can be determined colorimetrically according to Lowry's method and Gaudy's method, respectively [30,31].

2.5. Other Analysis

The pH and temperature of the suspension were measured using a pH meter (pH 700, Oakton, Charleston, SC, USA) and thermometer, respectively. The trans-membrane pressure (TMP) was monitored using a pressure gage. The growth of the microalgae was monitored through the determination of mixed liquor suspended solids (MLSS). The determination of MLSS was conducted following the standard method [32]. Total nitrogen (TN) and total phosphorus (TP) were measured following the methods previously adopted [26]. The analyses were conducted twice for each sample, and the average values were reported.

3. Results

3.1. Biomass Concentration and Chlorophyll-a Content

The contents of the microalgal biomass (represented by MLSS) and chlorophyll-a/MLSS are shown in Figure 2. During the period before the decay occurred, the microalgal biomass in the MPBR gradually increased and reached 3.48 g/L on day 22. Unlike the biomass concentration, the chlorophyll-a/MLSS content remained relatively stable and the average value was 34.44 ± 3.23 g/L. The gradual increase in microalgal biomass and relatively stable content of chlorophyll-a suggested that MPBR operated in a stable manner in the first 22 days. When the microalgae decayed on day 23, the biomass concentration and chlorophyll-a/MLSS dropped rapidly from 3.48 to 1.94 g/L and 34.56 to 10.71 mg/g, respectively, within 5 days. The rapid decrease in biomass concentration and chlorophyll-a/MLSS indicated that a large number of microalgae died in a short time.

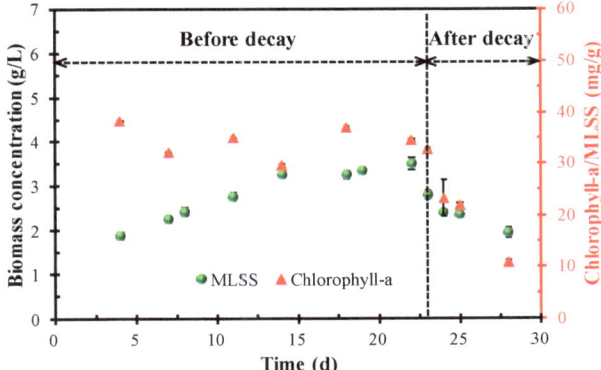

Figure 2. Variation in biomass concentration and chlorophyll-a/MLSS in the MPBR.

The occurrence of microalgae decay suggested that a high SRT greatly impacted the continued long-term operation of MPBR. In this system, SRT directly affects the biomass concentration, which has a trade-off relationship with the effective light transmittance (i.e., a high biomass concentration corresponds to a low effective light transmittance) [15,17]. In the current work, a high SRT of 50 d was adopted, and the higher initial biomass concentration enabled the system to achieve a high MLSS of 3.48 g/L in a short time. With the increase in biomass concentration, the intraspecific competition among the microalgae became increasingly fierce because of the significant decrease in light transmittance. On the other hand, untreated municipal wastewater was used as the influent. Unlike the secondary effluent in previous studies, the organic matter in the municipal wastewater will lead to the growth of bacteria, which will enhance the stress effect on microalgae growth [33,34]. Because of the above two reasons, a large number of microalgae died, which led to a significant reduction in the biomass and chlorophyll-a/MLSS in the MPBR system.

3.2. Nutrients Removal

Figure 3 shows the TN and TP concentrations in the feed and permeate. The real TN and TP in the feed were 46.7 ± 4.4 and 9.5 ± 0.4 mg/L, respectively. Under normal operating conditions (before the occurrence of decay), MPBR is a promising technique that can effectively remove TN and TP from the wastewater, although it requires a period of adaptation. In this study, the lowest TN and TP concentrations in the permeate were 10.9 and 3.2 mg/L, respectively, corresponding to the highest removal efficiency of 76.7% and 66.2%, respectively. However, once the decay occurred, both the TN and TP concentrations in the permeate significantly rose and even exceeded that of the feed. This result indicated that the released cytoplasm as a result of the microalgae decomposition would severely degrade the permeate quality. Although the microalgal decay phenomenon has seldom been studied in MPBR systems, such a phenomenon has been widely reported in natural systems such as lakes [35–38]. In addition, two days after the decay occurred, the TN and TP concentration in the permeate gradually decreased, suggesting that the destroyed MPBR system can self recover and the deteriorated permeate quality can also gradually improve. However, such a recovery process needs one week or even longer. Therefore, from the biological treatment performance, microalgae decay undoubtedly should be avoided and preferably prevented in advance in the practical operation and maintenance of MPBR.

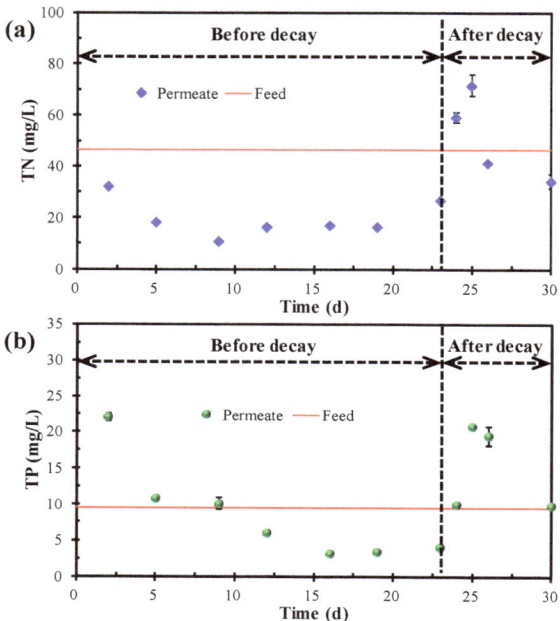

Figure 3. Variation in (**a**) TN and (**b**) TP in the feed and permeate.

3.3. Microalgae Properties and Membrane Fouling
3.3.1. PSD and Micromorphology

Figure 4 shows the PSD of the microalgae suspension before and after the occurrence of decay. It can be seen that the suspended flocs before the occurrence of decay had a double-peak shape, corresponding to a sharp primary peak ranging from 10–100 μm and a weak secondary peak in the range of 1–10 μm. In comparison, the microalgae liquor after the occurrence of decay had a perfect unimodal shape; the peak in the range of 1–10 μm disappeared and the proportion of the flocs in the range of 10–100 μm increased. The microscopic morphology of the microalgae in Figure 5 further demonstrates the variation in PSD for the microalgae suspension before and after the occurrence of decay. As shown in Figure 5, *Chlorella vulgaris* cells dispersed individually or combined as flocs before the occurrence of decay, while almost existing as flocs after the occurrence of decay. The PSD and microscopic observation jointly proved that decay shock had a great influence on the biological properties of the microalgae particles, especially for the small dispersed *Chlorella vulgaris* cells. Combined with the sudden decline of biomass (Figure 2) and the surge of nutrients in the effluent (Figure 3), it can be reasonably speculated that the *Chlorella vulgaris* cells in the system might be the first to decompose under the stress environment, and the released substances from the lysis could promote the aggregation of free algal cells.

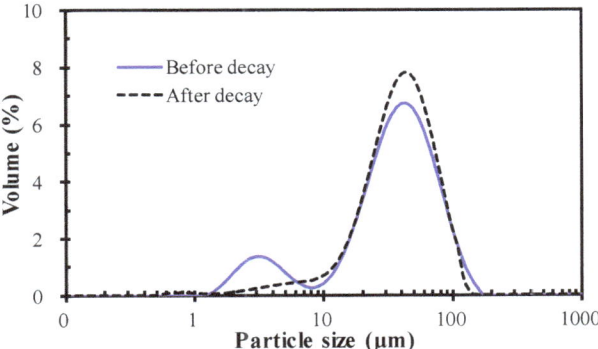

Figure 4. PSD of the microalgae in the MPBR before and after decay.

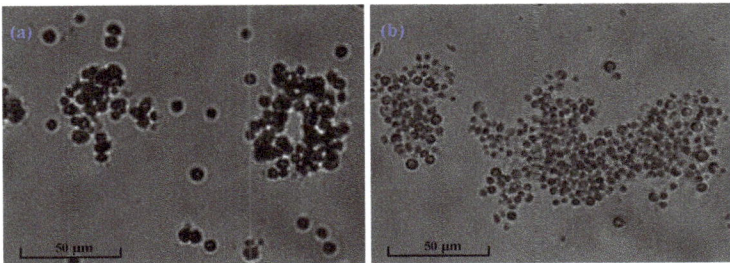

Figure 5. Microscopic morphology of the microalgae in the MPBR (**a**) before and (**b**) after decay.

For the MPBR system, particle size is very important for membrane fouling formation. It is generally believed that larger size flocs have a lower fouling potential because of their lower adhesive ability and looser cake layer that can be formed [39,40]. From this aspect, it seems that the occurrence of decay favors membrane fouling control because of the improved floc size. However, it should be noted that microalgae decay also leads to the release of the cytoplasm, which will increase the type and content of pollutants in the system and then may aggravate membrane fouling [41,42]. Therefore, the final effect of microalgae decay on membrane fouling mainly depends on the comprehensive influence degree of the above two opposite factors (increased floc size and foulants content).

3.3.2. EPS and SMP

Figure 6 compares the EPS and SMP values of the microalgal suspension before and after the occurrence of decay. It can be seen from Figure 6a that the amount of carbohydrates, proteins, and total EPS was comparable before the occurrence of decay, while gradually decreasing after the occurrence of decay. These results suggested that the microalgae flocs remained in a stable state before the occurrence of decay. However, after the decay occurred, a large number of microalgae cells died and decomposed suddenly, which led to the decomposition and corresponding decrease in EPS (from 26.18 ± 1.99 mg/g MLSS on day 23 to 11.61 ± 0.57 mg/g MLSS on day 25). Generally, EPS is considered a protective substance secreted by organisms to prevent them from being harmed in adverse environments [43,44]. As for membrane-related systems, a lower EPS is preferred because a higher EPS content would accelerate the formation of membrane fouling [45,46]. Therefore, according to the results obtained, it can be speculated that with the gradual increase in biomass concentration, the stress of photoinhibition and bacteria growth was too severe to be resisted by EPS secretion. On the other hand, it also suggests that the microalgae flocs after the occurrence of decay had better anti-fouling properties.

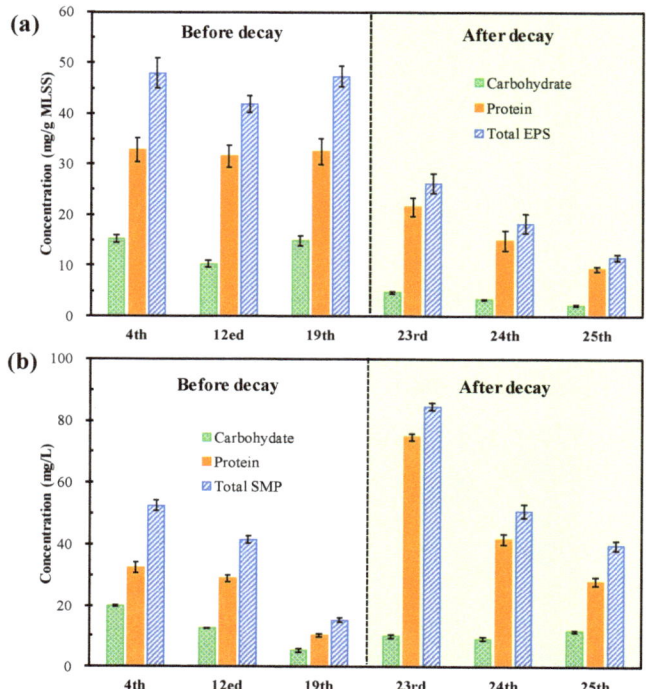

Figure 6. Comparison of the (**a**) EPS and (**b**) SMP content before and after decay.

Figure 3b shows that the contents of total SMP gradually decreased from 52.63 ± 1.66 mg/L on day 4 to 15.41 ± 0.79 mg/L on day 19 (before the occurrence of decay), suddenly sharply increased to 84.68 ± 1.14 mg/L on day 23, and then gradually decreased to 39.75 ± 1.50 mg/L on day 25 (after the occurrence of decay). The variation trend in the protein was the same as that of the total SMP both before and after decay. SMPs are generally defined as biomass-released biopolymers and EPS hydrolysis is an important source of SMPs [45,47]. As illustrated above, the sudden increase in SMP corresponded to the decrease in EPS, reasonably demonstrating the outlet of the decreased EPS. SMPs also play a vital role in membrane fouling, especially when the main form of membrane fouling is gel layer formation, the increase in SMP content will significantly promote the increase in filtration resistance [42,48–50]. Therefore, after the occurrence of decay, the variation trends in EPS and SMP exhibited the opposite effects on membrane fouling. The comprehensive effects of EPS and SMP will rely on the main format of membrane fouling in the system.

3.3.3. Membrane Fouling Performance

The variations in TMP and flux for the MPBR system are displayed in Figure 7. It can be seen that the flux remained relatively constant and the TMP gradually increased with little fluctuation during the whole experimental period. The TMP increased from 1.69 kPa to 3.72 kPa before microalgae decay and then gradually reached 4.74 kPa after microalgae decay. Overall, there is no significant increase in TMP after the occurrence of decay. As stated above, the flocs size and EPS content exhibited the opposite effect when compared with the SMP content for the membrane fouling formation. The small rise in TMP further demonstrated that membrane fouling formation is the comprehensive interaction results of various factors.

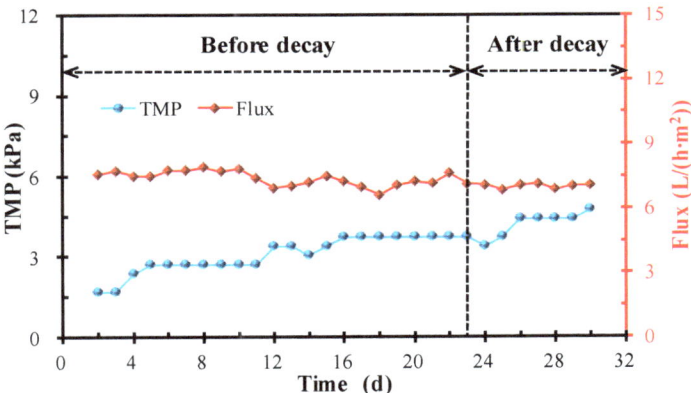

Figure 7. Variations in TMP and flux for the MPBR system.

3.4. Implications of High SRT for Long-Term Municipal Wastewater Treatment in MPBR

The occurrence of microalgae decay on the 23rd day suggests that the MPBR system cannot maintain long-term operation under a high SRT for municipal wastewater treatment. The occurrence of microalgae decay had a great impact on the MPBR system, which was reflected by the significant changes in the microalgae biomass, chlorophyll-a content, effluent quality, and the microalgae properties before and after the occurrence of decay.

Before the occurrence of decay, the microalgae biomass gradually increased, the chlorophyll-a and EPS content remained stable, and the TN and TP removal remained steady after a period of adaptation. The removal rate of TN and TP reached 10.54 and 2.14 mg/(L·d), respectively, which was comparable to or higher than that of most of the results that were previously reported [11,18,51]. From this aspect, it is feasible and promising to utilize microalgae for municipal wastewater treatment and microalgal biomass accumulation in the MPBR system under a high SRT.

However, the occurrence of microalgae decay on day 23 suggests that the above-mentioned feasibility is not always valid. That is, when the stress of the systematic environment caused by a high SRT exceeds the tolerance of the microalgae cells, the microalgae will start to die. In the current study, the high SRT and the application of municipal wastewater resulted in two main stresses on the microalgal cells. On the one hand, under high SRT operating conditions, the concentration of biomass gradually increased, which had a significant impact on light transmission. Ma et al. pointed out that all the light spectra attenuate exponentially with the light path based on a modified Cornet model for light transmission in the microalgal suspension [52]. In addition, the higher the microalgae concentration in the system, the faster the light attenuates at the same light path distance [52]. In this study, because of the separation effect of the membrane, the microalgae concentration (3.48 g/L) was very high compared with previous studies [14,18,22,53]. As a result, the photo masking effect was serious in the system. The severe insufficient light apparently will lead to fierce competition for light among the microalgae. On the other hand, unlike secondary effluents often reported in the literature, municipal wastewater has medium-strength organic matter, which provides a breeding ground for bacterial growth. In fact, the relationship between microalgae and bacteria is complex. They are cooperative and competitive [33,34]. Although suitable bacteria can provide CO_2 for microalgae and thus facilitate the growth of microalgae, bacteria can also secrete toxic substances that limit the growth of microalgae [37]. Moreover, the growth of bacteria would preempt the growth space of microalgae and enhance the photo masking effect, which indirectly reduces the light transmission rate and enhances light competition among the microalgae. In this work, the microalgae started to die when the biomass content increased to 3.48 g/L. This indicated that the stress of the microalgal cells had reached the limit, which eventually led to the lysis of a large number of microalgal cells. Therefore, for MPBR systems operating at a high SRT, there should be a critical biomass concentration above which microalgae decay would occur. If the biomass can be effectively removed and always maintained below the critical value, then the total environmental stress can be controlled within the tolerance range of the microalgae, and the long-term stable operation can be maintained in the MPBR system.

After the decay occurred, the microalgae biomass, chlorophyll-a content, and nutrients removal decreased. From the perspective of biological performance, the occurrence of decay will not only reduce the biomass accumulation, but also reduce the effluent quality. The recovery of effluent quality required one week or even longer, and thus must be prevented in practice. Otherwise, the unsatisfactory effluents will enter the water body and cause problems such as eutrophication. On the other hand, from the perspective of membrane fouling, the particles became larger and the EPS content decreased, indicating that the microalgae flocs after decay had a better antifouling performance. However, the increased SMP suggested that more colloid-like substances were released into the system because of the occurrence of decay, which is beneficial to membrane fouling formation. As membrane fouling is the result of the comprehensive interactions of various factors, no significant TMP increase was observed after the occurrence of decay (Figure 7).

The above results provide some implications for municipal wastewater treatment and membrane fouling control in MPBR systems. According to the results in this work, a critical biomass concentration above which microalgae decay would occur exists. The MPBR system cannot maintain long-term operation under a high SRT for municipal wastewater treatment because the biomass concentration that can be achieved under a high SRT is too high. Therefore, to avoid microalgae decay, periodic biomass removal is required to control the environmental stress within the tolerance range of the microalgae. In addition, for MPBR, microalgae cells of a small size can more easily adhere to the membrane than sludge to form a filter cake layer, which will lead to serious membrane fouling problems [39]. Based on the above results, if a method or operating conditions can help flocculate free microalgal cells into flocs and prevent the increase in SMP content, the problem of membrane fouling can be significantly reduced. As a result, the subsequent enrichment and collection of microalgal biomass will also be facilitated because of the enlarged floc size.

It should be noted that, despite the valuable inspiration provided by this work, a distinctive limitation of this study is the design of a single experimental run. Although the single experimental run setting is not a special case and has been extensively applied in previous studies regarding MPBR [20,22,23], a duplicate design would be better and can avoid misinterpretation. In addition, too much biomass and bacterial development were speculated as to the potential reasons for microalgae decay. Apparently, the speculation was based on previous literature and the results obtained in this study. Nevertheless, their respective effects were not independently demonstrated in the current work. Therefore, further studies can be conducted on the following aspects in the future. For instance, finding the optimal HRT value by setting up experimental groups with HRT as a single variable. Afterwards, under the optimized HRT, several MPBRs can be operated in parallel with the long-term treatment of municipal wastewater. The effect of biomass concentration on the occurrence of decay can be confirmed by setting different SRTs. Furthermore, as organic matter could provide a breeding ground for bacterial growth, two types of wastewater, with and without organic matter, can be used as feed to verify the effects of bacterial development on the occurrence of decay.

4. Conclusions

In the current study, a lab-scale submerged MPBR was operated to treat synthetic municipal wastewater at a long SRT of 50 d. It was found that serious microalgae decay occurred on day 23, which had a great impact on the MPBR performance and the biological properties of the microalgae particles. A comparison of the microalgae properties showed that the biomass concentration, chlorophyll-a/MLSS, and effluent quality sharply decreased. However, the floc size increased, the EPSs content decreased, and the SMPs content increased. This suggests that the biological performance of the MPBR deteriorated while the antifouling performance of the microalgae flocs improved. However, the filtration resistance had no significant increase due to the comprehensive interactions of the floc size, EPSs, and SMPs. The occurrence of microalgae decay suggested that the MPBR system cannot maintain long-term operation under a high SRT for municipal wastewater treatment. The occurrence of decay was attributed to the double stresses from the light shading and intraspecific competition under a high biomass concentration. As a result, to avoid microalgae decay, periodic biomass removal is suggested to control the environmental stress within the tolerance range of the microalgae. Further studies are required in order to explore the underlying mechanism of the occurrence of decay in the future.

Author Contributions: Methodology, H.Z. and N.C.R.; formal analysis, S.C.; investigation, H.Z., N.C.R. and S.C.; data curation, S.C.; writing—original draft preparation, H.Z. and N.C.R.; writing—review and editing, M.Z., H.L. and B.L.; supervision, M.Z.; project administration, M.Z. and H.L.; funding acquisition, M.Z. and H.L. All authors have read and agreed to the published version of the manuscript.

Funding: This research was funded by the Natural Science Foundation of Zhejiang Province (nos. LQ21E080010 and LD21E080001), the National Natural Science Foundation of China (nos. 52000160 and 51978628), the Key Research and Development Program of Zhejiang Province (no. 2022C03069), and the Zhejiang Provincial Ten Thousand Talent Program (ZJWR0302055).

Institutional Review Board Statement: Not applicable.

Data Availability Statement: Not applicable.

Conflicts of Interest: The authors declare no conflict of interest.

References

1. Bouju, H.; Buttiglieri, G.; Malpei, F. The use of microcalorimetry to compare the biological activity of a CAS and a MBR sludge-application to pharmaceutical active compounds. *Water Sci. Technol.* **2008**, *58*, 529–535. [CrossRef]
2. Munz, G.; Gualtiero, M.; Salvadori, L.; Claudia, B.; Claudio, L. Process efficiency and microbial monitoring in MBR (membrane bioreactor) and CASP (conventional activated sludge process) treatment of tannery wastewater. *Bioresour. Technol.* **2008**, *99*, 8559–8564. [CrossRef]
3. Cicek, N. A review of membrane bioreactors and their potential application in the treatment of agricultural wastewater. *Can. Biosyst. Eng.* **2003**, *45*, 37–49.
4. Lin, H.; Gao, W.; Meng, F.; Liao, B.-Q.; Leung, K.-T.; Zhao, L.; Chen, J.; Hong, H. Membrane bioreactors for industrial wastewater treatment: A critical review. *Crit. Rev. Env. Sci. Technol.* **2012**, *42*, 677–740. [CrossRef]
5. Radjenović, J.; Petrović, M.; Barceló, D. Fate and distribution of pharmaceuticals in wastewater and sewage sludge of the conventional activated sludge (CAS) and advanced membrane bioreactor (MBR) treatment. *Water Res.* **2009**, *43*, 831–841. [CrossRef]
6. Wu, M.; Chen, Y.; Lin, H.; Zhao, L.; Shen, L.; Li, R.; Xu, Y.; Hong, H.; He, Y. Membrane fouling caused by biological foams in a submerged membrane bioreactor: Mechanism insights. *Water Res.* **2020**, *181*, 115932. [CrossRef]
7. Abinandan, S.; Shanthakumar, S. Challenges and opportunities in application of microalgae (Chlorophyta) for wastewater treatment: A review. *Renew. Sust. Energ. Rev.* **2015**, *52*, 123–132. [CrossRef]
8. Chen, G.; Zhao, L.; Qi, Y. Enhancing the productivity of microalgae cultivated in wastewater toward biofuel production: A critical review. *Appl. Energy* **2015**, *137*, 282–291. [CrossRef]
9. Razzak, S.A.; Hossain, M.M.; Lucky, R.A.; Bassi, A.S.; de Lasa, H. Integrated CO2 capture, wastewater treatment and biofuel production by microalgae culturing—A review. *Renew. Sust. Energ. Rev.* **2013**, *27*, 622–653. [CrossRef]
10. Luo, Y.; Le-Clech, P.; Henderson, R.K. Simultaneous microalgae cultivation and wastewater treatment in submerged membrane photobioreactors: A review. *Algal Res.* **2017**, *24*, 425–437. [CrossRef]
11. Gao, F.; Yang, Z.-H.; Li, C.; Wang, Y.-J.; Jin, W.-H.; Deng, Y.-B. Concentrated microalgae cultivation in treated sewage by membrane photobioreactor operated in batch flow mode. *Bioresour. Technol.* **2014**, *167*, 441–446. [CrossRef] [PubMed]
12. Maity, J.P.; Bundschuh, J.; Chen, C.-Y.; Bhattacharya, P. Microalgae for third generation biofuel production, mitigation of greenhouse gas emissions and wastewater treatment: Present and future perspectives—A mini review. *Energy* **2014**, *78*, 104–113. [CrossRef]
13. Marbelia, L.; Bilad, M.R.; Passaris, I.; Discart, V.; Vandamme, D.; Beuckels, A.; Muylaert, K.; Vankelecom, I.F.J. Membrane photobioreactors for integrated microalgae cultivation and nutrient remediation of membrane bioreactors effluent. *Bioresour. Technol.* **2014**, *163*, 228–235. [CrossRef] [PubMed]

14. Boonchai, R.; Seo, G. Microalgae membrane photobioreactor for further removal of nitrogen and phosphorus from secondary sewage effluent. *Korean. J. Chem. Eng.* **2015**, *32*, 2047–2052. [CrossRef]
15. Xu, M.; Li, P.; Tang, T.; Hu, Z. Roles of SRT and HRT of an algal membrane bioreactor system with a tanks-in-series configuration for secondary wastewater effluent polishing. *Ecol. Eng.* **2015**, *85*, 257–264. [CrossRef]
16. Gao, F.; Li, C.; Yang, Z.-H.; Zeng, G.-M.; Feng, L.-J.; Liu, J.-Z.; Liu, M.; Cai, H.-W. Continuous microalgae cultivation in aquaculture wastewater by a membrane photobioreactor for biomass production and nutrients removal. *Ecol. Eng.* **2016**, *92*, 55–61. [CrossRef]
17. Honda, R.; Teraoka, Y.; Noguchi, M.; Yang, S. Optimization of Hydraulic Retention Time and Biomass Concentration in Microalgae Biomass Production from Treated Sewage with a Membrane Photobioreactor. *J. Water Environ. Technol.* **2017**, *15*, 1–11. [CrossRef]
18. Luo, Y.; Le-Clech, P.; Henderson, R.K. Assessment of membrane photobioreactor (MPBR) performance parameters and operating conditions. *Water Res.* **2018**, *138*, 169–180. [CrossRef]
19. Zhang, M.; Yao, L.; Maleki, E.; Liao, B.-Q.; Lin, H. Membrane technologies for microalgal cultivation and dewatering: Recent progress and challenges. *Algal Res.* **2019**, *44*, 101686. [CrossRef]
20. Van Thuan, N.; Thi Thanh Thuy, N.; Hong Hai, N.; Nguyen, N.C.; Bui, X.-T. Influence of microalgae retention time on biomass production in membrane photobioreactor using human urine as substrate. *Vietnam, J. Sci. Technol. Eng.* **2018**, *60*, 5. [CrossRef]
21. Zhang, M.; Leung, K.-T.; Lin, H.; Liao, B. Effects of solids retention time on the biological performance of a novel microalgal-bacterial membrane photobioreactor for industrial wastewater treatment. *J. Environ. Chem. Eng.* **2021**, *9*, 105500. [CrossRef]
22. Xu, M.; Bernards, M.; Hu, Z. Algae-facilitated chemical phosphorus removal during high-density Chlorella emersonii cultivation in a membrane bioreactor. *Bioresour. Technol.* **2014**, *153*, 383–387. [CrossRef] [PubMed]
23. Praveen, P.; Guo, Y.; Kang, H.; Lefebvre, C.; Loh, K.-C. Enhancing microalgae cultivation in anaerobic digestate through nitrification. *Chem. Eng. J.* **2018**, *354*, 905–912. [CrossRef]
24. Chen, X.; Li, Z.; He, N.; Zheng, Y.; Li, H.; Wang, H.; Wang, Y.; Lu, Y.; Li, Q.; Peng, Y. Nitrogen and phosphorus removal from anaerobically digested wastewater by microalgae cultured in a novel membrane photobioreactor. *Biotechnol. Biofuels* **2018**, *11*, 190. [CrossRef]
25. Chang, H.; Quan, X.; Zhong, N.; Zhang, Z.; Lu, C.; Li, G.; Cheng, Z.; Yang, L. High-efficiency nutrients reclamation from landfill leachate by microalgae Chlorella vulgaris in membrane photobioreactor for bio-lipid production. *Bioresour. Technol.* **2018**, *266*, 374–381. [CrossRef]
26. Zhang, M.; Leung, K.-T.; Lin, H.; Liao, B. The biological performance of a novel microalgal-bacterial membrane photobioreactor: Effects of HRT and N/P ratio. *Chemosphere* **2020**, *261*, 128199. [CrossRef]
27. Nautiyal, P.; Subramanian, K.A.; Dastidar, M.G. Production and characterization of biodiesel from algae. *Fuel Process. Technol.* **2014**, *120*, 79–88. [CrossRef]
28. Frølund, B.; Palmgren, R.; Keiding, K.; Nielsen, P.H. Extraction of extracellular polymers from activated sludge using a cation exchange resin. *Water Res.* **1996**, *30*, 1749–1758. [CrossRef]
29. Zhang, M.; Leung, K.-T.; Lin, H.; Liao, B. Membrane fouling in a microalgal-bacterial membrane photobioreactor: Effects of P-availability controlled by N:P ratio. *Chemosphere* **2021**, *282*, 131015. [CrossRef]
30. Lowry, O.H.; Rosebrough, N.J.; Farr, A.L.; Randall, R.J. Protein measurement with the folin phenol reagent. *J. Biol. Chem.* **1951**, *193*, 265–275. [CrossRef]
31. Gaudy, A.F. Colorimetric determination of protein and carbohydrate. *Ind. Water Wastes* **1962**, *7*, 17–22.
32. APHA. *Standard Methods for the Examination of Water and Wastewater*, 20th ed.; American Public Health Association; American Water Works Association; Water Environmental Federation: Washington, DC, USA, 2005.
33. Liu, J.; Wu, Y.; Wu, C.; Muylaert, K.; Vyverman, W.; Yu, H.-Q.; Muñoz, R.; Rittmann, B. Advanced nutrient removal from surface water by a consortium of attached microalgae and bacteria: A review. *Bioresour. Technol.* **2017**, *241*, 1127–1137. [CrossRef] [PubMed]
34. Muñoz, R.; Guieysse, B. Algal–bacterial processes for the treatment of hazardous contaminants: A review. *Water Res.* **2006**, *40*, 2799–2815. [CrossRef] [PubMed]
35. Han, C.; Ding, S.; Yao, L.; Shen, Q.; Zhu, C.; Wang, Y.; Xu, D. Dynamics of phosphorus–iron–sulfur at the sediment–water interface influenced by algae blooms decomposition. *J. Hazard. Mater.* **2015**, *300*, 329–337. [CrossRef]
36. Zhao, Y.; Zhang, Z.; Wang, G.; Li, X.; Ma, J.; Chen, S.; Deng, H.; Annalisa, O.-H. High sulfide production induced by algae decomposition and its potential stimulation to phosphorus mobility in sediment. *Sci. Total Environ.* **2019**, *650*, 163–172. [CrossRef] [PubMed]
37. Li, P.; Yao, Y.; Lian, J.; Ma, C. Effect of thermal stratified flow on algal blooms in a tributary bay of the Three Gorges reservoir. *J. Hydrol.* **2021**, *601*, 126648. [CrossRef]
38. Yao, Y.; Li, D.; Chen, Y.; Han, X.; Wang, G.; Han, R. High-resolution characteristics and mechanisms of endogenous phosphorus migration and transformation impacted by algal blooms decomposition. *Sci. Total Environ.* **2022**, *820*, 152907. [CrossRef]
39. Shen, L.-G.; Lei, Q.; Chen, J.-R.; Hong, H.-C.; He, Y.-M.; Lin, H.-J. Membrane fouling in a submerged membrane bioreactor: Impacts of floc size. *Chem. Eng. J.* **2015**, *269*, 328–334. [CrossRef]
40. Liu, J.; Zhao, Y.; Fan, Y.; Yang, H.; Wang, Z.; Chen, Y.; Tang, C.Y. Dissect the role of particle size through collision-attachment simulations for colloidal fouling of RO/NF membranes. *J. Membr. Sci.* **2021**, *638*, 119870. [CrossRef]
41. Zhang, M.; Hong, H.; Lin, H.; Shen, L.; Yu, H.; Ma, G.; Chen, J.; Liao, B.-Q. Mechanistic insights into alginate fouling caused by calcium ions based on terahertz time-domain spectra analyses and DFT calculations. *Water Res.* **2018**, *129*, 337–346. [CrossRef]

42. Teng, J.; Zhang, M.; Leung, K.-T.; Chen, J.; Hong, H.; Lin, H.; Liao, B.-Q. A unified thermodynamic mechanism underlying fouling behaviors of soluble microbial products (SMPs) in a membrane bioreactor. *Water Res.* **2019**, *149*, 477–487. [CrossRef] [PubMed]
43. Siddharth, T.; Sridhar, P.; Vinila, V.; Tyagi, R.D. Environmental applications of microbial extracellular polymeric substance (EPS): A review. *J. Environ. Manag.* **2021**, *287*, 112307. [CrossRef] [PubMed]
44. Martis, B.S.; Mohan, A.K.; Chiplunkar, S.; Kamath, S.; Goveas, L.C.; Rao, C.V. Bacterium isolated from coffee waste pulp biosorps lead: Investigation of EPS mediated mechanism. *Curr. Res. Microb. Sci.* **2021**, *2*, 100029. [CrossRef] [PubMed]
45. Lin, H.; Zhang, M.; Wang, F.; Meng, F.; Liao, B.-Q.; Hong, H.; Chen, J.; Gao, W. A critical review of extracellular polymeric substances (EPSs) in membrane bioreactors: Characteristics, roles in membrane fouling and control strategies. *J. Membr. Sci.* **2014**, *460*, 110–125. [CrossRef]
46. Meng, F.; Chae, S.-R.; Drews, A.; Kraume, M.; Shin, H.-S.; Yang, F. Recent advances in membrane bioreactors (MBRs): Membrane fouling and membrane material. *Water Res.* **2009**, *43*, 1489–1512. [CrossRef]
47. Liu, T.; Zheng, X.; Tang, G.; Yang, X.; Zhi, H.; Qiu, X.; Li, X.; Wang, Z. Effects of temperature shocks on the formation and characteristics of soluble microbial products in an aerobic activated sludge system. *Process Saf. Environ. Prot.* **2022**, *158*, 231–241. [CrossRef]
48. Tian, Y.; Li, Z.; Ding, Y.; Lu, Y. Identification of the change in fouling potential of soluble microbial products (SMP) in membrane bioreactor coupled with worm reactor. *Water Res.* **2013**, *47*, 2015–2024. [CrossRef]
49. Long, Y.; Yu, G.; Dong, L.; Xu, Y.; Lin, H.; Deng, Y.; You, X.; Yang, L.; Liao, B.-Q. Synergistic fouling behaviors and mechanisms of calcium ions and polyaluminum chloride associated with alginate solution in coagulation-ultrafiltration (UF) process. *Water Res.* **2021**, *189*, 116665. [CrossRef]
50. Pan, Z.; Zeng, B.; Lin, H.; Teng, J.; Zhang, H.; Hong, H.; Zhang, M. Fundamental thermodynamic mechanisms of membrane fouling caused by transparent exopolymer particles (TEP) in water treatment. *Sci. Total Environ.* **2022**, *820*, 153252. [CrossRef]
51. Gao, F.; Peng, Y.-Y.; Li, C.; Cui, W.; Yang, Z.-H.; Zeng, G.-M. Coupled nutrient removal from secondary effluent and algal biomass production in membrane photobioreactor (MPBR): Effect of HRT and long-term operation. *Chem. Eng. J.* **2018**, *335*, 169–175. [CrossRef]
52. Ma, S.; Zeng, W.; Huang, Y.; Zhu, X.; Xia, A.; Zhu, X.; Liao, Q. Revealing the synergistic effects of cells, pigments, and light spectra on light transfer during microalgae growth: A comprehensive light attenuation model. *Bioresour. Technol.* **2022**, *348*, 126777. [CrossRef] [PubMed]
53. Sheng, A.L.K.; Bilad, M.R.; Osman, N.B.; Arahman, N. Sequencing batch membrane photobioreactor for real secondary effluent polishing using native microalgae: Process performance and full-scale projection. *J. Clean. Prod.* **2017**, *168*, 708–715. [CrossRef]

Article

PAC-UF Process Improving Surface Water Treatment: PAC Effects and Membrane Fouling Mechanism

Tian Li [1,2,3], Hongjian Yu [2], Jing Tian [2], Junxia Liu [4,*], Tonghao Yuan [2], Shaoze Xiao [2], Huaqiang Chu [1,2,3,*] and Bingzhi Dong [1,2,3]

[1] Key Laboratory of Yangtze River Water Environment, Ministry of Education, College of Environmental Science and Engineering, Tongji University, Shanghai 200092, China; litian001@tongji.edu.cn (T.L.); dbz77@tongji.edu.cn (B.D.)
[2] State Key Laboratory of Pollution Control and Resource Reuse, College of Environmental Science and Engineering, Tongji University, Shanghai 200092, China; 1853855@tongji.edu.cn (H.Y.); tianjing1605@163.com (J.T.); 1952483@tongji.edu.cn (T.Y.); shaoze_xiao@163.com (S.X.)
[3] Shanghai Institute of Pollution Control and Ecological Security, Tongji University, Shanghai 200092, China
[4] School of Civil and Transportation Engineering, Guangdong University of Technology, Guangzhou 510006, China
* Correspondence: whjunxia@163.com (J.L.); chuhuaqiang@tongji.edu.cn (H.C.); Tel.: +86-20-39322515 (J.L.); +86-21-65985811 (H.C.)

Citation: Li, T.; Yu, H.; Tian, J.; Liu, J.; Yuan, T.; Xiao, S.; Chu, H.; Dong, B. PAC-UF Process Improving Surface Water Treatment: PAC Effects and Membrane Fouling Mechanism. *Membranes* **2022**, *12*, 487. https://doi.org/10.3390/membranes12050487

Academic Editor: Hongjun Lin

Received: 27 March 2022
Accepted: 26 April 2022
Published: 29 April 2022

Publisher's Note: MDPI stays neutral with regard to jurisdictional claims in published maps and institutional affiliations.

Copyright: © 2022 by the authors. Licensee MDPI, Basel, Switzerland. This article is an open access article distributed under the terms and conditions of the Creative Commons Attribution (CC BY) license (https://creativecommons.org/licenses/by/4.0/).

Abstract: In this study, the water purification effect and membrane fouling mechanism of two powdered activated carbons (L carbon and S carbon) enhancing Polyvinylidene Fluoride (PVDF) ultrafiltration (UF) membranes for surface water treatment were investigated. The results indicated that PAC could effectively enhance membrane filtration performance. With PAC addition, organic removal was greatly enhanced compared with direct UF filtration, especially for small molecules, i.e., the S-UF had an additional 25% removal ratio of micro-molecule organics than the direct UF. The S carbon with the larger particle size and lower specific surface area exhibited superior performance to control membrane fouling, with an operation duration of S-UF double than the direct UF. Therefore, the particle size and pore structure of carbon are the two key parameters that are essential during the PAC-UF process. After filtration, acid and alkaline cleaning of UF was conducted, and it was found that irreversible fouling contributed the most to total filtration resistance, while the unrecoverable irreversible resistance ratio with acid cleaning was greater than that with alkaline cleaning. With PAC, irreversible UF fouling could be relieved, and thus, the running time could be extended. In addition, the membrane foulant elution was analyzed, and it was found to be mainly composed of small and medium molecular organic substances, with 12% to 21% more polysaccharides than proteins. Finally, the hydrophilicity of the elution was examined, and it was observed that alkaline cleaning mainly eluted large, medium, and small molecules of hydrophilic and hydrophobic organic matter, while acid cleaning mainly eluted small molecules of hydrophilic organic matter.

Keywords: powdered activated carbon; ultrafiltration; particle size; cleaning; fouling

1. Introduction

Powdered activated carbon (PAC) adsorption is a powerful and easily adjustable technology due to the effective removal of many contaminants [1–3]. The process of PAC adsorption to remove aquatic contaminants from natural water can be described with a three-step transition, i.e., from water to carbon, then to the carbon surface, and finally to the binding sites [4].

Membrane filtration has been identified as a safe barrier to microorganisms, suspended particles, and colloids through size exclusion, and ultrafiltration (UF) can be such a way to remove contaminants from surface water for its compactness, easy automation, and high removal ratio of turbidity [5–7]. However, membrane fouling is an urgent problem in membrane technology applications for water and wastewater treatment [6,8–10]. Hybrid

PAC-microfiltration/ultrafiltration (MF/UF) has become an emerging water treatment technology for the thermos dynamically unstable surface and widely commercial availability of PAC [1,11]. Meanwhile, by reducing acid-base interaction energy, PAC can also control membrane fouling [12]. In the PAC-UF/MF process, PAC can effectively relieve membrane fouling since PAC can reduce the natural organic matter (NOM) deposition on the membrane surface or in the pores, thus extending the membrane filtration cycle and even enhancing organic matter adsorption and removal more than two-fold [13,14]. However, there are still some contradictory results regarding PAC influence on membrane filtration performance. Some researchers found that the higher the permeate flux, the longer the filtration duration or the lower the frequency of chemical washing [15,16]. In contrast, others reported that PAC-UF exhibited a similar flux to UF, and even flux decline occurred [17–19]. In the study of Shao et al., it was found that external fouling was caused by the deposition of biological PAC on the membrane surface, and during the operation of PAC-MBR, low flux and effective physical cleaning protocols were needed [20].

The effect of PAC on membrane fouling can be attributed to the membrane characteristics of hydrophobicity. Some authors have reported that PAC could reduce the flux decline of hydrophilic membranes to some extent, but it has almost no influence on the flux of the hydrophobic membranes [21,22]. Meanwhile, PAC has a mechanical effect on the performance of the membrane, as it is not usually separated in front of the membrane process. In other words, they influence the membrane process through the adsorption of solutes as well as through their properties as geometric bodies [18].

Since the different types of PAC with various properties can be responsible for the operation results of the PAC-UF system, the objective of this study is to investigate the operation process and the fouling control effect of two different types of PAC with different particle sizes and pore structures on the UF filtration process, especially for the long duration of combined operations for surface water treatment. In addition, the UF membrane cleaning effect associated with membrane resistance and elution characteristics is also analyzed systematically, such that the mechanism of PAC properties on UF performance could be elucidated. Before the detailed results are reported, the materials and methodologies are first reported in the following section.

2. Materials and Methodologies

2.1. PAC

Two commercial powdered activated carbons were used in this study, i.e., Li yuan carbon (L carbon, Fujian Yuanli Active Carbon Co., Ltd., Nanping, China) and Su carbon (S carbon, Suzhou Water Supply Co., Ltd., Suzhou, China). They were added during filtration to keep the mixed liquor suspended solids (MLSS) in the mixing reactor at about 4 g/L. At the beginning of the operation, an appropriate amount of glucose was artificially added to promote the growth of microorganisms. After each sampling, carbon was replenished to maintain the stability of the MLSS.

The two carbon size distributions were similar, mainly between 10 and 100 μm (Figure S1 and Table S1 in Supplementary Materials). The average particle size of the L carbon was smaller. The pore volume of the two types of carbons varies in different pore size ranges, but the pore volume of L carbon was generally larger than that of S carbon, especially in the pore size range of 0–15 nm (Figure S2). The specific surface area of the activated carbons was composed of micropore (d < 2.0 nm) and primary mesopore (d < 5.0 nm) (Figure S3), with L carbon much larger than S carbon. The content of acidic oxygen-containing functional groups on the surface of L carbon was higher, while the content of basic oxygen-containing functional groups on the surface of S carbon was larger (Table S2).

2.2. Source Water

The raw water was collected from Sanhaowu Lake at Tongji University, Shanghai, China. The pH, turbidity, DOC, UV_{254}, NH_4-N values of the water sample were 7.90 ± 0.06,

2.83 ± 1.18 NTU, 3.27 ± 0.63 mg/L, 0.068 ± 0.009 cm^{-1} and 0.50 ± 0.38 mg/L (n = 8), respectively.

2.3. PAC-UF Experimental Setup

Two reactors, A and B, were used in the experiment. Rector A was a direct UF filtration setup, and Reactor B was an activated carbon-reinforced setup. Both adopted immersed hollow fiber membrane filtration, with the setup diagrams shown in Figure S4.

The Polyvinylidene Fluoride (PVDF) ultrafiltration membrane used in this study has a membrane pore size of 0.05 μm and an effective surface area of about 0.12 m^2 in a single module. The membrane flux was set to 30 L/(m^2·h). The device adopted a negative suction pressure for filtration. After the initial sinking, the raw water was injected into the mixing tank with the feed water pump, and the liquid level of the mixing tank was controlled with a high-level floating ball valve. The mixing tank was connected with the submerged UF membrane tank, and the total volume of the mixed liquid of the two tanks was always maintained at about 16 L. Perforated pipe aerators were installed at the bottom of the mixing tank and the UF membrane tank. The membrane permeate was stored in a clean water tank, and excess water was discharged through the overflow pipe.

2.4. Membrane Cleaning Protocol

The contaminated membrane module was disassembled from the reactor after the experiment. In order to distinguish the source and size of reversible pollution from irreversible pollution, the physical and chemical cleaning of the contaminated membrane was carried out. In the chemical cleaning, HCl and NaOH solutions with the same mass concentration of 0.2 wt% were adopted. The cleaning setup diagram is shown in Figure S5.

The two cleaning tanks were connected in the middle, and a quantitative amount of ultrapure water (physical cleaning) or chemical cleaning liquid (chemical cleaning) was poured into the tank; after the fouled membrane module was fixed, water was supplied from pipeline 7 and discharged from pipeline 5. When the cross-flow cleaning mode was turned on, the cleaning liquid was supplied with water from pipeline 6; when the backwash mode was turned on, the water was supplied from pipeline 8. The cleaning cycle mode was automatically controlled with the Programmable Logic Controller (PLC). After these operations, the cleaning solution and membrane module were collected for subsequent analysis.

2.5. Analytical Methods

2.5.1. Extracellular Polymeric Substances (EPS)

The concentration of EPS in the water sample was quantitatively calculated by analyzing the content of carbohydrates (soluble polysaccharides) and proteins. Carbohydrates were determined with an anthrone colorimetric assay, and proteins were measured with the modified Lowry method [23,24].

2.5.2. Determination of Relative Molecular Weight Distribution

In this study, a high-performance size exclusion chromatography-UV detector-TOC detector (HPSEC-UV-TOC, with UV detector from Waters USA, and TOC detector from Shimadzu, Tokyo, Japan) was used to determine the molecular weight distribution (MW) of organic matter. The column was TSK-GEL G3000PWXL (Japan TOSOH Co., Ltd., Tokyo, Japan), measuring 7.8 mm × 300 mm, and the material was a methacrylate copolymer. The guard column was made of a TSK-GEL PWXL Guard column measuring 6.0 mm × 40 mm.

2.5.3. Three-Dimensional Excitation-Emission Matrix (EEM) Fluorescence Spectroscopy

A VARIAN Cary Eclipse fluorescence spectrophotometer (Agilent Technologies, Inc., La Jolla, CA, USA) was applied in this study. The excitation source was a xenon lamp, with an excitation wavelength of 200–400 nm, the emission wavelength of 250–550 nm, the excitation slit width of 10 nm, an emission slit width of 2 nm, and a scanning speed of

12,000 nm/min. The fluorescence intensity was reduced with decreasing pH. Therefore, the pH of the water sample was adjusted to about 7.0 before the measurement, and then tested at room temperature (25 °C) using a 1 cm fluorescent cuvette. To eliminate the effects of pure water, the blank experiment was measured with ultrapure water before scanning the sample. The resulting data was processed into contour plots using Origin 8.5 and Surfer 8.0 software (OriginLab, Northampton, MA, USA) to characterize the fluorescence information of the organics in the sample. The fluorescent region boundaries and characteristic substance types are displayed in Table S3 [25,26].

2.5.4. Separation of Hydrophilic and Hydrophobic Components

The DAX and XAD resin separation methods can be used to separate NOM into three components. DAX-8 and XAD-4 resins were adopted to adsorb and elute hydrophobic components (HPO) and transphilic components (TPI). If it was not retained by the above two resins, it was a hydrophilic component (HPI). The water sample was filtered through a 0.45 μm filter and 250 mL was collected for separation. The effluent pH was adjusted to 2.0 with a 5 mol/L HCl, and 50 mL passed through the DAX-8 and XAD-4 ion exchange columns trapped in series at a flow rate of 1.5 to 2.5 mL/min. Then, the two columns were eluted with 0.1 mol/L NaOH each at a flow rate of 0.5 to 1.0 mL/min to obtain the HPO and HPI components, respectively. The organic component that was not adsorbed through the two columns was the HPI component. After adjusting the pH of the HPO, TPI, and HPI solutions to about 7.0, the corresponding DOC, UV_{254} and MW distribution were measured.

3. Results and Discussion

3.1. Membrane Operating Conditions and Variations in Organic Matter

3.1.1. Effect of Filtration Performance and Variation in Organic Matter Removal

The effects of operating conditions (direct UF, S carbon, and L carbon) on the TMP variations are illustrated in Figure 1. It could be observed that the TMP increased slowly at different operating conditions in the initial stage of operation, and the TMP of the direct UF, L-UF (the enhanced reactor with L carbon addition was abbreviated as L-UF), and S-UF (the enhanced reactor with S carbon addition was abbreviated as S-UF) increased rapidly after 12, 16, and 26 days. The direct UF filtration, L-UF, and S-UF conditions lasted 15 d, 22 d, and 30 d, respectively, when serious membrane fouling occurred with the TMP above 70 kPa. Thus, after adding PAC, the operation cycle was prolonged, with the S-UF reactor exhibiting superior performance, which was probably attributed to that PAC could adsorb the organics in the reactor, thus reducing the UF membrane fouling.

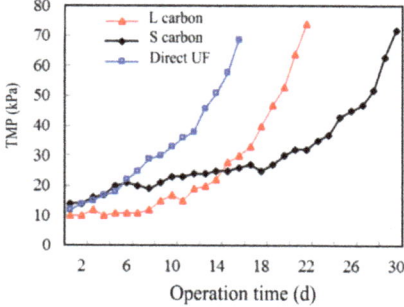

Figure 1. Effects of operating conditions on TMP performance.

Figure S6 provides the removal ratio of DOC and UV_{254} for all three operations. Clearly, after PAC addition, the removal effect of the reactor on DOC was significantly improved compared with direct UF filtration. The DOC removal ratio of direct UF filtration was only about 7%, while the removal ratio of L-UF was 55% and 47% for S-UF after

PAC strengthening. Similar to the DOC, the PAC-UF had a much higher UV_{254} removal ratio of above 50%, while the direct UF exhibited a much lower ratio of around 15%. This may be due to that PAC had a large specific surface area and could adsorb some organic matters [27]. At the same time, the PAC concentration in the mixing tank was stable at about 4 g/L, which provided an attachment carrier for the microorganisms. After the initial bioaugmentation, the microorganisms gradually adhered to the surface of the PAC to form a biofilm, which could degrade the organic matter together with the unattached microbial flocs [28].

The removal ratio of the UV_{254} in the reactor with S carbon addition was slightly higher than that with L carbon, which may be ascribed to the following reasons, i.e., the surface of the S carbon had more basic groups, and it was easier to adsorb the organic substances with ultraviolet response; the microorganisms presented in the S-UF mixing tank had a higher removal ratio of organic substances with ultraviolet response [29]. In addition, the average removal ratio of UV_{254} of L-UF and S-UF was higher than that of DOC, which may be due to that microorganisms in the mixing tank were more likely to degrade organic substances with strong absorption of ultraviolet rays containing unsaturated bonds. In addition, the DOC and UV_{254} of the effluent exhibited similar fluctuations with the raw water, extrapolating that the UF membrane itself could also intercept the organic matter, but the effect was greatly affected by the original water content [30].

3.1.2. EPS Variation

Figure 2 provides the variation of soluble EPS content in raw water, L-UF, and S-UF mixtures and Table S4 presents the ANOVA statistical analysis using Tukey's test of the comparison of EPS concentration in terms of both protein and polysaccharide from the raw water, the L-UF and S-UF reactors in the initial, middle and later stages. As shown in Figure 2, the protein content was lower than that of the polysaccharide for the three water qualities. After adding carbon, the protein and polysaccharide content of the mixture were significantly higher than the raw water during the whole operation. It could be deduced that there were active microorganisms in the mixture, and as the operation proceeded, the content of protein and polysaccharide in the mixture of L-UF and S-UF demonstrated a trend of increasing first and then decreasing. The variation of protein and polysaccharide contents in the raw water was smaller in the middle and later stages than in the mixture, surmising that there was an increase or decrease in the proliferation and metabolism of the cultured microorganisms. Meanwhile, the EPS content of the S-UF mixture was higher than that of L-UF, especially in the middle and later stages, indicating that the microbial metabolism in S-UF was more vigorous, which might result from the favored growth of microorganisms in the larger pore volume of the S carbon.

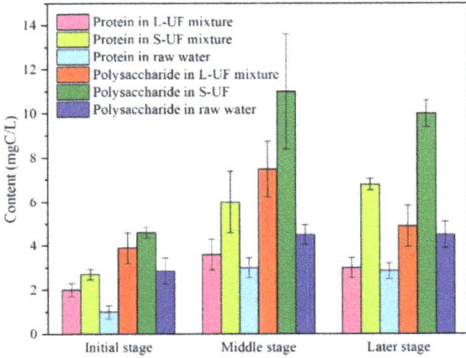

Figure 2. Variation in soluble EPS content in raw water and mixed liquor in L-UF and S-UF reactors.

In addition, the increase in the EPS content in the reactor could lead to an increment in the particle size of mixed liquor. Table S5 presents the variation in particle size of the mixed liquor during the operations. It can be observed that the particle size in both the L-UF and S-UF reactors increased when the filtration proceeded. Specifically, the size grading in the S-UF reactor is more significant, toward the right, i.e., a larger particle size (Figure S7). In the middle stage of the operation, the average particle size of the mixed liquor was over 200 μm in S-UF, which was much larger than that in the L-UF. Thus, the EPS in the S-UF was much higher than that in the L-UF (Figure 2).

3.1.3. Molecular Weight (MW) Distribution

The molecular weight distribution of the water samples in the raw water, effluent, and mixing tank were examined, and the results are shown in Figure 3. From Figure 3a,b the organic matter in raw water could be divided into three sections, i.e., macromolecule, medium molecule and small molecule. In the initial stage of operation, the peak of macromolecules in the mixture of L-UF and S-UF was very low, while it was very strong for medium and small molecules. Meanwhile, the medium and small molecular peaks of the organic matter in the reactor and effluent reduced as compared with that in the raw water, indicating the content of humic organic matter and small molecular protein in the incubator had a considerable decline. According to the analysis of the nature of the activated carbon, the decrease of L-UF was larger than that of S-UF. This may be due to that the adsorption capacity of L carbon was stronger, and at this time, the growth of L-UF microorganisms was earlier. In the middle of the operation, there was no significant change in L-UF, i.e., the peak of macromolecules was small, and the decrease of small and medium molecules was similar to that in the initial stage. However, the peak of the macromolecules was observed in the mixture of S-UF, with the intensity much larger than that in the raw water. The content of small molecules decreased significantly. This may be attributed to the fact that the microbes in S-UF proliferated and the hypermetabolism was enhanced. A large number of hydrophilic small molecules could be degraded or synthesized into macromolecular organics, but the microbial activity in L-UF showed no significant change. At the end of the operation, the reduction of organic matter in the molecular weight range of the two working conditions significantly decreased, and the removal effect was slightly worse than in the medium and early stages. The reason for the performance may be that the microbial activity decreased with the decline of the microbial organic matter, the degradation ability of the small molecule organic matter was reduced, and some sedimentation occurred.

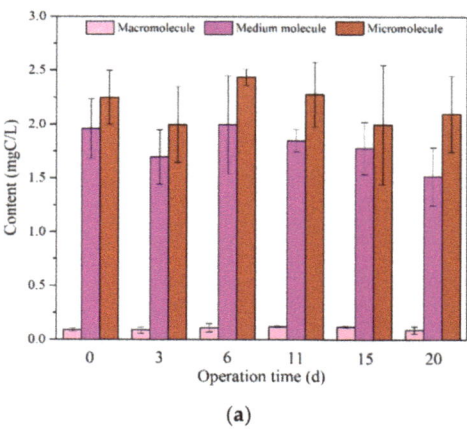

(a)

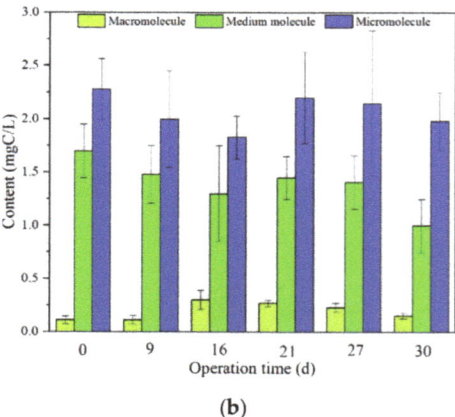

(b)

Figure 3. Cont.

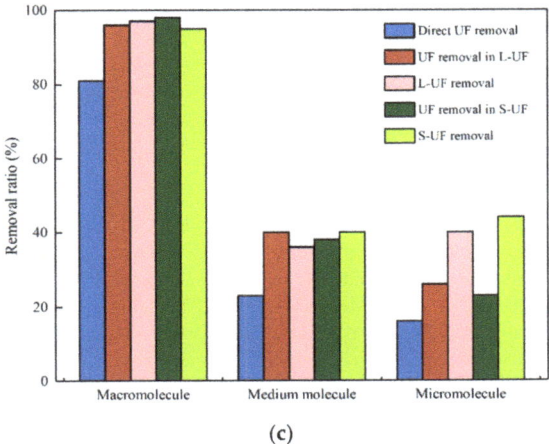

Figure 3. (**a**,**b**) MW variations of the mixed liquor in the L-UF reactor and S-UF reactor, and (**c**) removal ratio of different interval molecular weights.

The peak area of Peakfit was used to calculate the percentage of each molecular weight range. Comparing the molecular weight content of the mixture in the L-UF and S-UF mixing tanks (Figure 3a,b), it could be seen that the content of both macromolecules increased and then decreased. The variation of medium and small molecules was the same, and they were decreasing first, then increasing, and then decreasing. This phenomenon may be attributed to the microbial activity and the fluctuation of raw water quality. The medium molecular content of the S-UF mixture was lower than that of L-UF. The reason may be as follows, i.e., PAC showed an obvious adsorption effect at the initial stage of operation, especially for adsorbing small molecular substances, but the adsorption was quickly weakened, and the amount of carbon added per day was small, relying solely on PAC; the addition of PAC was beneficial to the growth of microorganisms, and in the process, microorganisms selectively consumed a higher proportion of smaller molecular weight organic matter, synthesized, and produced macromolecular substances during metabolism. The separation principle of the UF membrane made it easier to retain large molecules. Therefore, the macromolecular organic matter was accumulated in the mixing tank, and the decrease in the later stage may be due to the influence of the decrease of the microbial metabolic rate and the fluctuation of the raw water quality. It has been reported that PAC particle size significantly affects the growth environment of microorganisms in the system [31,32]. The smaller the PAC particle size, the more obvious the growth of microorganisms in the system was inhibited. As shown in Figures S1–S3, comparing with the L carbon, the S carbon has a larger average particle size and a relatively larger pore volume in 0–5 nm and over 30 nm. The larger particle size and pore volume of S carbon may provide a more favorable growth environment for microorganisms. Therefore, more microbes, more metabolism, and more synthetic macromolecules were produced in the S-UF reactor.

From the removal ratio results, as shown in Figure 3c, the direct UF filtration had a good retention effect on macromolecules, with a removal ratio of over 80%. After adding activated carbon, the removal ratio increased to 96% or even more. UF membranes have the ability to retain medium and small molecules, especially for small molecules (less than 20%). This was owing to that membrane filtration was one of the mechanical actions, which included adsorption, blockage, and mechanical retention [33]. Contaminants with particle size larger than the membrane pore size could be effectively retained [34]. Clearly, the addition of PAC enhanced the removal effect of NOM. Regardless of the carbon added, the total removal ratio of the effluent from large, medium, and small molecules relative to raw water was increased compared with direct UF filtration, and the removal ratio of small molecules was the largest. After being activated by PAC, it exceeded 40%. However, the removal ratio of small molecular substances exhibited a minimum value. This was attributed to the fact that the composition of the mixture after carbon addition had a significant reduction in small molecules, and some weakly polar and biodegradable small molecular organic substances were consumed by the activated carbon adsorption and biological metabolism [35].

3.1.4. Three-Dimensional Fluorescence (EEM) Spectroscopy

The EEM spectra of raw water, effluent, and mixed liquor in the reactor under different working conditions are shown in Figure 4. There were four peak regions, B, T, A, and C, in the Sanhaowu raw water. The protein and humic areas had an obvious response, but the A and C peaks were strong, and the B and T peaks were weak, indicating that the raw water in the filtration test contained more humic substances and protein substances. However, comparing the effluent of direct UF filtration, it could be observed that all four peaks of the effluent were weakened to varying degrees, and the protein response almost disappeared, but the humic response peak was still evident, indicating that the protein was the main pollution leading to membrane fouling. The UF membrane could hardly remove humic organic matter.

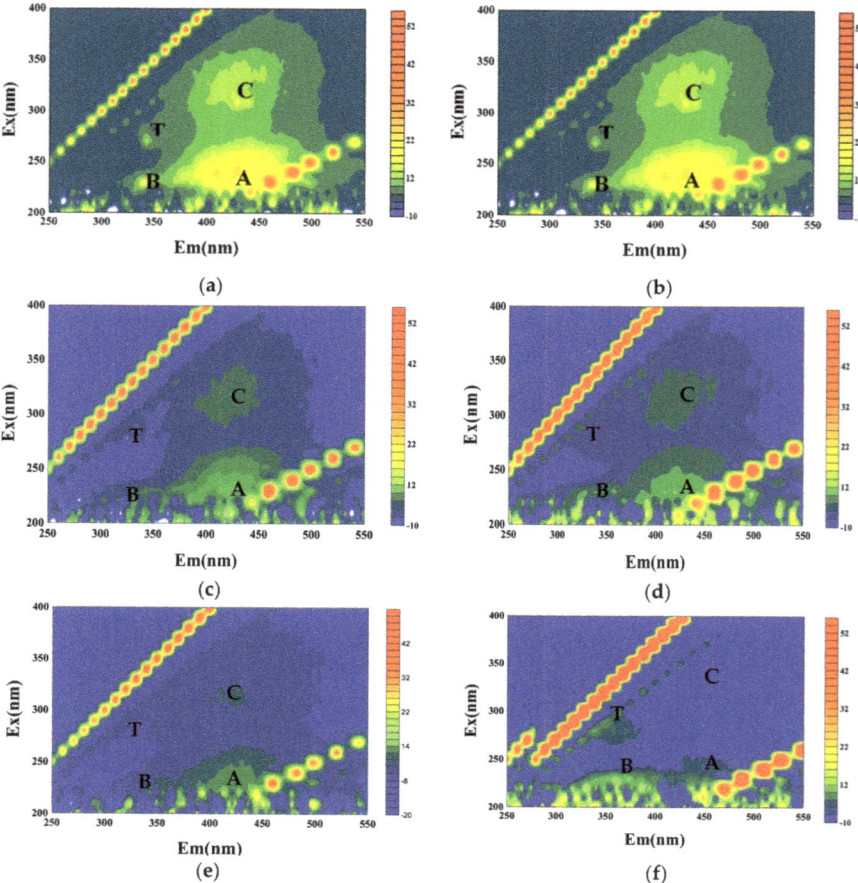

Figure 4. *Cont.*

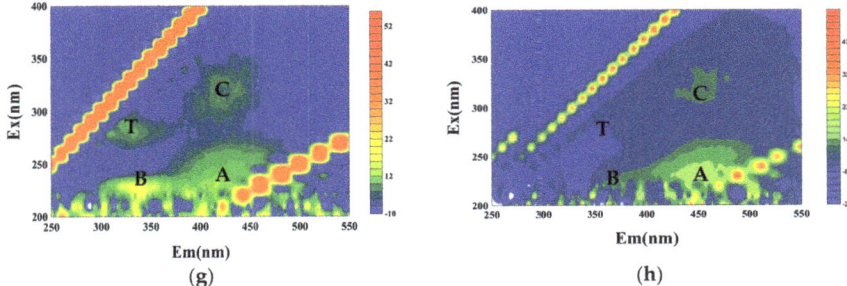

Figure 4. EEM spectra of the water sample of raw water, effluent, and mixing tank under different conditions: (**a**) raw water; (**b**) direct UF filtration, effluent; (**c**) L-UF, initial mixture; (**d**) L-UF, middle and later mixture; (**e**) L-UF, effluent; (**f**) S-UF, initial mixture; (**g**) S-UF, middle and later mixture; and (**h**) S-UF, effluent.

It can be seen from the fluorescence spectra of the L-UF mixture in different stages of operation that the four response peaks of the mixture were significantly weaker than the raw water in the initial stage of operation. Although the change in the middle and later stages was small, the A peak response could be observed, indicating that the addition of the L carbon contributed to the removal of organic matter, especially humic organic matter, by the UF membrane. Similar to L-UF, the peak intensity of the four peaks in the S-UF mixture was weakened at the beginning of the operation, and the C-peak response decreased more, suggesting that the humic organic matter was obviously decreased in the initial stage of the mixed liquid of S-UF. The response peak of proteinaceous material may be caused by the metabolites of raw water and initial microorganisms. In the middle and later stages, enhanced B, T, and C peaks were observed in the mixture. This was mainly attributed to the two sources, i.e., the part introduced by the fluctuation of raw water quality and the metabolism and synthesis of microorganisms.

3.2. Membrane Cleaning Characteristics under Different Conditions

3.2.1. Variation of Membrane Resistance

The membrane modules of the direct UF filtration, L-UF, and S-UF conditions were cleaned, and membrane resistance was observed as a function of TMP change. The resistance distributions are displayed in Figure 5 and Figure S3. Regardless of the TMP of 10 kPa, 15 kPa, or 30 kPa, the order of the resistance of the carbonization condition was irreversible (cleanable, acid cleaning) > irreversible (cleanable, alkaline cleaning) > inherent resistance of the membrane > reversible (alkaline cleaning) > reversible (acid cleaning) > concentration difference resistance. It can be seen that the highest ratio of the inherent resistance of the membrane was 35% of the total filtration resistance in the direct UF filtration, which was significantly higher than the PAC-UF. After the end of the three working conditions, the irreversible resistance ratio reached over 50%, which accounted for the highest proportion of membrane filtration resistance. This phenomenon illustrated that irreversible resistance in the membrane filtration process was the biggest contributor to the membrane resistance growth process. In addition, it was found that the irreversible resistance ratio of acid cleaning was greater than that of alkaline cleaning, which indicated that for irreversible fouling, acid cleaning had greater cleaning efficiency and produced higher membrane recovery. The proportion of reversible resistance was much lower in direct UF filtration than PAC-UF, which could be due to the fact that adding PAC could improve the porous structure of the membrane cake layer, increase reversible fouling, and alleviate irreversible fouling, thus prolonging the filtration duration.

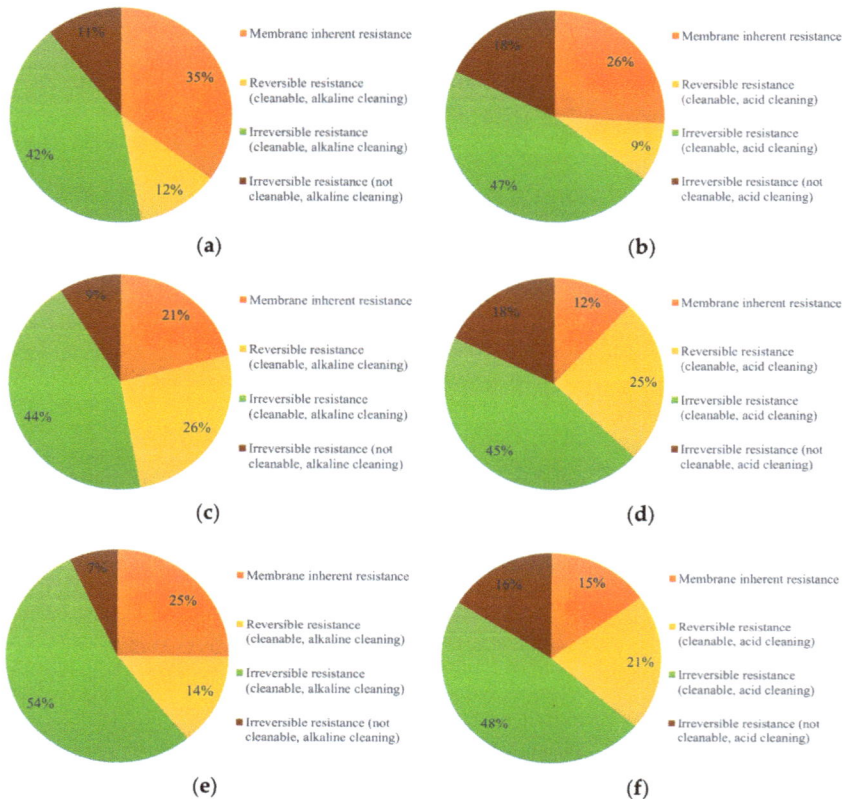

Figure 5. The ratio of the resistance of the membrane after cleaning: (**a**) direct UF filtration, alkaline cleaning; (**b**) direct UF filtration, acid cleaning; (**c**) L-UF, alkaline cleaning; (**d**) L-UF, acid cleaning; (**e**) S-UF, alkaline cleaning; and (**f**) S-UF, acid cleaning.

3.2.2. Analysis of Membrane Contaminants

Content of Organic Matter in Membrane Elution

The organic matter content of the membrane elution is shown in Table S6. It can be seen that the DOC and UV_{254} values of the alkaline cleaning liquid were the highest in each working condition, followed by acid cleaning, and the lowest physical washing, indicating that the alkaline cleaning could elute more. Comparing the organics in the alkaline cleaning and acid cleaning of the L-UF and S-UF conditions, it could be found that the DOC in the respective alkaline cleaning and acid cleaning was quite different, but the UV_{254} values were similar, indicating that the acid cleaning could elute more easily the substances that respond to ultraviolet light. The DOC in the alkaline cleaning and acid cleaning of L-UF were lower than that of S-UF, but the corresponding UV value was larger than that of S-UF, indicating that in the L-UF cleaning solution, UV responses, such as humic acid and aromatic organic compounds containing carbon-oxygen double bonds, accounted for a higher proportion.

The content of soluble EPS in the cleaning solution is shown in Figure 6, with the ANOVA statistical analysis using Tukey's test of the comparison of EPS concentration in terms of both protein and polysaccharide from the direct UF, L-UF, and S-UF reactors with the physical, acid, and alkaline cleanings are presented in Table S7. Comparing the EPS concentration in the membrane elution from L-UF and S-UF reactors with that from the direct UF reactor, the three cleaning protocols all behaved significantly differently.

While comparing the EPS concentration in the membrane elution from the L-UF reactor with that from the S-UF reactor, the alkaline cleaning exhibited significantly different, and the physical cleaning was not. While the acid cleaning behaved statistically significantly different for the proteins in the membrane elution but did not for polysaccharides. As shown in Figure 6, the content of polysaccharide and protein in the cleaning solution of each working condition was higher than that of the previously tested mixture and raw water, indicating that the polysaccharide and protein contributed greatly to membrane fouling. As depicted in Figure 6, the content of EPS in the direct UF filtration was the lowest, and the highest in the S-UF. Meanwhile, the EPS content of the cleaning solution after acid cleaning was slightly higher than that of the alkaline cleaning, indicating that the acid cleaning was more likely to elute polysaccharide and protein.

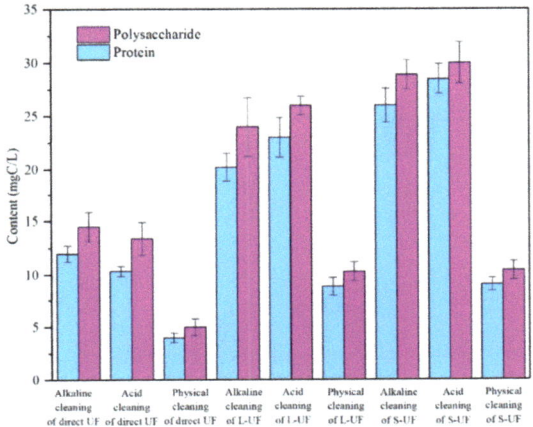

Figure 6. EPS content in membrane elution.

Molecular Weight Distribution

Figure 7 displays the peak separation analysis of the solution after membrane washing using Peakfit. It could be observed that the direct UF filtration, L-UF, and S-UF had the least proportion of macromolecules in the elution ($\leq 15\%$), which could be due to the fact that the organic matter in the elution was composed mainly of small and medium molecules.

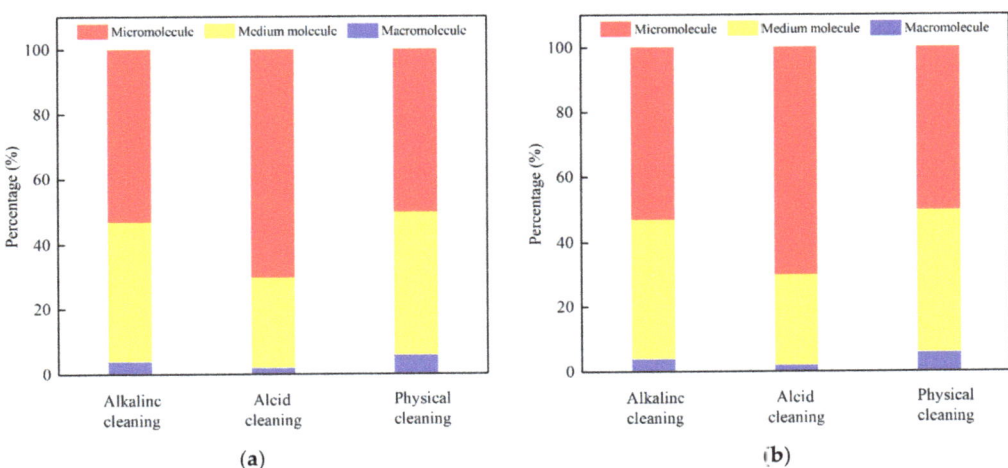

Figure 7. Cont.

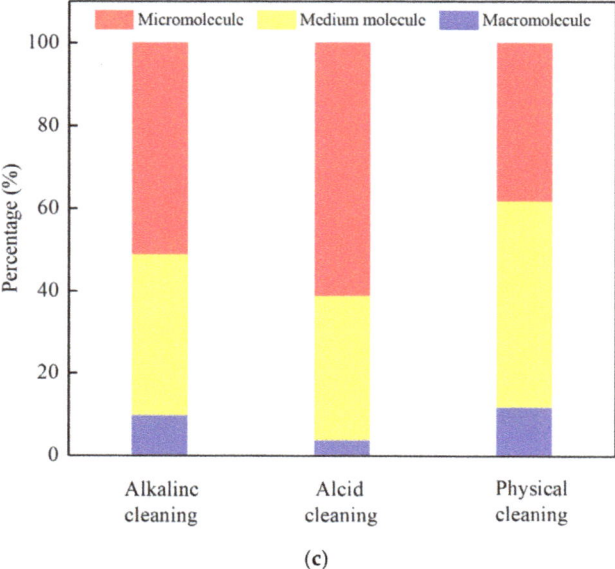

(c)

Figure 7. The MW of the membrane elution: (**a**) direct UF; (**b**) L-UF; and (**c**) S-UF.

Among the alkaline cleaning of L-UF, the proportion of medium molecules was the highest, and the other elution had the highest proportion of small molecules, indicating that the small molecules were more likely to cause irreversible fouling of the membrane. In addition, the proportion of small molecules in the acid cleaning solution was the highest among the three cleaning methods, again demonstrating that acid cleaning made it easier to elute small molecular organics [36].

Hydrophilic-Hydrophobic Property Analysis of Membrane Elution

The characteristics of the hydrophilic and hydrophobic separation organics of the membrane elution are shown in Figure 8. Clearly, the main organic substances of the elution were hydrophilic and strong hydrophobic components, and the proportion of weakly hydrophobic components was less than 5%. In the same working condition, the ratio of hydrophilic components was higher in the acid cleaning solution than in the alkaline and physical cleaning, indicating that acid cleaning was more prone to elute the hydrophilic components. The DOC of the alkaline cleaning liquid in the direct UF filtration condition was mainly composed of strong hydrophobic organic matter. The strong hydrophobic content of L-UF in the alkaline cleaning solution (43%) was higher than that of S-UF (33%), indicating that the strong hydrophobic matter had a greater influence on membrane fouling in L-UF. The content of hydrophilic substances in S-UF was the highest, no matter with the alkaline cleaning or the acid cleaning, indicating that hydrophilic organic compounds had a greater impact on membrane fouling in S-UF.

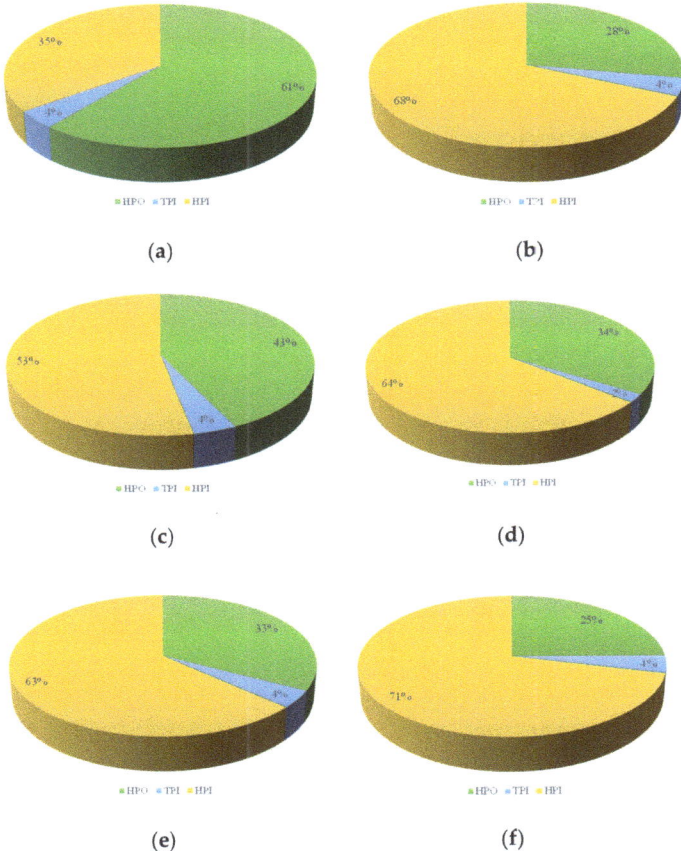

Figure 8. Proportion of hydrophilic and hydrophobic components: (**a**) direct UF filtration, alkaline cleaning; (**b**) direct UF filtration, acid cleaning; (**c**) L-UF, alkaline cleaning; (**d**) L-UF, acid cleaning; (**e**) S-UF, alkaline cleaning; (**f**) S-UF, acid cleaning.

3.3. Mechanisms of Different PACs' Effect on Membrane Filtration

Compared with the L carbon with the smaller particle size and the lower BET, the S carbon with the larger particle size and the higher BET exhibited superior performance in prolonging the UF filtration process, which suggests that the membrane fouling could be better controlled. Since the small PAC particle size could inhibit the growth of microorganisms in the mixed liquor, the S carbon with a larger particle size and pore volume (d > 30.0 nm) provided a more favorable growth environment for microorganisms and covered more EPS on their surface. Therefore, the particle size of the S carbon mixed liquor also increased as the filtration process proceeded. At the end of the filtration, the S carbon cake layer that formed on the UF membrane surface and contained the highest quantity of EPS (i.e., hydrophilic substances, polysaccharide, and protein) could alleviate irreversible fouling effectively. Thus, particle size and pore structure are the two key parameters that should be considered for the PAC addition during the UF filtration process, which can also be analyzed with the Hagen–Poiseuille equation, as shown in Figure 9.

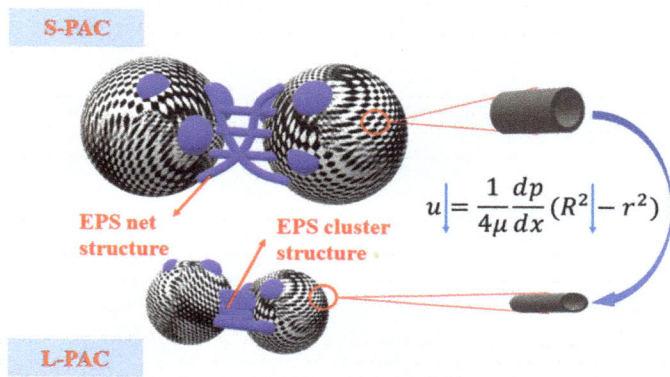

Figure 9. Mechanism of PAC addition on the UF filtration process.

4. Conclusions

In this study, the water purification effect and membrane fouling mechanism of two PAC (L carbon and S carbon)-enhanced PVDF UF membranes on micro-polluted water filtration were investigated. The results indicated that adding PAC could effectively enhance the removal ratio of large, medium, and small molecules on the membrane, and played an active role in the water purification effect and membrane fouling reduction. The S carbon with the larger particle size and pore volume exhibited superior performance in controlling membrane fouling. Particle size and pore structure are the two key parameters that are essential for the PAC-UF process. The composition of the carbon-added mixture was analyzed, and the results showed that with the increase of the operation time, the EPS content of the S-UF mixture increased significantly. The molecular weight distribution results showed that the content of macromolecules in the L-UF and S-UF mixtures increased first and then decreased as the S-UF increased more. With the two PACs, the total removal ratio of large, medium, and small molecular organic matter was increased as compared with direct UF filtration, with the removal ratio of small molecule the largest. After acid cleaning and alkaline cleaning, irreversible fouling played the most important role in the growth of membrane resistance. In addition, the irreversible resistance ratio of acid cleaning was greater than that of alkaline cleaning, indicating that for irreversible fouling, acid cleaning could produce greater cleaning efficiency and membrane recovery. The reversible resistance of direct UF filtration was much lower than PAC-UF, demonstrating that adding activated carbon could improve the structure of the membrane cake layer and increase reversible fouling, thus relieving irreversible fouling to some extent and prolonging the running duration. Finally, through the analysis of the solution after membrane cleaning, it was found that the elution was mainly composed of small and medium molecular organic substances, and polysaccharides were more likely to contribute to membrane fouling than proteins. Alkaline cleaning mainly eluted large, medium, and small molecules of hydrophilic and hydrophobic organic matter, while acid cleaning mainly eluted small molecules of hydrophilic organic matter.

Supplementary Materials: The following supporting information can be downloaded at: https://www.mdpi.com/article/10.3390/membranes12050487/s1, Figure S1: Particle size distribution of L carbon and S carbon; Figure S2: The pore structure analysis of L carbon and S carbon; Figure S3: The specific surface area analysis of the PAC pores; Table S1: The size distribution of the two different activated carbons (μm); Table S2: Surface oxygen-containing functional group of L carbon and S carbon; Figure S4: Schematic diagram of the experimental setup, (a) direct UF filtration; (b) PAC-UF filtration; Figure S5: Cleaning process diagram; Table S3: Fluorescent region boundaries and characteristic substance types; Figure S6: Removal effects of DOC, UV_{254} at different operating conditions (direct UF filtration, L-UF and S-UF filtration); Table S4: ANOVA statistical analysis of the comparison of EPS concentrations in the raw water, the L-UF and S-UF reactors in the initial, middle and later stages using Tukey's test; Table S5: The particle size variation of the mixed liquor during operations (μm); Figure S7: The particle size variation of mixed liquor during operations; Table S6: Organic content in the membrane elution; Table S7: ANOVA statistical analysis of the comparison of EPS in the membrane elution from direct UF, L-UF and S-UF reactors with physical, acid and alkaline cleanings using Tukey's test; Figure S8: The membrane resistance in different operations.

Author Contributions: Conceptualization, T.L. and J.L.; methodology, H.Y., J.T. and T.Y.; formal analysis, J.T.; investigation, H.Y. and S.X.; resources, J.L.; writing—original draft, T.L. and J.T.; writing—review and editing, T.L. and H.C.; supervision, B.D.; project administration, B.D.; funding acquisition, T.L., J.L., H.C. and B.D. All authors have read and agreed to the published version of the manuscript.

Funding: This research was funded by financial support from the National Natural Science Foundation of China (No. 52100012, 51922078 and 51708130), the Natural Science Foundation of Shanghai (No. 20ZR1460800), the Major Science and Technology Program for Water Pollution Control and Treatment (No. 2017ZX07201001), and the Foundation of Key Laboratory of Yangtze River Water Environment, Ministry of Education (Tongji University), China (No. YRWEF202104).

Institutional Review Board Statement: Not applicable.

Informed Consent Statement: Not applicable.

Data Availability Statement: The datasets analyzed during the current study are available from the corresponding author on reasonable request.

Acknowledgments: The authors would like to greatly acknowledge instruction on the statistical analysis from Jiaying Ma at Tongji University.

Conflicts of Interest: The authors declare that they have no known competing financial interest or personal relationships that could have appeared to influence the work reported in this paper.

References

1. Cheng, X.X.; Hou, C.S.; Li, P.J.; Luo, C.W.; Zhu, X.W.; Wu, D.J.; Zhang, X.Y.; Liang, H. The role of PAC adsorption-catalytic oxidation in the ultrafiltration performance for treating natural water: Efficiency improvement, fouling mitigation and mechanisms. *Chemosphere* **2021**, *284*, 131561. [CrossRef] [PubMed]
2. Schwaller, C.; Hoffmann, G.; Hiller, C.X.; Helmreich, B.; Drewes, J.E. Inline dosing of powdered activated carbon and coagulant prior to ultrafiltration at pilot-scale-Effects on trace organic chemical removal and operational stability. *Chem. Eng. J.* **2021**, *414*, 128801. [CrossRef]
3. Sun, L.H.; Zhu, J.Y.; Shi, P.F.; Ding, Y.; Feng, C.M. Preoxidation Combined with Powdered Activated Carbon and Ultrafiltration to Remove Antibiotic Resistance Genes in Secondary Effluent. *Environ. Eng. Sci.* **2021**, *38*, 822–831. [CrossRef]
4. Sheng, C.; Nnanna, A.G.A.; Liu, Y.; Vargo, J.D. Removal of Trace Pharmaceuticals from Water using coagulation and powdered activated carbon as pretreatment to ultrafiltration membrane system. *Sci. Total Environ.* **2016**, *550*, 1075–1083. [CrossRef]
5. Yu, H.K.; Li, X.; Chang, H.Q.; Zhou, Z.W.; Zhang, T.T.; Yang, Y.L.; Li, G.B.; Ji, H.J.; Cai, C.Y.; Liang, H. Performance of hollow fiber ultrafiltration membrane in a full-scale drinking water treatment plant in China: A systematic evaluation during 7-year operation. *J. Membr. Sci.* **2020**, *613*, 118469. [CrossRef]
6. Gao, K.; Li, T.; Zhao, Q.Q.; Liu, W.; Liu, J.X.; Song, Y.L.; Chu, H.Q.; Dong, B.Z. UF fouling behavior of allelopathy of extracellular organic matter produced by mixed algae co-cultures. *Sep. Purif. Technol.* **2021**, *261*, 118297. [CrossRef]
7. Peters, C.D.; Rantissi, T.; Gitis, V.; Hankins, N.P. Retention of natural organic matter by ultrafiltration and the mitigation of membrane fouling through pre-treatment, membrane enhancement, and cleaning-A review. *J. Water Process. Eng.* **2021**, *44*, 102374. [CrossRef]
8. Pourbozorg, M.; Li, T.; Law, A.W.K. Effect of turbulence on fouling control of submerged hollow fibre membrane filtration. *Water Res.* **2016**, *99*, 101–111. [CrossRef]
9. Wu, S.Q.; Hua, X.; Ma, B.W.; Fan, H.W.; Miao, R.; Ulbricht, M.; Hu, C.Z.; Qu, J.H. Three-Dimensional Analysis of the Natural-Organic-Matter Distribution in the Cake Layer to Precisely Reveal Ultrafiltration Fouling Mechanisms. *Environ. Sci. Technol.* **2021**, *55*, 5442–5452. [CrossRef]
10. Meng, S.; Wang, R.; Zhang, K.; Meng, X.; Xue, W.; Liu, H.; Liang, D.; Zhao, Q.; Liu, Y. Transparent exopolymer particles (TEPs)-associated protobiofilm: A neglected contributor to biofouling during membrane filtration. *Front. Environ. Sci. Eng.* **2021**, *15*, 64. [CrossRef]
11. Huang, W.; Wang, L.; Zhou, W.; Lv, W.; Hu, M.; Chu, H.; Dong, B. Effects of combined ozone and PAC pretreatment on ultrafiltration membrane fouling control and mechanisms. *J. Membr. Sci.* **2017**, *533*, 378–389. [CrossRef]

12. Liu, J.; Tian, J.; Wang, Z.; Zhao, D.; Jia, F.; Dong, B. Mechanism analysis of powdered activated carbon controlling microfiltration membrane fouling in surface water treatment. *Colloid. Surf. A* **2017**, *517*, 45–51. [CrossRef]
13. Campinas, M.; Rosa, M.J. Assessing PAC contribution to the NOM fouling control in PAC/UF systems. *Water Res.* **2010**, *44*, 1636–1644. [CrossRef] [PubMed]
14. Khan, M.M.; Takizawa, S.; Lewandowski, Z.; Habibur Rahman, M.; Komatsu, K.; Nelson, S.E.; Kurisu, F.; Camper, A.K.; Katayama, H.; Ohgaki, S. Combined effects of EPS and HRT enhanced biofouling on a submerged and hybrid PAC-MF membrane bioreactor. *Water Res.* **2013**, *47*, 747–757. [CrossRef]
15. Konieczny, K.; Klomfas, G. Using activated carbon to improve natural water treatment by porous membranes. *Desalination* **2002**, *147*, 109–116. [CrossRef]
16. Parameshwaran, K.; Fane, A.G.; Cho, B.D.; Kim, K.J. Analysis of microfiltration performance with constant flux processing of secondary effluent. *Water Res.* **2001**, *35*, 4349–4358. [CrossRef]
17. Matsui, Y.; Sanogawa, T.; Aoki, N.; Mima, S.; Matsushita, T. Evaluating submicron-sized activated carbon adsorption for microfiltration pretreatment. *Water Treat.* **2006**, *6*, 149–155. [CrossRef]
18. Meier, J. Mechanical influence of PAC particles on membrane processes. *J. Membr. Sci.* **2010**, *360*, 404–409. [CrossRef]
19. Yiantsios, S.G.; Karabelas, A. An experimental study of humid acid and powdered activated carbon deposition on UF membranes and their removal by backwashing. *Desalination* **2001**, *140*, 195–209. [CrossRef]
20. Shao, S.; Qu, F.; Liang, H.; Chang, H.; Yu, H.; Li, G. Characterization of membrane foulants in a pilot-scale powdered activated carbon–membrane bioreactor for drinking water treatment. *Process Biochem.* **2014**, *49*, 1741–1746. [CrossRef]
21. Mozia, S.; Tomaszewska, M.; Morawski, A.W. Studies on the effect of humic acids and phenol on adsorption-ultrafiltration process performance. *Water Res.* **2005**, *39*, 501–509. [CrossRef] [PubMed]
22. Shi, X.; Tal, G.; Hankins, N.P.; Gitis, V. Fouling and cleaning of ultrafiltration membranes: A review. *J. Water Process Eng.* **2014**, *1*, 121–138. [CrossRef]
23. Li, T.; Law, A.W.K.; Jiang, Y.S.; Harijanto, A.K.; Fane, A.G. Fouling control of submerged hollow fibre membrane bioreactor with transverse vibration. *J. Membr. Sci.* **2016**, *505*, 216–224. [CrossRef]
24. Zhao, F.C.; Zhang, Y.L.; Chu, H.Q.; Jiang, S.H.; Yu, Z.J.; Wang, M.; Zhou, X.F.; Zhao, J.F. A uniform shearing vibration membrane system reducing membrane fouling in algae harvesting. *J. Clean. Prod.* **2018**, *196*, 1026–1033. [CrossRef]
25. Her, N.; Amy, G.; McKnight, D.; Sohn, J.; Yoon, Y. Characterization of DOM as a function of MW by fluorescence EEM and HPLC-SEC using UVA, DOC, and fluorescence detection. *Water Res.* **2003**, *37*, 4295–4303. [CrossRef]
26. Sim, L.N.; Chong, T.H.; Taheri, A.H.; Sim, S.T.V.; Lai, L.; Krantz, W.B.; Fane, A.G. A review of fouling indices and monitoring techniques for reverse osmosis. *Desalination* **2018**, *434*, 169–188. [CrossRef]
27. Wang, H.; Qu, F.; Ding, A.; Liang, H.; Jia, R.; Li, K.; Bai, L.; Chang, H.; Li, G. Combined effects of PAC adsorption and in situ chlorination on membrane fouling in a pilot-scale coagulation and ultrafiltration process. *Chem. Eng. J.* **2016**, *283*, 1374–1383. [CrossRef]
28. Gao, W.; Liang, H.; Ma, J.; Han, M.; Chen, Z.-L.; Han, Z.-S.; Li, G.-B. Membrane fouling control in ultrafiltration technology for drinking water production: A review. *Desalination* **2011**, *272*, 1–8. [CrossRef]
29. Davini, P. Adsorption and Desorption of So2 on Active-Carbon-the Effect of Surface Basic Groups. *Carbon* **1990**, *28*, 565–571. [CrossRef]
30. Chew, C.M.; Aroua, M.K.; Hussain, M.A. Advanced process control for ultrafiltration membrane water treatment system. *J. Clean. Prod.* **2018**, *179*, 63–80. [CrossRef]
31. Ozkaya, B.; Kaksonen, A.H.; Sahinkaya, E.; Puhakka, J.A. Fluidized bed bioreactor for multiple environmental engineering solutions. *Water Res.* **2019**, *150*, 452–465. [CrossRef] [PubMed]
32. Tan, L.C.; Lens, P.N.L. Addition of granular activated carbon during anaerobic oleate degradation overcomes inhibition and promotes methanogenic activity. *Environ. Sci.-Water Res.* **2021**, *7*, 762–774. [CrossRef]
33. Thongmak, N.; Sridang, P.; Puetpaiboon, U.; Héran, M.; Lesage, G.; Grasmick, A. Performances of a submerged anaerobic membrane bioreactor (AnMBR) for latex serum treatment. *Desalin. Water Treat.* **2015**, *57*, 20694–20706. [CrossRef]
34. Zhao, F.; Chu, H.; Yu, Z.; Jiang, S.; Zhao, X.; Zhou, X.; Zhang, Y. The filtration and fouling performance of membranes with different pore sizes in algae harvesting. *Sci. Total Environ.* **2017**, *587–588*, 87–93. [CrossRef] [PubMed]
35. Xing, L.; Fabris, R.; Chow, C.W.K.; van Leeuwen, J.; Drikas, M.; Wang, D. Prediction of DOM removal of low specific UV absorbance surface waters using HPSEC combined with peak fitting. *J. Environ. Sci.* **2012**, *24*, 1174–1180. [CrossRef]
36. Zularisam, A.W.; Ismail, A.F.; Salim, R. Behaviours of natural organic matter in membrane filtration for surface water treatment—A review. *Desalination* **2006**, *194*, 211–231. [CrossRef]

Article

A Membrane with Strong Resistance to Organic and Biological Fouling Using Graphene Oxide and D-Tyrosine as Modifiers

Jiarui Guo [1,2], Yan Zhang [1,2,*], Fenghua Chen [2] and Yuman Chai [2]

[1] Key Laboratory of Water Quality Science and Water Environment Recovery Engineering, Beijing University of Technology, Beijing 100124, China; guojiarui@emails.bjut.edu.cn

[2] Faculty of Architecture, Civil and Transportation Engineering, Beijing University of Technology, Beijing 100124, China; chenfh@cnpe.cc (F.C.); chaiyuman@163.com (Y.C.)

* Correspondence: yzhang@bjut.edu.cn

Abstract: Membrane fouling markedly influences the service life and performance of the membrane during the using process. Herein, hydrophilic polyvinylidene fluoride (PVDF) nanocomposite (P-GO-DAA) membranes with antifouling and anti-biofouling characteristics were fabricated by employing graphene oxide (GO) and different concentrations of D-Tyrosine. The structural properties of the prepared nanocomposite membranes as well as pure PVDF membranes were characterized using FTIR, XPS, SEM, AFM, and contact angle analysis. It was found that the introduction of GO fillers made an excellent antifouling performance compared to pure PVDF indicated by the pure water flux, flux recovery rate, and rejection rate during ultrafiltration experiments as a result of the formation of the hydrophilic and more porous membrane. In particular, the nanocomposite membranes showed an increased flux of 305.27 L/(m^2·h) and the rejection of 93.40% for the mixed pollutants solution (including Bull Serum Albumin, Sodium Alginate, and Humic Acid). Besides, the outstanding anti-biofouling activity was shown by the P-GO-DAA membrane with the properties of D-Tyrosine for inhibiting biofilm formation during the bacterial adhesion experiments. Furthermore, the adhesion ratio of bacteria on the membrane was 26.64% of the P-GO-DAA membrane compared to 84.22% of pure PVDF. These results were confirmed by CLSM.

Keywords: membrane fouling; mixed matrix membranes; D-amino acid; graphene oxide (GO); biofouling control

1. Introduction

Ultrafiltration (UF) membranes have been considered as a promising strategy for wastewater treatment [1–3]. Nevertheless, a massive quantity of foulants in the environment such as heavy metal ions, suspended solids, natural organic matters, microorganisms, and others, is likely to accumulate on the membrane surface leading to the organic fouling and biofouling of the membrane [4,5].

Nanomaterials are commonly added as a modifier to the membrane matrix or surface to enhance the permeability and the antifouling property of the membranes [6–9]. Among them, GO and functionalized GO have been recognized as attractive materials to prepare nanocomposite membranes owing to their high hydrophilic, porous structure, the abundance of oxygen-containing functional groups, and strong mechanical properties [10–12]. Zwitterionic modified GO sheets were used to prepare PVDF nanocomposite membranes for improving the hydrophilic property, which reduced the water contact angle to 65.1° [13]. A nanocomposite forward osmosis (FO) membrane was synthesized by using polyvinylpyrrolidone (PVP) modified GO to enhance the membrane permeability, with a water flux of 33.2 L/(m^2·h), which was 3.3 times higher compared with the pristine FO membranes [14]. Khan et al. fabricated a membrane blended hybrid nanosheets by using the aid of the functional groups of GO, which induced covalent organic frameworks (COFs) to GO, exhibiting the highest water flux of 226.3 L/(m^2·h) [15].

GO contributes significantly to the improvement of membrane resistance to organic foulants in the nanocomposite membranes. Recently, a novel polysulfide (PSF) membrane used vanillin-modified GO nanosheets as pore formers was reported with a rejection rate of 99%, and the membrane showed a flux recovery rate of 88.55% when filtering BSA as a pollutant [16]. Similarly, polyethersulfone (PES)/sulfonated polysulfone (SPSf) nanocomposite membranes were prepared by using a very low GO content of 0.012 wt% as a modifier, with a high flux recovery rate of 92.4% for BSA filtration [17]. Wang et al. fabricated a self-assembly polyacrylonitrile membrane with GO to improve the antifouling capability of the membrane, achieving the flux recovery rate up to 91.2% for HA [18]. GO makes the membrane more resistant to organic fouling, but it has a limited effect on the biofouling of the membrane surface.

Biofouling is usually inevitable due to the inherent properties of microorganisms as well as the accumulation of biologically active organisms along with extracellular polymeric substances (EPS) on the membrane surface [19]. That brings up the tricky problem which is, even if 99.9% of them are eliminated, there are still enough cells remaining that could continue to grow by using biodegradable substances in the water [20]. Consequently, biofouling is inescapable unless it is sustained in sterile surroundings or no nutrients are present at all [20–22].

Presently, the widely used strategy to deal with biofouling is inducing Ag, Cu, and heavy metal materials to the matrix of the membrane, to achieve their effect through the lethal action on microorganisms. Sun et al. used GO-Ag nanoparticles to develop the biofouling resistant membranes, which found the presence of GO-Ag nanoparticles on the membrane led to an inactivation of 86% E. coli after contact with the membrane [23]. However, the excessive release of biocidal materials is unfriendly to non-target microorganisms and even deteriorates the surroundings [24,25]. Hence, natural materials as eco-friendly substitutions for antimicrobial agents have become a huge focus, such as chitosan [26], D-amino acid (DAA) [27], and pancreatic enzymes [28].

D-Tyrosine (a typical DAA), a newly discovered green substance, was demonstrated to mitigate bacterial fouling by prompting self-disintegrating of biofilm effectively at extremely low concentrations [29–33]. Auto-inducers (AIs) are known as signal molecules that elicit quorum sensing (QS) of microorganisms to coordinate their collective behavior including biofilm formation [34]. A series of works in the available literature has confirmed that DAA could prevent the synthesis of bacteria by altering their cell wall composition, leading to the loss of the ability of bacteria to secrete AIs normally, which triggers the disassembly of biofilms [35–37]. Based on the explorations above, D-Tyrosine has been applied in membrane technology to suppress bacterial adhesion to the membrane. Yu et al. developed an anti-biofouling membrane by incorporating D-Tyrosine onto a membrane using zeolite nanoparticles that inhibited biofilm formation without inactivating the bacteria [36]. Guo et al. prepared DAA-modified PVDF nanocomposite membranes supported by PDA and halloysite nanotubes (HNTs), and a stable anti-biofouling over 10 days was obtained for the prepared membrane [38].

The DAA-modified membranes usually introduce DAA through adhesion on the surface of the membrane [38–40]. Khan et al. modified an anti-biofouling membrane by using alginate dialdehyde (ADA) to graft D-Tyrosine on the membrane surface, and this membrane could achieve nearly 80% bacterial inhibition [39]. Jiang et al. modified the membrane by adhering D-Tyrosine through polydopamine (PDA) on the membrane surface, and the bacterial attachment rate of the membrane decreased by about 10% after modification [40]. The membranes mentioned above exhibited superior anti-biofouling performance, but the water permeability of the membranes may be partly influenced.

Herein, in this study, GO incorporated DAA-modified PVDF membranes (P-GO-DAA) with strong resistance to organic and biological fouling were prepared by blending GO nanosheets into the PVDF matrix, and followed by the introduction of D-Tyrosine through hydrogen bonding. The characteristics and performances of the membranes were analyzed through FTIR, XPS, SEM, AFM, and contact angles. The antifouling activity of the prepared membranes was investigated by filtrating multiple typical organic foulants including bull serum albumin (BSA), sodium alginate (SA), and humic acid (HA). The anti-biofouling activity was evaluated through the bacterial adhesion tests and cyclic filtration tests using E. coli as the model microorganism.

2. Materials and Methods

2.1. Materials and Chemicals

All reagents and chemicals are of analytical grade and used as received. To prepare membranes, Polyvinylidene fluoride (PVDF), polyvinylpyrrolidone (PVP), and N, N-dimethylacetamide (DMAC) were all obtained from Shanghai Maclin Biochemical Technology Company. GO (Sheet diameter is 0.5~5 μm and thickness is 0.8~1.2 nm) used as membrane matrix modifier was purchased from Nanjing Xianfeng Nanomaterials Technology Company. D-Tyrosine as an antibacterial agent was supplied by Perfemiker. BSA (Shanghai Maclin Biochemical Technology Company, Shanghai, China), SA, and HA (Tianjin Fuchen Chemical Reagent Company, Tianjin, China) were used as typical organic foulants. E. coli used as a biological foulant was purchased from Guangzhou Strain Preservation Center. LB broth and LIVE/DEAD® BacLight Bacterial Viability Kits were supplied by Thermo Fisher Technology (Shanghai, China) Company. DI water was purified by a Millipore Direct-Q3 in this paper.

2.2. Preparation of PVDF, P-GO, P-GO-DAA Membranes

PVDF and P-GO membranes were prepared via the non-solvent induced phase separation (NIPS) technique and the percentages of all of the components in the membrane casting solutions are shown in Table S1. For the preparation of the PVDF membrane, a homogeneous cast solution was prepared by dissolving PVP as the pore-forming agent and PVDF in DMAC with the aid of a magnetic stirrer at 45 °C for 12 h. After the solution placed in an oven at 60 °C for more than 6 h to release the trapped gas bubbles, it was cast using a 250 μm scraper on clean glass plates at room temperature. The cast membrane was left for 30 s to partially evaporate the solvent and immersed in DI water that served as a non-solvent bath. The resulting membrane was washed with DI water and soaked in DI water for later use when it was detached from the glass surface within tens of seconds. Similarly, the P-GO membrane was fabricated by the same process but GO was dispersed in DMAC in advance by sonication for 30 min at room temperature.

The P-GO-DAA membranes were modified by separately immersing the P-GO membranes in D-tyrosine solutions (DI water as solvent) at different concentrations of 50, 100 and 150 mg/L at 45 °C for 24 h. for modification. The samples were referred to as P-GO-50, P-GO-100 and P-GO-150, respectively.

The synthesis schematic was shown in Figure 1.

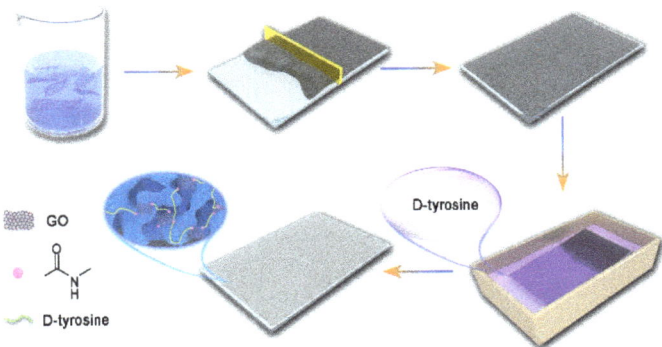

Figure 1. Synthesis schematic of membranes.

2.3. Characteristics of Membranes

The morphology of membrane surface and cross-section was probed by scanning electron microscope (SEM; SU8020, Hitachi, Tokyo, Japan). The cross-section samples were pretreatment by brittle fracturing with the aid of liquid nitrogen and the surface was covered with a thin gold sputtering layer employing an ion sputter device (Buehler, Lake Bluff, IL, USA) prior to the SEM analysis. Atomic force microscopy (AFM; Dimension Icon, Bruker AXS, Karlsruhe, Germany) was implemented using a three-direction closed-loop scanner to obverse the roughness of the membrane surface.

The element composition and functional groups were identified through X-ray photo-electron spectroscopy (XPS; ESCALAB 250Xi, Thermo, Waltham, MA, USA) and Fourier transform infrared (FTIR; V70, Bruker, Karlsruhe, Germany) spectroscopy with a diamond attenuated total reflection (ATR) accessory.

The hydrophilicity was investigated via static contact angle conducted by goniometer (OCA50, Dataphysics, Filderstadt, Germany) to reflect the wettability properties of the membrane surface.

The average pore size of membranes was analyzed with an absorption and desorption instrument (Autosorb iQ, Quantachrome Instruments, Boynton Beach, FL, USA). The porosity (ε) of the membrane was measured by the gravimetric method as reported before and calculated as [41]:

$$\varepsilon = \frac{\frac{w_w - w_d}{\rho_w}}{\frac{w_w - w_d}{\rho_w} + \frac{w_d}{\rho_p}} \times 100\% \tag{1}$$

where the w_w and w_d denote the weight (g) of the wet and dry membrane, respectively, ρ_w and ρ_p represent the densities (g/cm³) of DI water and PVDF, respectively.

2.4. The Permeability of the Membranes

The performance of the pristine and modified membranes was systemically assessed by a lab-scale dead-end filtration unit (Figure 2) at a stable trans-membrane pressure of 0.1 MPa supplied by a nitrogen cylinder. The membranes were initially pre-pressed with DI water at the pressure of 0.2 MPa for 20 min before testing. For the DI water permeation test, the pure water through the membrane was collected every 5 min for 30 min and the flux J_w was calculated by Equation (2):

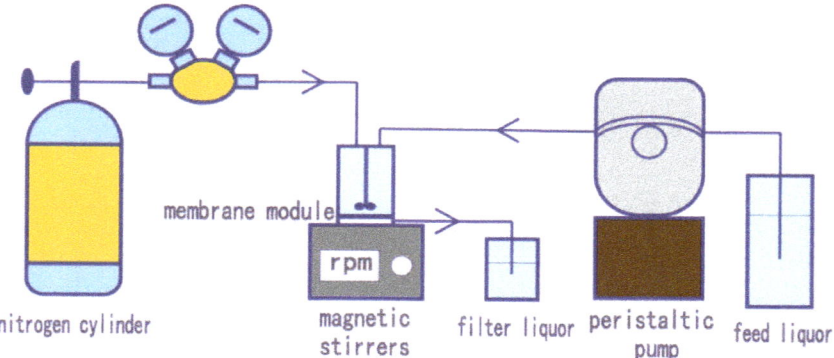

Figure 2. Lab-scale dead-end filtration unit.

$$J_w = \frac{V}{A \times \Delta_t} \tag{2}$$

Here V denotes the volume of permeated water (L); A denotes the effective filtration membrane area in m² and Δt represents the filtration time in hours (h).

2.5. Separate Performance and Antifouling Activity

To determine the separation and antifouling capacity of the membranes, the ultrafiltration experiments were performed again with several kinds of feed water solutions instead of DI water, and the contents and concentrations of these solutions are shown in Table 1.

Table 1. Composition of pollutant solutions.

Name of the Solution	Contents and Concentrations
BSA (Bull Serum Albumin)	BSA, 100 mg/L
SA (Sodium alga Acid)	SA, 50 mg/L
HA (Humic Acid)	HA, 50 mg/L
Bi (Binary pollutants)	BSA, 100 mg/L SA, 50 mg/L
Tri (Triple pollutants)	BSA, 100 mg/L SA, 50 mg/L HA, 50 mg/L
AS (Actual Sewage)	Effluent from a secondary sedimentation tank in a sewage plant in Beijing, TOC: 182 mg/L
Synthetic pollutants solution	BSA, 100 mg/L SA, 100 mg/L HA, 50 mg/L E. coli, 3.78×10^7 CFU

The pure water flux (J_0) was recorded continuously for 30 min after the pressure stabilized, and the permeate (J_F) passing through the membrane was also recorded for 30 min when the DI water was replaced with the polluted feed water. Later, the coupons were backwashed by DI water completely prior to the measure of pure water flux (J_R) of the backwashed membranes. From these experiments, five parameters were calculated: *rejection rate*, flux recover rate (*FRR*, %), reversible fouling rate (R_r), irreversible fouling rate (R_{ir}), and total fouling rate (R_t). *Rejection rate* and *FRR* were calculated as follows:

$$rejection\ rate = \left(1 - \frac{C_P}{C_F}\right) \times 100\% \quad (3)$$

$$FRR = \frac{J_R}{J_0} \times 100\% \quad (4)$$

where the C_F and C_P denote the organic content measured with a TOC meter (Vario TOC, Element, Munich, Germany) of feed water and permeate water (mg/L) collected from filtration experiments.

While R_r, R_{ir} and R_t were calculated using Equations (5)–(7), respectively:

$$R_r = \frac{J_R - J_F}{J_0} \times 100\% \quad (5)$$

$$R_{ir} = \frac{J_0 - J_R}{J_0} \times 100\% \quad (6)$$

$$R_t = R_r + R_{ir} = \left(1 - \frac{J_F}{J_0}\right) \times 100\% \quad (7)$$

2.6. Anti-Biofouling Activity

The antibacterial efficiency of the membranes was evaluated by employing *E. coli* as the model microorganism. The test bacterium was cultivated in a Luria-Bertani (LB) liquid medium using a modified method reported before [42]. The cultivated suspension with *E. coli* 10 mL was poured into a conical flask containing a membrane sample sterilized and then incubated at 37 °C for 5 d. The concentration of the *E. coli* suspension was measured with a UV spectrophotometer at wavelength 650 nm, and the membranes were cut into coupons and then flushed with DI water for further testing.

In order to visualize the bacterial adhesion and distinguish the activity of bacteria on the membrane surface, the bacteria on the membranes were colored by the LIVE/DEAD Bacterial Viability Kit and the stained membrane samples were observed under the confocal laser scanning electron microscopy (CLSM, LSM800, Zeiss, Jena, Germany).

2.7. Characterization of Comprehensive Fouling and Biofouling Resistance Activity

To determine the antifouling and anti-biofouling activities of the membranes, the flux, concentration of TOC and the adhesion of *E. coli* on the membrane surface were measured with the synthetic heavily polluted sewage containing *E. coli* suspension by the three cycles of ultrafiltration experiments. Subsequently, the adhesion and growth of bacteria on the membrane surface were investigated by the CLSM and SEM to characterize the membrane's resistance to biological foulants.

3. Results and Discussion

3.1. Characterization of PVDF Membranes

The FTIR spectra were performed to investigate the functional groups of the prepared membranes. Figure 3 shows the FTIR spectra for the pristine PVDF, P-GO, and P-GO-DAA membranes from 500 cm^{-1} to 4000 cm^{-1}. The infrared absorption peaks with corresponding chemical bonds are shown in Table S5. Compared with the PVDF membrane, a higher level of O-H and C=O for the carboxyl group observed in the P-GO membrane at the peak of 3385 cm^{-1} and 1720 cm^{-1}, respectively, indicated the successful introduction of GO in the P-GO membrane. The increase in intensity at 1413 cm^{-1} and 860 cm^{-1} for the P-GO-DAA membrane corresponding to the deformation and stretching of N-H and C-N [43], confirmed the existence of DAA on the P-GO-DAA membrane. After the incorporation of DAA with the P-GO membrane, the peaks at 1615 cm^{-1} and 1720 cm^{-1} for the P-GO membrane that ascribed to the bending vibration of carboxyl and hydroxyl groups [44], shifted to a broad vibration band at 1653 cm^{-1} for the P-GO-DAA membrane. Likewise, the peaks of 1320 cm^{-1} that reflected O-H bond of phenol present in D-Tyrosine shifted to 1280 cm^{-1} for the P-GO-DAA membrane. These changes in peak position may be due to intermolecular hydrogen bonding between D-Tyrosine and GO [44,45]. To further clarify the connection between GO and DAA, the mixture of DAA and GO in solution was tested and the same changes in peak positions were observed in the GO-DAA by the FTIR spectra (Figure S2). It was these peaks shifted that speculated the presence of intermolecular hydrogen bonding between D-Tyrosine and GO. Besides, the peak at 1037 cm^{-1} represents C-O-C vibration of GO shifted to 1069 cm^{-1} in the GO-DAA, which speculated the possible intermolecular hydrogen bonding between the epoxy group of GO and D-Tyrosine. The suggestions for interaction are shown in Figure S1.

XPS provided further confirmation of the transformations of the functional groups and elemental compositions on the surface of the membranes. As shown in Figure 4, the C1s narrow sweep spectrum for different membranes could be distinguished into C-C, C-F, C=C and C-O/C-N species, which peaks located at 284.8, 289.3, 283.6, and 286.2 eV, respectively. The increased peak area of C=C compared with the PVDF membrane was testified to the presence of GO in the membrane matrix. The slight enhancement of the peak area of C-O/C-N was also probably owing to the limited distribution of GO on the membrane surface. For the P-GO-DAA membrane, the increased proportion of O and N for the P-GO-DAA membrane (Tables S2 and S3) compared to the P-GO membrane confirmed that the P-GO membrane was successfully modified with DAA. The increased peak area of C-O/C-N was also gained by incorporating the DAA since DAA possessed amino, hydroxyl and carboxyl groups.

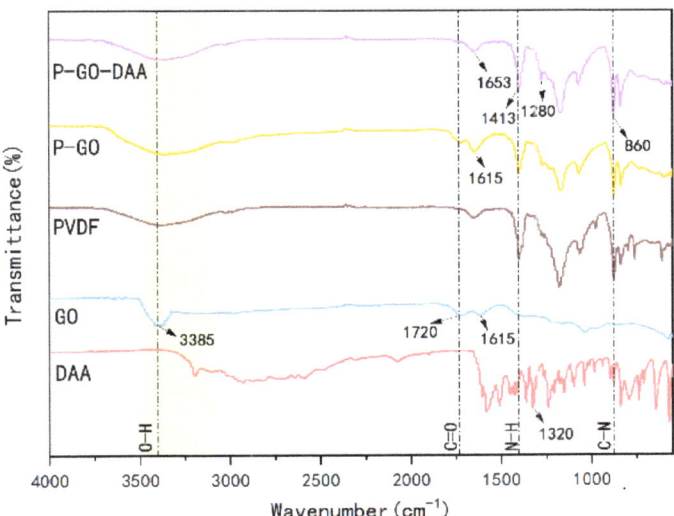

Figure 3. FTIR spectra of DAA, GO, PVDF, P-GO, and P-GO-DAA membranes.

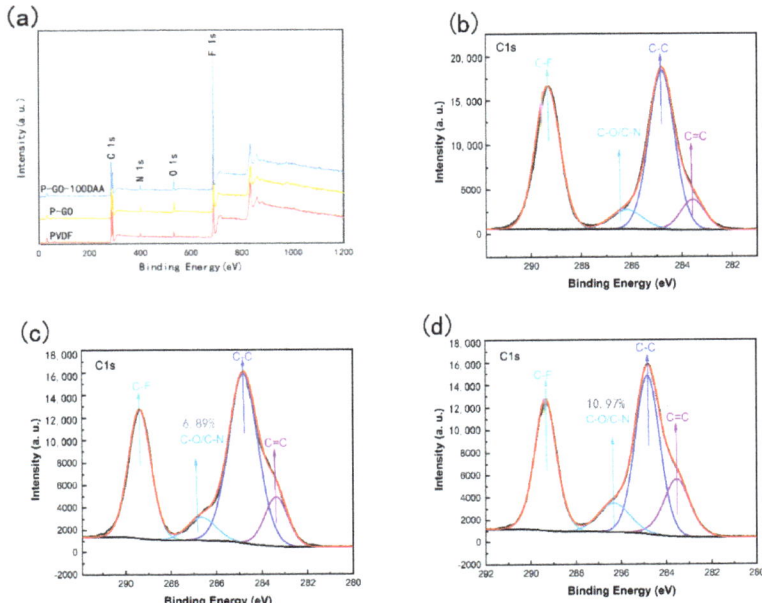

Figure 4. XPS characterization of nanocomposites. (**a**) XPS spectra of PVDF, P-GO, and P-GO-DAA membranes; (**b**) C1s XPS spectra of PVDF; (**c**) C1s XPS spectra of P-GO; (**d**) C1s XPS spectra of P-GO-DAA membranes.

The surface and cross-sectional images of the prepared membranes were taken by SEM and presented in Figure 5. The SEM images showed the PVDF and nanocomposite membranes had a porous structure on the surface as well as an asymmetric cross-section structure consisting of a compact surface layer and a sponge-like porous structure beneath. The pristine PVDF membrane possessed fewer pores with a minimal number of disconnected voids possibly due to its high hydrophobicity [46]. After GO was induced to the membrane matrix, a much more porous surface with relatively interconnected large pores in the bulk and a thinner surface layer (Figure S3) were observed in the P-GO membrane. For the DAA-modified membranes, a denser and compressed layer on top of the membrane was observed which seemed to be increased with the concentration of D-Tyrosine in the membrane. This may be related to the molecular weight and conformation of the D-Tyrosine immobilized on the membrane surface, and this kind of morphological feature was supposed to promote membrane separation performance [47].

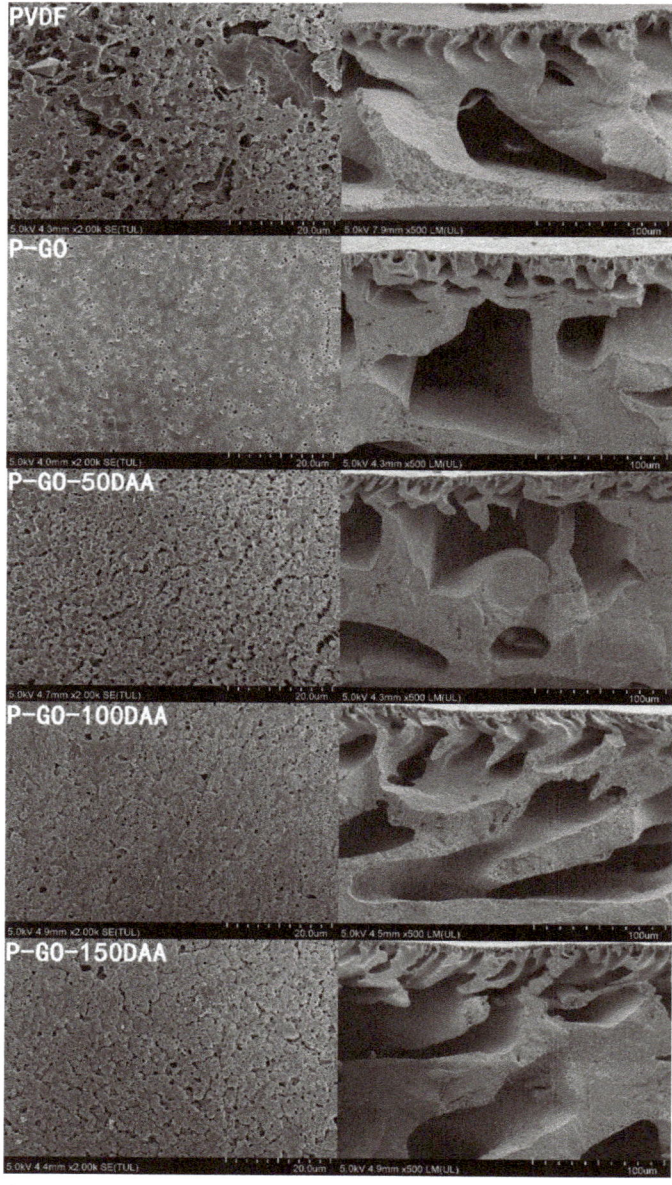

Figure 5. SEM images of the surface and cross-sectional morphologies of different membranes.

The roughness has a relatively great influence on the water permeability and antifouling property of the membrane. High roughness would lead to easier deposition of pollutions on the membrane surface, and very low roughness might diminish the tendency of water molecules through the membrane [48,49]. The surface roughness of each fabricated membrane was determined using AFM images (Figure 6). The P-GO membrane displayed the lowest roughness, while the P-GO-DAA membrane possessed higher roughness than the P-GO membrane and lower roughness than the PVDF membrane. The addition of hydrophilic GO could facilitate the formation of a smoother surface during the membrane casting process when using the DI water as the non-solvent bath. The presence of D-Tyrosine changed the morphology of the nanocomposite membrane surface in a way leading to higher surface roughness that depended on the number of D-Tyrosine on the membrane surface. As expected, the P-GO-DAA membrane had a more appropriate surface roughness in contrast with other membranes, which would better balance the permeability and antifouling ability of the membrane.

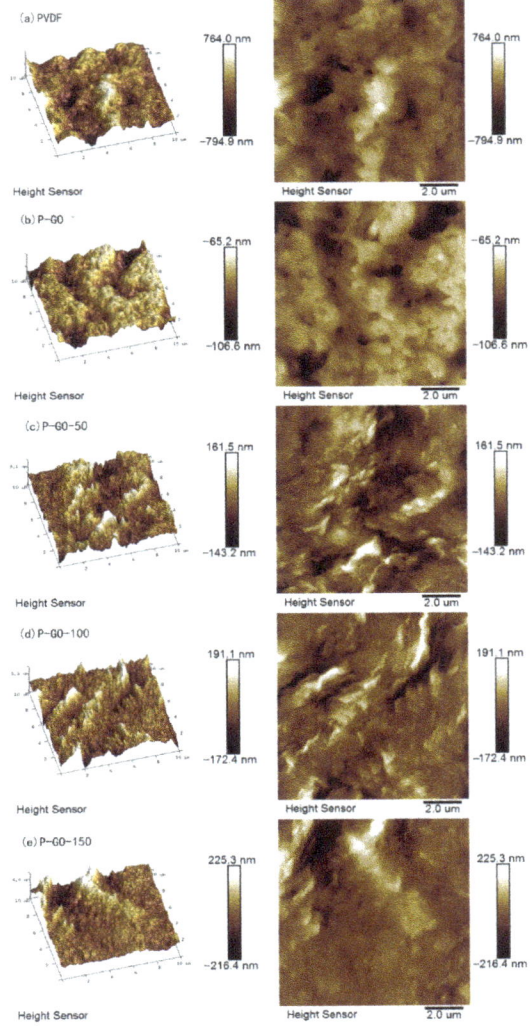

Figure 6. AFM images of various membranes.

The average pore size and porosity were analyzed and the results were listed in Table 2. It was found that the average pore size of the P-GO membrane reduced after the addition of GO, but the porosity of the membranes had improved inversely. This may be since the existence of GO in the P-GO membrane might disrupt the continuity of the membrane matrix resulting in more voids in the membrane. With the introduction of D-Tyrosine, the average pore size and porosity of the P-GO-DAA membranes showed a very minor difference in comparison with the P-GO membrane, which could be presumed that the addition of D-Tyrosine did not have a noticeable impact on the membrane pores.

Table 2. Average pore size and porosity of membranes.

Membrane	Average Pore Size (nm)	Porosity (%)
PVDF	16.9 (±2.1)	74.5 (±1.4)
P-GO	16.2 (±1.6)	79.1 (±1.8)
P-GO-50	15.8 (±1.3)	80.3 (±2.5)
P-GO-100	15.7 (±1.8)	81.0 (±2.3)
P-GO-150	14.4 (±2.0)	79.6 (±2.7)

To evaluate the wettability of the membranes, water contact angles analyses (Figure 7) of membranes were carried out by the static contact angle test. The highest contact angle was exhibited by the pure PVDF with 82.4°, while the P-GO membrane was 72.27°. The contact angle of the P-GO-DAA membranes decreased even more to 67.77°, 64.67°, and 60.27° for the P-GO-50, P-GO-100, and P-GO-150 membranes, respectively. The presence of numerous hydrophilic groups (such as the hydroxyl and carboxyl groups) around the boundaries of GO nanosheets endowed the PVDF membrane with higher hydrophilicity [18]. There is a positive correlation between the hydrophilic of the membrane and the DAA content that may not only be attributed to the increased polar functional groups on the membrane surface, but also to the increased surface roughness.

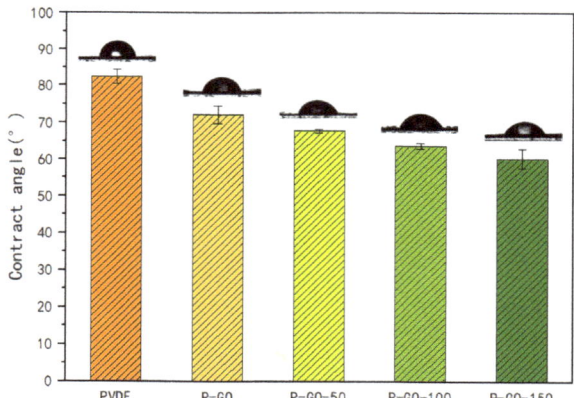

Figure 7. The contact angle of PVDF, P-GO, P-GO-50, P-GO-100, P-GO-150 membrane. The CA data and images were obtained by dropping 2.0 µL DI water on five different locations for each membrane surface sample at room temperature.

3.2. Pure Water Permeability of the Membranes

The permeation flux of the membranes plays a crucial role in the membrane application. The filtrations with pure water were performed, and the results were depicted in Figure 8. The addition of GO to the PVDF casting solution could increase the pure water flux due to the improved porous network and surface hydrophilicity of the membrane. 0.2 wt% of GO in the casting solution was enough to raise the pure water flux from 138.33 L/(m^2·h) (for pure PVDF) to 282.08 L/(m²·h). A more increase in permeation flux was observed for the P-GO-DAA membranes, with the maximum reaching 305.27 L/(m²·h) for the P-GO-100 membrane, which was more than twice compared with the pure PVDF. This higher permeability was probably a combined result that the improved hydrophilicity of the membrane surface by adding GO and DAA to the membrane, as well as a higher surface roughness produced by the modification using DAA, allowing water molecules to pass through the membrane more easily [46]. Nevertheless, a small decrease in the pure water flux for the P-GO-150 membrane could be attributed to a small amount of pore narrowing that occurred due to the relatively high concentration of DAA added. This phenomenon revealed that the water permeation would decline beyond a certain DAA content in the membranes. Overall, the contribution of GO to the increased flux of the membranes was more, and similar kinds of results that nanoparticles contributed more to the water flux of membranes compared to DAA have been reported [26].

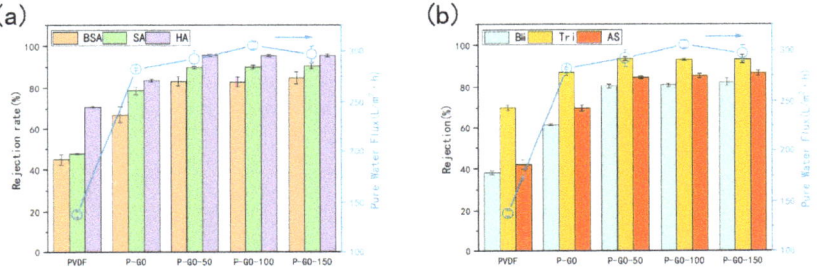

Figure 8. Pure water permeability and *rejection rate* for various foulants of pristine and functioned PVDF membranes: (**a**) *rejection rate* for BSA, SA, and HA; (**b**) *rejection rate* for Bi (binary foulants: BSA and SA), Tri (triple foulants: BSA, SA, and HA), and AS (Actual sewage: effluent of secondary sedimentation tank).

3.3. Separate Performance of the Membranes

In order to study the separate performance of the prepared membranes, filtrations with different aqueous solutions containing single and mixed foulants were performed (Figure 8). For single foulants, it was seen from Figure 8a that the P-GO membrane had good *rejection rates* with 67.25%, 78.87%, and 83.75% compared with the pure PVDF with 44.98%, 47.70% and 71.05% for BSA, SA, and HA, respectively. This increase suggested the separate capacity of the membrane could be contributed to the reduction of the average pore size caused by the blending of GO. After the addition of D-Tyrosine, the highest rejections were observed in the P-GO-DAA membranes for BSA, SA, and HA around 84%, 90% and 95%, respectively, and this increase was due to a dense layer created by the immobilized of DAA on the P-GO membrane surface. The rejection capacity of the membrane for different pollutants had a great relationship with the membrane pore size and the size of pollutant molecules [50]. The membrane rejection for BSA gained the lowest result could be owing to the smaller molecule size of BSA used in this experiment than SA and HA.

As expected, the separation performance of membranes was manifested to be better while employing the solutions comprising complex foulants as feed water during the filtration tests (Figure 8b). The *rejection rates* of the P-GO-150 membrane for Bi, Tri, and AS reached 82.53%, 93.40%, and 86.82%, respectively, which were superior to those of the pure PVDF. This result might be strongly related to a more complex interaction between pollutants leading to the formation of larger agglomerates, so the *rejection rate* of the membranes for the triple foulants solution was higher than the binary foulants solution. The *rejection rate* of membrane dropped for actual sewage which was likely due to the presence of smaller molecule foulants in actual sewage [51,52]. These favorable results indicated that the P-GO-DAA membranes possess an enhanced separate performance for various organic foulants. The difference in these *rejection rates* among the DAA-modified membranes was not significant, which could be ascribed to the addition of D-Tyrosine having little effect on the membrane pores (Table 2), leading to a similar rejection capacity. Similar results were found by Khan et al. [39].

3.4. Antifouling Performance of the Membranes

Flux recovery rate is one of the most crucial measures to reveal the antifouling properties of membranes. The FRR of the prepared membranes was tested through filtrations with single or compound foulants and the results were shown in Figure 9. The P-GO-DAA membranes exhibited better flux recovery performance for single foulants, especially the P-GO-150 membrane, having the highest FRR for BSA, SA and HA with 76.61%, 87.87%, and 61.65%, respectively. The FRR of the P-GO-150 membrane for Bi, Tri, and AS reached 74.58%, 88.72%, and 86.67%, respectively, which were about 28%, 38%, and 39% higher than the P-GO membrane, and about 64%, 64%, and 70% higher than the PVDF membrane. Except for the more hydrophilic membrane surface induced by adding GO and DAA, this improvement could also be a result that DAA reduced the interaction between membrane and foulants leading to less adsorption of foulants by the membrane [17,53,54]. FRR of the P-GO-DAA membranes for all pollutions increased with increasing DAA content, which denoted the antifouling ability might have a positively correlated with the content of DAA in the membrane.

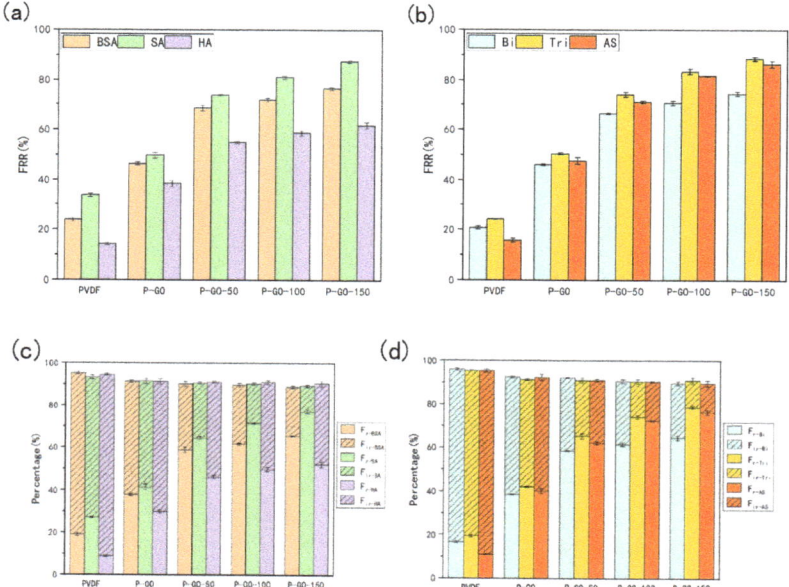

Figure 9. Flux recovery ratio (FRR), reversible flux decline ratio (R_r) and the irreversible flux decline ratio (R_{ir}) for different foulants during filtration experiments of original and prepared membranes. (**a**,**b**) FRR of the membranes; (**c**,**d**) R_r and R_{ir} of the membranes.

Meanwhile, R_r and R_{ir} were also employed as evaluation indexes of membrane antifouling activity (Figure 9), and the accumulation of irreversible fouling would directly descend the performance and service life of the membrane. For all DAA-modified membranes, the percentage of irreversible fouling by various pollutants was obviously reduced. The R_r for AS of the P-GO-150 membrane was 13.85%, which decreased markedly compared with that of PVDF (84.31%) and P-GO (52.33%) membrane. It could be speculated that various interactions between foulants and foulants or foulants and membrane of complex multicomponent systems retarded the fouling instead. As most foulants were hydrophobic, the interaction between the foulants and the membrane was weakened with the assistance of hydrophilic GO and DAA, which greatly improved the membrane's ability to resist organic fouling. A higher proportion of reversible fouling was obtained with increasing DAA content. This might be attributed to the fact that the higher the roughness, the easier it is to remove the foulants attached to the membrane surface, resulting in less fouling remaining in the membrane.

3.5. Anti-Biofouling Performance of the Membranes

To investigate the effect of DAA on bacterial fouling, a five-day static adhesion experiment was conducted using *E. coli* as the model bacteria and the CLSM images (Figures 10 and 11) of the membranes were taken during the experiment. It can be observed that the P-GO-DAA membranes possess strong antibacterial activity as the less bacterial adhesion extent on the membrane surfaces. On the first day, the pure PVDF membrane had already exhibited worse resistance to biofouling. In comparison almost no bacteria on the P-GO-DAA membrane surface. As time passed, the bacteria adhering to the origin membrane surface showed a rapid growth trend. Until the fifth day, a large number of bacteria adhered to the surface of the PVDF membrane (as the stronger the fluorescence in the images, the greater the number of bacteria), as well as the live and dead bacteria on the surface of the membrane co-formed the biofilm. On the contrary, only a small area of the modified membrane surface was fouled and almost no biofilm formation during the period. It was apparent that the number of bacteria on the membrane surface decreased after D-Tyrosine grafted. Hence, the P-GO-DAA membrane exhibited better performance to biofouling, possibly due to the fact that the presence of D-Tyrosine not only prevented bacteria from adhering to the membrane but also promoted the detachment of biofilm from the membrane surface by influencing the secretion of QS signals of the bacteria [36].

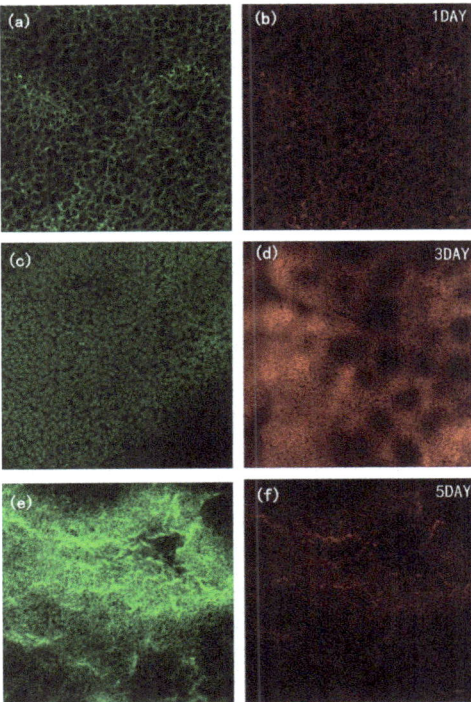

Figure 10. CLSM images of PVDF membrane during five days of bacterial culture cycle. (**a,c,e**) Live bacteria on PVDF membrane; (**b,d,f**) dead bacteria on PVDF membrane.

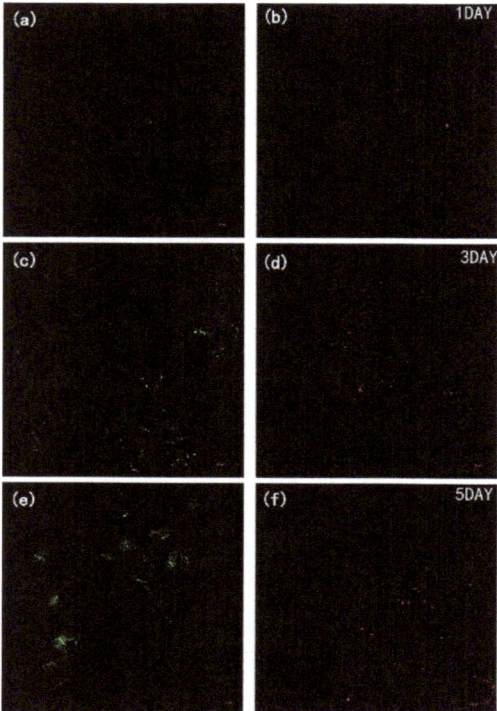

Figure 11. CLSM images of P-GO-DAA membrane during five days of bacterial culture cycle. (**a,c,e**) live bacteria on PVDF membrane; (**b,d,f**) dead bacteria on PVDF membrane.

3.6. Antifouling and Anti-Biofouling Performance of the Membranes through Cycle Test

The antifouling and anti-biofouling performance of the membranes were studied by the cyclic filtration experiments using a heavily polluted solution containing BSA, SA, HA, and *E. coli* (Figure 12). The flux (Figure 12a) of the modified membranes decreased slowly during filtration using the polluted solution as the feed water, while the PVDF membrane was already reduced dramatically after the first cycle, even down to 2 L/(m^2·h). On the contrary, the P-GO-100 membrane exhibited a higher flux than the origin membrane and maintained the flux of 158 L/(m^2·h) after three cycles, even higher than the initial flux of the PVDF membrane (138.5 L/(m^2·h)). The *FRR* (Figure 12b) of the P-GO-DAA membranes was relatively higher during the whole cycles, especially for the P-GO-100 and P-GO-150 membranes. Among the prepared membranes, the *rejection rate* of the P-GO-DAA membrane was maintained at a higher level during the filtration process. The difference in *rejection rates* between the DAA-modified membranes was not significant, which was similar to the results observed before in Figure 9. It was seen from Table S4 that the average pore size of the pure PVDF dropped considerably after the test, indicating the serious pore blockage of the origin membrane, while the pores of the DAA-modified membrane did not change much.

The outstanding anti-biofouling capacity of the modified membranes was also showed in Figure 12c. In contrast with the P-GO-DAA membranes, the PVDF membrane and the P-GO membrane seemed to have no bacterial inhibition with the bacteria adhering ratio after three cycles reaching 84.22% and 80.98%, respectively. The bacteria adhering ratio of the P-GO-100 membrane was only 26.64%, about 60% and 56% lower than the PVDF and P-GO membranes, respectively, indicating an excellent resistance to bacterial adhesion on the D-Tyrosine modified membranes. It was demonstrated that the addition of D-Tyrosine played a greater role in the antibacterial property. The CLSM images of the membranes (the two membranes circled in red in Figure 12c) after the experiment (Figure 13) also demonstrated the conclusions above mentioned. Both live and dead bacteria that adhered to the pure PVDF were significantly more than the modified membranes. Fewer bacteria adhered to the surface of the modified membrane. It was also indicated that probably due to the presence of DAA with the inhibiting effect of QS signals diminishing the bacterial attachment on the membrane surface.

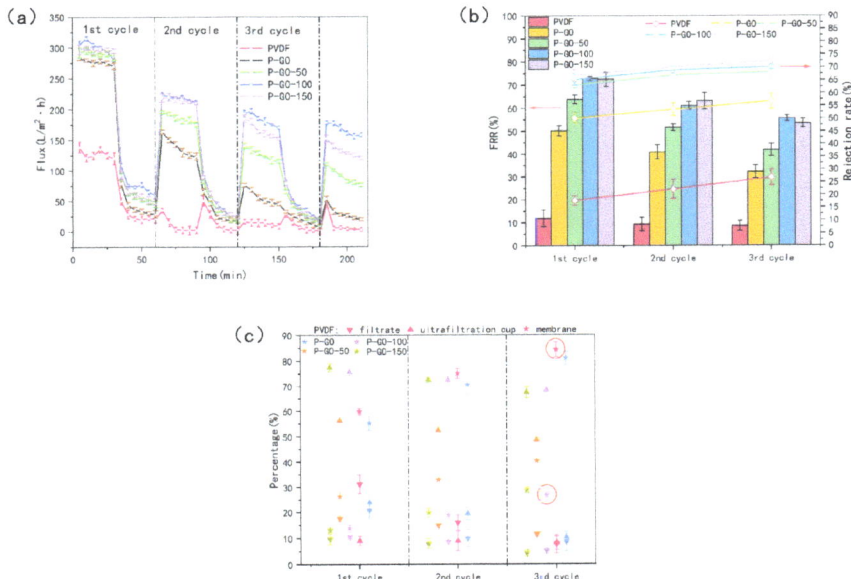

Figure 12. The results of three cycles ultrafiltration experiment. (**a**) The membrane flux change; (**b**) flux recovery rate and *rejection rate* of membranes; (**c**) the bacterial distribution in the test.

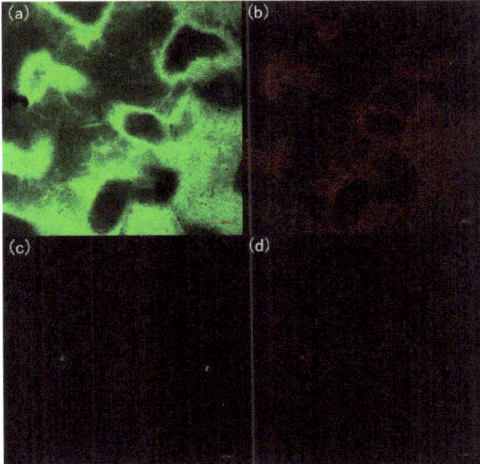

Figure 13. CLSM images of the membranes after the three cycles ultrafiltration experiment. (**a**) Live bacteria on the origin membrane; (**b**) dead bacteria on the origin membrane; (**c**) live bacteria on the P-GO-DAA membrane; (**d**) dead bacteria on the P-GO-DAA membrane.

The improved antifouling and anti-biofouling performances of the P-GO-DAA membranes were verified by SEM images (Figure 14). The surface roughness of the DAA-modified membrane did not have much change after the experiment comparing to the origin membrane.

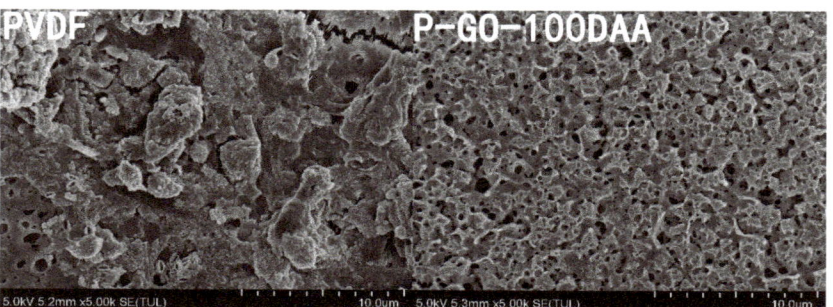

Figure 14. SEM images of the membranes after the three cycles ultrafiltration experiment.

To sum up, the addition of GO and D-Tyrosine to the membrane could weaken the interaction between the foulants and membrane surface due to the improved hydrophilic of the membrane, so that pollutants on the membrane surface are more easily carried away by the shear force of water. On the other hand, the addition of DAA could effectively inhibit bacterial adhesion and induce biofilm self-decomposition.

4. Conclusions

In this study, a P-GO-DAA membrane with strong resistance to organic and biological fouling was developed via the phase inversion technique by using GO and D-Tyrosine as modifiers. The FTIR results suggested that DAA might be introduced into the membrane by forming hydrogen bonds with GO. The addition of DAA to membrane formed a dense layer on the membrane surface observed from SEM. The porous structure and high hydrophilic nature of GO led to enhanced pore structure and surface hydrophilicity of the nanocomposite membranes, which were verified by the increasing porosity and decreasing water contact angle. By adding GO combined with DAA, the P-GO-DAA membrane exhibited a further improvement in pure water flux up to 305.27 L/(m^2·h) that was more than two folds of the PVDF membrane, and a more hydrophilic surface was confirmed by the decline in water contact angle compared with the PVDF membrane. The modified membranes displayed better separate ability contrast with the pure PVDF. The P-GO-DAA membrane had the best separation performance with a rejection rate of 95% especially for HA. The antifouling ability of the P-GO-DAA membrane became stronger evidenced by its significantly increased FRR for various foulants, especially for actual sewage by 70% compared to the origin membrane. For the anti-biofouling activity of the membranes, the addition of DAA played a key role with the results of 60% and 56% less surface bacterial adhesion than of the PVDF and P-GO membranes, respectively, and almost no bacteria adhered to the P-GO-DAA membrane also observed by CLSM. It was supposed that DAA could diminish the interaction between the membrane and bacteria by impacting the secretion of QS signals. Consequently, it could be concluded that the GO blended DAA-modified nanocomposite membranes possess a synergistic effect on the antifouling and anti-biofouling activity of the membrane.

Supplementary Materials: The following supporting information can be downloaded at: https://www.mdpi.com/article/10.3390/membranes12050486/s1. Figure S1: Schematic diagram of the Intermolecular hydrogen bonding between GO and D-Tyrosine in the membrane matrix, Figure S2: FTIR spectra of GO and GO-DAA, Figure S3: The surface layer thickness of the membranes, (a) PVDF, (b) P-GO, (c) P-GO-50, (d) P-GO-100, (e) P-GO-150, Table S1: Components in the polymer casting solutions, Table S2: Elemental composition of three membranes' surface, Table S3: Carbon chemical bond composition of three membranes' surface, Table S4: Average pore size and porosity of fouled membranes, Table S5: Infrared absorption peaks with corresponding chemical bonds.

Author Contributions: Conceptualization, J.G. and Y.Z.; methodology, J.G.; software, J.G.; validation, J.G., F.C., and Y.C.; formal analysis, J.G.; investigation, J.G.; resources, J.G.; data curation, J.G., F.C. and Y.C.; writing—original draft preparation, J.G.; writing—review and editing, J.G. and Y.Z.; visualization, J.G.; supervision, F.C. and Y.C.; project administration, Y.Z.; funding acquisition, Y.Z. All authors have read and agreed to the published version of the manuscript.

Funding: This research was funded by National Natural Science Foundation of China, grant number 51478015.

Institutional Review Board Statement: Not applicable.

Informed Consent Statement: Not applicable.

Data Availability Statement: Not applicable.

Conflicts of Interest: The authors declare no conflict of interest.

References

1. Wang, X.; Sun, M.; Zhao, Y.; Wang, C.; Ma, W.; Wong, M.S.; Elimelech, M. In Situ Electrochemical Generation of Reactive Chlorine Species for Efficient Ultrafiltration Membrane Self-Cleaning. *Environ. Sci. Technol.* **2020**, *54*, 6997–7007. [CrossRef] [PubMed]
2. Zhou, M.Y.; Zhang, P.; Fang, L.F.; Zhu, B.K.; Wang, J.L.; Chen, J.H.; Abdallah, H. A positively charged tight UF membrane and its properties for removing trace metal cations via electrostatic repulsion mechanism. *J. Hazard. Mater.* **2019**, *373*, 168–175. [CrossRef] [PubMed]
3. Vatanpour, V.; Haghighat, N. Improvement of polyvinyl chloride nanofiltration membranes by incorporation of multiwalled carbon nanotubes modified with triethylenetetramine to use in treatment of dye wastewater. *J. Environ. Manag.* **2019**, *242*, 90–97. [CrossRef] [PubMed]
4. Chen, G.; Lin, Q.; Chen, S.; Chen, X. In-plane biaxial ratcheting behavior of PVDF UF membrane. *Polym. Test.* **2016**, *50*, 41–48. [CrossRef]
5. Tang, W.; Zhang, Y.; Bai, J.; Li, J.; Wang, J.; Li, L.; Zhou, T.; Chen, S.; Rahim, M.; Zhou, B. Efficient denitrification and removal of natural organic matter, emerging pollutants simultaneously for RO concentrate based on photoelectrocatalytic radical reaction. *Sep. Purif. Technol.* **2020**, *234*, 116032. [CrossRef]
6. Esfahani, M.R.; Koutahzadeh, N.; Esfahani, A.R.; Firouzjaei, M.D.; Anderson, B.; Peck, L. A novel gold nanocomposite membrane with enhanced permeation, rejection and self-cleaning ability. *J. Membr. Sci.* **2019**, *573*, 309–319. [CrossRef]
7. Liu, Q.; Li, L.; Pan, Z.; Dong, Q.; Xu, N.; Wang, T. Inorganic nanoparticles incorporated in polyacrylonitrile-based mixed matrix membranes for hydrophilic, ultrafast, and fouling-resistant ultrafiltration. *J. Appl. Polym. Sci.* **2019**, *136*, 47902. [CrossRef]
8. Deng, H.; Zheng, Q.; Chen, H.; Huang, J.; Yan, H.; Ma, M.; Xia, M.; Pei, K.; Ni, H.; Ye, P. Graphene oxide/silica composite nanofiltration membrane: Adjustment of the channel of water permeation. *Sep. Purif. Technol.* **2022**, *278*, 119440. [CrossRef]
9. Fan, G.; Chen, C.; Chen, X.; Li, Z.; Bao, S.; Luo, J.; Tang, D.; Yan, Z. Enhancing the antifouling and rejection properties of PVDF membrane by Ag3PO4-GO modification. *Sci. Total Environ.* **2021**, *801*, 149611. [CrossRef]
10. Joshi, R.K.; Carbone, P.; Wang, F.C.; Kravets, V.G.; Su, Y.; Grigorieva, I.V.; Wu, H.A.; Geim, A.K.; Nair, R.R. Precise and ultrafast molecular sieving through graphene oxide membranes. *Science* **2014**, *343*, 752–754. [CrossRef]
11. Nair, R.R.; Wu, H.A.; Jayaram, P.N.; Grigorieva, I.V.; Geim, A.K. Unimpeded permeation of water through helium-leak-tight graphene-based membranes. *Science* **2012**, *335*, 442–444. [CrossRef] [PubMed]
12. Shen, L.; Xie, Q.; Hong, Z.; Wu, C.; Yu, T.; Fang, H.; Xiong, Y.; Zhang, G.; Lu, Y.; Shao, W. Facile Strategy to Construct High-Performance Nanofiltration Membranes by Synergy of Graphene Oxide and Polyvinyl Alcohol. *Ind. Eng. Chem. Res.* **2020**, *59*, 19001–19011. [CrossRef]
13. Rahimi, A.; Mahdavi, H. Zwitterionic-functionalized GO/PVDF nanocomposite membranes with improved anti-fouling properties. *J. Water Process Eng.* **2019**, *32*, 100960. [CrossRef]
14. Wu, X.; Field, R.W.; Wu, J.J.; Zhang, K. Polyvinylpyrrolidone modified graphene oxide as a modifier for thin film composite forward osmosis membranes. *J. Membr. Sci.* **2017**, *540*, 251–260. [CrossRef]
15. Khan, N.A.; Yuan, J.; Wu, H.; Cao, L.; Zhang, R.; Liu, Y.; Li, L.; Rahman, A.U.; Kasher, R.; Jiang, Z. Mixed Nanosheet Membranes Assembled from Chemically Grafted Graphene Oxide and Covalent Organic Frameworks for Ultra-high Water Flux. *ACS Appl. Mater. Interfaces* **2019**, *11*, 28978–28986. [CrossRef]
16. Yadav, S.; Ibrar, I.; Samal, A.K.; Altaee, A.; Deon, S.; Zhou, J.; Ghaffour, N. Preparation of fouling resistant and highly permselective novel PSf/GO-vanillin nanofiltration membrane for efficient water purification. *J. Hazard. Mater.* **2022**, *421*, 126744. [CrossRef] [PubMed]

17. Hu, M.; Cui, Z.; Li, J.; Zhang, L.; Mo, Y.; Dlamini, D.S.; Wang, H.; He, B.; Li, J.; Matsuyama, H. Ultra-low graphene oxide loading for water permeability, antifouling and antibacterial improvement of polyethersulfone/sulfonated polysulfone ultrafiltration membranes. *J. Colloid. Interface Sci.* **2019**, *552*, 319–331. [CrossRef]
18. Wang, L.; Wang, N.; Li, J.; Li, J.; Bian, W.; Ji, S. Layer-by-layer self-assembly of polycation/GO nanofiltration membrane with enhanced stability and fouling resistance. *Sep. Purif. Technol.* **2016**, *160*, 123–131. [CrossRef]
19. Khan, M.M.T.; Stewart, P.S.; Moll, D.J.; Mickols, W.E.; Burr, M.D.; Nelson, S.E.; Camper, A.K. Assessing biofouling on polyamide reverse osmosis (RO) membrane surfaces in a laboratory system. *J. Membr. Sci.* **2010**, *349*, 429–437. [CrossRef]
20. Flemming, H.-C.; Griebe, T.; Schaule, G.; Schmitt, J.; Tamachkiarowa, A. Biofouling—The Achille's heel of membrane processes. *Desalination* **1997**, *113*, 215–225. [CrossRef]
21. Wang, Q.; Hu, M.; Wang, Z.; Hu, W.; Cao, J.; Wu, Z.-C. Uniqueness of biofouling in forward osmosis systems: Mechanisms and control. *Crit. Rev. Environ. Sci. Technol.* **2018**, *48*, 1031–1066. [CrossRef]
22. Kochkodan, V.; Hilal, N. A comprehensive review on surface modified polymer membranes for biofouling mitigation. *Desalination* **2015**, *356*, 187–207. [CrossRef]
23. Sun, X.-F.; Qin, J.; Xia, P.-F.; Guo, B.-B.; Yang, C.-M.; Song, C.; Wang, S.-G. Graphene oxide–silver nanoparticle membrane for biofouling control and water purification. *Chem. Eng. J.* **2015**, *281*, 53–59. [CrossRef]
24. Hu, J.Y.; Song, L.F.; Ong, S.L.; Phua, E.T.; Ng, W.J. Biofiltration pretreatment for reverse osmosis (RO) membrane in a water reclamation system. *Chemosphere* **2005**, *59*, 127–133. [CrossRef] [PubMed]
25. Asha, A.B.; Chen, Y.; Narain, R. Bioinspired dopamine and zwitterionic polymers for non-fouling surface engineering. *Chem. Soc. Rev.* **2021**, *50*, 11668–11683. [CrossRef]
26. Guo, X.; Yang, H.; Liu, Q.; Liu, J.; Chen, R.; Zhang, H.; Yu, J.; Zhang, M.; Li, R.; Wang, J. A chitosan-graphene oxide/ZIF foam with anti-biofouling ability for uranium recovery from seawater. *Chem. Eng. J.* **2020**, *382*, 122850. [CrossRef]
27. Gao, Z.; Yu, Z.; Zhang, X.; Fan, S.; Gao, H.; Liu, C.; Zhou, Q.; Shao, H.; Wang, L.; Guo, X. Exploration on Optimized Control Way of D-Amino Acid for Efficiently Mitigating Membrane Biofouling of Membrane Bioreactor. *Membranes* **2021**, *11*, 612. [CrossRef]
28. Li, Y.; Wang, H.; Wang, S.; Xiao, K.; Huang, X. Enzymatic Cleaning Mitigates Polysaccharide-Induced Refouling of RO Membrane: Evidence from Foulant Layer Structure and Microbial Dynamics. *Environ. Sci. Technol* **2021**, *55*, 5453–5462. [CrossRef]
29. Kolodkin-Gal, I.; Cao, S.; Chai, L.; Bottcher, T.; Kolter, R.; Clardy, J.; Losick, R. A self-produced trigger for biofilm disassembly that targets exopolysaccharide. *Cell* **2012**, *149*, 684–692. [CrossRef]
30. Kolodkin-Gal, I.; Romero, D.; Cao, S.; Clardy, J.; Kolter, R.; Losick, R. D-amino acids trigger biofilm disassembly. *Science* **2010**, *328*, 627–629. [CrossRef]
31. Xing, S.F.; Sun, X.F.; Taylor, A.A.; Walker, S.L.; Wang, Y.F.; Wang, S.G. D-amino acids inhibit initial bacterial adhesion: Thermodynamic evidence. *Biotechnol. Bioeng.* **2015**, *112*, 696–704. [CrossRef] [PubMed]
32. Yu, C.; Wu, J.; Contreras, A.E.; Li, Q. Control of nanofiltration membrane biofouling by Pseudomonas aeruginosa using D-Tyrosine. *J. Membr. Sci.* **2012**, *423–424*, 487–494. [CrossRef]
33. Yu, C.; Li, X.; Zhang, N.; Wen, D.; Liu, C.; Li, Q. Inhibition of biofilm formation by D-Tyrosine: Effect of bacterial type and D-Tyrosine concentration. *Water Res.* **2016**, *92*, 173–179. [CrossRef] [PubMed]
34. Xiong, Y.; Liu, Y. Biological control of microbial attachment: A promising alternative for mitigating membrane biofouling. *Appl. Microbiol. Biotechnol.* **2010**, *86*, 825–837. [CrossRef] [PubMed]
35. Xu, H.; Liu, Y. D-Amino acid mitigated membrane biofouling and promoted biofilm detachment. *J. Membr. Sci.* **2011**, *376*, 266–274. [CrossRef]
36. Yu, C.; Wu, J.; Zin, G.; Di Luccio, M.; Wen, D.; Li, Q. D-Tyrosine loaded nanocomposite membranes for environmental-friendly, long-term biofouling control. *Water Res.* **2018**, *130*, 105–114. [CrossRef]
37. Lam, H.; Oh, D.C.; Cava, F.; Takacs, C.N.; Clardy, J.; de Pedro, M.A.; Waldor, M.K. D-amino acids govern stationary phase cell wall remodeling in bacteria. *Science* **2009**, *325*, 1552–1555. [CrossRef]
38. Guo, X.; Fan, S.; Hu, Y.; Fu, X.; Shao, H.; Zhou, Q. A novel membrane biofouling mitigation strategy of D-amino acid supported by polydopamine and halloysite nanotube. *J. Membr. Sci.* **2019**, *579*, 131–140. [CrossRef]
39. Khan, R.; Khan, M.K.; Wang, H.; Xiao, K.; Huang, X. Grafting d-amino acid onto MF polyamide nylon membrane for biofouling control using biopolymer alginate dialdehyde as a versatile platform. *Sep. Purif. Technol.* **2020**, *231*, 115891. [CrossRef]
40. Jiang, B.-B.; Sun, X.-F.; Wang, L.; Wang, S.-Y.; Liu, R.-D.; Wang, S.-G. Polyethersulfone membranes modified with D-Tyrosine for biofouling mitigation: Synergistic effect of surface hydrophily and anti-microbial properties. *Chem. Eng. J.* **2017**, *311*, 135–142. [CrossRef]
41. Abdel-Karim, A.; Gad-Allah, T.A.; El-Kalliny, A.S.; Ahmed, S.I.A.; Souaya, E.R.; Badawy, M.I.; Ulbricht, M. Fabrication of modified polyethersulfone membranes for wastewater treatment by submerged membrane bioreactor. *Sep. Purif. Technol.* **2017**, *175*, 36–46. [CrossRef]
42. Miller, D.J.; Araujo, P.A.; Correia, P.B.; Ramsey, M.M.; Kruithof, J.C.; van Loosdrecht, M.C.; Freeman, B.D.; Paul, D.R.; Whiteley, M.; Vrouwenvelder, J.S. Short-term adhesion and long-term biofouling testing of polydopamine and poly(ethylene glycol) surface modifications of membranes and feed spacers for biofouling control. *Water Res.* **2012**, *46*, 3737–3753. [CrossRef] [PubMed]
43. Xie, W.; Han, Y.; Tai, S. Biodiesel production using biguanide-functionalized hydroxyapatite-encapsulated-γ-Fe_2O_3 nanoparticles. *Fuel* **2017**, *210*, 83–90. [CrossRef]

44. Yadav, S.; Ibrar, I.; Altaee, A.; Déon, S.; Zhou, J. Preparation of novel high permeability and antifouling polysulfone-vanillin membrane. *Desalination* **2020**, *496*, 114759. [CrossRef]
45. Khalili, D. Graphene oxide: A promising carbocatalyst for the regioselective thiocyanation of aromatic amines, phenols, anisols and enolizable ketones by hydrogen peroxide/KSCN in water. *New J. Chem.* **2016**, *40*, 2547–2553. [CrossRef]
46. Sri Abirami Saraswathi, M.S.; Rana, D.; Divya, K.; Gowrishankar, S.; Nagendran, A. Versatility of hydrophilic and antifouling PVDF ultrafiltration membranes tailored with polyhexanide coated copper oxide nanoparticles. *Polym. Test.* **2020**, *84*, 106367. [CrossRef]
47. Sukitpaneenit, P.; Chung, T.-S. Molecular design of the morphology and pore size of PVDF hollow fiber membranes for ethanol–water separation employing the modified pore-flow concept. *J. Membr. Sci.* **2011**, *374*, 67–82. [CrossRef]
48. Al-Jeshi, S.; Neville, A. An investigation into the relationship between flux and roughness on RO membranes using scanning probe microscopy. *Desalination* **2006**, *189*, 221–228. [CrossRef]
49. Feng, S.; Yu, G.; Cai, X.; Eulade, M.; Lin, H.; Chen, J.; Liu, Y.; Liao, B.Q. Effects of fractal roughness of membrane surfaces on interfacial interactions associated with membrane fouling in a membrane bioreactor *Bioresour. Technol.* **2017**, *244*, 560–568. [CrossRef]
50. Monash, P.; Pugazhenthi, G. Effect of TiO_2 addition on the fabrication of ceramic membrane supports: A study on the separation of oil droplets and bovine serum albumin (BSA) from its solution. *Desalination* **2011**, *279*, 104–114. [CrossRef]
51. Bai, Z.; Zhang, R.; Wang, S.; Gao, S.; Tian, J. Membrane fouling behaviors of ceramic hollow fiber microfiltration (MF) membranes by typical organic matters. *Sep. Purif. Technol.* **2021**, *274*, 118951. [CrossRef]
52. Tian, J.-Y.; Ernst, M.; Cui, F.; Jekel, M. Effect of different cations on UF membrane fouling by NOM fractions. *Chem. Eng. J.* **2013**, *223*, 547–555. [CrossRef]
53. Zhao, X.; Zhang, R.; Liu, Y.; He, M.; Su, Y.; Gao, C.; Jiang, Z. Antifouling membrane surface construction: Chemistry plays a critical role. *J. Membr. Sci.* **2018**, *551*, 145–171. [CrossRef]
54. Khalid, A.; Abdel-Karim, A.; Ali Atieh, M.; Javed, S.; McKay, G. PEG-CNTs nanocomposite PSU membranes for wastewater treatment by membrane bioreactor. *Sep. Purif. Technol.* **2018**, *190*, 165–176. [CrossRef]

Article

Hydrophilic and Positively Charged Polyvinylidene Fluoride Membranes for Water Treatment with Excellent Anti-Oil and Anti-Biocontamination Properties

Zirui Wang [1], Shusu Shen [1,2,*], Linbin Zhang [1], Abdessamad Ben Hida [1] and Ganwei Zhang [1,2]

[1] School of Environmental Science and Engineering, Suzhou University of Science and Technology, 1 Kerui Road, Suzhou 215009, China; 15956940790@139.com (Z.W.); lb_zhang2022@163.com (L.Z.); abdessamadbenhida@gmail.com (A.B.H.); zhangganwei214036@126.com (G.Z.)
[2] Jiangsu Engineering Research Center for Separation and Purification Materials and Technology, 1 Kerui Road, Suzhou 215009, China
* Correspondence: shusushen@usts.edu.cn

Abstract: Membrane fouling limits the rapid development of membrane separations. In this study, a blend membrane containing polycationic liquid (P(BVImBr-co-PEGMA$_1$)) is presented that can improve the antifouling performance of polyvinylidene fluoride (PVDF) membranes. By mixing the polycationic liquid into PVDF, an improved membrane-surface hydrophilicity and enlarged membrane porosity were detected. The water contact angle decreased from 82° to 67°, the porosity enlarged from 7.22% to 89.74%, and the pure water flux improved from 0 to 631.68 L m^{-2} h^{-1}. The blend membrane surfaces were found to be always positively charged at pH 3~10. By applying the membranes to the filtration of oil/water emulsion and bovine serum albumin (BSA) solution, they showed a very high rejection rate to pollutants in wastewater (99.4% to oil droplets and 85.6% to BSA). The positive membrane surface charge and the increased membrane hydrophilicity resulted in excellent antifouling performance, with the flux recovery rates of the dynamic filtration tests reaching 97.3% and 95.5%, respectively. Moreover, the blend membranes demonstrated very low BSA adhesion and could even kill *S. aureus*, showing excellent antifouling properties.

Keywords: blend membrane; water treatment; antifouling modification; hydrophilic; polycationic liquid

1. Introduction

Membrane separation technology has great potential in water and wastewater treatment. Based on the pore size diameter, membranes are often classified into microfiltration membranes, ultrafiltration membranes, nanofiltration membranes, and reverse osmosis membranes [1]. According to different membrane materials, membranes can be divided into inorganic membranes, organic membranes, or organic-inorganic hybrid membranes. Among various matrix membrane materials, polyvinylidene fluoride (PVDF) is the most-used organic polymer membrane material due to its thermal stability, excellent chemical resistance, and good membrane-forming ability [2].

However, PVDF is a semi-crystalline polymer with repeating units of –CH$_2$–CF$_2$– that form hydrophobic structures that make the membrane prone to fouling, which leads to a reduction in membrane performance by reducing permeability and imposing additional costs for membrane cleaning or replacement [3]. Membrane fouling leads to easy blockage of membrane pores, increases the transmembrane pressure, and decreases the water flux. In addition, it increases the cleaning cost of the membrane and, finally, shortens the service life of the membrane. It is generally believed that a hydrophilic membrane could somehow overcome membrane fouling problems [4–7], because the hydrophilic membrane easily forms a hydration layer on the membrane surface that can effectively reduce the adhesion between pollutants and the membrane surface. In addition, the charge on the membrane surface is also an important factor that slows down membrane fouling [8], because some pollutants in water are charged. The electrostatic action between the membrane surface and the charged pollutants in water can reduce the adhesion of pollutants, and even improve the interception effect of the membrane on charged pollutants [9,10].

To achieve antifouling PVDF membranes, different kinds of modifications have been developed [11,12]. The modification methods for PVDF membranes mainly include membrane surface modifications such as surface coating [13,14] and surface grafting [15,16], and membrane bulk modifications such as copolymerization modification [17,18] or blending modification [19,20]. Among them, blending modification can afford uniform membrane structure, consistent chemical composition, good separation effect, etc.

Polyionic liquids, also known as polymer ionic liquids [21], have been reported to improve the hydrophilicity of membranes and charge the membranes, so they are quite applicable in solving the membrane fouling problem [22,23]. Many of these zwitterions have been employed as polymer brushes to modify the surface of materials with enhanced hydrophilicity and antifouling ability [24,25]. By blending a polyionic liquid, P(MMA-co-BVIm-Br) with PVDF, the obtained membranes showed a reduced fouling rate from 68% to 40% [26].

In our previous report [27], a polycationic liquid P(BVImBr$_1$-co-PEGMA$_2$), was blended with PVDF. The prepared membranes showed good repellence (up to 99%) against positively charged BSA, and the flux recovery rate was improved to 76%. However, the pollutant object of membrane treatment was limited, and the research was not deep enough.

For our ongoing study, blend PVDF membranes with a newly synthesized polycationic liquid, P(BVImBr$_1$-co-PEGMA$_1$) (P11), were fabricated. The major difference in this work is that the monomer ratio of the copolymer was switched from 1/2 to 1/1. However, the membrane performance was found to be greatly improved. The properties of the modified membranes were carefully examined and the separation efficiency of different organic pollutants, including the oil droplets and the typical protein, BSA, were explored in this work. In addition, the antibacterial properties of the membrane were also tested.

2. Materials and Methods

2.1. Materials

PVDF (FR 904, >99.5%, Mw 400,000) was purchased from 3F New Materials Co. Ltd., Shanghai, China. Poly(ethylene glycol) methyl ether methacrylate (PEGMA, average Mn 950, contains 300 ppm BHT and 100 ppm MEHQ as inhibitor) was purchased from Aldrich (St. Louis, MO, USA). 1-Vinylimidazole ($C_5H_6N_2$, 99%), 1-bromobutane (C_4H_9Br, >99%), N,N-dimethylformamide (DMF, AR) and azobisisobutyronitrile (AIBN, 98%), n-hexadecane (AR, 98%) were supplied by Macklin Biochemical Technology Co., Ltd., Beijing, China. BSA (Mn 67 kDa) was obtained from Aladdin Chemical Co., Ltd., Shanghai, China. Other chemicals utilized in this study were purchased with analytical quality and purified before use. Deionized (DI) water (18.2 MΩ) purified with a Milli-Q system from Millipore (Burlington, MA, USA) was used to prepare all solutions as needed in the work.

2.2. Synthesis of Polycationic Liquid P(BVImBr$_1$-co-PEGMA$_1$) (P11)

The utilized polycationic liquid, or so-called cationic polyionic liquid, (P(BVImBr$_1$-co-PEGMA$_1$)) was synthesized by following the similar protocol in our previous research [27], but in a different monomer ratio (1/1). The product is denoted as P11 in later description. It was characterized by ^{1}H NMR (Bruker Avance 300) and IR analysis. Figure 1 shows the ^{1}H NMR spectra (using CDCl$_3$ as internal standard) of the obtained P11 and the starting materials. The disappearance of the peaks a' originated from BVImBr and the peaks j' originated from PEGMA indicates that the product P11 was successfully synthesized via the RAFT reaction, where the ratio of BVImBr/PEGMA was around 1/1. The IR analysis can be found in the latter Figure 2a.

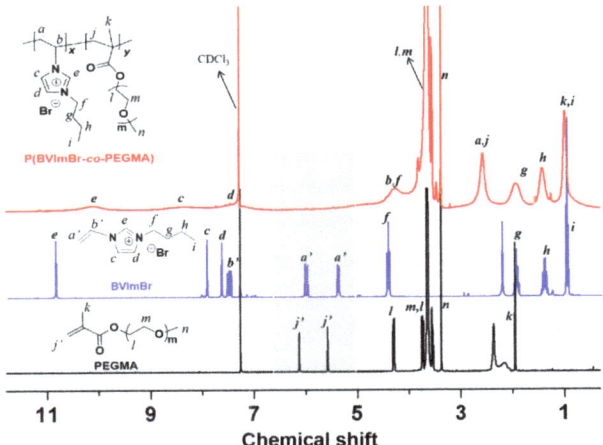

Figure 1. The ^{1}H NMR spectra compare of the obtained P11 and the starting materials.

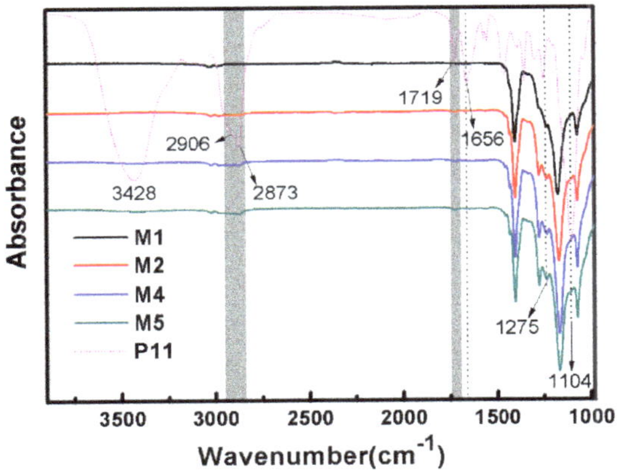

Membranes	Atomic %				
	F1s	N1s	O1s	Br3d	C1s
M0	43.62	0.78	2.96	0.06	52.59
M1	39.02	1.22	4.32	0.07	55.37
M4	39.5	0.93	5.19	0.07	54.31
M5	35.8	1.09	7.34	0.14	55.63

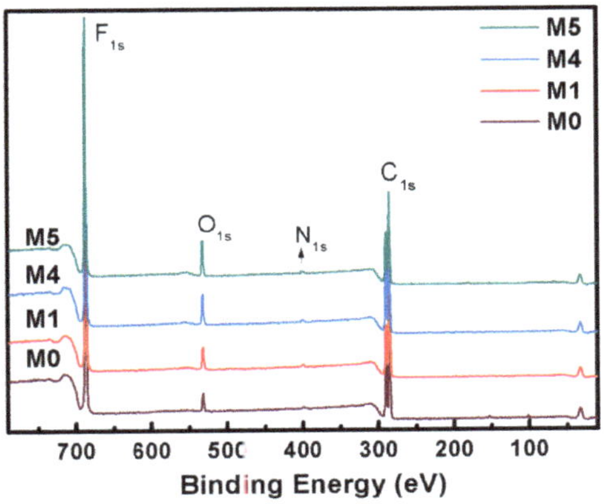

Figure 2. (**a**) FT-IR spectra of P11 and membranes; (**b**) full XPS spectra of membranes; (**c**) the elemental composition of membrane surface; and (**d**) the C core layer XPS spectrum of M5.

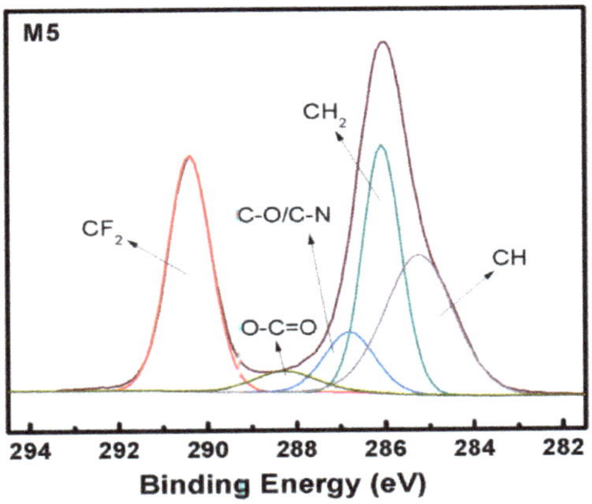

2.3. Preparation of Membranes

The flat sheet membranes were prepared by blending P11 with PVDF in different weight ratios and the polymer concentrations were 20 wt% and 22 wt%, respectively, via a non-solvent-induced phase separation (NIPS) method [27]. Six different membranes were prepared in this study; the composition of the casting solution and some of the properties are summarized in Table 1. The polymer concentrations were 20 wt% (M0, M1, M2) and 22 wt% (M3, M4, M5), and for different membrane samples, the ratio of PVDF/P11 was 10/0 (M0, M3, pure PVDF membranes), 9/1 (M1, M4) and 8/2 (M2, M5), respectively.

Table 1. The casting solution composition and general properties of the membranes.

Membranes	M0	M1	M2	M3	M4	M5
Polymer mixture	PVDF	PVDF/P11 = 9/1	PVDF/P11 = 8/2	PVDF	PVDF/P11 = 9/1	PVDF/P11 = 8/2
Polymer concentration/wt%	20	20	20	22	22	22
Porosity/%	7.22 ± 1.75	75.31 ± 1.24	89.74 ± 2.01	5.18 ± 0.95	70.33 ± 1.56	82.65 ± 2.20
Mean pore size/nm	/	29.91 ± 1.02	35.20 ± 1.31	/	24.75 ± 1.14	27.32 ± 1.57
Pure water flux/L $m^{-2}\ h^{-1}$	/	427.47 ± 10.32	631.68 ± 15.79	/	193.24 ± 7.39	238.63 ± 6.50
Thickness/μm	30.4 ± 0.28	113.6 ± 1.08	121.2 ± 0.92	39.2 ± 0.83	92.3 ± 1.02	149.0 ± 0.71
Mechanical strength/MPa	1.82 ± 0.24	0.83 ± 0.36	0.48 ± 0.15	2.01 ± 0.69	0.94 ± 0.34	0.54 ± 0.28

2.4. Characterization of Membranes

2.4.1. Fourier Transform Infra-Red (FTIR) Spectroscopy

FT-IR model (Nicolet 6700) was used to analyze the surface functional groups of the membranes, with a spectral range of 500~4000 cm^{-1} and a resolution of 2 cm^{-1}.

2.4.2. Morphological Analysis

XPS measurement was carried out on the membrane surface using Thermo Scientific, ESCALAB 250Xi. The elements tested on the membrane surface were C, F, O, N and Br. SEM (Phenom Pro, USA) was used to observe the surface, cross-sectional structure, and thickness of membrane samples. AFM (Bruker Dimension Icon, Tucson, AZ, USA) was used to measure the surface morphology and roughness of the membranes; the sample tested was 5 μm × 5 μm.

2.4.3. Contact Angles

The static water contact angle and underwater oil contact angle of the membranes were measured by the static hanging drop method, with a membrane surface contact angle tester (Ramé-Hart 500). The contact angle was measured at 5 different positions on each sample, the average value was calculated and recorded with the obtained data, and the accepted error range was less than 3.

2.4.4. Mechanical Characterization

The mechanical strength was tested by a tensile strength tester (5944, Instron, Norwood, MA, USA); 3–5 strips of 5 cm × 1 cm were measured from different positions, and the average value was recorded.

2.4.5. Zeta Potential and Thermogravimetric Analysis

The membrane surface charge properties were measured with a membrane solid sample flow field potential analyzer (Surpass 3). An SDT 2960 analyzer was used for thermogravimetric analysis (TGA).

2.4.6. Porosity and Pore Size

Aperture measurement involved randomly measuring 100 membrane pores on the surface image of electron micrograph with Nano Measure, and the average pore size D (nm) was calculated. The membrane porosity was measured by cutting each membrane sample into circular slices with a diameter of 2.5 cm, which were washed in ethanol and then soaked in deionized water for 24 h. The soaked membranes were removed and residual water on the membrane surface was gently wiped with dust-free paper. The mass of the wet sample was weighed and recorded as m_1. The wet membranes were dried in a vacuum drying oven (VD115, BINDER, Tuttlingen, Germany) at 60 °C until a consistent mass was obtained, m_2. The membrane porosity ε was calculated by Equation (1):

$$\varepsilon = \frac{m_1 - m_2}{\rho \cdot A \cdot d} \times 100\% \tag{1}$$

where ε represents the porosity (%), m_1 and m_2 are the wet weight and dry weight (kg), ρ represents the density of water (kg m^{-2}), A is the area (m^2), and d is the thickness (m) of the membrane.

2.5. Membrane Performance: Pure Water Flux and Dynamic Filtration Tests

A pure water flux test was carried out by using a dead-end filtration system. The membrane sample was cut into a certain size wafer and fixed in an ultrafiltration cup, with an effective filtration area of 8.55 cm^2. At first, the membrane was pretreated under 0.12 MPa pressure for 30 min, and was then filtered by pure water under 0.1 MPa pressure. The filtered pure water was collected every 10 min and the volume (L) of the collected water was measured and recorded. Until the effluent volume reached a stable value, the pure water flux J_0 was calculated by Equation (2):

$$J_0 = \frac{V}{S \cdot t} \tag{2}$$

where J_0 is pure water flux (L m^{-2} h^{-1}), V is the volume of pure water passing through the membrane (L), S is the effective area of water passing through (m$_2$), and t is the time of each water intake (h).

Oil/water emulsion separation: The oil/water emulsion (0.1 g L^{-1}) was prepared by using 0.5 g of n-hexadecane diluted into 5 L deionized water. The operation for the oil/water separation was carried out by following the pure water flux test. After the stable flux J_0 of pure water was calculated, the oil/water emulsion was continuously filtered by the same membrane sample for another 120 min. The flux of oil/water emulsions for 1 min was recorded every 10 min. The permeation flux after 120 min was recorded as J_p. Then, the fouled membrane was taken out and placed in deionized water for 10 min for ultrasonic cleaning. The cleaned membrane was put back into the dead-end filtration system for testing with another pure water flux and the data were recorded as J_c. During the filtration of oil/water emulsion, the oil concentration was detected with an organic carbon analyzer (TOC-L CPH, Shimadzu, Japan). The dynamic antifouling properties were evaluated by the flux decay rate (RFD) and the relative flux recovery rate (RFR) according to Equations (3) and (4) below:

$$RFD = \frac{J_c}{J_0} \times 100\% \tag{3}$$

$$RFR = \left(1 - \frac{J_p}{J_0}\right) \times 100\% \tag{4}$$

where J_0 is the initial pure water flux (L m^{-2} h^{-1}), J_p is the permeation flux (L m^{-2} h^{-1}) after running for 120 min, and J_c is the pure water flux (L m^{-2} h^{-1}) measured after membrane cleaning.

BSA solution separation: 1.0 g L^{-1} of BSA solutions at different pH were prepared by using phosphate buffer or acetic acid/sodium acetate buffer. The filtration operation of the BSA solution was almost the same as the oil/water emulsion separations. The concentrations of BSA before and after the filtration were measured by a UV-vis spectrometer. The rejection rate (R) was obtained by Equation(5):

$$R = \left(1 - \frac{C_p}{C_r}\right) \times 100\% \tag{5}$$

where C_p represents pollutant concentration in the filtrate (mg L^{-1}) and C_r represents pollutant concentration in feed liquid (mg L^{-1}).

2.6. Static Antifouling Tests

The BSA adsorption experiment proceeded as follows: 0.1 g of dried membrane sample was placed into 50 mL of 1.0 g L^{-1} BSA solution (pH 7.0) and shaken for 24 h. After that, the membrane was taken out of the solution and the BSA concentration of the residual solution was measured by UV-vis spectrometer. The adsorption capacity (mg g^{-1}) of the membrane was then calculated.

The antibacterial test of the membranes was carried out via *S. aureus* suspension by first cutting the sterilized membrane sample into a circular piece with a diameter of 15 mm and laying it on the bottom of the 24-well plate. 100 μL of *S. aureus* suspension (10^6 CFU mL^{-1}) was extracted and dropped onto the membrane sample, and the 24-well plate was placed in the shaking table at 37 °C and 80 rpm. After incubation for 2 h, the bacterial solution was resuspended with 900 μL PBS. Then, 100 μL of bacterial solution after resuspension was extracted and spread evenly onto a solid medium. The coated solid medium was put in the shaking table and incubated upside-down for 24 h at 37 °C. Finally, the bacterial culture dish was photographed and the number of the remaining *S. aureus* was counted using Image J software. Each experiment was repeated at least three times.

3. Results and Discussion

3.1. Surface Chemical Composition of Membranes

FT-IR analysis was conducted to determine the surface chemical composition of the membranes. The IR spectrum of P11 is also shown in Figure 2a, with the peaks at 1104 cm^{-1} and 1275 cm^{-1} attributed to stretching vibration and bending vibration of the C-O bond, and the peak at 1719 cm^{-1} assigned to the C=O double bond in the PEGMA segment. In addition, 1656 cm^{-1} can be assigned to C=C and C=N double bonds in the BVIm-Br segment, and the peaks at 2873 and 2906 cm^{-1} must be the saturated C-H bonds that present in P11. Compared with pure PVDF membrane M0, the blend membranes M1, M4, and M5 showed obvious characteristic peaks at 1719 cm^{-1} and 2873~2906 cm^{-1}, with the peak intensity enhanced with increased blending ratios of P11.

XPS analysis of the membrane surface further proved the successful blending of P11. Figure 2b shows the full XPS spectra of the pure PVDF membrane (M0) and the blend membranes M1, M4, and M5. Compared with M0, the blend membranes containing P11 had a new peak of N1s at 401 eV, and the peak area enhanced with the increasing content of P11. The peak area of O1s at 532 eV also increased due to the greater content of P11 in the polymer mixture. The elemental composition of the membrane surface is shown in Figure 2c; by increasing P11 in the membrane from M0 to M5, the atomic ratio of F atom on the membrane surface decreased and the atomic ratio of N, O, and Br atoms improved gradually. Figure 2d is the C core layer XPS spectrum of the blend membrane M5; the peak analysis indicated characteristic peaks at 288 eV and 287 eV, which can be corresponded to O-C=O and C-O/C-N bonds which exist in the blended P11.

3.2. SEM Images of the Membranes

As shown in Figure 3a, no obvious pore can be found on the surface of pure PVDF membranes (M0 or M3). Comparably, more pores appeared on the blend membrane surface. The porosity data are summarized in Table 1; for example, the porosity of the pure PVDF membrane M0 was 7.21%, and the porosity of blend membrane M2 became 89.24%, almost 13 times more than M0. This is mainly due to the hydrophilic polyionic liquid (P11) that may accelerate the mass transfer rate of the solvent and non-solvent during the phase transfer stage, thus accelerating the transient liquid–liquid phase separation [28], producing a more porous and loose membrane structure. It can be seen from the cross-sectional view (Figure 3b) that the blend membranes (M1, M2, M4, M5) all showed typical asymmetric structure. With an increased ratio of the polycationic liquid mixed into the membrane, the membrane became thicker. As summarized in Table 1, the thicknesses of membranes M0, M1, and M2 were 30.4 µm, 113.6 µm and 121.2 µm, respectively. This accords with the observation of a more loose structure. In addition, by adding more P11 into the membrane, more finger-like macropores can be observed on the cross-section of the membranes.

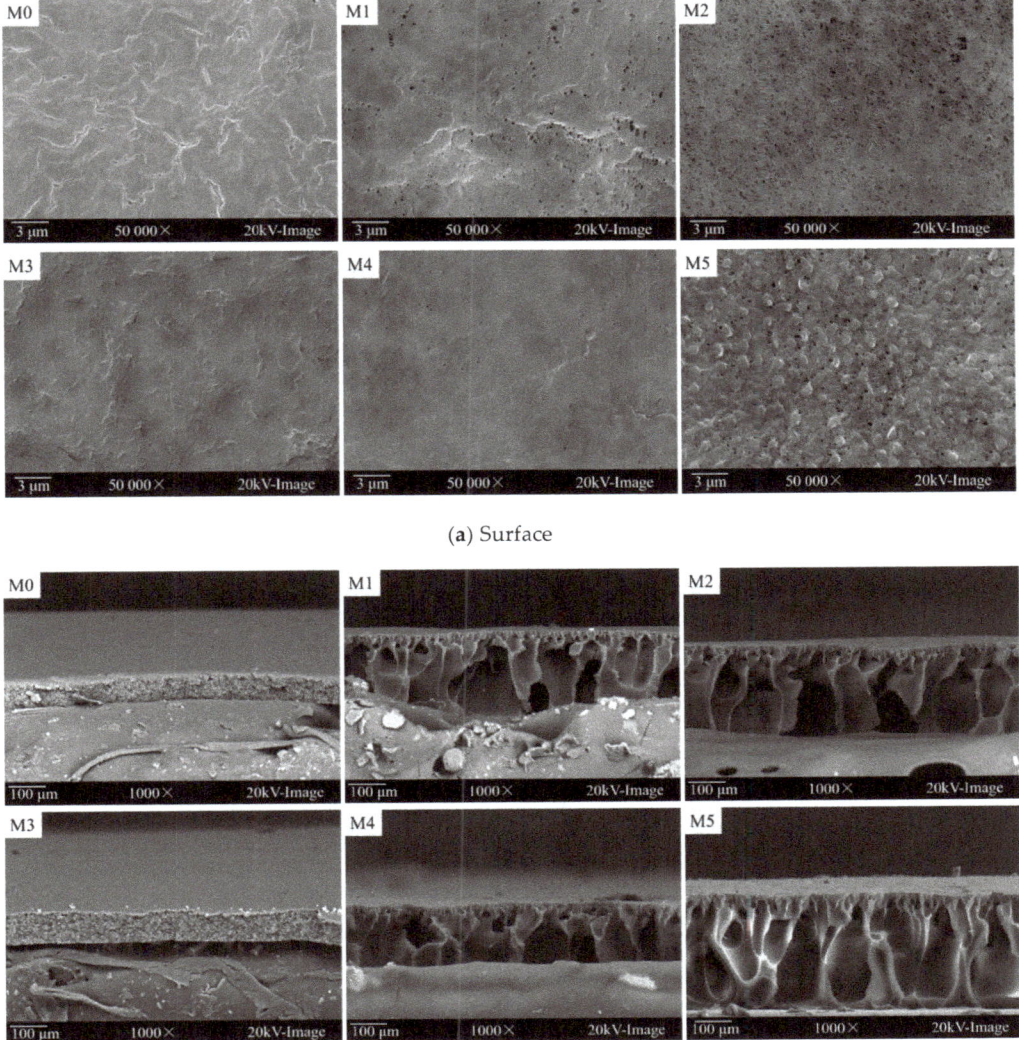

Figure 3. The SEM images of all membranes. (a) Surface; (b) Cross-section.

The mean pore size of the membranes are summarized in Table 1; the pore size and the pure water flux were enlarged. For example, the blend membrane M2 gave the largest mean pore size (35.2 nm) and water flux (631.68 L m^{-2} h^{-1}). By increasing the polymer concentration from 20 wt% to 22 wt%, the obtained membranes became denser, and the pore size or the pure water flux decreased (such as for membrane M5, reduced to 27.32 nm and 238.63 L m^{-2} h^{-1}, respectively).

As detected from the SEM images, the thickness of membrane M5 was the thickest (Table 1, M5: 149.0 μm); this is due to the higher polymer concentration of the casting solution and the higher ratio of the copolymer P11 (2/8 of P11/PVDF) in the membrane. Correspondingly, the mechanical strength was found to be stronger (Table 1, M5: 0.54 MPa) than membrane M2, which has a lower polymer concentration (Table 1, M2: 0.48 MPa). Although the blend membranes' mechanical strength was smaller than that of the pure PVDF membranes (Table 1, M0: 1.82 MPa; M3: 2.01 MPa), they can survive under ultrafiltration pressure, which is usually lower than 0.30 MPa.

3.3. Surface Wettability of the Membranes

The water contact angles and underwater oil contact angles of the membranes can be found in Figure 4a. The hydrophilicity and the oleophobicity of the PVDF membranes were improved by adding the polycationic liquid P11. For example, the water contact angle of the pure PVDF membrane M3 was 81.9°, its under-water oil contact angle was 113.3°, and the same data points for the blend membrane M5 were 67.0° and 138.1°, respectively. This is mainly caused by the introduction of polycationic liquid, P11, which is both hydrophilic and positively charged. The PEGMA segments in P11 helped the hydrophilicity; meanwhile, the cationic liquid (BVImBr) segments may improve both the positive charge and hydrophilicity.

The dynamic changes in water contact angles of the membranes M3, M4, and M5 are also recorded in Figure 4b, which can reflect the surface wettability of the membrane more intuitively, and the faster the wetting speed, the better the surface wettability. With the increase of immersion time, the contact angles of all membranes decreased. Compared with pure membrane M3 and blend membrane M4, the wetting speed of blend membrane M5 was much faster: the water contact angle dropped from 67.07° to 30.34° in 300 s. This is mainly because membrane M5 has a higher blending of hydrophilic P11 in the membrane.

In addition, the AFM images (Figure 4c) show that the surface roughness of M5 was higher than M4, where the average roughness (Ra) for M4 was 34.9 nm and the Ra of M5 was 38.6 nm. There are many protrusions on the surface of the membrane M5; this phenomenon can also be found in the surface SEM images of M5 (Figure 4a, M5) as more pores appear on the surface face. Therefore, the rougher and more porous surface accelerated the membrane wetting speed.

3.4. Thermal Stability and Zeta Potentials of the Membranes

TGA curves of blend membranes and the copolymer P11 are described in Figure 4d. The copolymer P11 showed obvious weight loss at 250 °C, which indicated the decomposition of the polymer. The pure PVDF membrane M3 showed the only thermal weight loss at 414 °C because the pure membrane had strong thermal stability. Compared with M3, the blend membranes M4 and M5 gave the first obvious weight loss at a temperate 250 °C, which can be attributed to the decomposition of P11; M5 gave more weight loss than M4 because M5 was fabricated from a higher content of P11 in the polymer mixture.

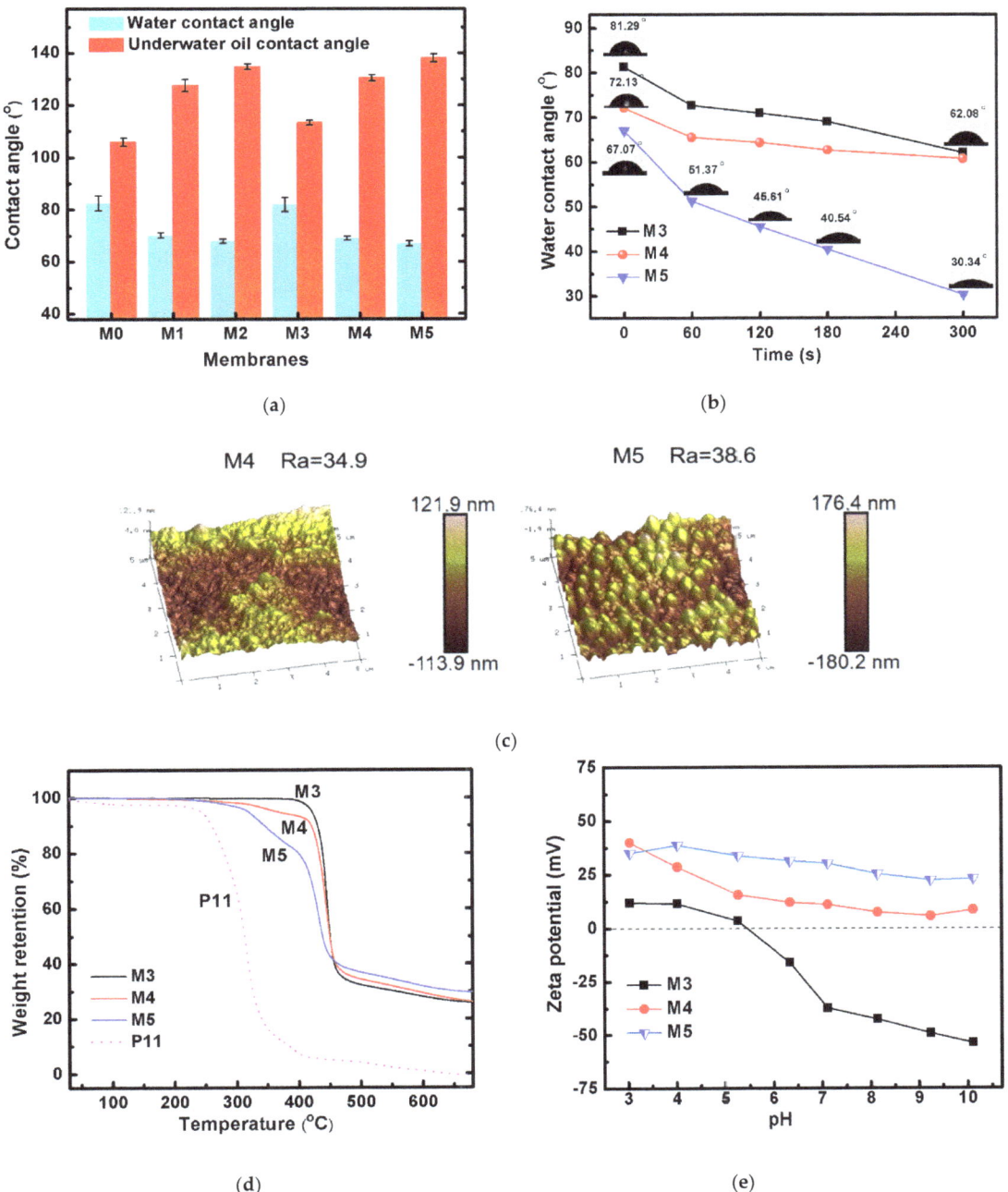

Figure 4. (**a**) The static water/underwater oil contact angles of all membranes; (**b**) the dynamic water contact angles of membranes M3, M4, and M5; (**c**) AFM images of blend membranes M4 and M5; (**d**) TGA diagram of the polymer P11 and membranes; and (**e**) Zeta potentials of membranes M3, M4, and M5.

Figure 4e shows that by blending cationic P11 into the PVDF membrane, the membranes M4 and M5 were always positively charged at pH 3~10, while the isoelectric point of pure PVDF membrane M3 is 5.4. Moreover, with the higher content of P11, the membrane M5 (2/8) appeared to be more positively charged than M4 (1/9). This enrichment of positive electricity on the membrane surface may help to enhance the electrostatic interaction between the membrane surface and pollutants in wastewater, which is quite beneficial for antifouling performance and the interception effects of the membrane on some charged pollutants.

3.5. Filtration of Oil/Water Emulsions

Figure 5a and Table 2 describe the filtration performance of blend membranes M1, M2, M4, and M5 in separating 0.1 g L^{-1} of oil/water emulsion. All the membranes showed a rapid flux decay within 30~60 min, which was attributed to the fast aggregation of oil droplets on the membrane surface and even blockage of internal membrane pores driven by pressure. Rates of higher than 91% in oil rejection were obtained for all the membranes, with membrane M2 giving the lowest R at 91.6% and membrane M4 showing the highest R (up to 99.4%). This indicates that the rejection mechanism of the membranes is mainly due to the pore size screening effect. It was mentioned in a previous section that M4 had the smallest pore size and M2 had the largest pore size; the data are shown in Table 1.

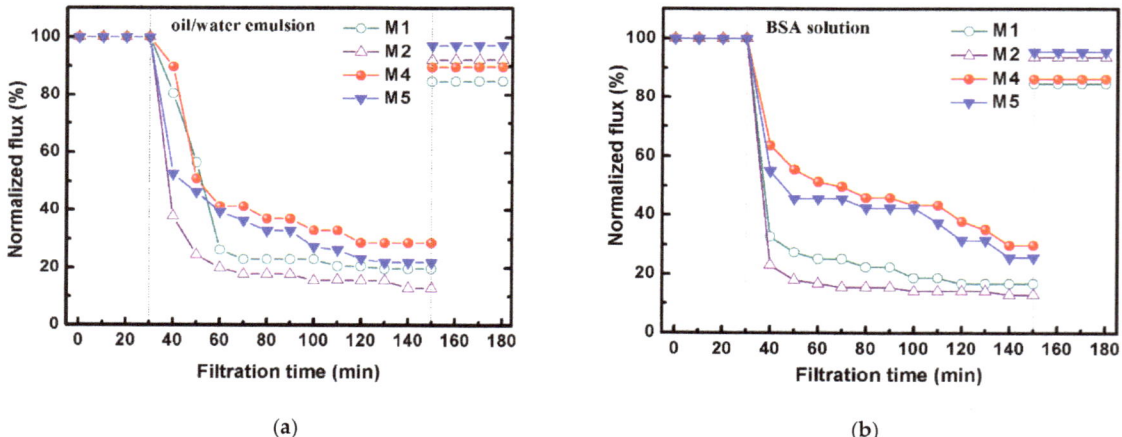

Figure 5. The normalized flux during the filtration of: (**a**) oil/water emulsions (0.1 g L^{-1}, pH 7) and (**b**) BSA solutions (1.0 g L^{-1}, pH 3.6).

Table 2. Parameters in the filtration of oil/water emulsions.

Membranes	J_0 /L m^{-2} h^{-1}	J_p /L m^{-2} h^{-1}	J_c /L m^{-2} h^{-1}	RFD /%	RFR /%	R /%	RFR/%	R/%	RFR/%	R/%
							2nd Cycle		3rd Cycle	
M1	427.47	84.18	362.35	80.3	84.8	96.4	77.3	93.2	61.4	85.7
M2	631.68	81.49	582.41	87.1	92.2	91.6	87.8	87.4	83.4	79.6
M4	193.24	51.21	173.34	73.5	89.7	99.4	82.6	96.5	73.2	88.2
M5	238.63	50.35	232.19	78.9	97.3	98.5	92.4	94.1	88.7	87.3

Despite the high flux decay, the blend membranes demonstrated good antifouling properties after an ultrasonic cleaning by DI water; up to 97.3% of RFR can be found for membrane M5. The RFR for M1 was 84.8%; by adding more polycationic liquid P11 into the membrane, a better RFR was obtained. This can be explained by the fact that on a more hydrophilic membrane surface, a denser hydrated layer can be formed, which can effectively prevent the adhesion of oil droplets.

The R and RFR of membranes M1, M2, M4, and M5 in the repeated filtration cycles of oil/water emulsion are summarized in Table 2. Both the flux recovery and the rejection rate of each membrane sample decline after three repeated filtrations. For example, the rejection rate of M5 in each cycle were 98.5%, 94.1%, and 87.3%. The decrease in the membrane rejection may be caused by the clean method utilized in this experiment, such that the ultrasonic clean might damage the membrane pores and therefore affecting the pore size screening effect. Although the RFR was also reduced in each cycle, membrane M5 showed the smallest decline while the RFR of M5 in the third cycle remained as high as 89%.

3.6. Filtration of BSA Solutions

A similar operation was employed in the filtration of BSA solutions (Figure 5b, 1.0 g L^{-1}, pH 3.6). During the BSA separation, the changing trend was quite similar to the oil/water emulsion separation (Figure 5a). The blend membranes M4 and M5 showed reduced flux decay (Table 3, 70.3% and 74.4%) as compared to membranes M1 and M2 (Table 3, 83.4% and 87.1%); M5 gave the best relative flux recovery rate (RFR) at 95.5%, indicating the improved antifouling property. The explanation follows the previous discussion on the oil/water separation. As shown in Table 3, the rejection rates of M4 and M5 were much higher than that of M1 and M2 due to the decreased membrane pore size.

Table 3. Parameters in the filtration of BSA solutions.

Membranes	J_0 /L m^{-2} h^{-1}	J_p /L m^{-2} h^{-1}	J_c /L m^{-2} h^{-1}	RFD /%	RFR /%	R /%	RFR/%	R/%	RFR/%	R/%
							2nd Cycle		3rd Cycle	
M1	439.47	73.15	371.45	83.4	84.5	38.4	not tested			
M2	547.45	70.19	512.36	87.1	93.6	31.0				
M4	186.24	55.27	160.35	70.3	86.1	87.2	82.9	79.3	76.3	65.6
M5	220.47	56.34	210.45	74.4	95.5	85.6	93.2	77.9	88.3	62.4

By using the optimum membranes M4 and M5, which have narrower pore sizes and good BSA rejections, three repeated filtration cycles were carried out to evaluate the membrane performance. As demonstrated in Table 3, despite the slight decreased R and RFR in every cycle, the RFR of M5 was still as high as 88.3%, with the blend membranes with the highest content of P11 showing good anti-protein-fouling properties.

It is noted that the pH value of the BSA solution was 3.6 and the isoelectric point of BSA is 4.7, and thus, the BSA molecules at pH 3.6 appears to be positively charged. The membranes M4 and M5 have been confirmed to have positively charged surfaces: the electrostatic repulsion between the positive BSA and the positive membrane is suggested to affect the pollutant rejection and the enhanced antifouling performance.

A test of BSA rejection at different levels of pH was carried out by using the membrane M5, which possessed the higher ratio of P11 in the polymer mixture, but in different polymer concentrations. Rejections of M5 were 85.6% (pH 3.6), 81.5% (pH 4.7) and 86.4% (pH 7.0), respectively. Given the lowest rejection of BSA at pH 4.7 and the fact that the BSA molecules were neutral and so no specific electrostatic interaction between BSA and membrane presented there, the pore size screening effect is the major rejection mechanism. Comparably, membrane M5 showed higher rejections when the pH values of the BSA solution were 3.6 and 7.0. They are almost the same, but higher than the neutral conditions. It proved that the retention of pollutants can be assisted by the electrostatic effects, and that both electrostatic repulsion and electrostatic attraction between the charged pollutant and the charged membrane surface help the retention improvement.

The data of our previous work [27]—in which the different polycationic liquid P(BVImBr$_1$-co-PEGMA$_2$) was used in a 1/2 ratio between the two monomers and the obtained PVDF membranes were also applied into the BSA filtration—are summarized in Table 4. As mentioned at the very beginning of this report, we newly synthesized P11 and the monomer ratio was changed to 1/1. It can be seen in Table 4 that the pure water flux of the membranes were similar and the membrane hydrophilicity enhanced gradually. All of them showed ~99% BSA rejection; however, the RFR was greatly improved from 76.2% to 97.3%. This is mainly due to the increased monomer ratio of cationic liquid (BVImBr) in the synthesized polymer, which enhanced the positive charge of the membrane, and due to the higher content of cationic P11 in the polymer mixture, the membranes became more positively charged and hydrophilic, which finally helped the antifouling ability.

Table 4. Parameters of different membranes during the filtration of BSA solution (pH 3.6).

Membranes	Pure Water Flux	Water Contact Angle	R	RFR	Ref.
Optimum membrane in previous work: P(BVImBr$_1$-co-PEGMA$_2$)/PVDF = 6/15	200.8 ± 9.0 L m^{-2} h^{-1}	76.3°	99.1%	76.2%	[27]
M4: P(BVImBr$_1$-co-PEGMA$_1$)/PVDF = 1/9	193.24 ± 7.39 L m^{-2} h^{-1}	72.1°	99.4%	89.7%	this work
M5: P(BVImBr$_1$-co-PEGMA$_1$)/PVDF = 2/8	238.63 ± 6.50 L m^{-2} h^{-1}	67.0°	98.5%	97.3%	this work

3.7. The Static Antifouling Performance

Since the blend membranes demonstrated good antifouling performance in the dynamic filtrations of typical organic pollutants (oil and protein), the static antipollution of the membranes was also evaluated by the adsorption test of BSA [29] and antibacterial tests.

Pure PVDF membrane M3 showed no adsorption capacity of BSA molecule (0 mg g^{-1}), the blend membranes showed an increased adsorption capacity from M4 (0.013 mg g^{-1}) to M5 (0.017 mg g^{-1}). The main reason should be the increased membrane porosity that may increase the specific surface area and adsorption sites of the membranes. The electrostatic interaction between the positively charged surface of the blend membranes and negatively charged BSA (at pH 7.0) was also considered to facilitate the adsorption process, but should not be the dominant reason since the adsorption capacity data is very small. Such low adsorption ability indicated good anti-protein adhesion performance of the blend membranes, and this is mainly attributed to the good hydrophilicity.

The membranes M3, M4, and M5 were tested for the antibacterial activity against *S. aureus*. The results in Figure 6 showed the pure PVDF membrane M3 had almost no antibacterial ability, while the blend membranes M4 and M5 showed good bactericidal ability. This is due to the blending of polycationic liquid P11 in the membranes, which can interact with the negatively charged cell membrane of the bacteria, destroying the structure of bacteria and killing them [30,31]. With the increase of polycationic liquid, the number of bacteria (Figure 6a) and the bacterial activity (Figure 6b) were greatly reduced. For example, the bacterial viability of M5 was as low as 5.74%: the blend membranes showed excellent bactericidal ability.

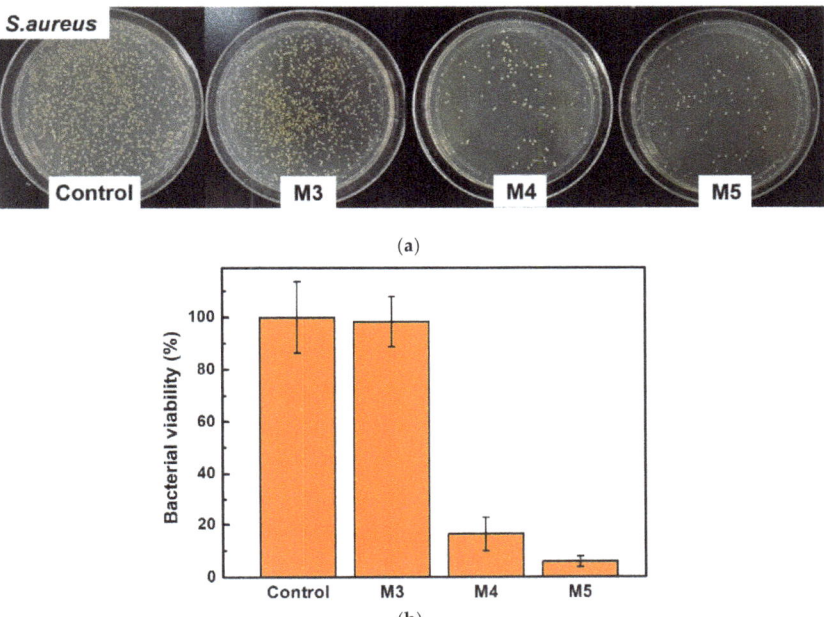

Figure 6. (a) Photographs and (b) the bacterial viabilities of the membranes M3, M4, and M5 against *S. aureus*.

4. Conclusions

In this study, a positively charged and hydrophilic PVDF membrane has been well-developed by blending with a novel synthetic polycationic liquid, P11, via the NIPS method. By using the membranes M4 and M5, fabricated from 22 wt% of polymer concentration in the casting solution, the blend PVDF membranes showed excellent antifouling properties due to the cationic structure of and the hydrophilic segments in P11. The pure water flux has been improved to 238.63 L m^{-2} h^{-1} because of the increased membrane pores. During the filtration of oil/water emulsion, the membranes showed a very high R at 99.4%, and up to 97.3% of RFR was achieved after the ultrasonic cleaning of the fouled membranes. When applied to BSA solution separation, 87.2% R and 95.5% RFR can be detected for the membranes. After three repeated filtration cycles, although the rejection rates and the relative flux recovery rate declined slightly, the blend membranes still exhibited good separation and antifouling effects. Moreover, the static antifouling ability of the membranes was so strong that extremely low BSA adsorption capacity (0.013 mg g^{-1}) was found, and the bacterial viability of *S. aureus* was reduced greatly from 98.9% to 5.74%. In summary, an efficient anti-pollution separation membrane has been developed which has good application prospects for water and wastewater treatment.

Author Contributions: Conceptualization, S.S.; methodology, Z.W. and L.Z.; investigation, Z.W. and L.Z.; data curation, Z.W. and A.B.H.; writing—original draft preparation, Z.W.; writing—review and editing, S.S. and G.Z.; visualization, S.S.; supervision, S.S.; funding acquisition, S.S. and Z.W. All authors have read and agreed to the published version of the manuscript.

Funding: This research is financial supported by the National Natural Science Foundation of China (No. 51608342) and the Postgraduate Research & Practice Innovation Program of Jiangsu Province (KYCX21-3029).

Institutional Review Board Statement: Not applicable.

Informed Consent Statement: Not applicable.

Data Availability Statement: Not applicable.

Acknowledgments: The authors acknowledge the support from the Jiangsu Collaborative Innovation Center for Technology and Material of Water Treatment. We would also like to thank Jie Huang (Suzhou Institute of nano-tech and nano-bionics, Chinese Academy of Sciences) for the assistance on the antibacterial analysis.

Conflicts of Interest: The authors declare they have no competing interests.

References

1. Yao, M.; Tijing, L.D.; Naidu, G.; Kim, S.-H.; Matsuyama, H.; Fane, A.G.; Shon, H.K. A review of membrane wettability for the treatment of saline water deploying membrane distillation. *Desalination* **2020**, *479*, 114312. [CrossRef]
2. Samree, K.; Srithai, P.-U.; Kotchaplai, P.; Thuptimdang, P.; Painmanakul, P.; Hunsom, M.; Sairiam, S. Enhancing the antibacterial properties of PVDF membrane by hydrophilic surface modification using titanium dioxide and silver nanoparticles. *Membranes* **2020**, *10*, 289. [CrossRef] [PubMed]
3. van den Berg, T.; Ulbricht, M. Polymer nanocomposite ultrafiltration membranes: The influence of polymeric additive, dispersion quality and particle modification on the integration of zinc oxide nanoparticles into polyvinylidene difluoride membranes. *Membranes* **2020**, *10*, 197. [CrossRef] [PubMed]
4. Zhu, X.; Loo, H.E.; Bai, R. A novel membrane showing both hydrophilic and oleophobic surface properties and its non-fouling performances for potential water treatment applications. *J. Membr. Sci.* **2013**, *436*, 47–56. [CrossRef]
5. Shen, S.; Hao, Y.; Zhang, Y.; Zhang, G.; Zhou, X.; Bai, R.B. Enhancing the antifouling properties of poly(vinylidene fluoride) (PVDF) membrane through a novel blending and surface-grafting modification approach. *ACS Omega* **2018**, *3*, 17403–17415. [CrossRef]
6. Liu, L.; Huang, L.; Shi, M.; Li, W.; Xing, W. Amphiphilic PVDF-g-PDMAPMA ultrafiltration membrane with enhanced hydrophilicity and antifouling properties. *J. Appl. Polym. Sci.* **2019**, *136*, 48049. [CrossRef]
7. Deng, Y.; Zhang, G.; Bai, R.; Shen, S.; Zhou, X.; Wyman, I. Fabrication of superhydrophilic and underwater superoleophobic membranes via an in situ crosslinking blend strategy for highly efficient oil/water emulsion separation. *J. Membr. Sci.* **2019**, *569*, 60–70. [CrossRef]
8. Ghiasi, S.; Behboudi, A.; Mohammadi, T.; Khanlari, S. Effect of surface charge and roughness on ultrafiltration membranes performance and polyelectrolyte nanofiltration layer assembly. *Colloids Surf. A Physicochem. Eng. Asp.* **2019**, *580*, 123753. [CrossRef]
9. Guo, J.; Farid, M.U.; Lee, E.-J.; Yan, D.Y.-S.; Jeong, S.; An, A.K. Fouling behavior of negatively charged PVDF membrane in membrane distillation for removal of antibiotics from wastewater. *J. Membr. Sci.* **2018**, *551*, 12–19. [CrossRef]
10. Zhang, L.; Shen, S.; Zhang, Y.; Zhou, X.; Bai, R. Modification of polyvinylidene fluoride membrane by blending with cationic polyionic liquid. *Desalin. Water Treat.* **2020**, *189*, 119–125. [CrossRef]
11. Kang, G.; Cao, Y. Application and modification of poly(vinylidene fluoride) (PVDF) membranes—A review. *J. Membr. Sci.* **2014**, *463*, 145–165. [CrossRef]
12. Cheng, K.; Zhang, N.; Yang, N.; Hou, S.; Ma, J.; Zhang, L.; Sun, Y.; Jiang, B. Rapid and robust modification of PVDF ultrafiltration membranes with enhanced permselectivity, antifouling and antibacterial performance. *Sep. Purif. Technol.* **2021**, *262*, 118316. [CrossRef]
13. Li, N.; Chen, H.-D.; Lu, Y.-Z.; Zhu, M.-C.; Hu, Z.-X.; Chen, S.-W.; Zeng, R.J. Nanoscale zero-valent iron-modified PVDF membrane prepared by a simple filter-press coating method can robustly remove 2-chlorophenol from wastewater. *Chem. Eng. J.* **2021**, *416*, 127701. [CrossRef]
14. Xing, J.; Zhang, G.; Jia, X.; Liu, D.; Wyman, I. Preparation of Multipurpose Polyvinylidene Fluoride Membranes via a Spray-Coating Strategy Using Waterborne Polymers. *ACS Appl. Mater. Inter.* **2021**, *13*, 4485–4498. [CrossRef]
15. Nayak, K.; Tripathi, B.P. Molecularly grafted PVDF membranes with in-air superamphiphilicity and underwater superoleophobicity for oil/water separation. *Sep. Purif. Technol.* **2021**, *259*, 118068. [CrossRef]
16. Lim, S.J.; Shin, I.H. Graft copolymerization of GMA and EDMA on PVDF to hydrophilic surface modification by electron beam irradiation. *Nucl. Eng. Technol.* **2020**, *52*, 373–380. [CrossRef]
17. Li, C.; Wang, L.; Wang, X.; Kong, M.; Zhang, Q.; Li, G. Synthesis of PVDF-g-PSSA proton exchange membrane by ozone-induced graft copolymerization and its application in microbial fuel cells. *J. Membr. Sci.* **2017**, *527*, 35–42. [CrossRef]
18. Wang, S.; Li, T.; Chen, C.; Liu, B.; Crittenden, J.C. PVDF ultrafiltration membranes of controlled performance via blending PVDF-g-PEGMA copolymer synthesized under different reaction times. *Front. Env. Sci. Eng.* **2018**, *12*, 3. [CrossRef]
19. Du, J.; Li, N.; Tian, Y.; Zhang, J.; Zuo, W. Preparation of PVDF membrane blended with graphene oxide-zinc sulfide (GO-ZnS) nanocomposite for improving the anti-fouling property. *J. Photoch. Photobio. A* **2020**, *400*, 112694. [CrossRef]
20. Zhao, J.; Han, H.; Wang, Q.; Yan, C.; Li, D.; Yang, J.; Feng, X.; Yang, N.; Zhao, Y.; Chen, L. Hydrophilic and anti-fouling PVDF blend ultrafiltration membranes using polyacryloylmorpholine-based triblock copolymers as amphiphilic modifiers. *Reac. Funct. Polym.* **2019**, *139*, 92–101. [CrossRef]
21. Yuan, J.; Mecerreyes, D.; Antonietti, M. Poly(ionic liquid)s: An update. *Prog. Polym. Sci.* **2013**, *38*, 1009–1036. [CrossRef]

22. Yu, L.; Zhang, Y.; Wang, Y.; Zhang, H.; Liu, J. High flux, positively charged loose nanofiltration membrane by blending with poly(ionic liquid) brushes grafted silica spheres. *J. Hazard. Mater.* **2015**, *287*, 373–383. [CrossRef] [PubMed]
23. Cheng, Y.-Y.; Du, C.-H.; Wu, C.-J.; Sun, K.-X.; Chi, N.-P. Improving the hydrophilic and antifouling properties of poly(vinyl chloride) membranes by atom transfer radical polymerization grafting of poly(ionic liquid) brushes. *Polym. Adv. Technol.* **2018**, *29*, 623–631. [CrossRef]
24. Chen, S.; Zheng, J.; Li, L.; Jiang, S. Strong resistance of phosphorylcholine self-assembled monolayers to protein adsorption: Insights into nonfouling properties of zwitterionic materials. *J. Am. Chem. Soc.* **2005**, *127*, 14473–14478. [CrossRef]
25. Zhu, Y.-L.; Lu, Z.-Y.; Li, Z.-W.; Sun, Z.-Y.; Liu, X. Effect of the self-assembled structures of hydrated polyzwitterionic and polyanionic brushes on their self-cleaning capabilities. *Langmuir* **2019**, *35*, 6669–6675. [CrossRef]
26. Du, C.-H.; Ma, X.-M.; Wu, C.-J.; Cai, M.-Q.; Hu, M.-X.; Wang, T. Polymerizable ionic liquid copolymer P(MMA-co-BVIm-Br) and its effect on the surface wettability of PVDF blend membranes. *Chin. J. Polym. Sci.* **2015**, *33*, 857–868. [CrossRef]
27. Shen, S.; Zhang, L.; Zhang, Y.; Zhang, G.; Yang, J.; Bai, R. Fabrication of antifouling membranes by blending poly(vinylidene fluoride) with cationic polyionic liquid. *J. Appl. Polym. Sci.* **2020**, *137*, 48878. [CrossRef]
28. Hashim, N.; Liu, F.; Li, K. A simplifified method for preparation of hydrophilic PVDF membranes from an amphiphilic graft copolymer. *J. Membr. Sci.* **2009**, *345*, 134–141. [CrossRef]
29. Mockel, D.; Staude, E.; Guiver, M.D. Static protein adsorption, ultrafiltration behavior and cleanability of hydrophilized polysulfone membranes. *J. Membr. Sci.* **1999**, *158*, 63–75. [CrossRef]
30. Ye, Q.; Gao, T.; Wan, F.; Yu, B.; Pei, X.; Zhou, F.; Xue, Q. Grafting poly(ionic liquid) brushes for anti-bacterial and anti-biofouling applications. *J. Mater. Chem.* **2012**, *22*, 13123–13131. [CrossRef]
31. Cihanoğlu, A.; Altinkaya, S.A. A facile route to the preparation of antibacterial polysulfone-sulfonated polyethersulfone ultrafiltration membranes using a cationic surfactant cetyltrimethylammonium bromide. *J. Membr. Sci.* **2020**, *594*, 117438. [CrossRef]

Article

Novel Sandwich-Structured Hollow Fiber Membrane for High-Efficiency Membrane Distillation and Scale-Up for Pilot Validation

Marn Soon Qua [1], Yan Zhao [1], Junyou Zhang [1], Sebastian Hernandez [1], Aung Thet Paing [1], Karikalan Mottaiyan [1], Jian Zuo [2], Adil Dhalla [1], Tai-Shung Chung [3,4,*] and Chakravarthy Gudipati [1,*]

1. Separation Technologies Applied Research and Translation Centre (START), Nanyang Technological University–NTUitive Pte Ltd., Nanyang Technological University, Singapore 637141, Singapore; ananda.qua@ntu.edu.sg (M.S.Q.); zhao_yan@ntu.edu.sg (Y.Z.); junyou.zhang@century-water.com (J.Z.); sebastian.hs@ntu.edu.sg (S.H.); aungthetpaing@ntu.edu.sg (A.T.P.); karikalan@ntu.edu.sg (K.M.); adil.dhalla@ntu.edu.sg (A.D.)
2. Food, Chemical and Biotechnology Singapore Institute of Technology, Singapore 637141, Singapore; zuolezl@gmail.com
3. Department of Chemical and Biomolecular Engineering, National University of Singapore, Singapore 637141, Singapore
4. Graduate Institute of Applied Science and Technology, National Taiwan University of Science and Technology, Taipei 10607, Taiwan
* Correspondence: chencts@mail.ntust.edu.tw (T.-S.C.); chakra@ntu.edu.sg (C.G.); Tel.: +886-2-2730-1158 (T.-S.C.); +65-6908-2275 (C.G.)

Citation: Qua, M.S.; Zhao, Y.; Zhang, J.; Hernandez, S.; Paing, A.T.; Mottaiyan, K.; Zuo, J.; Dhalla, A.; Chung, T.-S.; Gudipati, C. Novel Sandwich-Structured Hollow Fiber Membrane for High-Efficiency Membrane Distillation and Scale-Up for Pilot Validation. *Membranes* **2022**, *12*, 423. https://doi.org/10.3390/membranes12040423

Academic Editors: Hongjun Lin and Meijia Zhang

Received: 1 April 2022
Accepted: 12 April 2022
Published: 14 April 2022

Publisher's Note: MDPI stays neutral with regard to jurisdictional claims in published maps and institutional affiliations.

Copyright: © 2022 by the authors. Licensee MDPI, Basel, Switzerland. This article is an open access article distributed under the terms and conditions of the Creative Commons Attribution (CC BY) license (https://creativecommons.org/licenses/by/4.0/).

Abstract: Hollow fiber membranes were produced from a commercial polyvinylidene fluoride (PVDF) polymer, Kynar HSV 900, with a unique sandwich structure consisting of two sponge-like layers connected to the outer and inner skin layers while the middle layer comprises macrovoids. The sponge-like layer allows the membrane to have good mechanical strength even at low skin thickness and favors water vapor transportation during vacuum membrane distillation (VMD). The middle layer with macrovoids helps to significantly reduce the trans-membrane resistance during water vapor transportation from the feed side to the permeate side. Together, these novel structural characteristics are expected to render the PVDF hollow fiber membranes more efficient in terms of vapor flux as well as mechanical integrity. Using the chemistry and process conditions adopted from previous work, we were able to scale up the membrane fabrication from a laboratory scale of 1.5 kg to a manufacturing scale of 50 kg with consistent membrane performance. The produced PVDF membrane, with a liquid entry pressure (LEPw) of >3 bar and a pure water flux of >30 L/m^2·hr (LMH) under VMD conditions at 70–80 °C, is perfectly suitable for next-generation high-efficiency membranes for desalination and industrial wastewater applications. The technology translation efforts, including membrane and module scale-up as well as the preliminary pilot-scale validation study, are discussed in detail in this paper.

Keywords: PVDF; hollow fiber membranes; vacuum membrane distillation; flux; liquid entry pressure; wastewater treatment; desalination

1. Introduction

Freshwater scarcity is becoming a major challenge for meeting requirements toward basic human needs for agriculture and industry, impeding efforts to meet the global water demand due to an increase in population and industrialization, particularly in coastal countries due to the lack of sufficient fresh water sources or storage capacity [1–5]. Only 3% of the water on Earth is considered fresh water and only 1.2% has potential use as drinking water because the rest is locked in glaciers, ice caps, and permafrost. To deal with water scarcity and freshwater shortage, seawater desalination processes are being widely used.

The current solutions utilized for producing drinking water from seawater largely rely on thermal methods such as multistage flash and multi-effect distillation (MSF and MED, respectively) [5–8] as well as pressure-driven seawater reverse osmosis (SWRO) [8,9], which comprises about 70% of the desalination processes worldwide. However, both thermal and pressure-driven RO technologies are extremely intensive in terms of energy consumption and, currently, they strongly depend on fossil fuels for their operation [8,9]. In addition, although the brine resulting from the SWRO process still has the potential for additional water recovery, it is very difficult to treat [10,11].

Another potential freshwater source is found through wastewater reclamation. In this process, the use of reverse osmosis (RO) by itself is not enough and using conventional processes or a combination of these with RO systems tends to be costly and the purification ineffective [11–13]. Due to these reasons, alternative desalination technologies and the use of renewable energies are being researched and developed to reduce the energy consumption and improve overall process productivity [14–18].

Membrane distillation (MD) is a promising technology that can potentially compete with the existing MSF/MED and SWRO solutions and could be capable of overcoming the issues previously described [17,19,20]. The key benefits of using MD technology include: (1) less stringent operation conditions compared to conventional desalination, including lower vacuum, or pumping pressures, (2) higher rejection of salts (theoretically approaching 100%), (3) larger contact areas in smaller modular footprints, and (4) ability to treat extremely high salinities of feed water beyond the SWRO tolerance limit. Salts and non-volatile compounds are rejected by the membrane and could produce an even more concentrated brine than the one typically obtained by the RO processes, facilitating efforts toward zero liquid discharge [21–23]. This latter characteristic is an advantage because, in theory, one can have higher recoveries using MD compared to RO. Finally, (5) the use of low-grade heat sources, which opens the possibility of the use of renewable energy sources [10,18,20,24,25].

In membrane distillation, the membrane is one of the key factors that govern the separation process in different applications. The MD membrane works as a barrier separating a feed water solution from the permeate vapor stream and simultaneously transfers the vapor produced from the feed stream through the membrane wall to the permeate stream. The permeate stream is condensed on the permeate side or externally using different techniques. MD is a process based on the change in phase due to a thermal gradient allowing the separation of the volatiles, in this case, water from the feed solution, and applying the principles of vapor–liquid equilibrium and heat and mass transfer [24,26]. An ideal MD membrane should be highly porous and hydrophobic, with a very tight pore size distribution and small pore size, and it should have cheap and easy fabrication conditions for large-scale production. Many technology breakthroughs using hydrophobic materials in both flat sheet and hollow fiber membrane configurations have been accomplished, including polyvinylidene fluoride (PVDF), polytetrafluoro-ethylene (PTFE), and polypropylene (PP) [27,28].

The techniques using MD to recover water vary according to what is used in the permeate side to drive the separation; these include Direct Contact Membrane Distillation (DCMD), where the permeate side is in contact with cold pure water coming from the DCMD process itself. In Air Gap Membrane Distillation (AGMD) and Sweeping Gas Membrane Distillation (SGMD), air or an inert gas is used to collect water vapor that will be condensed in situ or externally, respectively. In vacuum membrane distillation (VMD), water vapor is removed by applying vacuum on the permeate side and is subsequently condensed externally, similar to the SGMD process [12,17,21]. Other liquids and materials have been used in MD configurations, which are combinations of the above techniques [25].

From these configurations, VMD is one of the most attractive MD processes for water reclamation purposes due to its lower costs in operation. These costs are related to fewer stringent mechanical properties needed for the membrane material given that VMD needs lower temperatures and pressures to work with, including pumping and vacuum pressures [29,30]. Using vacuum boosts the water flux due to the gradient in vapor pressure across the membrane and improves heat and mass transfer as there are no separation media involved compared to DCMD, for example [31,32]. These characteristics make VMD suitable for decreasing energy consumption by replacing SWRO or complementing it using the SWRO brines to extract even more freshwater [29].

The main shortcomings of using vacuum are that the vacuum pressure cannot exceed the liquid entry pressure (LEP_w) to avoid membrane wetting and breaking. Another issue is the energy consumption related to the heating of the feed solution and the condensation of the permeate [24,32]. To assess these drawbacks, several works using hollow fiber membranes have addressed different methods to increase the efficiency of the membranes. The most relevant is membrane fabrication through a spinning process that includes the formulation of the dope composition and the spinning process itself, conducted by thermally induced phase separation (TIPS) or non-solvent induced phase separation (NIPS) [25].

The dope composition coupled with the preparation are the main obstacles that need to be overcome for the transition from a bench-scale to a commercial product. PVDF has been employed for MD membrane manufacturing via NIPS due to its processing versatility, hydrophobicity, and resistance to a wide range of chemical products [25]. However, the LEP_w values and the mechanical strength of such PVDF membranes have proven to be unsuitable for scaling up as commercial products. A good hollow fiber PVDF membrane must have a spongy internal structure with a reduced presence of macrovoids, which are important to avoid membrane malfunction when they are working under high stresses from the feed pressure and the vacuum. However, there must be a balance, as a more open structure (low tortuosity, higher porosity including macrovoids, and pore size) increases flux, whereas the opposite is important to prevent membrane wetting and failure [33,34].

To design a PVDF hollow fiber membrane for MD with greater mechanical strength and excellent hydrophobic properties, it is necessary to tackle some intrinsic characteristics of the dope, specifically the viscosity, which affects the phase inversion and the general structure of the membranes [25,35]. The concentration of the polymer as well as the solvents and additives must be formulated to have the correct balance as previously described, and this has to be coupled with an optimized set of phase inversion conditions [33–37].

The present article is primarily focused on scaling up a patented MD technology for manufacturing PVDF hollow membranes, which have a unique sandwich structure, from laboratory scale to commercial scale and producing commercial size modules for pilot validation [19,20,28,37]. Furthermore, the effects in the final product characteristics of the PVDF source in the dope formulation, the coagulation bath composition and temperature, the bore fluid temperature, and the high-speed spinning are investigated in this work. After the membranes were successfully manufactured, they were used for module fabrication at various scales, namely 0.5-inch modules (bench scale) and 2-inch modules (large scale). Two-inch modules were further tested for $\geq$100 h in a VMD pilot plant, simulating a seawater desalination process using NaCl solutions ($\approx$35 g/L).

2. Materials and Methods
2.1. Materials

All chemicals used during the membrane fabrication and scale-up were of industrial and reagent grade and used without further purification. Polyvinylidene fluoride (PVDF 1, PVDF 2) (Kynar HSV 900 PWD resin, Arkema, Calvert City, KY, USA and Changshu, China, respectively); lithium chloride (LiCl) (GCE Laboratory Chemicals–TACT Chemie S.E.A. Pte. Ltd., Singapore); N-Methyl-2-Pyrrolidone (NMP) (Puyang Guangming Chemicals Co., Ltd., Puyang city, China); ethylene glycol (EG) (TACT Chemie S.E.A. Pte. Ltd., Singapore); methanol (MegaChem Ltd., Singapore); hexane HPLC grade (Fisher Scientific, Fairlawn, NJ, USA); sodium chloride (NaCl) (Pure Dried Vacuum Salt, INEOS Enterprises, Runcorn, UK). Deionized water was acquired from a PURELAB Option-Q DV 25 unit from ELGA with a resistivity of 18.2 MΩ·cm.

2.2. Polymer Characterization

The molecular weights of the two PVDF 1 and PVDF 2 polymers were not provided by the manufacturer. However, the gel permeation chromatography (GPC) of the commercial Kynar® HSV900 has been reported to contain two peaks corresponding to the number-average molecular weights (M_n); one at ~92,840 kDa (24.92%) and another one at ~1367 kDa (75.08%) [38]. It is possible that the molecular weights of the two polymers may differ from each other due to process variations at different locations. The potential differences in the molecular weights of PVDF 1 and PVDF 2 are reflected in the slightly different solution viscosities measured at the spinning temperature (~50 °C), as shown in Table 1.

Table 1. Characterization of PVDF and dope samples.

Polymer	PVDF 1	PVDF 2
Melting point (°C)	162.65	162.64
Crystallization point (°C)	127.68	126.81
Max. thermal degradation (°C)	472.49	472.66
Melting enthalpy (J/g)	36.08	33.92
Crystallization enthalpy (J/g)	41.41	36.64
Dope viscosity (Pa·s)	101.93 (@50.1 °C)	167.27 (@ 51.7 °C)

These samples were analyzed using differential scanning calorimetry (DSC), thermogravimetric analysis (TGA), and pyrolysis–gas chromatography–mass spectrometry (pyrolysis-GCMS).

The DSC analysis (Q20, TA Instruments, New Castle, DE, USA) was performed in a dry nitrogen atmosphere. Around 5–10 mg of powder was tightly encapsulated into an aluminum pan. The melting behavior of polymer/diluent samples was analyzed after equilibrating the sample at 40 °C and then heating it at a rate of 10 °C/min until reaching a temperature of 250 °C, and subsequently sustaining this value for 2 min. The crystallization curve was later obtained by cooling the sample to 40 °C at a rate of 10 °C/min after equilibrating at 250 °C for 2 min. The thermogravimetric analysis (TGA) was conducted using a thermal analyzer (SDT Q600, TA Instruments, New Castle, DE, USA) under a nitrogen flow at 100 mL/min. The samples were tested after equilibrating the sample at 40 °C and then heating it in a temperature range of 40–700 °C at a rate of 20 °C/min with an isothermal treatment at the end point for 5 min.

To perform the pyrolysis-GCMS tests, a GCMS-Pyrolyzer (Agilent Technologies 7890B GC, Agilent Technologies 5977A MSD, Frontier Lab Multi-Shot Pyrolyzer EGA/PY3030D) was used. The analysis cup containing a 0.2 mg sample was inserted into the Multi-Shot Pyrolyzer EGA/PY3030D. Samples were pyrolyzed at 600 °C for 1 min. Pyrolysis products were injected with a split of 50 using the Agilent Technologies 7890B GC (equipped with an Ultra ALLOY-5 column (30 m, 0.25 mm, 0.25 mm film of 5% diphenyl–95% dimethylpolysiloxane) (Frontier Lab). The temperatures of the pyrolizer interface and the injection port were both set at 300 °C. Helium was used as a carrier gas with a constant flow of 1 mL/min. The initial oven program was set as follows: 40 °C for 2 min, then increased to 320 °C at 20 °C/min and then maintained for 13 min. Mass spectra were obtained by the Agilent Technologies 5977A MSD. The interface temperature was set at 300 °C, the ion source temperature was set at 230 °C, the ionization voltage was set at 70 eV, and a mass range from 33 to 600 m/z was scanned at a scan speed of 1526 µ/s.

2.3. Fabrication of Hollow Fiber Membranes

The PVDF hollow fiber membranes were fabricated with a formulation of the polymer dope and spinning conditions developed by Zuo and Chung [28,37]. The spinneret used is a dual-layer spinneret with a bore output of 0.44 mm and an inner channel between 0.6 and 1.14 mm. The bore fluid was fed from the top and the dope from the side of the spinneret. Table 2 summarizes the spinning parameters, such as line speed, air-gap distance, dope flowrate, and bore fluid flowrate, and includes temperatures of the dope, bore liquid solution, and the coagulation bath. However, some of the conditions for the mix of the dope and the spinning of the membranes were modified to adapt the process to large-scale production and have consistent results. Briefly, PVDF 1 and PVDF 2 were mixed separately in each batch for 24 h at 65 °C. Then, (1) each dope was degassed for another 48 h in the reactor to guarantee complete dissolution of the polymer and removal of entrapped air bubbles in the mix; (2) the take-up speed (line speed) of the fiber and the temperature of the coagulation bath were optimized during the spinning process; (3) dope and bore flowrates were adjusted to the line speed and to obtain similar results as the baseline work cited; (4) after spinning, the new fibers were stored in water for 3 days to remove the residual solvents; (5) the membranes were post-treated with alternate baths of methanol followed by hexane to remove the water from the fibers and increase hydrophobicity; (6) the membranes were dried in a dry room at room temperature (RT) at least two days before being inspected and selected for testing and module production.

Table 2. Spinning conditions for production of PVDF MD hollow fibers. Based on conditions from Zuo and Chung [28,37]. Design of experiments (DOE) using speed line and coagulation bath temperature as variables. Bore flowrate was adjusted to the DOE parameters.

Batch Number	B1	B2	B3	B4	B5	B6	B7	B8 Onwards
Dope (wt. %)				PVDF/LiCl/EG/NMP: 13/5/5/77				
Bore solution (wt. %)				NMP/Water: 50/50				
Scale-up (kg Dope)	1.5	1.5	1.5	1.5	20	20	50	50
Air gap (mm)	30	30	30	30	30	30	30	30
PVDF source	1	2	1	1	2	1	2	1
Coagulation bath, tap water (°C)	RT and 40	40	40	40	40	40	40	40
Line speed (m/min)	3	9	3	9	9	9	9	9
Dope flowrate (mL/min)	4.5	13.5	4.5	13.5	13.5	13.5	13.5	13.5
Bore flowrate (mL/min)	1.5–4.5	4.5–9.0	1.5–3.0	4.5–6.8	4.5–6.8	4.5–6.8	4.5–6.8	4.5

As part of the initial scale-up trials from a lab-scale fabrication line to pilot-scale production, two different PVDF (PVDF1 and PVDF 2) sources were identified based on the prior data, cost of materials, and ease of availability for large-scale production. Different spinning conditions were employed to optimize the membrane fabrication process, which could be scaled-up from small 1.5 kg batch sizes to 50 kg batch sizes. The coagulation bath temperatures and the bore fluid flowrates were varied for both the PVDF materials employed, as shown in Table 3.

Table 3. Hollow fiber casting conditions for small 1.5 kg batches using PVDF 1 and PVDF 2.

PVDF	Batch No	Bore Fluid Flowrate (mL/min)	Coagulation Bath Temperature (°C)	Outer Diameter (mm)	Inner Diameter (mm)	Contact Angle (°)
1	B1-a	1.5	≈24	1.13 ± 0.01	0.73 ± 0.01	69.3
1	B1-b	3	≈24	1.29 ± 0.02	0.95 ± 0.01	77.2
1	B1-c	4.5	≈24	1.38 ± 0.01	1.09 ± 0.01	72.3
1	B1-d	1.5	38.3	1.12 ± 0.01	0.70 ± 0.00	66.6
1	B1-e	3	38.3	1.25 ± 0.03	0.89 ± 0.01	64.1
2	B2-a	4.5	38.6	1.07 ± 0.02	0.66 ± 0.03	70.6
2	B2-b	6.8	38.6	1.16 ± 0.01	0.79 ± 0.01	72.6
2	B2-c	9	38.6	1.25 ± 0.01	0.90 ± 0.01	n/a

The temperatures of the polymer dope were constantly monitored during mixing, degassing, and spinning. The spinning required up to three working days for batch sizes of

≥20 kg of dope, which required degassing at the end of each working day. The viscosity of the different dopes was measured close to the spinning temperature of 55 °C using a viscometer (Cole-Palmer VCPL 340015, Vernon Hills, IL, USA).

The membranes were characterized with a Field Emission Scanning Electron Microscope (FESEM) (JEOL JSM-7200F) operated at 5.0 kV of accelerating voltage. A goniometer (OCA15EC, DataPhysics Instruments, Filderstadt, Germany) was used to test the static water contact angle of each membrane using the sessile drop method. A droplet of deionized water was mechanically pipetted onto the membrane surface and a static image of the droplet on the membrane surface after the equilibrium was taken. This was repeated five times at different locations of the membrane and the average results were reported. The optical images of hollow fibers were obtained using a Leica DVM6 optical microscope.

The pore size distribution was determined by a capillary flow porometer (CFP 1500AEX, Porous Material. Inc., Ithaca, NY, USA), whose working principle was based on the bubble-point and gas permeation tests. The hollow fiber samples were potted into the sample holder and soaked by the wetting fluid (Galwick, with surface tension 15.9×10^{-3} N/m) until completely wet. During the test, the gas flowrate was increased stepwise and passed through the saturated sample until the applied pressure exceeded the capillary attraction of the fluid in the pores. By comparing the gas flowrates of both wet and dry samples at the same pressures, the percentage of flow passing through the pores larger than or equal to the specified size can be calculated from the pressure–size relationship. The mechanical properties of hollow fiber membranes were examined using a universal tensile tester (Instron 3342, Norwood, MA, USA). Each specimen was firmly clamped by the testing holder and pulled longitudinally at an elongation rate of 50 mm/min at room temperature. The corresponding mechanical properties were determined by the built-in software.

In another method, the contact angle was determined using a tensiometer (DCAT11 Dataphysics, Filderstadt, Germany). The contact angle quantifies the wettability of a solid surface by a liquid. The sample was inserted into an electro balance for cyclical immersion into DI water. The contact angle was calculated from the wetting force using Wihelmy's method. The overall porosity of membranes was determined by the gravimetric method with the following Equation (1):

$$Porosity = 1 - \frac{Volume_{Polymer}}{Volume_{total}} = 1 - \frac{Membrane\ weight / Membrane\ volume}{Polymer\ density} \quad (1)$$

where the PVDF density was 1.78 g/cm^3 and the membrane volume was calculated based on OD and ID of the fibers.

LEP$_w$ was determined using dead-end hollow fiber modules containing a single membrane fiber. LEP$_w$ measures the pressure required to force water through the pores of a dried membrane and is an indication of how easily a hydrophobic membrane could be wetted. Water was gradually pressurized at a 0.5 bar increment. As water pressure was increased, water could be pushed out of the membrane pores, and the pressure at which water droplets were visible on the outer surface of hollow fibers was recorded as the LEP$_w$ of the membranes.

2.4. Membrane Module Testing

The hollow fibers provided were assembled into 0.5-inch diameter or 2-inch diameter modules, as shown in Figure 1, and tested at the Environment & Water Innovation Centre of Innovation (EWTCOI) and our facility, Separation Technologies Applied Research and Translation Centre (START), respectively. For the 0.5-inch modules, after the target temperature of the feed was reached, temperature sensors for the feed inlets and outlets were calibrated. The feed water was recirculated through the lumen side of the hollow fibers. The liquid feed entered the module in an upward direction to minimize air bubbles in the module. Once the feed inlet temperature in the membrane module reached a steady state, the vacuum pump was switched on to create a vacuum in the shell side of the hollow fibers. The timer for permeate collection was started and permeate was collected by condensing the water vapor either in an ice chip bath, which was periodically refilled with ice chips (0.5-inch modules) or using a chiller at 15 °C (2-inch modules). The amount of permeate collected was gravimetrically determined using a weighing scale and the electrical conductivity (EC) was measured. Table 4 shows the conditions for each of the tests and Figure 2 presents a process flow schematic for the in-to-out setup of VMD used in this study, which was a semi-continuous operation with variation in EC. The feedwater was filled with a NaCl solution whenever the EC value was nearly doubled or the tank was at half capacity, whichever came first. Each time the feed was filled, the vacuum was switched off until the target temperature of the feed was reached again.

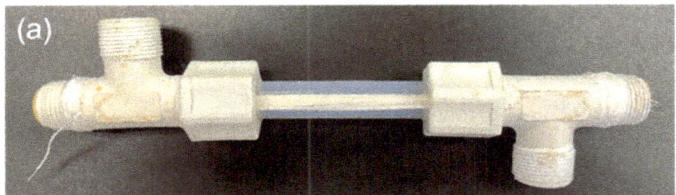

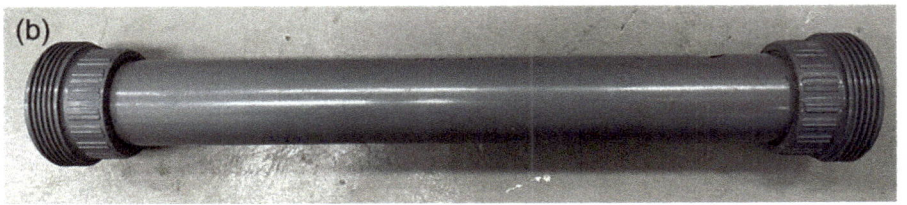 

Figure 1. Scale-up of membrane distillation modules. (**a**) Lab-scale testing module (0.5-inch diameter); (**b**) pilot-scale testing module (2-inch diameter).

Table 4. Characteristics of testing modules and operating conditions for vacuum membrane distillation (VMD) of water from a NaCl solution. † Estimated values.

Testing Site	Lab-Scale Module	Pilot-Scale Module
Module (nominal inches)	0.5	2
VMD configuration	in-to-out	in-to-out
Number of fibers	15	560
Effective length (mm)	120	370
Effective membrane area (m^2)	0.0035–0.0051	0.456
Packing density (%)	≈13 †	35
Feed flowrate (L/min)	0.5	8.5–9.5
Feed temperature (°C)	88	≥70
Vacuum (bar)	−0.80	−0.85
Test duration (hr)	≥1	>100
Feed concentration (g/L NaCl)	35	≈35.7

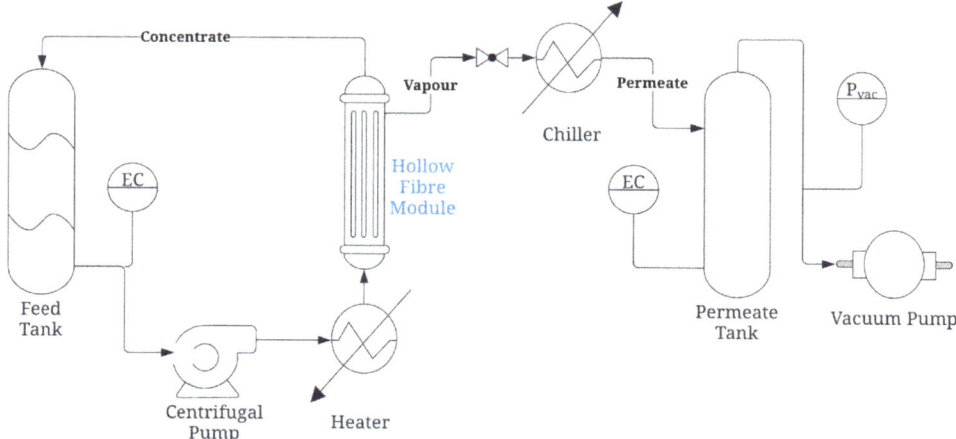

Figure 2. Process flow schematic for the vacuum membrane distillation (VMD) used for testing 2-inch modules at START.

3. Results and Discussion

3.1. Characterization of Polymers and Dopes

The two samples of PVDF1 and PVDF 2 were similar according to their melting, crystallization, thermal degradation temperatures, and pyrolysis–GCMS chromatograms, as shown in Table 1 and in Figure S1 in the Supplementary Materials, respectively. Melting transition temperatures differed from those reported for pure PVDF (177–179 °C) but were close to the reported values for commercial samples, from 159 °C to 173 °C [39–41]. Rapid crystallization was demonstrated by sharp and narrow peaks with a degree of supercooling of about 35 °C difference from the melting points for both samples, as seen in Figure S1. These crystallization points were lower than the ones reported for pure PVDF, which are between 139 °C and 141 °C, at the same rate of cooling [42].

Based on the TGA results, differences with pure PVDF were also observed in its maximum temperature for thermal degradation [43,44]. However, there were no differences among the samples examined, as shown in Figure S2 for the TGA thermograms. In the case of the pyrolysis–GCMS, the peaks on the chromatograms differed by the number of counts, but the times of separation were the same. The repeating unit of PVDF, vinylidene fluoride, was separated after 2.5 min for both samples PVDF 1 and PVDF 2. On the other hand, melting enthalpy, crystallization enthalpy, and the viscosity of the dope showed a clear difference between the two PVDF samples, as well as with the reported value of pure PVDF (104.7 J/g) [45]. These differences have an impact on the performance of the final product characteristics and could be correlated to differences in molecular weight, polydispersity, or the branching of the PVDF chains [39].

3.2. Characterization of PVDF Hollow Fiber Membranes

The bore fluid flowrate and coagulation temperature were the initial factors that were chosen to conduct an experimental design based on the previous work [28,37]. The first and second batches were conducted using a polymer dope of 1.5 kg. The dope B1 and B2 were made using PVDF 1 and PVDF 2, respectively. For the batch B1, the dope flowrate was kept constant at 4.5 mL/min and the take-up line speed at 3.0 m/min. The process conditions for the 1.5 kg batch sizes were optimized by varying the coagulation bath temperature and the bore fluid flowrate, as shown in Table 3.

The membranes fabricated under these conditions were visually examined using an optical microscope, and the fiber images are displayed in Figure 3 (B1-a to B1-e). The membranes were visually examined and their performance in handling during spinning was also checked. From B1-a to B1-c, the coagulation bath was kept at room temperature (≈24 °C). Membrane B1-a showed the highest strength (i.e., did not break nor collapse during spinning) with the highest thickness and the smallest dimensions (i.e., internal diameter, ID, and outer diameter, OD). Membrane B1-b was in the middle of B1-a and B1-c in terms of dimensions and showed the characteristic sandwich structure, with a thin layer of small-size macrovoids that was present in all subsequent batches. B1-c possessed a softer structure than B1-a as it easily collapsed due to handling during spinning; therefore, B1-c spinning conditions were not retested. Membranes B1-d and B1-e had the coagulation bath at a higher temperature. B1-d had a similar dimension to B1-a, while the structure changed to a more porous one than B1-a due to the higher coagulation bath temperature. The results are consistent with numerous similar observations reported in the literature [46–48]. It has been well established that an increase in coagulation bath temperature results in a faster solvent–non-solvent exchange. Consequently, it leads to a more porous structure, while a slower de-mixing at lower temperatures results in a denser film [49]. Membranes B1-d and B1-e were fabricated at a higher coagulation bath temperature of ~40 °C and bore fluid flowrates of 1.5 and 3.0 mL/min, respectively. As shown in Table 3, the increase in bore fluid flowrate resulted in an increase in ID and OD of the membrane B1-e. A similar trend was also observed for membranes B1-a, B1b and B1-c, where the diameter increased as the bore fluid rate increased from 1.5 to 4.5 mL/min. also leading to a reduced wall thickness. Membrane B1-e started to lose its round shape and the membrane strength significantly decreased when elevating the bath temperature. From these results, it is revealed that the membrane dimensions increase and thicknesses reduce when the bore flowrate is boosted [25,50].

As part of our efforts to scale-up the membrane fabrication process from a lab scale to a pilot scale, the batches with PVDF 2 were run at a higher take-up speed of 9 m/min and, consequently, the dope flowrate had to be adjusted to 13.5 mL/min to be consistent with the lower line speeds used for PVDF 1 batches. As shown in Table 3, membrane B2-a had the highest strength and the highest thickness of the batch. Membrane B2-b had similar features as B2-a but with a slightly higher ID. Membrane B2-c appeared deformed due to its small membrane wall thickness and, therefore, these spinning conditions were not considered for further experiments. Like the previous batch results, when the bore flowrate was boosted, the membrane dimensions increased and the thicknesses reduced. As evident from Table 3, the increased bore fluid flowrate resulted in larger diameters (ID and OD) as well as a lower wall thickness, eventually leading to the loss of mechanical integrity (for B1-c and B2-c). As the bore fluid flowrate increased, the solvent–non-solvent exchange rate increased, leading to higher mass transfer and faster polymer de-mixing. The higher bore fluid flowrate also radially expanded the fiber dimensions and thinned the fiber wall, thus reducing the overall mechanical strength [50–52].

The microscopy images and the very feasible optimization of the membranes suggested that membranes B2-a and B2-b had the potential to be scaled-up to 20 kg and subsequently 50 kg batches. Based on the performance results, it was determined that the conditions used for the membrane B2-a were most suitable for the final scale-up stage when using PVDF 1 in the polymer dope instead of PVDF 2.

Figure 4 shows the FESEM images of the membrane samples from a small batch size (1.5 kg, B1-a) and a large production-scale batch size (50 kg, B8). The SEM images confirm the formation of the novel sandwich-like structure with a porous inner layer filled with macrovoids between two thin, denser outer layers. They are consistent with the previously reported literature [19,33,37].

The sandwich structure, with the two sponge-like layers, improves the mechanical properties, and increases the evaporation area and the vapor transport during the VMD process in an in-to-out configuration. The rapid de-mixing in the outer layer is due to the use of water as a non-solvent, which produces a closer porous structure than the inner layer that helps to avoid membrane wetting due to condensation in the permeate side. The inner surface is more porous due to the use of an NMP/water solution (50/50 wt./wt.)

as the bore fluid, which delays the phase inversion. LiCl and EG are used to decrease the miscibility of the solvent in the dope, allowing a more controlled liquid–liquid extraction of the dope. LiCl, by increasing the dope viscosity, also helps to reduce the size of the macrovoids, thereby increasing the strength of the membranes. While the inner surface of B8 is still very porous, it is less porous than the smaller scale batches B2-a and B2-b. This small change in the structure is probably due to the increase in dope viscosity shown in Table 1 when using PVDF 1 as the base polymer.

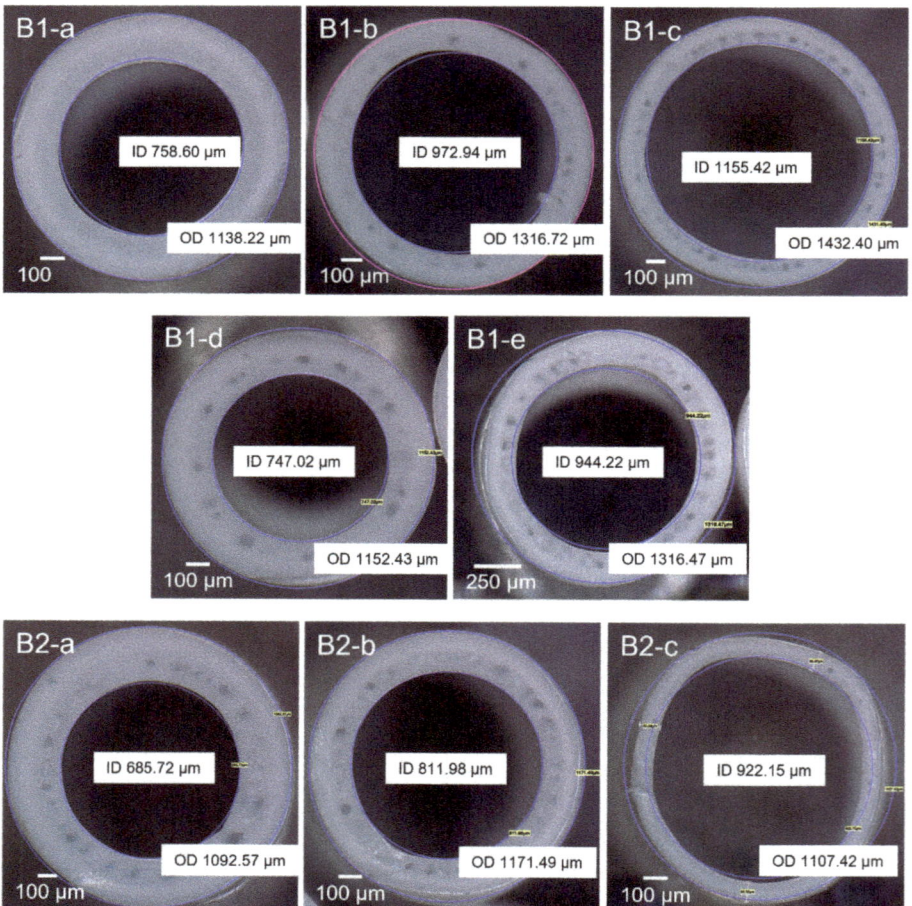

Figure 3. Changes in membrane morphology with changes in spinning conditions. B1: samples from PVDF 1 dope. B2: samples from PVDF 2 dope.

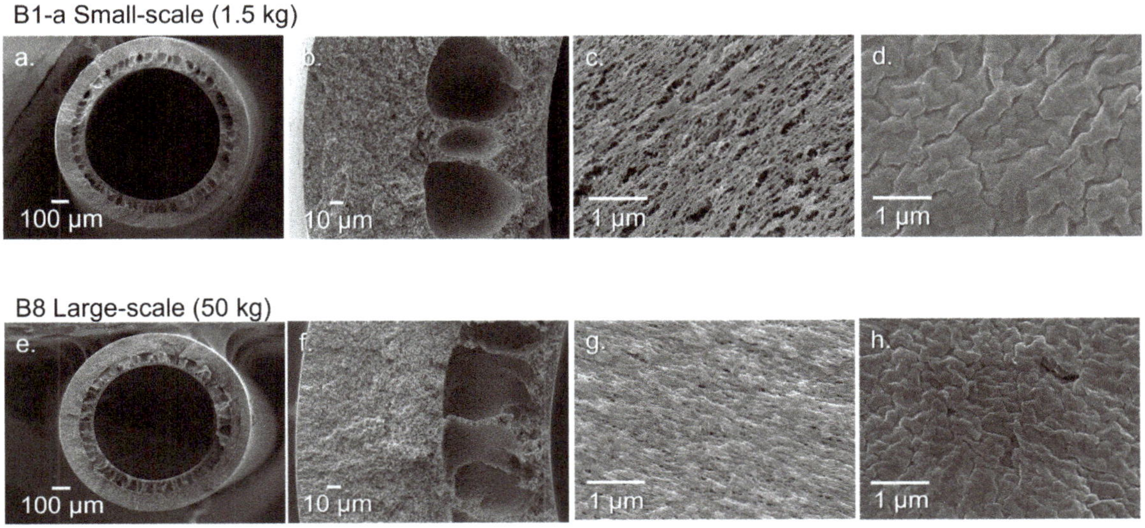

B1-a Small-scale (1.5 kg)

B8 Large-scale (50 kg)

Figure 4. FESEM of PVDF hollow fiber membranes. (**a**,**e**) Cross-section of membrane; (**b**,**f**) zoom-in of membrane's cross-section; (**c**,**g**) inner surface of membrane; (**d**,**h**) outer surface of membrane.

The porosity, contact angle, and thickness of the samples increased when the polymer in the dopes was changed from PDVF 2 to PDVF 1, as shown in Figure 5. The membranes made from PVDF 1, prepared using similar spinning and dope conditions as the membrane B-2a (Table 3), showed higher contact angle and porosity values than the ones from PVDF 2 for batch sizes of 1.5 kg and 20 kg, as shown in Figure 5.

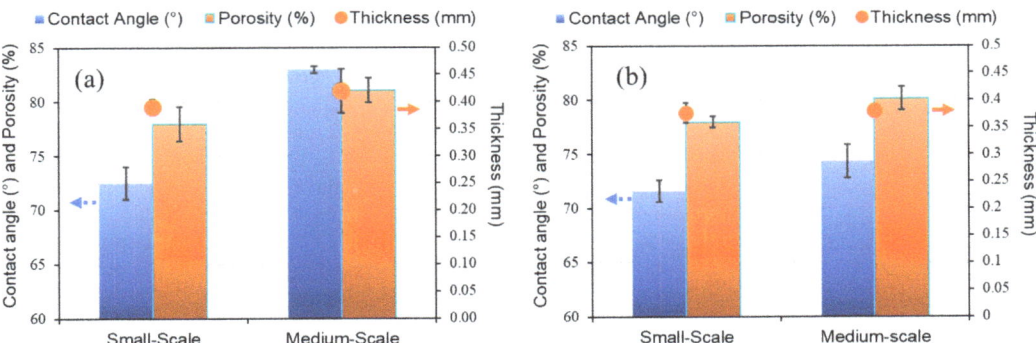

Figure 5. Porosity, contact angle, and thickness of produced hollow fiber membranes in each batch. Small scale: 1.5 kg batch; medium scale: 20 kg batch. (**a**) Fibers made with PVDF 1 dope; (**b**) fibers made with PVDF 2 dope.

On the other hand, the tensile strength decreased when the production was scaled-up, as shown in Figure 6. This behavior could be explained by the constant tension to which the membranes were subjected during the long continuous fabrication process. The spinning process was performed for up to three days due to the larger quantities of dopes, and these conditions could have subtle changes in dope compositions from one day to the next. Here, the tensile strength was proportionally higher when using PVDF 1 in the dope than when PVDF 2 was used, as shown in Figure 6a. The tensile strain also decreased when increasing the batch size using PVDF 1 (Figure 6a). On the contrary, the tensile strain values seemed to increase when using PVDF 2 in the dope. It is worth noting that from batch B3 onwards, the conditions of the spinning process were adopted at larger scales of 20 kg and 50 kg (see Table 2); thus, the values obtained by the large-scale batch (50 kg) show that there was an optimization of conditions that led to an increase in the mechanical properties. In addition, the LEP$_w$ values showed more consistency between batches, which is a very important feature for obtaining better results in the VMD process. These findings suggest that large-scale reproducibility is commercially achievable with small changes aiming to increase production.

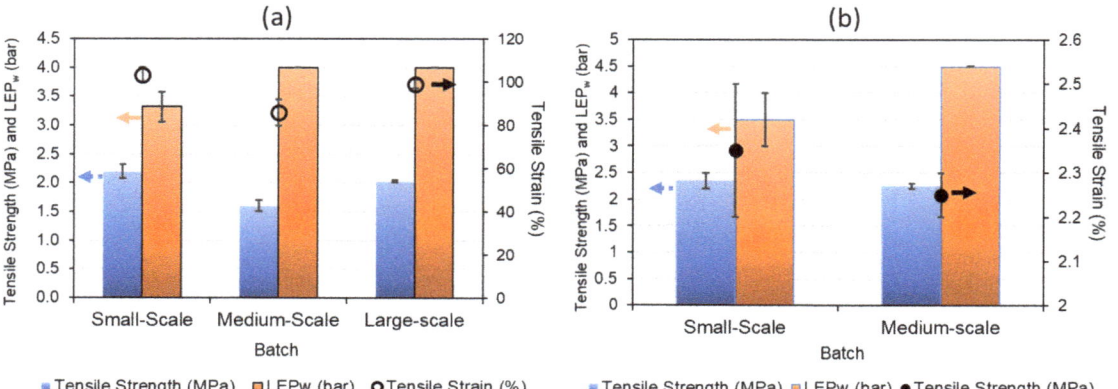

Figure 6. Tensile strength and strain and liquid entry pressure (LEP$_w$) of water in produced hollow fiber membranes in each batch. Small scale: 1.5 kg batch; medium scale: 20 kg batch; large scale: 50 kg. (**a**) Fibers made with PVDF 1 dope; (**b**) fibers made with PVDF 2 dope.

3.3. VMD Tests

Once assembled, the small 0.5-inch modules prepared using PVDF 1 and 2 dopes were placed in a vacuum membrane distillation unit at the EWTCOI facility. These modules were tested by treating a 35 g/L synthetic NaCl feed solution, which was used to simulate seawater to validate the membrane modules for desalination application, as per the operating parameters outlined in Table 4. The modules were prepared using membranes spun using the conditions described in Tables 2 and 3 in a 1.5 kg batch size, and characteristics such as salt rejection and flux were evaluated in VMD mode (Figure S4). The VMD tests performed on the 0.5-inch modules showed a higher flux for hollow fibers produced from the dope 1 than from the dope 2, with differences of about 20 L/m^2.h for the same spinning conditions (B3 vs. B5). The salt rejection, based on electrical conductivity (EC) measurements, remained consistent with values close to 100% for all modules tested, as seen in Figure S4. In comparison to the research reported in the literature, the membranes produced in this work showed higher flux values under similar operating conditions [29,32]. It is worth noting that with each batch iteration, the consistency in the membrane characteristics increased and was maintained, especially for the batches that used PVDF 1 (Figure S4a). These VMD results of the 0.5-inch modules from smaller-scale batches confirmed the suitability of PVDF 1 for the full-scale spinning process, as discussed in the previous section.

Once the production conditions were selected, 2-inch modules were assembled (Figure 1b) and tested using a custom-built MD unit capable of operating in VMD and DCMD modes. The process flow diagram for the VMD operation using the skid is shown in Figure 2. Figure 7 depicts the VMD test results for the 0.5-inch and 2-inch modules at EWTCOI and START facilities, respectively. In order to evaluate the reproducibility of the membrane characteristics as a function of dope batch sizes, two sets of 0.5-inch modules were assembled with membranes prepared from small- and medium-size batches (i.e., 1.5 kg and 20 kg dope sizes) and then compared with 2-inch modules assembled with membranes prepared from a large batch size of 50 kg. As shown in Figure 7, the fluxes of small 0.5-inch modules remained high at 47 L/m².h and 60 L/m².hr for membranes prepared from 1.5 kg and 20 kg batch sizes, respectively. However, as the module size is increased to 2-inch, the flux drops significantly to ~10 L/m².h.while the salt rejection remains >90%. The high salt rejection indicates that the membrane's microporous structure is still intact and reproducible at different batch sizes; the decline in flux may be attributed to module characteristics such as flow pattern, flow distribution, and temperature polarization. This flux decline phenomenon tends to be higher in an in-to-out configuration, thus diminishing the mass and heat transfer efficiencies in 2-inch modules [53–56].

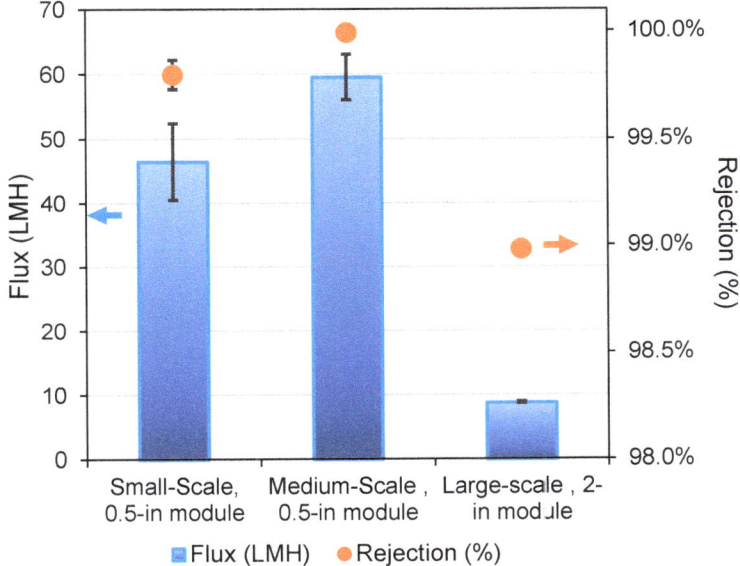

Figure 7. Flux and rejection of VMD tests in each batch using PVDF 1. All tests were performed for time ≥ 1 h using 0.5-inch modules and time ≥ 100 h using 2-inch modules. Small scale: 1.5 kg batch; medium scale: 20 kg batch; large scale: 50 kg.

Due to the limitations of the existing VMD unit, the effects of some operating parameters such as feed flowrate, temperature gradient, pressure differential across the membranes, and temperature polarization coefficient (TPC) were not thoroughly evaluated in the current study. A larger 5000 L/day capacity pilot unit with the requisite engineering design to study the effect of the above-mentioned operating parameters on permeate flux is under construction and the results from VMD testing of 16 4-inch modules will be the subject of a subsequent publication.

In order to evaluate the long-term performance and to assess failure modes such as pore wetting under the given test conditions, the 2-inch modules were tested with synthetic seawater prepared with a 35 g/L NaCl, in a batch mode previously described, for over 100 h. The pilot unit was operated for 5–6 h per day with the feed water replenished at the beginning of the day. The flux and the salt rejection data for the 2-inch modules are summarized in Figure 8. Throughout the test duration, the salt rejection and the permeate flux remained consistent at ~100% and within 8–9 L/m²·h, respectively, despite the variations in feed concentrations due to the batch mode operations previously described. The stable permeate flux through the test duration and under the given conditions indicates that the membrane pore structure remained intact with no pore wetting, which would have otherwise caused a spike in the permeate conductivity, not seen in this study.

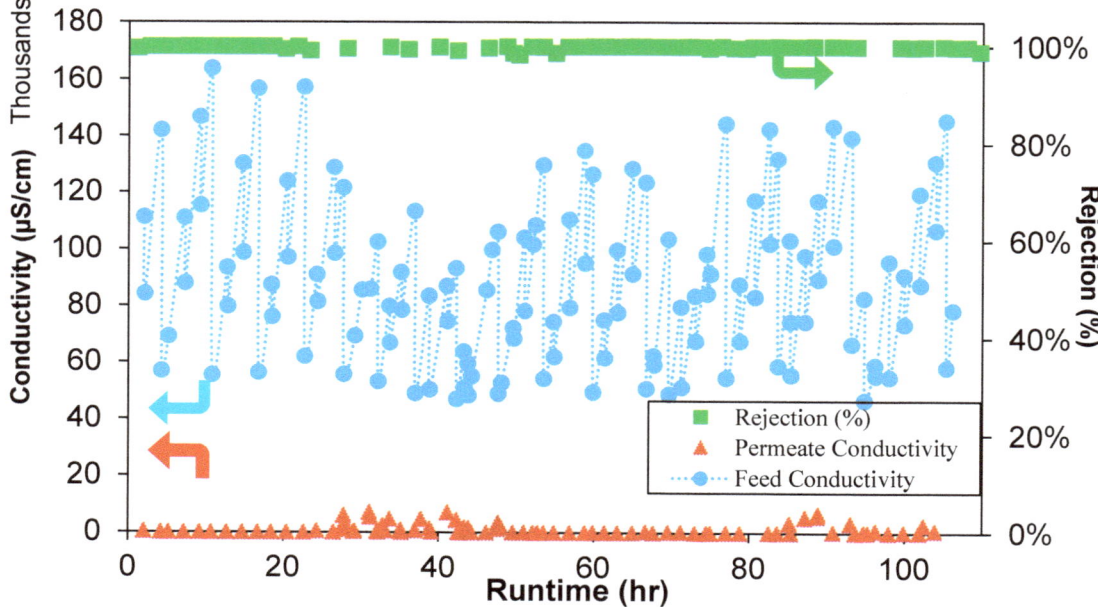

Figure 8. Rejection and conductivity of the feed and permeate as a function of runtime from pilot tests using a 2-inch module.

It is important to highlight that the flux of the 2-inch module was nearly six times lower than the highest flux previously reported for the 0.5-inch modules (Figure 7). The significant drop in flux with an increase in module size can be attributed to several factors such as (a) multi-fold increase in the membrane area as well as a much tighter packing density in a larger module, leading to a decrease in the residence time at the feed flowrates employed, (b) a high conductive heat loss leading to the loss of driving force for water vapor transport, and (c) sub-optimal flow distribution either in laminar flow regime or the transition flow regime, all resulting in less efficient mass transfer across the membrane [55,56]. Despite this reduction in flux, the product water flux using the 2-inch module falls within the range of previously reported works where the tests were carried out on a pilot scale [24,54]. These results confirm the long-term effectiveness and high performance of the sandwich-structured hollow fibers developed in this study.

In addition, test conditions such as feed temperatures and flowrates, as well as the test duration, have an impact on the flux (see Table 4 and Figure 9) [55]. For example, the lower feed flowrate used in the 0.5-inch modules increases the residence time, leading to a higher flux. However, these conditions are not suitable for use in industrial or commercial settings because of the very low productivity rates. Nevertheless, the results of the 2-inch modules show fluxes almost four times higher than and comparable rejections to SWRO systems, which are typically in the range of 2.5–3 L/m^2.h bar and ~99.7% salt rejection, respectively, making these produced hollow fibers suitable alternatives for desalination.

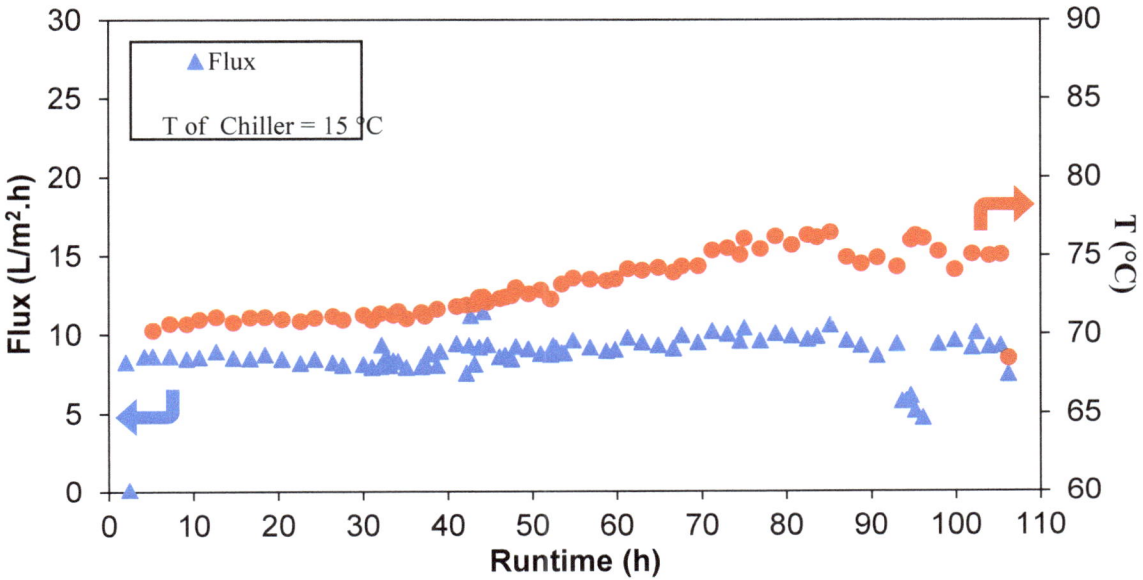

Figure 9. Flux and feed temperature profile from pilot tests using a 2-inch module.

4. Conclusions

In this study, we successfully scaled-up a lab-scale membrane fabrication process to produce a novel sandwich-like structure comprising an inner porous layer with controllable macrovoids between two thin layers of sponge-like dense layers. The membrane morphology was optimally designed for membrane distillation applications. The membrane fabrication processes were scaled-up from 1.5 kg batch sizes to 20 kg and 50 kg, clearly demonstrating the feasibility of translating the chemistry and process to a manufacturing set up. The membrane properties such as porosity, mechanical strength, and morphology were optimized by careful control of the spinning conditions. The scaled-up membranes prepared using the optimized conditions were assembled into small 0.5-inch diameter modules with 15 fibers (i.e., an effective membrane area of ~0.0035–0.0051 m^2) as well as 2-inch diameter modules (i.e., an effective membrane area of 0.46 m^2), which were tested using MD testing units in VMD mode against a synthetic feed water simulating seawater concentration (35 g/L NaCl). The small modules showed a very high flux of >40 L/m^2·h under the operating conditions, while the flux drops to ≤10 L/m^2·h as the module size is increased to 2-inch. Nevertheless, the 2-inch modules tested for over 100 h demonstrated the long-term efficiency of the membranes with a flux maintained at ~8.8 L/m^2·h while the salt rejection remains close to 100%. These results validate the morphological design employed for the novel PVDF membranes that imparts high mechanical integrity as well as optimal pore structure for highly efficient vapor transport. While the observations are highly encouraging and stand testimony to the suitability of these membranes in applications such as seawater desalination and high-strength industrial wastewater treatment for recycle and reuse, challenges with retaining the flux still linger as the modules are further scaled to a commercial industrial scale of 4-inch or 8-inch diameters. The future efforts of our group are to be extensively focused on module scale-up and field validation using a 5000 L/day pilot unit against actual seawater or industrial wastewater. The results from the pilot validation will be the subject of subsequent publication.

Supplementary Materials: The following supporting information can be downloaded at: https://www.mdpi.com/article/10.3390/membranes12040423/s1. Figure S1. DSC of different PVDF powders used for hollow fiber membrane manufacture. (a) PVDF 1, USA origin; (b) PVDF 2, China origin; Figure S2. DSC-TGA of different PVDF powders used for hollow fiber membrane manufacture. (a) PVDF 1, USA origin; (b) PVDF 2, China origin; Figure S3. Pyrolysis–GCMS chromatograms of PVDF at 600 °C. Orange—PVDF 1; green—PVDF 2; Table S1. Characteristics of manufactured hollow fibers. FR: flowrate; CB: coagulation bath; ID: internal diameter; OD: outer diameter; Figure S4. Flux and rejection of VMD tests in each batch. All tests were performed for time ≥ 1 h and using 0.5-inch modules. Descending values based on flux (LMH). (a) Fibers made with PVDF 1 dope; (b) fibers made with PVDF 2 dope.

Author Contributions: Conceptualization, T.-S.C., J.Z. (Jian Zuo) and C.G.; methodology, T.-S.C. and C.G.; validation, J.Z. (Junyou Zhang), A.T.P., K.M., M.S.Q. and Y.Z.; formal analysis, Y.Z.; investigation, S.H.; resources, C.G.; data curation, S.H. and Y.Z.; writing—original draft preparation, S.H.; writing—review and editing, C.G. and T.-S.C.; visualization, C.G. and T.-S.C.; supervision, C.G.; project administration, J.Z. (Junyou Zhang) and S.H.; funding acquisition and technology guidance, A.D. All authors have read and agreed to the published version of the manuscript.

Funding: This research was funded by the Economic Development Board (Singapore) through grant number S15-1068-NRF EWI-RCFS.

Institutional Review Board Statement: Not applicable.

Informed Consent Statement: Not applicable.

Data Availability Statement: Not applicable.

Acknowledgments: We would like to acknowledge Lv Yunbo, Lam Mei Shan, and She Ka Keng from NEWRI Analytics Cluster, Nanyang Technological University, Singapore, for sample testing on SEM, Goniometer, Porometer, Pyrolysis GC-MS, TGA, and DSC. We also want to acknowledge EWTCOI, Ngee Ann Polytechnic, Singapore, for sample characterization and vacuum membrane distillation testing of hollow fiber membranes. The sample characterization includes tensile stress and strain, contact angle, LEP_w, porosity.

Conflicts of Interest: The authors declare no conflict of interest.

References

1. Escobar, I.C. Chapter 14 Conclusion: A Summary of Challenges still Facing Desalination and Water Reuse. In *Sustainable Water for the Future: Water Recycling versus Desalination*; Escobar, I.C., Schafer, A.I., Eds.; Elsevier: Amsterdam, The Netherlands, 2010; Volume 2, pp. 389–397.
2. Christidis, N.; Scott, P.A. The influence of anthropogenic climate change on wet and dry summers in Europe. *Sci. Bull.* **2021**, *66*, 813–823. [CrossRef]
3. World Health Organization; UNICEF. *Progress on Drinking Water, Sanitation and Hygiene: 2017 Update and SDG Baselines*; World Health Organization: Geneva, Switzerland, 2017; ISBN 9789241512893.
4. Ahmed, F.E.; Khalil, A.; Hilal, N. Emerging desalination technologies: Current status, challenges and future trends. *Desalination* **2021**, *517*, 115183. [CrossRef]
5. Zapata-Sierra, A.; Cascajares, M.; Alcayde, A.; Manzano-Augliaro, F. Worldwide research trends on desalination. *Desalination* **2022**, *519*, 115305. [CrossRef]
6. Li, B.-J.; Choi, S.-M.; Cho, S.-H.; Master, G.-R.L.; Park, C.-D. Design and performance analysis of vertical multi-effect diffusion solar distiller: A review. *Desalination* **2022**, *527*, 115572. [CrossRef]
7. Jhon Jairo Feria-Díaz, J.J.; López-Méndez, M.C.; Rodríguez-Miranda, J.P.; Sandoval-Herazo, L.C.; Correa-Mahecha, F. Commercial Thermal Technologies for Desalination of Water from Renewable Energies: A State of the Art Review. *Processes* **2021**, *9*, 262. [CrossRef]
8. Curto, D.; Franzitta, V.; Guercio, A. A review of the water desalination technologies. *Appl. Sci.* **2021**, *11*, 670. [CrossRef]
9. Lim, Y.J.; Goh, K.; Kurihara, M.; Wang, R. Seawater desalination by reverse osmosis: Current development and future challenges in membrane fabrication—A review. *J. Membr. Sci.* **2021**, *629*, 119292. [CrossRef]
10. Edwie, F.; Chung, T.S. Development of simultaneous membrane distillation-crystallization (SMDC) technology for treatment of saturated brine. *Chem. Eng. Sci.* **2013**, *98*, 160–172. [CrossRef]
11. Choudhury, M.R.; Anwar, N.; Jassby, D.; Rahaman, M.S. Fouling and wetting in the membrane distillation driven wastewater reclamation process—A review. *Adv. Colloid Interface Sci.* **2019**, *269*, 370–399. [CrossRef]

12. Yan, Z.; Jiang, Y.; Liu, L.; Li, Z.; Chen, X.; Xia, M.; Fan, G.; Ding, A. Membrane distillation for wastewater treatment: A mini Review. *Water* **2021**, *13*, 3480. [CrossRef]
13. Deshmukh, A.; Boo, C.; Karanikola, V.; Lin, S.; Straub, A.P.; Tong, T.; Warsinger, D.M.; Elimelech, M. Membrane distillation at the water-energy nexus: Limits, opportunities, and challenges. *Energy Environ. Sci.* **2018**, *11*, 1177–1196. [CrossRef]
14. Ibrar, I.; Yadav, S.; Naji, O.; Alanezi, A.A.; Ghaffour, N.; Deon, S.; Subbiah, S.; Altaee, A. Development in forward Osmosis-Membrane distillation hybrid system for wastewater treatment. *Sep. Purif. Technol.* **2022**, *286*, 120498. [CrossRef]
15. Meng, L.; Mansouri, J.; Li, X.; Liang, J.; Huang, M.; Lv, Y.; Wang, Z.; Chen, V. Omniphobic membrane via bioinspired silicification for the treatment of RO concentrate by membrane distillation. *J. Membr. Sci.* **2022**, *647*, 120267. [CrossRef]
16. Kharraz, J.A.; Khanzada, N.K.; Farid, M.U.; Kim, J.; Jeong, S.; An, A.K. Membrane distillation bioreactor (MDBR) for wastewater treatment, water reuse, and resource recovery: A review. *J. Water Proc. Eng.* **2002**, *47*, 102687. [CrossRef]
17. Parani, S.; Oluwafemi, O.S. Membrane Distillation: Recent Configurations, Membrane Surface Engineering, and Applications. *Membranes* **2021**, *11*, 934. [CrossRef]
18. Abdelkareem, M.A.; El Haj Assad, M.; Sayed, E.T.; Soudan, B. Recent progress in the use of renewable energy sources to power water desalination plants. *Desalination* **2018**, *435*, 97–113. [CrossRef]
19. Zuo, J.; Chung, T.S. PVDF hollow fibers with novel sandwich structure and superior wetting resistance for vacuum membrane distillation. *Desalination* **2017**, *417*, 94–101. [CrossRef]
20. Zhong, W.; Guo, L.; Ji, C.; Dong, G.; Li, S. Membrane distillation for zero liquid discharge during treatment of wastewater from the industry of traditional Chinese medicine: A review. *Environ. Chem. Lett.* **2021**, *19*, 2317–2330. [CrossRef]
21. Julian, H.; Nurgirisia, N.; Sutrisna, P.D.; Wenten, I.G. Advances in seawater membrane distillation (SWMD) towards stand-alone zero liquid discharge (ZLD) desalination. *Rev. Chem. Eng.* **2021**, 000010151520200073. [CrossRef]
22. Yadav, A.; Labhasetwar, P.K.; Shahi, V.K. Membrane distillation crystallization technology for zero liquid discharge and resource recovery: Opportunities, challenges and futuristic perspectives. *Sci. Total Environ.* **2022**, *806*, 150692. [CrossRef]
23. Zuo, J.; Bonyadi, S.; Chung, T.S. Exploring the potential of commercial polyethylene membranes for desalination by membrane distillation. *J. Membr. Sci.* **2016**, *497*, 239–247. [CrossRef]
24. Francis, L.; Ahmed, F.E.; Hilal, N. Advances in Membrane Distillation Module Configurations. *Membranes* **2022**, *12*, 81–93. [CrossRef] [PubMed]
25. Pagliero, M.; Khayet, M.; García-Payo, C.; García-Fernández, L. Hollow fibre polymeric membranes for desalination by membrane distillation technology: A review of different morphological structures and key strategic improvements. *Desalination* **2021**, *516*, 115235–115263. [CrossRef]
26. Camacho, L.; Dumée, L.; Zhang, J.; Li, J.; Duke, M.; Gomez, J.; Gray, S. Advances in Membrane Distillation for Water Desalination and Purification Applications. *Water* **2013**, *5*, 94–196. [CrossRef]
27. Chiam, C.K.; Sarbatly, R. Vacuum membrane distillation processes for aqueous solution treatment-A review. *Chem. Eng. Process. Process Intensif.* **2013**, *74*, 27–54. [CrossRef]
28. Zuo, J.; Chung, T.S.; O'Brien, G.S.; Kosar, W. Hydrophobic/hydrophilic PVDF/Ultem® dual-layer hollow fiber membranes with enhanced mechanical properties for vacuum membrane distillation. *J. Membr. Sci.* **2017**, *523*, 103–110. [CrossRef]
29. Abu-Zeid, M.A.E.R.; Zhang, Y.; Dong, H.; Zhang, L.; Chen, H.L.; Hou, L. A comprehensive review of vacuum membrane distillation technique. *Desalination* **2015**, *356*, 1–14. [CrossRef]
30. Sorour, M.H.; Hani, H.A.; Shaalan, H.F.; El-Toukhy, M. Fabricatio nand characterization of hydrophobic PVDF-based hollow fiber membranes for vacuum membrane distillation of seawater and desalination brine. *Egypt. J. Chem.* **2021**, *64*, 4889–4899. [CrossRef]
31. Albloushi, A.; Giwa, A.; Mukherjee, D.; Calabro, V.; Cassano, A.; Chakraborty, S.; Hasan, S.W. Chapter 7—Renewable Energy-Powered Membrane Systems for Water Desalination. In *Current Trends and Future Developments on (Bio-) Membranes*; Basile, A., Cassano, A., Figoli, A., Eds.; Elsevier: Amsterdam, The Netherlands, 2019; pp. 153–177, ISBN 978-0-12-813545-7.
32. Baghel, R.; Upadhyaya, S.; Singh, K.; Chaurasia, S.P.; Gupta, A.B.; Dohare, R.K. A review on membrane applications and transport mechanisms in vacuum membrane distillation. *Rev. Chem. Eng.* **2017**, *34*, 73–106. [CrossRef]
33. Wang, P.; Teoh, M.M.; Chung, T.S. Morphological architecture of dual-layer hollow fiber for membrane distillation with higher desalination performance. *Water Res.* **2011**, *45*, 5489–5500. [CrossRef]
34. Peng, N.; Chung, T.S.; Wang, K.Y. Macrovoid evolution and critical factors to form macrovoid-free hollow fiber membranes. *J. Membr. Sci.* **2008**, *318*, 363–372. [CrossRef]
35. Sukitpaneenit, P.; Chung, T.S. Molecular elucidation of morphology and mechanical properties of PVDF hollow fiber membranes from aspects of phase inversion, crystallization and rheology. *J. Membr. Sci.* **2009**, *340*, 192–205. [CrossRef]
36. Fontananova, E.; Jansen, J.C.; Cristiano, A.; Curcio, E.; Drioli, E. Effect of additives in the casting solution on the formation of PVDF membranes. *Desalination* **2006**, *192*, 190–197. [CrossRef]
37. Zuo, J.; Chung, T.S.N. A Hollow Fiber Membrane. WO2018080398A1, 3 May 2018.
38. Chen, Z.; Rana, D.; Matsuura, T.; Meng, D.; Lan, C.Q. Study on structure and vacuum membrane distillation performance of PVDF membranes: II. Influence of molecular weight. *Chem. Eng. J.* **2015**, *276*, 174–184. [CrossRef]
39. Liu, Z.; Maréchal, P.; Jérôme, R. Melting and crystallization of poly(vinylidene fluoride) blended with polyamide 6. *Polymer* **1997**, *38*, 5149–5153. [CrossRef]
40. Marega, C.; Marigo, A. Influence of annealing and chain defects on the melting behaviour of poly(vinylidene fluoride). *Eur. Polym. J.* **2003**, *39*, 1713–1720. [CrossRef]

41. Ji, D.; Xiao, C.; Chen, K.; Zhou, F.; Gao, Y.; Zhang, T.; Ling, H. Solvent-free fabrication of PVDF hollow fiber membranes with controlled pore structure via melt spinning and stretching. *J. Membr. Sci.* **2021**, *621*, 118593. [CrossRef]
42. Dikshit, A.K. Effect of solvent on thermal transitions and conductivity of poly(vinylidene fluoride) gel electrolytes. *Polym. Plast. Technol. Mater.* **2020**, *59*, 822–834. [CrossRef]
43. Rosenberg, Y.; Siegmann, A.; Narkis, M.; Shkolnik, S. The sol/gel contribution to the behavior of γ-irradiated poly(vinylidene fluoride). *J. Appl. Polym. Sci.* **1991**, *43*, 535–541. [CrossRef]
44. Silva, A.J.D.J.; Contreras, M.M.; Nascimento, C.R.; da Costa, M.F. Kinetics of thermal degradation and lifetime study of poly(vinylidene fluoride) (PVDF) subjected to bioethanol fuel accelerated aging. *Heliyon* **2020**, *6*, e04573. [CrossRef]
45. Xu, B.; Hou, S.; Chu, M.; Cao, G.; Yang, Y. An activation-free method for preparing microporous carbon by the pyrolysis of poly(vinylidene fluoride). *Carbon* **2010**, *48*, 2812–2814. [CrossRef]
46. Liu, S.-H.; Liu, M.; Xu, Z.-L.; Wei, Y.-M. A polyethersulfone–bisphenol sulfuric acid hollow fiber ultrafiltration membrane fabricated by a reverse thermally induced phase separation process. *RSC Adv.* **2018**, *8*, 7800–7809. [CrossRef]
47. Sathiya, S.; Norlisa, H.; Syed, M.S. Effect of coagulation bath temperature during preparation of PES hollow fiber supported liquid membrane for acetic acid removal. *Chem. Eng. Res. Bull.* **2017**, *19*, 118–122.
48. Xu, J.; Tang, Y.; Wang, Y.; Shan, B.; Yu, L.; Gao, C. Effect of Coagulation Bath Conditions on the Morphology and Performance of PSf Membrane Blended with a Capsaicin-Mimic Copolymer. *J. Membr. Sci.* **2014**, *455*, 121–130. [CrossRef]
49. Blanco, F.J.; Sublet, J.; Nguyen, Q.T.; Schaetzel, P. Formation and Morphology Studies of Different Polysulfones-Based Membranes Made by Wet Phase Inversion Process. *J. Membr. Sci.* **2006**, *283*, 27–37. [CrossRef]
50. Bildyukevich, A.V.; Plisko, T.V.; Usosky, V.V. The formation of polysulfone hollow fiber membranes by the free fall spinning method. *Pet. Chem.* **2016**, *56*, 379–400. [CrossRef]
51. Bonyadi, S.; Chung, T.S.; Krantz, W.B. Investigation of corrugation phenomenon in the inner contour of hollow fibers during the non-solvent induced phase-separation process. *J. Membr. Sci.* **2007**, *299*, 200–210. [CrossRef]
52. Li, G.; Kujawski, W.; Knozowska, K.; Kujawa, J. The Effects of PEI Hollow Fiber Substrate Characteristics on PDMS/PEI Hollow Fiber Membranes for CO_2/N_2 Separation. *Membranes* **2021**, *11*, 56. [CrossRef]
53. Yang, X.; Wang, R.; Fane, A.G. Novel designs for improving the performance of hollow fiber membrane distillation modules. *J. Membr. Sci.* **2011**, *384*, 52–62. [CrossRef]
54. Shin, Y.; Choi, J.; Park, Y.; Choi, Y.; Lee, S. Influence of operation conditions on the performance of pilot-scale vacuum membrane distillation (VMD). *Desalin. Water Treat.* **2017**, *97*, 1–7. [CrossRef]
55. Han, F.; Bian, Y.; Zhang, G. Conductive heating vacuum membrane distillation for brine desalination: Study on operational conditions, temperature polarization and energy consumption. *Desalination* **2022**, *531*, 115726. [CrossRef]
56. Anvari, A.; Yancheshme, A.A.; Kekre, K.M.; Ronen, A. State-of-the-art methods for overcoming temperature polarization in membrane distillation process: A review. *J. Membr. Sci.* **2020**, *616*, 118413. [CrossRef]

Article

Comparison between Thermophilic and Mesophilic Membrane-Aerated Biofilm Reactors—A Modeling Study

Duowei Lu [1,2], Hao Bai [2] and Baoqiang Liao [1,*]

[1] Department of Chemical Engineering, Lakehead University, 955 Oliver Road, Thunder Bay, ON P7B 5E1, Canada; dlu5@lakeheadu.ca
[2] Department of Mechanical Engineering, Lakehead University, 955 Oliver Road, Thunder Bay, ON P7B 5E1, Canada; hbai@lakeheadu.ca
* Correspondence: bliao@lakeheadu.ca; Tel.: +1-807-343-8437; Fax: +1-807-343-8928

Abstract: The concept of thermophilic membrane-aerated biofilm reactor (ThMABR) is studied by modeling. This concept combines the advantages and overcomes the disadvantages of conventional MABR and thermophilic aerobic biological treatment and has great potential to develop a new type of ultra-compact, highly efficient bioreactor for high-strength wastewater and waste gas treatments. Mathematical modeling was conducted to investigate the impact of temperature (mesophilic vs. thermophilic) and oxygen partial pressure on oxygen and substrate concentration profiles, membrane–biofilm interfacial oxygen concentration, oxygen penetration distance, and oxygen and substrate fluxes into biofilms. The general trend of oxygen transfer and substrate flux into biofilm between ThAnMBR and MMABR was verified by the experimental results in the literature. The results from modeling studies indicate that the ThMABR has significant advantages over the conventional mesophilic MABR in terms of improved oxygen and pollutant flux into biofilms and biodegradation rates, and an optimal biofilm thickness exists for maximum oxygen and substrate fluxes into the biofilm.

Keywords: membrane-aerated biofilm reactor; thermophilic membrane-aerated biofilm reactor; thermophilic biological treatment; biofilm; mass transfer; modeling

Citation: Lu, D.; Bai, H.; Liao, B. Comparison between Thermophilic and Mesophilic Membrane-Aerated Biofilm Reactors—A Modeling Study. *Membranes* **2022**, *12*, 418. https://doi.org/10.3390/membranes12040418

Academic Editor: Gaetano Di Bella

Received: 8 March 2022
Accepted: 8 April 2022
Published: 12 April 2022

Publisher's Note: MDPI stays neutral with regard to jurisdictional claims in published maps and institutional affiliations.

Copyright: © 2022 by the authors. Licensee MDPI, Basel, Switzerland. This article is an open access article distributed under the terms and conditions of the Creative Commons Attribution (CC BY) license (https://creativecommons.org/licenses/by/4.0/).

1. Introduction

Treatment of high-strength chemical oxygen demand (COD) industrial wastewater and waste gases has posed a significant challenge to engineers and scientists. Novel technologies for wastewater COD removal and waste gas treatment are highly desirable for sustainable development and pollution control. Recently, two promising approaches have emerged as competitive alternatives for process intensification in wastewater treatment facilities that can handle larger substrate loads and achieve higher effluent quality without increasing the footprint [1,2]. These two approaches are membrane-aerated biofilm reactor (MABR) technology [3,4] and thermophilic aerobic biological treatment (TABT) [5], which has a high oxygen transfer rate in MABR and a high biodegradation rate in TABT, and the synergy of these two technologies will develop a highly efficient and compact biological treatment system.

In an MABR system, the biofilm is immobilized on the outside of a gas-permeable membrane where the oxygen and gas pollutants are supplied for biodegradation, while the nutrients and wastewater pollutants are transported into the biofilm from the opposite direction [6]. The use of gas-permeable membranes to deliver oxygen and gas pollutants can achieve bubble-free aeration as well as extremely high removal efficiency for gas pollutants. This novel design represents a high energy efficiency compared to conventional biological treatment processes. In addition, the average TN removal in the biofilm membrane reactor was increased by around 6% compared with conventional membrane bioreactor [7]. Moreover, MABR technology is particularly suitable for the treatment of wastewaters containing volatile organic compounds (VOCs) and waste gases containing hydrophobic compounds, which are challenging to conventional aerated biological wastewater treatment and biofiltration technologies [8,9].

Nevertheless, the development of the MABR technology has been mainly a laboratory curiosity and only a few full-scale applications have been reported [3,10]. A common observation from most researchers is excessive biofilm formation (mm thickness) and decreasing pollutant flux rate with time [11]. Consequently, strategies for controlling biofilm thickness and porosity and increasing the penetration depth of oxygen, pollutants and nutrients in biofilms are crucial to improve the performance of the MABR. Unfortunately, only limited work [11] has been done in this area. It is believed that a breakthrough in biofilm structure control, particularly with respect to thickness and porosity, will lead to the development of commercial MABR technologies.

Findings from the literature review also indicate that optimization of MABR technology suffers from a lack of detailed fundamental knowledge about biofilm structure (thickness, density, porosity, diffusivity, microbial populations and their spatial distributions across biofilm depth) [12–15]. These fundamental properties have a dramatic influence on biofilm formation, transport, and reactions within biofilms. Previous studies assumed that the biofilm on gas-permeable membranes was homogeneous [16,17]. Moreover, a past work on conventional fixed biofilms suggests that there is a constantly changing population mixture and physical properties inside the biofilm [18]. Therefore, more realistic models to describe reactions and transport in biofilms will require a better understanding of biofilm structure. The full potential of MABR technology will only be realized when strategies for biofilm structure control and the relationship between biofilm structure and activity are properly understood.

The other emerging technology for waste abatement is the TABT process. It is a unique and relatively new process characterized by rapid biodegradation rates, low sludge yields, and excellent process stability [19]. Under thermophilic conditions (45–65 °C), substrate utilization rates are 3–10 times higher than those observed in mesophilic processes (25–35 °C) [20,21], and the sludge yield is similar to that of anaerobic processes [22]. These advantages have made Thermophilic MABR (ThMABR) extremely suitable for the treatment of high-strength industrial wastewater, such as pulp and paper mill effluent and food processing wastewater. However, low oxygen solubility combined with the high oxygen transfer rate required to sustain rapid biodegradation makes the selection of aeration equipment one of the most critical processes at thermophilic temperatures [23]. In addition, the poor flocculation potential and foaming problem of thermophilic bacteria represent other unique challenges for biomass separation in the suspended growth process.

In this paper, the concept of ThMABR technology is proposed and studied by theoretical analyses and modeling. Coupling the advantages of conventional MMABR technology with TABT overcomes their disadvantages and represents an innovative approach to the treatment of high-strength industrial wastewater and waste gases. On the one hand, the gas-permeable membrane is the ideal piece of aeration equipment for the delivery of the high-rate oxygen transfer required for rapid biodegradation in the ThMABR process; such rates are not achievable with conventional aeration technologies. On the other hand, the low yield and dispersing growth nature of thermophilic microorganisms represent a unique strategy for controlling the excessive growth of biofilms on the gas-permeable membrane. In addition, thermophilic treatment increases the penetration distance of oxygen, pollutants and nutrients in biofilms significantly due to increased diffusivities and decreased viscosities at thermophilic temperatures. It is anticipated that an ultra-compact, highly efficient bioreactor will be developed for high-strength wastewater and waste gas treatment through the ThMABR concept.

This communication presents theoretical analyses and modeling results of ThMABR and MMABRs. The particular interest are the differences between ThMABRs and MMABRs in terms of oxygen and pollutant flux and penetration distances and biodegradation rate.

2. Materials and Methods

2.1. Theoretical Analysis of the Impact of Temperature on Biofilm, Water and Mass Transfer Characteristics

As a biological treatment system, the ThMABR is mainly composed of membranes for oxygen delivery, and biofilms formed on membrane surfaces for biodegradation. Oxygen, pollutants and nutrients are transferred into the biofilm for biodegradation with a counter-diffusion manner. Among various factors that affect the performance of MABR, temperature plays a dominant role [24]. The various temperatures resulted in changes in biofilm characteristics (thickness, density, porosity, growth and detachment rates, microbial community, biodegradation rate, etc.), water and gas properties (viscosity, surface tension, density, etc.), membrane properties (pore size, tortuosity, solubility) and transport properties (diffusivity, flux, permeability). In return, these properties have a profound effect on the overall performance of ThMABR.

2.1.1. Impact of Temperature on Biofilm Properties

As shown in Figure 1, biofilm is the layer between the membrane surface and the bulk water phase, and mainly consists of microorganisms, extracellular polymeric substances (EPS), which are excreted by the cells, and which immobilize these cells and entrap particles within the matrix of biofilm. Biofilm is one of the most important components in MABR, as physical, chemical and biological properties of biofilms determine diffusion and biodegradation rates within the biofilm. Although extensive studies have been conducted on biofilms, the literature review indicates that most temperature-related studies focus on the formation of biofilm and very little attention has been paid to the impact of temperature on physical and chemical properties. Zhang and Bishop [25] found that the freezing technique in preparing biofilm samples for micro-slicing had no obvious adverse effects on biofilm properties (density, pore size, etc.) compared to that of the control samples. Overall, there is a lack of fundamental information on the temperature impact. However, it is clear that when the temperature is changed from the mesophilic (25–35 °C) to the thermophilic (45–65 °C) range, different microbial communities will be expected [25]. Thermophiles will survive at thermophilic temperatures and mesophiles will grow at mesophilic temperatures.

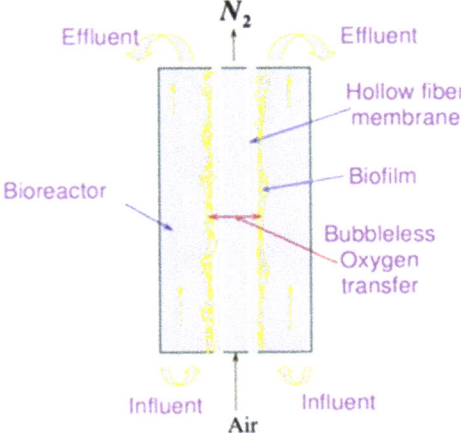

Figure 1. Schematics diagram of membrane-aerated biofilm reactor (MABR).

It is generally assumed that substrate consumption rate r_s within a biofilm can be described by Monod growth kinetics for two limiting substrates (oxygen and organic substrate) (C_s and C_o):

$$r_s = \mu_{max} \left[\frac{C_s}{(K_s + C_s)}\right]\left[\frac{C_o}{(K_o + C_o)}\right] \quad (1)$$

where K_s is the substrate half-saturation constant and K_o is the oxygen half-saturation constant.

In general, biodegradation rates are doubled for every 10 °C increase, in the range of 5–30 °C. A comparison of the biodegradation rates between the mesophilic and the thermophilic temperature may be difficult, owing to changes in microbial communities. However, it is generally accepted that biodegradation rates in thermophilic temperatures are much higher (3–10 times) than those in the mesophilic temperature range. Lapara and Alleman summarized the available biokinetic constants for the temperature range from 20 to 58 °C [2]. According to these figures of biokinetic constants against temperature [2], the maximum specific rate of microbial growth, maximum specific rate of substrate utilization and endogenous decay rate are strong functions of temperature. Although these data are obtained from the suspended growth biomass, it is believed, in principle, that similar trends will be observed for attached growth biomass. The oxygen transfer is a limiting factor in thermophilic treatment, due to the low oxygen solubility, high biodegradation rate, poor flocculation of biomass, and foaming issues. Thus, thermophilic treatment would negatively affect bacteria activities and may reduce process stability. Therefore, aeration must be precisely controlled to promote microbial activity and optimize organic removal and process stability.

Diffusion in biofilms is a complicated process, due to the heterogeneous nature of the biofilm structure. The pore size of channels and the porosity, tortuosity and thickness of biofilm affected the diffusivity of the oxygen and substrate. Past work assumed that the diffusivity in biofilms is equal to that in water, considering the majority of biofilm is water [23], while others consider the diffusivity in biofilm an effective diffusivity D_{eff}, which is equal to the diffusivity in water times the physical parameters of biofilm (porosity, tortuosity, pore size) [26]. López and coworkers explained the following equation to estimate the effective diffusivity in biofilms [27].

$$D_{eff} = (\varepsilon D_w)/\tau \quad (2)$$

where ε is the porosity of biofilms, τ is the tortuosity factor, and D_w is the diffusivity of water.

A change in temperature affects not only the physical properties of the bulk solution but also the physical properties of biofilms. As a result, the effective diffusivity in biofilms increases with an increase in temperature.

The impact of temperature on biofilm growth rates is generally well understood. However, very limited information is available in terms of the influence of temperature on detachment rates. It is generally believed that thermophiles have a poorer flocculating ability than mesophiles, e.g., the thermophiles have a dispersing growth nature. In addition, more substrate is converted to carbon dioxide and water instead of cell mass at thermophilic temperatures. Consequently, it is reasonable to believe that the growth rate of thermophilic biofilm thickness will be lower than that of the mesophilic biofilms under similar testing conditions.

2.1.2. Impact of Temperature on Water and Gas Properties

It is well known that the physical properties of water and gas are strong functions of temperature.

Empirical equations are as follows to correlate physical properties of water and gas with temperature.

The viscosity of water equation that is accurate to within 2.5% from 0 °C to 370 °C is shown below [28]:

$$\mu(w) = 2.414 \times 10^{-5} \times 10^{2.478/(T-140)} \quad (3)$$

where T has units of Kelvin, and $\mu(w)$ is the water viscosity which has units of Pa.s.

Sutherland's formula can be used to derive the dynamic viscosity of an ideal gas as a function of the temperature [29].

$$\mu(g) = \mu_0 (T_0 + C)(T/T_0)^{1.5}/(T+C) \quad (4)$$

where $\mu(g)$ is the dynamic viscosity of gas (Pa·s or µPa·s) at input temperature T, μ_0 is reference viscosity (in the same units as μ) at a reference temperature T_0, T is input temperature (K), T_0 is reference temperature (298 K), and C is Sutherland's constant for the gaseous material in question.

Lapara and coworkers provide an excellent summary of the physical properties of water at thermophilic temperatures [30]. It is concluded that an increase in temperature from the mesophilic to the thermophilic temperature range reduces the viscosity and surface tension of water and increases mixing and colloid solubility in water, which will improve oxygen, pollutants and nutrient transfer rates. In addition, the increase in temperature reduces the saturation oxygen concentration in water and thus increases oxygen driving force across the membrane and enhances oxygen transfer.

In bulk liquid solution, diffusivities of oxygen and substrates are proportional to T/μ [31], that is

$$\frac{D_{WT1}}{D_{WT2}} = \frac{T_1}{T_2} \frac{\mu T_2}{\mu T_1} \quad (5)$$

where D_w is the diffusion coefficient in water, T_1 and T_2 are the corresponding absolute temperatures, and μ is the dynamic viscosity of the solvent. An increase in temperature results in a decrease in bulk liquid solution viscosity. Accordingly, diffusivities of oxygen and substrates in biofilms are proportional to T^n ($n > 1$) (e.g., an increase in temperature leads to the increase in diffusivities in bulk liquid solution). The diffusivity of oxygen in the bulk liquid solution is increased from 2.1×10^{-5} cm^2/s at 25 °C to 4.67×10^{-5} cm^2/s at 60 °C [31].

In the lumen side of membranes, oxygen transfer to the biofilm involves adsorption, diffusion and desorption processes. According to the Chapman–Enskog kinetic theory, the diffusivity of oxygen in the bulk gas solution is proportional to $T^{1.5}/\mu$ [31], that is

$$\frac{D_{AB(T1)}}{D_{AB(T2)}} = \left(\frac{T_1}{T_2}\right)^{1.5} \left(\frac{\mu T_2}{\mu T_1}\right) \quad (6)$$

An increase in temperature results in a decrease in viscosity. Consequently, the diffusivity of oxygen in the bulk gas phase is proportional to T^m ($m > 1.5$). Estimation indicates that the oxygen diffusion coefficient in water is strongly affected by temperature. This effect is even stronger in the case in air, more than doubling as temperatures increase from 20 to 60 °C [32].

2.1.3. Impact of Temperature on Membrane Properties

An increase in temperature results in an increase in pore size, due to the impact of swelling [33]; thus, a high flux or permeability will be anticipated at a higher temperature. In addition, an increase in temperature leads to a lower solubility and higher diffusivity of oxygen in membranes.

Empirical correlations based on previous research data [34–36] are regressed using the Arrhenius equation as follows:

Oxygen solubility in Polydimethylsiloxane (PDMS) membrane:

$$S_{O_g} = 3.88014 \times 10^{-11} \times e^{-58322.13/RT} \quad (7)$$

(Gas–PDMS membrane interface, T = 293–313 K)

$$S_{O_w} = S_{O_g} \times H \qquad (8)$$

(Water–PDMS membrane interface, H-Henry's constant 0.0635, T = 273–333 K)
Oxygen permeability in PDMS membrane:

$$P_{O_g} = 1.1042 \times 10^{-11} \times e^{-47601/RT} \qquad (9)$$

(Gas–PDMS–Gas, T = 293–313 K)

Effective diffusivity of oxygen in the membrane is a function of pore diffusivity, the porosity of membrane, and the solubility of oxygen in membrane and is expressed as follows [36]:

$$D_{eff} = \frac{D_{AB}\varepsilon}{\varepsilon + (1-\varepsilon)S_{Og}} \qquad (10)$$

Temperature is an important factor that has significant degradative effects on membrane filtration because of the nature of seasonal changes in the temperature of raw water.

2.2. Mathematical Modelling of the Impact of Temperature on the Performance of MABR

Based on theoretical analyses and the fundamental equations that correlate the temperature and parameters mentioned above, a counter-diffusion and reaction mathematical model was developed, with the temperature impact incorporated, to study the transport and reaction processes in ThMABRs. Of particular interest is the comparison of the performance between MMABR and ThMABR.

The following set of equations was developed and used for cylindrical hollow fiber membranes.

Oxygen flux to bulk water solution without biofilms on membrane surface [37]:

$$J = \left(\frac{P_m * H}{L_e}\right)\left(\frac{32 * P_O}{H} - C_O|r = r_{bf-in}\right) \qquad (11)$$

where P_m is the permeability of oxygen; H is Henry's constant of oxygen; L_e is the effective thickness of silicone membrane; and Po is the partial pressure of oxygen gas.

Oxygen flux cross membrane can be further expressed according to oxygen concentrations in the gas phase and in the biofilm at the membrane–biofilm interface [18]:

$$J = K_d\left(\frac{C_{o,g}}{H} - C_{o,0}\right) \qquad (12)$$

where $C_{o,0}$ and $C_{o,g}$ are the dissolved oxygen concentrations in the membrane and biofilm bottom (g O_2 m^{-3}), k_d is the overall mass transfer coefficient of oxygen (m day^{-1}), and Henry's constant is H.

Under steady-state conditions, diffusion and reaction of oxygen and substrate within biofilms can be described using the following equations based on Fick's first law and Monod equation [31,37,38]:

$$D_{Oeff}\left[\frac{d^2C_O}{d_{r^2}} + \left(\frac{1}{r}\right)\frac{dC_O}{dr}\right] - \left[\frac{\mu_m S}{K_S+S}\right]\left[\frac{C_O}{K_O+C_O}\right]\frac{X_{bf}}{Y_{XO}} = 0 \qquad (13)$$

$$D_{Seff}\left[\frac{d^2S}{d_{r^2}} + \left(\frac{1}{r}\right)\frac{dS}{dr}\right] - \left[\frac{\mu_m S}{K_S+S}\right]\left[\frac{C_O}{K_O+C_O}\right]\frac{X_{bf}}{Y_{XS}} = 0 \qquad (14)$$

where D_{Seff} and D_{Oeff} are the effective diffusivity of substrate and oxygen in biofilm at temperature T, respectively; K_S and K_O are the half-saturation constant of substrate and oxygen at temperature T, respectively; μ_m is the maximum specific growth rate at temperature T; Y_{XS}, and Y_{XO} are the biofilm yield based on substrate utilization and oxygen consumption for biofilm growth, respectively; X_{bf} is the density of biofilm.

Boundary conditions can be calculated based on mass balance [31,37,39]:

$r = r_{bf-in}$,

$$D_{Oeff}\frac{dC_O}{dr}\bigg|r = r_{bf-in} = -\left(\frac{P_m * H}{L_e}\right)\left(\frac{32 * P_O}{H} - C_O\big|r = r_{bf-in}\right) \quad (15)$$

$$D_{Seff}\frac{dS}{dr}\bigg|r = r_{bf-in} = 0 \quad (16)$$

$r = r_{bf-out}$,

$$D_{Oeff}\frac{dC_O}{dr}\bigg|r = r_{bf-out} = \left(\frac{D_{OW}}{L_S}\right)\left(C_b - C\big|r = r_{bf-out}\right) \quad (17)$$

$$D_{Seff}\frac{dS}{dr}\bigg|r = r_{bf-out} = \left(\frac{D_{SW}}{L_S}\right)\left(S_b - S\big|r = r_{bf-out}\right) \quad (18)$$

where D_{SW} is the substrate diffusivity in water, L_S is the thickness of the stagnant layer of liquid. In order to simplify computations, the linear Finite-Difference Method is introduced. In this paper, MATLAB2021a (9.10.0.1710857) was used for data calculation and analysis.

2.3. Model Validation

The experiment data were collected from past literature as the input in this modeling work, as shown in Table 1. The diffusion coefficients were estimated based on Equation (5) and past literature [31,40–43]. Other kinetic parameters, such as Ko and Ks, are from literature [44–46].

Table 1. Parameters for numerical modeling of diffusion and reaction in membrane-attached biofilm, MMABR and ThMABR.

Parameters	Symbol	Unit	Typical Value MMABR (25 °C)	Typical Value ThMABR (60 °C)	Typical Value ThMABR (55 °C)
Oxygen diffusivity in biofilm	D_{oeff}	m²/s	1.67 × 10⁻⁹ [41]	3.37701 × 10⁻⁹ (Equation (5))	3.32632 × 10⁻⁹ (Equation (5))
Substrate diffusivity in biofilm	D_{seff}	m²/s	1 × 10⁻⁹ [42]	2.00216 × 10⁻⁹ (Equation (5))	1.99181 × 10⁻⁹ (Equation (5))
Oxygen half-saturation constant	K_O	g/m³	0.2 [44]	0.2 [44]	0.2 [44]
Substrate half-saturation constant	K_S	g/m³	20 [44]	20 [44]	20 [44]
Maximum growth rate	μ_m	1/s	2.3148 × 10⁻⁵ [2]	1.1574 × 10⁻⁴ [2]	1.1574 × 10⁻⁴ [2]
Biomass yield based on oxygen	Y_{xo}	/	0.2 [45]	0.2 [45]	0.2 [45]
Biomass yield based on substrate	Y_{xs}	mg/mg substrate	0.45 [2]	0.35 [2]	0.35 [2]
Biofilm density	X_{bf}	g/m³	55,000 [31]	55,000 [31]	55,000 [31]
Permeability	P_m	gmole*m/(m²*s*pa)	1.65 × 10⁻¹³ [45]	2.81 × 10⁻¹³ [Equation (9)]	2.73 × 10⁻¹³ [Equation (9)]
Effective thickness of hollow fiber membrane	L_e	m	7.52 × 10⁻⁵ [37]	7.52 × 10⁻⁵ [37]	7.52 × 10⁻⁵ [37]
Substrate diffusivity in water	D_{sw}	m²/s	1.26 × 10⁻⁹ [43]	2.54792 × 10⁻⁹ (Equation (5))	2.37 × 10⁻⁹ [Equation (5)]
oxygen diffusivity in water	D_{ow}	m²/s	2.41 × 10⁻⁹ [40]	5.15 × 10⁻⁹ [40]	4.76 × 10⁻⁹ [40]
Outside radius of hollow fiber membrane	r_0	m	3.18 × 10⁻⁴ [37]	3.18 × 10⁻⁴ [37]	3.18 × 10⁻⁴ [37]
Outside radius of biofilm	r_b	m	8.18 × 10⁻⁴ (This study)	8.18 × 10⁻⁴ (This study)	8.18 × 10⁻⁴ (This study)
Henry's constant	H	atm*m³/mole	0.769 [46]	1.15761 [46]	1.09767 [46]

The operation conditions and information of membrane modules in literature were shown as follows. The influent was composed by a mixture of sodium acetate solution and glucose (50% glucose COD/50% sodium acetate COD in distilled water) with 1200 mg/L COD [20]. The experimental system was sequencing batch reactor MMABR and ThMABR system operated at room temperature and 55 °C, respectively. At the beginning of each reaction cycle, each batch of MABR was manually added to 1.5 L of synthetic wastewater and the reaction time was 1 day [20]. The composition details of the nutrient feed could be found in Liao and Liss's work [20]. The membranes of MMABR and ThAnMBR are hollow fiber silicone (Model: M60-130W-200L-FC8, 13 cm wide × 20 cm long, supplied by Nagayanagi Co., Ltd., Yashio, Japan) [20].

In order to maximize the modeling results effectively, it can be used to compare the modeling results with the experimental results and examine the overall impact of reactor design and biofilm properties and operating conditions on overall MABR performance. Therefore, the past experiment comparison work about COD removal efficiency [20] in MMABR and ThMABR could be considered a validation for the present model.

3. Results and Discussion

The results are organized for discussion in terms of model validation using literature data, oxygen and substrate concentration profiles, biological activity profiles, membrane–biofilm interfacial oxygen concentration, oxygen penetration distance, and oxygen and substrate fluxes into biofilms under thermophilic and mesophilic conditions.

3.1. Model Validation

Liao and Liss [20] found out that MABR running at a thermophilic temperature (ThMABR) was more effective than MMABR in COD removal and biofilm thickness controlling for a synthetic high-strength organic wastewater treatment. Therefore, with the same experiment parameters as the inputs at 55 °C (biofilm thickness of MMABR is 1080 μm and biofilm thickness of ThMABR is 280 μm), the general trend of model prediction on substrate removal rates could be validated by Liao and Liss' [20] investigation. The comparison between COD removal rate in this model and literature was shown in Table 2.

Table 2. The comparison between modeling predictions and experiment results from literature [20].

Biofilm Reactor	Outside Radius of Hollow Fiber (μm)	Inner Radius of Hollow Fiber (μm)	Biofilm Thickness (μm)	Simulate COD Removal Rate (g/d)	Experiment COD Removal Rate (g/d)	Relative Error
MMABR (air 4 psi)	320	200	1080 [20]	2.5780	1.1625 [20]	121.7%
MMABR (air 6 psi)	320	200	1080 [20]	2.6466	1.2375 [20]	113.8%
ThMABR (air 4 psi)	320	200	280 [20]	8.5929	1.6532 [20]	419.8%
ThMABR (air 6 psi)	320	200	280 [20]	9.0763	1.6826 [20]	439.4%

The experimental results from the literature [20] verified the general trend of the higher COD removal efficiency in the ThMABR system similar to the present model. The variation of COD removal rate in the literature [20] from MMABR and ThMABR was not as significant as that predicted by the modeling study, which shows the notable change between MMABR and ThMABR. The deviation between the modeling study and experimental results could be explained by the following reasons: (1) First, the experimental data were from a sequencing batch reactor MMABR and ThMABR study and, unfortunately, the COD profile (decrease) with respect to reaction time (in one reaction cycle) was not monitored and only the COD level at the end of the reaction (24 h) was determined and used for the COD removal rate calculations. It is very likely that the majority of COD was biodegraded and reached a flat residual COD in a shorter period of time much less than 24 h (particularly for the ThMABR), and in this case, the experimental COD removal rates could be many times higher than the one reported here and much closer to the modeling results. (2) The difference between the modeled results and experimental results could also be partially caused by the back diffusion of water vapor into the lumen side of the hollow fibers, which caused additional mass transfer resistance of oxygen to biofilm. It was noted that much more water condensate was observed from the ThMABR system, due to the higher back diffusion of water vapor at the thermophilic temperature [20]. Even with this significant difference, the general tendency in both still showed that the ThMABR provided better COD removal efficiency than that of MMABR. The more rigorous validation process is still required in future work.

As Table 2 shows, the modeling and experimental results both show that an increase in the oxygen partial pressure led to an improved COD removal efficiency. These results clearly show the advantages of the ThMABR system. The ThMABR system showed a higher substrate flux or COD removal in both the modeling and experimental results. Thermophilic biofilms were much thinner than mesophilic biofilms, which implied that operating at thermophilic temperatures might be an effective approach to controlling biofilm thickness. This explains why the ThMABR performed better than the MMABR—because a thicker biofilm in the millimeter thickness range deteriorated the performance of the MMABR. Similarly, when the oxygen pressure changed to 6 psi, the substrate flux was still higher than the flux in the MMABR system. According to the experimental results, the simulated results are reasonable. The pollutant removal efficiency of ThMABR is higher than the removal in MMABR. The experimental results from the literature [20] verified the general trend of the higher COD removal efficiency in the ThMABR system.

3.2. Impact of Temperature (Thermophilic vs. Mesophilic) on Oxygen and Substrate Concentration Profiles

Figures 2 and 3 show the concentration profiles of oxygen and substrate within biofilms. The results suggest that the penetration distance of both oxygen and substrate strongly depends on the membrane–biofilm interfacial oxygen concentration. For a low substrate concentration (Sb = 50 mg/L), substrate transfer is the rate-limiting step; for a medium substrate concentration (Sb = 100 mg/L), a dual limitation (both oxygen and substrate transfer limitation) is observed in biofilms; for a high substrate concentration (Sb = 200 mg/L), oxygen transfer is the rate-limiting step. In both situations (thermophilic and mesophilic conditions), substrate either fully or partially penetrates the biofilm, while oxygen always partially penetrates the biofilms.

In most cases for municipal and industrial wastewater treatment, oxygen transfer is the rate-limiting step. Therefore, an increase in interfacial oxygen concentration is required to accommodate biological reactions in biofilms. This can be achieved by using pure oxygen for oxygen transfer. The use of pure oxygen for replacing air can increase the interfacial oxygen concentration and thus increase the penetration distance significantly [47]. Simulating the oxygen and substrate transport process in biofilm can be used to predict the pollutant removal efficiency and oxygen utilization rate. Figure 2 shows the oxygen transport process at different substrate concentrations in an MMABR and a ThMABR with air and pure oxygen supply. Compared with previous modeling studies [31,48], this profile is more reasonable, as the impact of dissolved oxygen concentration in the bulk water phase on oxygen and substrate transfer is considered here. The dissolved oxygen concentration was around 2 and 8 g/m^3 at the end of biofilm in the bulk water phase under thermophilic and mesophilic temperatures, respectively. The oxygen profile in this simulation is similar to the result of Ntwampe et al. [49] and Matsumoto et al. [50].

In the oxygen concentration profile of the ThMABR system, the substrate concentration had a positive impact on oxygen utilization rate in both biofilm reactors. With increasing substrate concentration, the oxygen utilization rate increased. This increase stimulated the activity of microbial communities on the biofilm, which increased the reaction rate. Compared with MMABR, the oxygen concentration in the ThMABR system displayed a faster reaction rate and better oxygen utilization rate. The biofilm thickness in the ThMABR system is thinner than biofilm in the MMABR system as well. This explains why the performance of ThMABR is better than MMABR, because thicker biofilms in the millimeter thickness can degrade MMABR performance. These results also proved that the ThMABR system has more advanced points than the MMABR system. Thermophilic biofilms were much thinner than mesophilic biofilms, implying that operation at thermophilic temperatures could be an effective method to control biofilm thickness. This result is similar to Liao and Liss [20].

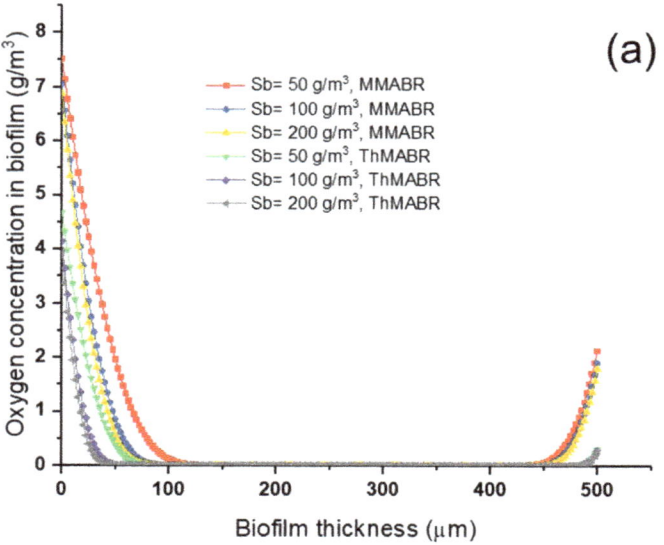
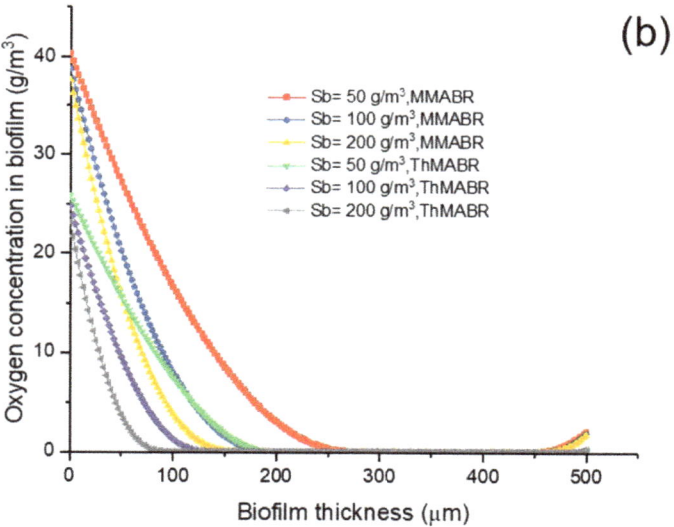

Figure 2. Oxygen concentration profile in MMABR and ThMABR: (**a**) air supplying; (**b**) pure oxygen supplying.

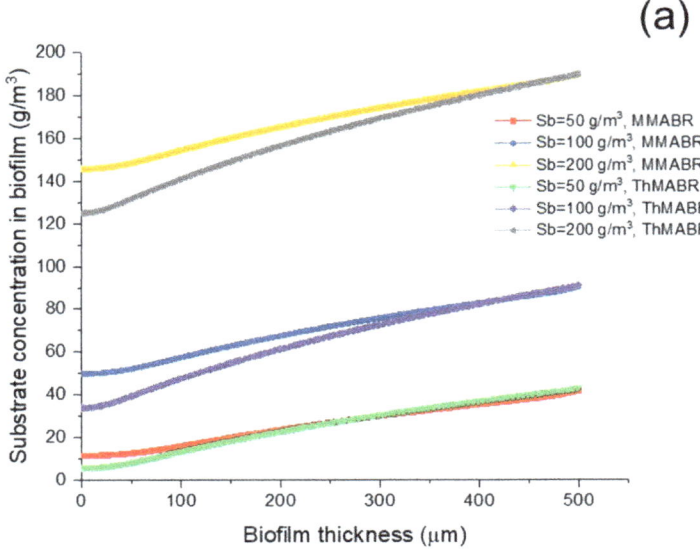

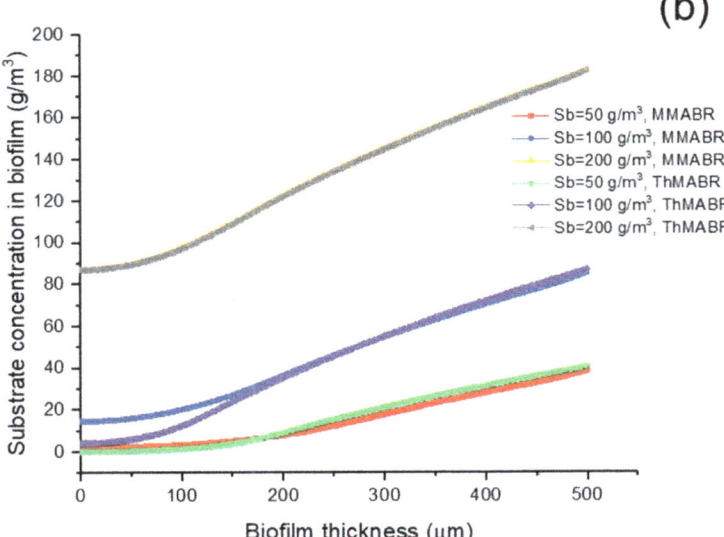

Figure 3. Substrate concentration profile in ThMABR and MMABR: (**a**) air supplying; (**b**) pure oxygen supplying.

The substrate concentration decreased with a decreased biofilm thickness, which means a decline in the substrate utilization rate as biofilm thickness increased [43]. As shown in Figure 3a, when the substrate concentration was increased to 200 g/m^3, a more significant difference in substrate removal could be found. A more significant substrate concentration decrease was observed in the ThMABR system. The ThMABR system had a better oxygen utilization performance, which supported that ThMABRs would provide more advanced performance on pollutant removal than the MMABR system.

If oxygen supply from the air was changed to pure oxygen, the oxygen concentration profiles under different operation conditions were totally similar. The modeling results of ThMABR still showed its outstanding removal abilities, especially for high-strength wastewater (Figure 3b). These results show that increasing oxygen partial pressure would increase reactor performance. However, the present model only considered the aerobic process in the MABR system. As the anaerobic parameters are still limited in the literature, the anaerobic process requires further study.

3.3. Impact of Temperature on Oxygen Penetration Distance into Biofilms

For a high-strength wastewater treatment, oxygen transfer is usually the limiting rate step. Therefore, it is important to know the penetration distance of oxygen within biofilms in order to control the biofilm thickness. The penetration distance of oxygen in a ThMABR and MMABR is shown in Figure 2a,b. The penetration distance of oxygen in MMABR is larger than that in ThMABR. This is probably not surprising, as the interfacial oxygen concentration in MMABR is always higher than that in ThMABRs. In addition, the consumption rate of oxygen in ThMABRs is higher than that in MMABRs. With substrate concentration increased, the distance of oxygen penetrated into biofilm distance was reduced. As shown in Figure 2b, when the air was replaced by pure oxygen, the penetration distance of oxygen almost doubled. This phenomenon is similar to that found by Wang and coworkers [51]. The penetrated distance in MABR was still higher than the distance in ThMABR. These results also indicate the advanced oxygen utilization of the ThMABR system.

3.4. Impact of Temperature on Membrane–Biofilm Interfacial Oxygen Concentration

The membrane–biofilm interfacial oxygen concentration is important in determining the penetration distance of oxygen in biofilms. Usually, a high membrane–biofilm interfacial concentration is associated with a larger penetration distance of oxygen in biofilms. A comparison of interfacial oxygen concentration between ThMABR and MMABR is shown in Figure 4. The results suggest that interfacial oxygen concentration in MMABR is higher than that in ThMABR under similar conditions. Of particular interest is the presence of a minimum interfacial oxygen concentration in terms of biofilm thickness. The presence of the minimum interfacial oxygen concentration may suggest the presence of an optimal biofilm thickness for maximum oxygen fluxes into biofilms. When the biofilm thickness is thinner than the optimal biofilm thickness, an increase in biofilm thickness results in increased consumption of oxygen and thus reduces the interfacial oxygen concentration. When the biofilm thickness is thicker than the optimal biofilm thickness, a further increase in biofilm thickness introduces more transport resistance for both oxygen and substrate and thus reduces the availability of substrate concentration at the membrane–biofilm interface, which corresponds to an increase in interfacial oxygen concentration. When the biofilm thickness is thinner than the optimal biofilm thickness, an increase in biofilm thickness results in increased consumption of oxygen and thus reduces the interfacial oxygen concentration. On the other hand, when the biofilm thickness is thicker than the optimal biofilm thickness, a further increase in biofilm thickness introduces more transport resistance for both oxygen and substrate and thus reduces the availability of substrate concentration at the membrane–biofilm interface, which corresponds to an increase in interfacial oxygen concentration. An optimization point of biofilm thickness can be observed in this paper. The profile of interfacial oxygen concentration in both biofilm reactors had the lowest point at certain biofilm thicknesses, which means the highest oxygen flux could be obtained at an optimal biofilm thickness. It provided a new design idea for future lab-scale research.

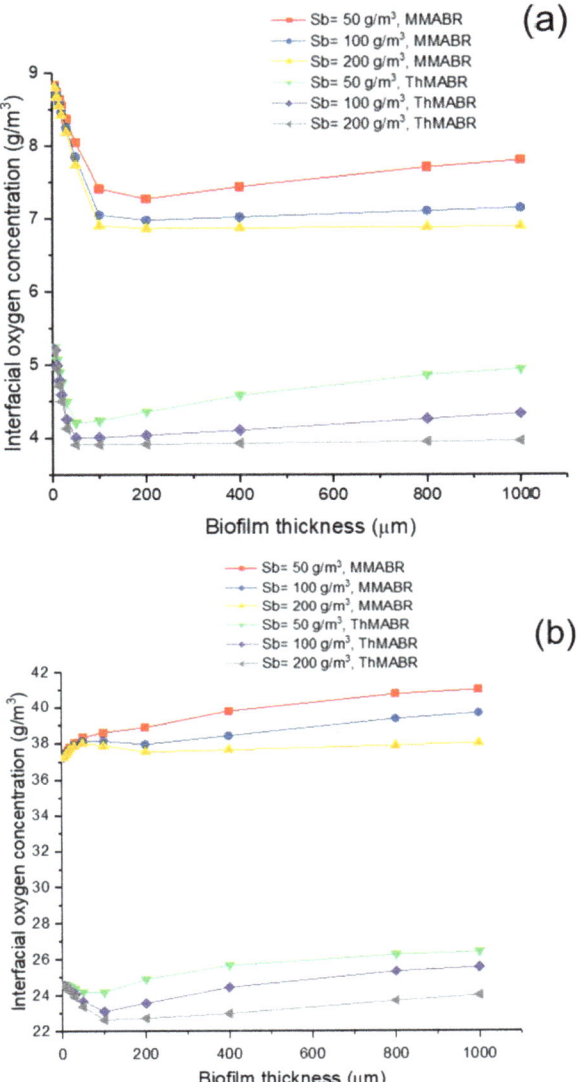

Figure 4. Interfacial oxygen concentration profile in ThMABR and MMABR: (**a**) air supplying; (**b**) pure oxygen supplying.

Figure 4b shows that the use of pure oxygen for replacing air can increase the interfacial oxygen concentration from about 6.5–8 to 36–38 g/m^3 in the MMABR system but only from 3.75–5.3 to 22–25 g/m^3 for ThMABR. Thus, the pure oxygen increased the penetration distance significantly. The use of sealed hollow fibers to deliver oxygen can achieve 100% utilization of oxygen. The optimal biofilm thickness in MMABR is hard to observe. However, the optimal thickness in ThMABR increased to double. It indicated that using pure oxygen to operate the ThMABR system needs thicker thickness.

3.5. Impact of Temperature on Oxygen and Substrate Fluxes into Biofilms

Figures 5 and 6 show the oxygen and substrate fluxes into biofilm in MMABR and ThMABR, respectively. The results suggest that the presence of a thin layer of biofilm could enhance the flux of oxygen into biofilms. This can be explained by the fact that the presence of a thin layer of biofilm would consume oxygen and thus reduce interfacial oxygen concentration, which led to an increase in oxygen flux into a biofilm. However, a further increase in biofilm thickness resulted in a minimum interfacial oxygen concentration, which corresponded to a maximum oxygen flux into a biofilm. The result indicates that an optimal biofilm thickness exists for a maximum oxygen flux into biofilms. After the optimal biofilm thickness, any further increase in biofilm thickness will introduce excessive transport resistance for oxygen and substrate transport and thus reduce the oxygen and substrate fluxes into biofilms. The optimal biofilm thickness strongly depends on the intracellular oxygen pressure [52].

A comparison of oxygen and substrate fluxes into biofilms between ThMABRs and MMABRs indicates that ThMABRs have advantages over MMABRs in terms of oxygen and substrate fluxes into biofilms. In a biofilm thickness close to the range of optimal biofilm thickness, the oxygen and substrate fluxes into biofilms in ThMABRs are about 30% higher than those in MMABRs. However, the advantages of fluxes in ThMABRs are reduced when biofilm thickness is further increased. The advantages of fluxes in ThMABRs totally disappear if the biofilm thickness is large enough. These results suggest that precise control of biofilm thickness at the range of optimal biofilm thickness is essential for achieving the advantages of ThMABRs.

According to Figures 5b and 6b, the pure oxygen increased the peak of oxygen flux, which improved substrate fluxes as well. Thus, by increasing the oxygen pressure inside the membranes, we can further increase the flux of oxygen and the substrate removal rate [53]. The peak of the high-strength (S_b = 200 g/m^3) oxygen flux decreased non-significantly in ThMABRs. It also showed thinner biofilm thickness more obviously. In both operation conditions (air and pure oxygen supplying), ThMABRs always displayed advanced removal abilities for the pollutant, which have already been applied in full-scale water treatment by their advantages. The thermophilic membrane biofilm system plants have been successfully used for pulp and papermaking wastewater treatment and food processing wastewater treatment. Both systems prove that there are many advantages compared to mesophilic bacteria. Compared to mesophilic bacteria, the biological properties of ThMABRs may be better, comparable or worse. The use of TABTs for high-temperature industrial wastewater treatment and sludge digestion significantly saves energy and enables energy-neutral or actively processed plants [54].

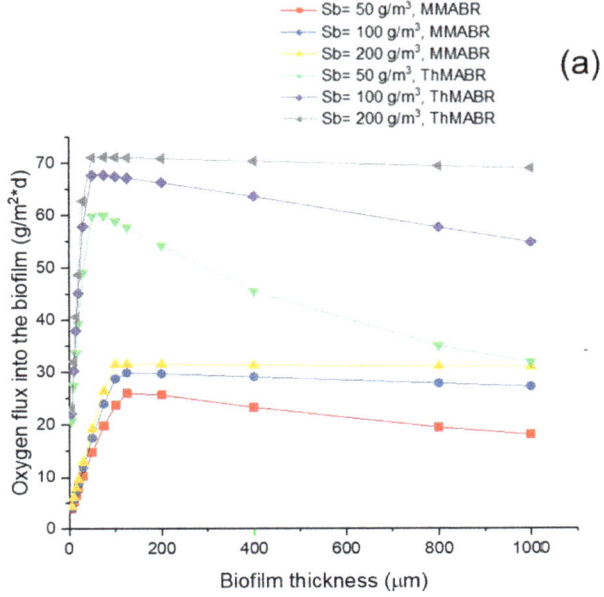

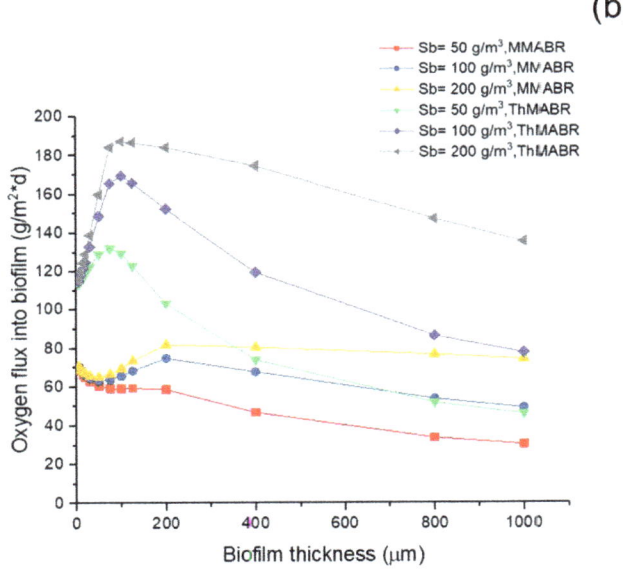

Figure 5. Oxygen flux comparison at different substrate concentrations: (**a**) air supplying; (**b**) pure oxygen supplying.

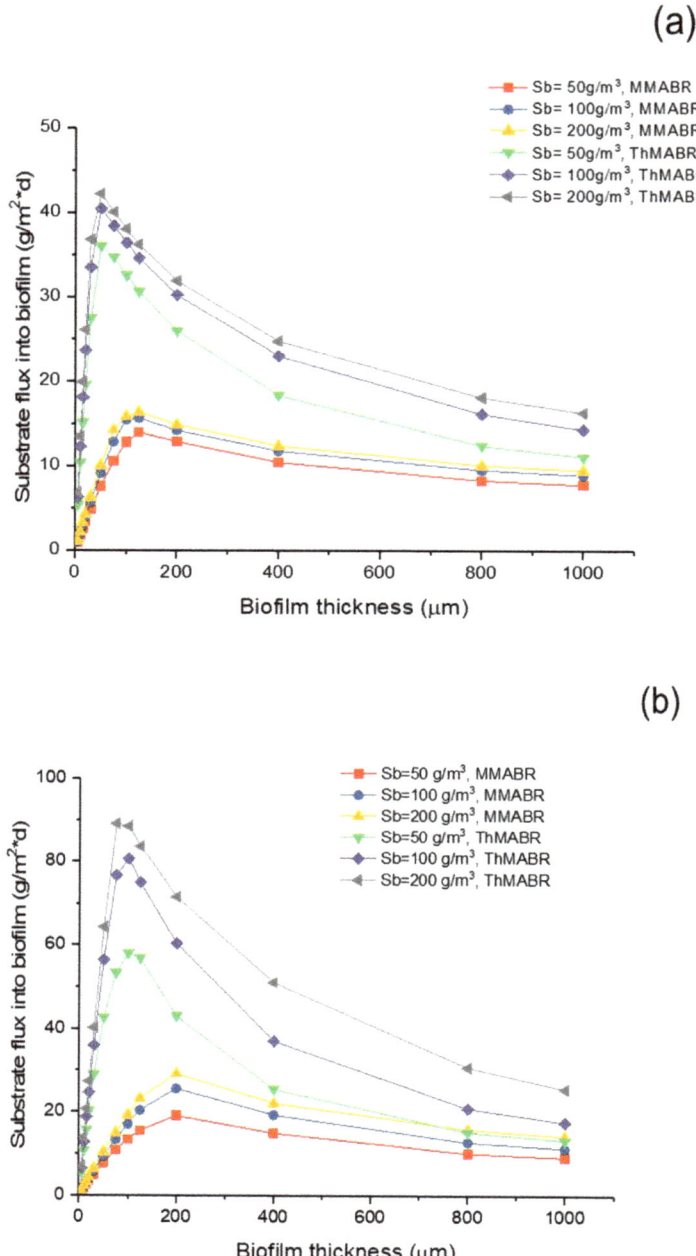

Figure 6. Substrate flux comparison at different substrate concentrations: (**a**) air supplying; (**b**) pure oxygen supplying.

3.6. Limitations of the Present Study

Based on the modeling results discussed above, it is evident that the biofilm may have aerobic, anoxic, and anaerobic zones co-existing on the membrane surface. The current model used in this study only considered the aerobic process for COD removal and has ignored the anoxic and anaerobic processes for COD and nutrient removal. Thus, the current models are more applicable for high-strength COD industrial wastewater treatments with the need of adding nutrients based on the biological reaction stoichiometry. For more comprehensive models that account for the contributions of anaerobic COD and nutrient removals, nitrifications should be developed and integrated into the current models for a comprehensive modeling and prediction of the MMABR and ThMABR processes in the future.

The current experimental results [20] only partially validate the general trend of the modeling results in terms of COD removal rates between the MMABR and ThMABR processes. The single hollow fiber MMABR and ThMABR system and experiments should be designed to precisely validate the modeling results, such as oxygen and substrate profiles among biofilm thickness. In this case, oxygen and substrate microsensors and the biofilm thickness monitoring technique are needed to get the needed information to validate the modeling results. This should be conducted in future studies.

4. Conclusions

The concept of ThMABR was proposed for high-strength wastewater and gas treatments. Theoretical analyses and modeling were conducted to elucidate the advantages and disadvantages of the ThMABR, as compared to the MMABR. The main conclusions are drawn below:

(1) An increase in temperature from the mesophilic to the thermophilic range results in a significant increase in the oxygen and substrate fluxes into biofilms. The oxygen and substrate flux into biofilms at 60 °C is 2–3 times higher than that at 25 °C, respectively.
(2) Under similar operating conditions, the oxygen penetration distance of ThMABRs is smaller than that of the MMABRs, implying that the control of biofilm thickness in ThMABRs is even more important than in MMABRs.
(3) Under similar operating conditions, the membrane–biofilm interfacial oxygen concentration in ThMABR is lower than that in MMABRs.
(4) An increase in oxygen partial pressure demonstrates that the advantages of the ThMABR are even superior to that of the MMABRs in treating high-strength wastewaters.
(5) The general trend of the higher substrate removal rates observed in the modeling study of the ThMABR was partially verified by the literature experimental results, although they were not perfect. Well-controlled single-fiber MABR experiments should be designed together with biofilm microsensor techniques to verify the modeling results in the future.

Author Contributions: Conceptualization, D.L., B.L. and H.B.; writing—review and editing, D.L., B.L. and H.B. All authors have read and agreed to the published version of the manuscript.

Funding: This research was funded by Natural Science and Engineering Research Council of Canada (NSERC) grant number RGPIN-2014-03727 and The APC was funded by NSERC.

Institutional Review Board Statement: Not applicable.

Informed Consent Statement: Not applicable.

Data Availability Statement: Data are available upon reasonable request.

Acknowledgments: The authors thank the financial support of the Natural Science and Engineering Research Council of Canada (NSERC).

Conflicts of Interest: The authors declare no conflict of interest.

Abbreviations

MMABR	mesophilic membrane-aerated biofilm reactor
ThMABR	thermophilic membrane-aerated biofilm reactor
TABT	thermophilic aerobic biological treatment
PDMS	Polydimethylsiloxane
J	flux (g/m^2*d)
K_{la}	overall mass transfer coefficient (min^{-1})
T	absolute temperature of liquid under testing (K)
E	modulus of elasticity of water at temperature T, (kNm^{-2})
μ	dynamic viscosity of the solvent
ρ	density of water at temperature T, (kg m^{-3})
σ	interfacial surface tension of water at temperature T, (N m^{-1})
Po	saturation pressure at the equilibrium position (atm).
$C_{O,g}$	dissolved oxygen concentrations in the membrane (g O$_2$ m^{-3})
$C_{O,0}$	dissolved oxygen concentrations in the biofilm bottom (g O$_2$ m^{-3})
K_d	the overall mass transfer coefficient of oxygen (m day^{-1})
K_O	oxygen half-saturation constant (mg/L)
K_S	substrate half-saturation constant (mg/L)
H	Henry's constant (atm*m^3/mole)
$\mu(w)$	viscosity of water (Pa·s)
$\mu(g)$	viscosity of gas (Pa·s)
S_{Og}	oxygen solubility in gas phase (g/L)
S_{Ow}	oxygen solubility in liquid phase (g/L)
P_{Og}	oxygen permeability in PDMS membrane (Barrer)
D_w	diffusion coefficient in water (m^2/s)
D_{AB}	diffusion coefficient in air (m^2/s)
ε	porosity of biofilms
τ	tortuosity factor
COD	chemical oxygen demand
μ_m	maximum growth rate (1/s)
Y_{xo}	biomass yield based on oxygen
Y_{xs}	biomass yield based on substrate
X_{bf}	biofilm density (g/m^3)
P_m	Permeability of oxygen gas (gmole*m/(m^2*s*Pa)
Le	effective thickness of hollow fiber membrane (m)
Ls	stagnant layer of liquid (m)
D_{sw}	substrate diffusivity in water (m^2/s)
D_{ow}	oxygen diffusivity in water (m^2/s)
$r_{bf\text{-}in}$	outside radius of hollow fiber membrane (m)
$r_{bf\text{-}out}$	outside radius of biofilm (m)

References

1. Kunetz, T.E.; Oskouie, A.; Poonsapaya, A.; Peeters, J.; Adams, N.; Long, Z.; Côté, P. Innovative membrane-aerated biofilm reactor pilot test to achieve low-energy nutrient removal at the Chicago MWRD, WEFTEC 2016—89th Water Environ. *Proc. Water Environ. Fed.* **2016**, *2016*, 2973–2987. [CrossRef]
2. LaPara, T.M.; Alleman, J.E. Thermophilic aerobic biological wastewater treatment. *Water Res.* **1999**, *33*, 895–908. [CrossRef]
3. Uri-Carreño, N.; Nielsen, P.H.; Gernaey, K.V.; Flores-Alsina, X. Long-term operation assessment of a full-scale membrane-aerated biofilm reactor under Nordic conditions. *Sci. Total Environ.* **2021**, *779*, 146366. [CrossRef]
4. He, H.; Wagner, B.M.; Carlson, A.L.; Yang, C.; Daigger, G.T. Recent progress using membrane aerated biofilm reactors for wastewater treatment. *Water Sci. Technol.* **2021**, *84*, 2131–2157. [CrossRef] [PubMed]
5. LaPara, T.M.; Alleman, J.E. Autothermal thermophilic aerobic waste treatment systems: A state-of-the-art review. In *Proceedings of the 52nd Purdue Industrial Waste Conference, Purdue University, West Lafayette, Indiana, 5–7 May 1997*; Ann Arbor Press: West Lafayette, IN, USA, 1998.
6. Shaowei, H.; Fenglin, Y.; Cui, S.; Zhang, J.; Tonghua, W. Simultaneous removal of COD and nitrogen using a novel carbon-membrane aerated biofilm reactor. *J. Environ. Sci.* **2008**, *20*, 142–148.
7. Subtil, E.; Mierzwa, J.; Hespanhol, I. Comparison between a conventional membrane bioreactor (C-MBR) and a biofilm membrane bioreactor (BF-MBR) for domestic wastewater treatment. *Braz. J. Chem. Eng.* **2014**, *31*, 683–691. [CrossRef]

8. Liao, B.; Zheng, M.; Ratana-Rueangsri, L.J.W.S. Treatment of synthetic kraft evaporator condensate using thermophilic and mesophilic membrane aerated biofilm reactors. *Water Sci. Technol.* **2010**, *61*, 1749–1756. [CrossRef]
9. Das, J.; Ravishankar, H.; Lens, P.N. Biological biogas purification: Recent developments, challenges and future prospects. *J. Environ. Manag.* **2022**, *304*, 114198. [CrossRef]
10. Syron, E. Innovative Energy Efficient Aerobic Bioreactors for Sewage Treatment. In *Sewage Treatment Plants: Economic Evaluation of Innovative Technologies for Energy Efficiency*; Stamatelatou, K., Tsagarakis, K.P., Eds.; IWA Publishing: London, UK, 2015.
11. Sen, D.; Randall, C.W.; Brink, W.; Farren, G.; Pehrson, D.; Flournoy, W.; Copithorn, R.R. Understanding the importance of aerobic mixing, biofilm thickness control and modeling on the success or failure of IFAS systems for biological nutrient removal. *Proc. Water Environ. Fed.* **2007**, *2*, 1098–1126. [CrossRef]
12. Li, M.; Li, P.; Du, C.; Sun, L.; Li, B. Pilot-scale study of an integrated membrane-aerated biofilm reactor system on urban river remediation. *Ind. Eng. Chem. Res.* **2016**, *55*, 8373–8382. [CrossRef]
13. Hou, F.; Li, B.; Xing, M.; Wang, Q.; Hu, L.; Wang, S. Surface modification of PVDF hollow fiber membrane and its application in membrane aerated biofilm reactor (MABR). *Bioresour. Technol.* **2013**, *140*, 1–9. [CrossRef] [PubMed]
14. Li, P.; Zhao, D.; Zhang, Y.; Sun, L.; Zhang, H.; Lian, M.; Li, B.J.C.E.J. Oil-field wastewater treatment by hybrid membrane-aerated biofilm reactor (MABR) system. *Chem. Eng. J.* **2015**, *264*, 595–602. [CrossRef]
15. Tian, H.; Zhang, H.; Li, P.; Sun, L.; Hou, F.; Li, B. Treatment of pharmaceutical wastewater for reuse by coupled membrane-aerated biofilm reactor (MABR) system. *RSC Adv.* **2015**, *5*, 69829–69838. [CrossRef]
16. Osborne, M.; Aryasomayajula, A.; Shakeri, A.; Selvaganapathy, P.R.; Didar, T.F. Suppression of biofouling on a permeable membrane for dissolved oxygen sensing using a lubricant-infused coating. *ACS Sens.* **2019**, *4*, 687–693. [CrossRef]
17. Desmond, P.; Huisman, K.T.; Sanawar, H.; Farhat, N.M.; Traber, J.; Fridjonsson, E.O.; Johns, M.L.; Flemming, H.-C.; Picioreanu, C.; Vrouwenvelder, J.S. Controlling the hydraulic resistance of membrane biofilms by engineering biofilm physical structure. *Water Res.* **2022**, *210*, 118031. [CrossRef]
18. Casey, E.; Glennon, B.; Hamer, G. Review of membrane aerated biofilm reactors. *Resour. Conserv. Recycl.* **1999**, *27*, 203–215. [CrossRef]
19. Zheng, M.; Liao, B. Membrane aerated biofilm reactors for thermomechanical pulping pressate treatment. *Int. J. Chem. React. Eng.* **2016**, *14*, 1017–1024. [CrossRef]
20. Liao, B.; Liss, S.N. A comparative study between thermophilic and mesophilic membrane aerated biofilm reactors. *J. Environ. Eng. Sci.* **2007**, *6*, 247–252. [CrossRef]
21. Ding, Y.; Liang, Z.; Guo, Z.; Li, Z.; Hou, X.; Jin, C. The performance and microbial community identification in mesophilic and atmospheric anaerobic membrane bioreactor for municipal wastewater treatment associated with different hydraulic retention times. *Water* **2019**, *11*, 160. [CrossRef]
22. Wijekoon, K.C.; Visvanathan, C.; Abeynayaka, A. Effect of organic loading rate on VFA production, organic matter removal and microbial activity of a two-stage thermophilic anaerobic membrane bioreactor. *Bioresour. Technol.* **2011**, *102*, 5353–5360. [CrossRef]
23. Reij, M.W.; Keurentjes, J.T.; Hartmans, S. Membrane bioreactors for waste gas treatment. *J. Biotechnol.* **1998**, *59*, 155–167. [CrossRef]
24. Lu, D.; Bai, H.; Kong, F.; Liss, S.N.; Liao, B. Recent advances in membrane aerated biofilm reactors. *Rev. Environ. Sci. Biotechnol.* **2021**, *51*, 649–703. [CrossRef]
25. Zhang, T.C.; Bishop, P.L. Density, porosity, and pore structure of biofilms. *Water Res.* **1994**, *28*, 2267–2277. [CrossRef]
26. Bonilla, M.R.; Bhatia, S.K. Diffusion in pore networks: Effective self-diffusivity and the concept of tortuosity. *J. Phys. Chem. C* **2013**, *117*, 3343–3357. [CrossRef]
27. López, L.G.; Veiga, M.; Nogueira, R.; Aparicio, A.; Melo, L. A technique using a membrane flow cell to determine average mass transfer coefficients and tortuosity factors in biofilms. *Water Sci. Technol.* **2003**, *47*, 61–67. [CrossRef]
28. Al-Shemmeri, T. *Engineering Fluid Mechanics*; Ventus Publishing ApS: Copenhagen, Denmark, 2012; pp. 17–18. ISBN 978-87-403-0114-4.
29. Smits, A.J.; Dussauge, J.-P. *Turbulent Shear Layers in Supersonic Flow*; Springer Science & Business Media: Berlin/Heidelberg, Germany, 2006.
30. LaPara, T.M.; Nakatsu, C.H.; Pantea, L.; Alleman, J.E. Phylogenetic analysis of bacterial communities in mesophilic and thermophilic bioreactors treating pharmaceutical wastewater. *Appl. Environ. Microbiol.* **2000**, *66*, 3951–3959. [CrossRef]
31. Essila, N.J. *Contrasting Behavior of Biofilms Grown on Gas-Permeable Membranes with Those Grown on Solid Surfaces: A Model Study*; University of Minnesota: Minneapolis, MN, USA; St. Paul, MI, USA, 1997.
32. Baker, R.; Wilkinson, D.P.; Zhang, J. Electrocatalytic activity and stability of substituted iron phthalocyanines towards oxygen reduction evaluated at different temperatures. *Electrochim. Acta* **2008**, *53*, 6906–6919. [CrossRef]
33. Simon, A.; McDonald, J.A.; Khan, S.J.; Price, W.E.; Nghiem, L.D. Effects of caustic cleaning on pore size of nanofiltration membranes and their rejection of trace organic chemicals. *J. Membr. Sci.* **2013**, *447*, 153–162. [CrossRef]
34. Robb, W. Thin silicone membranes-their permeation properties and some applications. *Ann. N. Y. Acad. Sci.* **1968**, *146*, 119–137. [CrossRef]
35. Merkel, T.; Bondar, V.; Nagai, K.; Freeman, B.; Pinnau, I. Gas sorption, diffusion, and permeation in poly (dimethylsiloxane). *J. Polym. Sci. B Polym. Phys.* **2000**, *38*, 415–434. [CrossRef]
36. Ismail, A.F.; Khulbe, K.C.; Matsuura, T. *Gas Separation Membranes*; Springer: Dordrecht, The Netherlands, 2015; Volume 10, pp. 973–978. ISBN 3319010956.

37. Casey, E.; Glennon, B.; Hamer, G. Oxygen mass transfer characteristics in a membrane-aerated biofilm reactor. *Biotechnol. Bioeng.* **1999**, *62*, 183–192. [CrossRef]
38. Essila, N.J.; Semmens, M.J.; Voller, V.R. Modeling Biofilms on Gas-Permeable Supports: Concentration and Activity Profiles. *J. Environ. Eng.* **2000**, *126*, 250–257. [CrossRef]
39. Syron; Kelly, H.; Casey, E. Studies on the effect of concentration of a self-inhibitory substrate on biofilm reaction rate under co-diffusion and counter-diffusion configurations. *J. Membr. Sci.* **2009**, *335*, 76–82. [CrossRef]
40. Xing, W.; Yin, M.; Lv, Q.; Hu, Y.; Liu, C.P.; Zhang, J.J. Oxygen solubility, diffusion coefficients, and solution viscosty. In *Rotating Electrode Methods and Oxygen Reduction Electrocatalysts*; Xing, W., Yin, G.P., Zhang, J.J., Eds.; Elsiver Publisher: New York, NY, USA, 2014; Chapter 1.
41. Rittmann, B. Comparative performance of biofilm reactor types. *Biotechnol. Bioeng.* **1982**, *24*, 1341–1370. [CrossRef] [PubMed]
42. Wanner, O.; Debus, O.; Reichert, P. Modelling the spatial distribution and dynamics of a xylene-degrading microbial population in a membrane-bound biofilm. *Water Sci. Technol.* **1994**, *29*, 243. [CrossRef]
43. Chen, G.; Ozaki, H.; Terashima, Y. Modelling of the simultaneous removal of organic substances and nitrogen in a biofilm. *Water Sci. Technol.* **1989**, *21*, 791–804. [CrossRef]
44. Tanase, C.; Chirvase, A.A.; Ungureanu, C.; Caramihai, M.; Muntean, O. Study of double-substrate limited growth of Pseudomonas aeruginosa in aerobic bioprocess. *Rev. Roum. Chem.* **2011**, *56*, 1147–1153.
45. Nicolella, C.; Pavasant, P.; Livingston, A.G. Substrate counterdiffusion and reaction in membrane-attached biofilms: Mathematical analysis of rate limiting mechanisms. *Chem. Eng. Sci.* **2000**, *55*, 1385–1398. [CrossRef]
46. Metcalf, L.; Eddy, H.P.; Tchobanoglous, G. *Wastewater Engineering: Treatment, Disposal, and Reuse*, 3rd ed.; McGraw-Hill: New York, NY, USA, 1991.
47. Stewart, P.S.; Zhang, T.; Xu, R.; Pitts, B.; Walters, M.C.; Roe, F.; Kikhney, J.; Moter, A. Reaction–diffusion theory explains hypoxia and heterogeneous growth within microbial biofilms associated with chronic infections. *NPJ Biofilms Microbiomes* **2016**, *2*, 16012. [CrossRef]
48. Li, S.; Chen, G.H. Modeling the organic removal and oxygen consumption by biofilms in an open-channel flow. *Water Sci. Technol.* **1994**, *30*, 53. [CrossRef]
49. Ntwampe, S.K.O.; Sheldon, M.S.; Volschenk, H. Oxygen mass transfer for an immobilised biofilm of Phanerochaete chrysosporium in a membrane gradostat reactor. *Braz. J. Chem. Eng.* **2008**, *25*, 649–664. [CrossRef]
50. Matsumoto, S.; Terada, A.; Tsuneda, S. Modeling of membrane-aerated biofilm: Effects of C/N ratio, biofilm thickness and surface loading of oxygen on feasibility of simultaneous nitrification and denitrification. *Biochem. Eng. J.* **2007**, *37*, 98–107. [CrossRef]
51. Wang, R.; Xiao, F.; Wang, Y.; Lewandowski, Z. Determining the optimal transmembrane gas pressure for nitrification in membrane-aerated biofilm reactors based on oxygen profile analysis. *Appl. Microbiol. Biotechnol.* **2016**, *100*, 7699–7711. [CrossRef] [PubMed]
52. Syron, E.; Casey, E.J.B. Bioengineering, Model-based comparative performance analysis of membrane aerated biofilm reactor configurations. *Biotechnol. Bioeng.* **2008**, *99*, 1361–1373. [CrossRef]
53. Motlagh, A.R.A.; Voller, V.R.; Semmens, M.J. Advective flow through membrane-aerated biofilms: Modeling results. *J. Membr. Sci.* **2006**, *273*, 143–151.
54. Duncan, J.; Bokhary, A.; Fatehi, P.; Kong, F.; Lin, H.; Liao, B. Thermophilic membrane bioreactors: A review. *Bioresour. Technol.* **2017**, *243*, 1180–1193. [CrossRef]

Article

Asymmetric Cellulose/Carbon Nanotubes Membrane with Interconnected Pores Fabricated by Droplet Method for Solar-Driven Interfacial Evaporation and Desalination

Zhiyu Yang [1], Linlin Zang [2], Tianwei Dou [1], Yajing Xin [1], Yanhong Zhang [1,*], Dongyu Zhao [1,*] and Liguo Sun [1]

[1] School of Chemical Engineering and Materials, Heilongjiang University, Harbin 150080, China; yangzy818yzy@163.com (Z.Y.); doutw1992@126.com (T.D.); xinjinzya@163.com (Y.X.); sunliguo1975@163.com (L.S.)

[2] School of Environmental Science and Engineering, Southern University of Science and Technology, Shenzhen 518055, China; zangll423@163.com

* Correspondence: zhangyanhong1996@163.com (Y.Z.); zhaody@hlju.edu.cn (D.Z.); Tel.: +86-188-4512-8078 (Y.Z.); +86-158-0461-1506 (D.Z.)

Abstract: Solar-driven interfacial water purification and desalination have attracted much attention in environmentally friendly water treatment field. The structure design of the photothermal materials is still a critical factor to improve the evaporation performance such as evaporation rate and energy conversion efficiency. Herein, an asymmetric cellulose/carbon nanotubes membrane was designed as the photothermal membrane via a modified droplet method. Under 1 sun irradiation, the evaporation rate and energy efficiency of pure water can reach up to 1.5 kg m^{-2} h^{-1} and 89%, respectively. Moreover, stable reusability and desalination performance made the cellulose/carbon nanotubes membrane a promising photothermal membrane which can be used for solar-driven desalination.

Keywords: droplet method; carbon nanotubes; porous materials; interfacial evaporation; solar energy materials

1. Introduction

In recent years, solar-driven interfacial evaporation has attracted widespread attention [1–4]. Many studies have shown that this emerging water treatment technology has great application prospects in seawater desalination and water purification [5–8]. Compared with traditional desalination technologies such as reverse osmosis (RO), multi-stage flash distillation (MSF), multi-effect distillation (MED) and vapor compression distillation (VCD), solar-driven interfacial desalination can maximize water recovery and reduce fossil fuel consumption by inducing various materials with light-to-heat conversion properties [7,9–11]. Therefore, it is considered to be one of the most promising technologies for water purification and desalination [12].

To improve the overall evaporation performance, photothermal membranes with vertically oriented, wrinkled or layered structure can be constructed by freeze-drying and hydrothermal methods to enhance the light absorption capacity and accelerate the vapor diffusion [13–19]. Although these membranes of different structure can achieve great performance of photothermal evaporation, their photothermal-vapor conversion capacities are relatively low, because the efficient photothermal steam generation is also somewhat related to the porous structure. Solar-driven interfacial desalination used porous structure membranes exhibits more excellent performance in terms of optical absorption, photothermal conversion and photothermal interfacial evaporation [20,21]. At present, the reported preparation methods of porous membranes include template method, stretching and electrospinning [22–24]. These preparation methods are expected to be great strategies for preparing photothermal membranes with excellent interfacial evaporation performance, but most of them are complicated in operation, which are not conductive to large-scale application [25–28]. Therefore, a simple method to synthesize photothermal evaporation membranes with abundant pore structures is urgently needed.

It has been reported that utilizing the gas–liquid interface can fabricate the ordered porous membranes. For example, the porous poly (lactic acid) membrane obtained by polymerizing or precipitating at the gas–liquid interface using the template of monolayer colloidal crystal floating on the liquid surface [29]. Adopting the two-phase anisotropy at the gas–liquid interface, asymmetric membranes with porous structures can be obtained [30–34].

Recently, inspired by membrane formation on the gas–liquid interface, droplet methods have been adopted in preparing porous membranes due to its uncomplicated operation. The typical process is summarized as follows: (1) Preparation of polymer solution for droplets formation. The droplets are homogenous solution composed of polymer, solvent and additions which improve the structure and performance of membranes. (2) Preparation of coagulation bath. The coagulation bath is composed of nonsolvent and solvent. Changing the proportion of solvent and nonsolvent could affect the structure of the prepared membranes. (3) Droplets dropping from the fixed nozzle. The droplets drop from nozzle and fall into the coagulation bath at a certain height to form membrane. There are some factors affecting the formation of membranes such as solution concentration, droplet height [35–37]. However, the key point is the selection of solvent and nonsolvent. The solvent must completely dissolve the polymer while the nonsolvent cannot dissolve the polymer, and the solvent should be miscible with the nonsolvent. When the droplet of homogenous solution is in contact with coagulation bath, the exchange of solvent and nonsolvent at the gas–liquid interface leads to phase separation, forming a polymer-rich phase (to form dense skin layer) and polymer-poor phase (to form porous structure). Subsequently, phase separation continues until the polymer-rich phase completely solidifies, forming an asymmetric porous membrane [38–41].

Herein, an asymmetric cellulose/carbon nanotubes membrane with porous structure was prepared by modified droplet, in which a mixed polymer solution of cellulose, polymethacrylic acid (PMAA), ionic liquid and carbon nanotubes (CNTs) was dropped into the coagulation bath. Under the combined effects of gravity, buoyancy and surface tension, droplet of polymer solution dropping into the coagulation bath would form a membrane at the gas–liquid interface of the coagulation bath. PMAA increased hydrophilicity as a polymer electrolyte and ionic liquid was used to dissolve cellulose. CNTs were used as photothermal materials to convert the absorbed light energy [42,43]. At the same time, it also acts as a supporting network together with cellulose to stabilize the porous structure. The three-dimensional (3D) interconnected pores structure of the membrane can effectively improve the evaporation performance by reducing light reflection and accelerating vapor escape.

2. Materials and Methods

2.1. Materials

Carbon nanotube were multi-walled CNT with a length of 1~2 μm and outer diameter of 20~40 nm obtained from Shenzhen Nanotech Port Co., Ltd. Cellulose was received from Shanghai Aladdin Bio-Chem Technology Co., LTD. (Shangai, China) 1-Butyl-3-methylimidazolium chloride ([Bmim] Cl, >99%) was obtained from Shanghai Yiji Industrial Co., Ltd. (Shangai, China) Methacrylic acid (MMA, AR), dimethylformamide (DMA, AR), sodium chloride (NaCl, AR), methylene blue (biological dye, BS), methyl orange (biological dye, BS) were purchased from Sinopharm Chemical Reagent Co., Ltd. (Shangai, China). Distilled water was received from Harbin Wenjing Distilled Water Factory.

2.2. Acidification of Carbon Nanotubes

First, 5 g pristine multi -walled CNTs with lengths of 1~2 μm were added to 500 mL of the mixed strong acid solution with H_2SO_4/HNO_3 which was a volume ratio of 1/3. The mixture was uniformly mixed under magnetic stirring, then heated to 333 K and stirred at reflux for 3 h. After cooling to room temperature, the solution was poured into a beaker and diluted with deionized water, and then filtered under reduced pressure until the sample became neutral. Finally, the acidified carbon nanotubes were prepared by drying at 323 K in a vacuum drying oven. Through the modification of acidification, oxygen-containing functional groups were added to improve the hydrophilicity of materials so that they can be fully dispersed in solvents.

2.3. Fabrication of Cellulose/CNTs Membrane

In this experiment, the cellulose/CNTs membrane was prepared by a droplet method. First, 0.15 g cellulose powder and 0.15 g PMAA were added to 3 g 1-Butyl-3-methylimidazolium chloride ([Bmim] Cl) at 80 °C and stirred for 2 h until they were completely dissolved. Then, 5 mL of N, N-Dimethylformamide (DMF) solution containing 20 mg acidified CNTs was poured into the above solution and mixed thoroughly. After that, the obtained solution was transferred to a 10 mL syringe and dropped into a coagulation bath with DMF/water (volume ratio = 1/1) through a micro-syringe pump at a height of 10 cm above the liquid level, thereby forming a composite membrane on the surface of the coagulation bath with DMF/water solution (Figure 1). Subsequently, the composite membrane was repeatedly washed with water to remove [Bmin] Cl and PMAA, and then freeze-dried using liquid nitrogen. The resulting cellulose/CNTs membrane was named as CCM. As a control, cellulose membrane without any CNTs were also prepared according to the above procedure, which was named as CM. Cellulose membrane without porous structure by freeze-drying prepared solution directly for another control experiment was named as CCM-N.

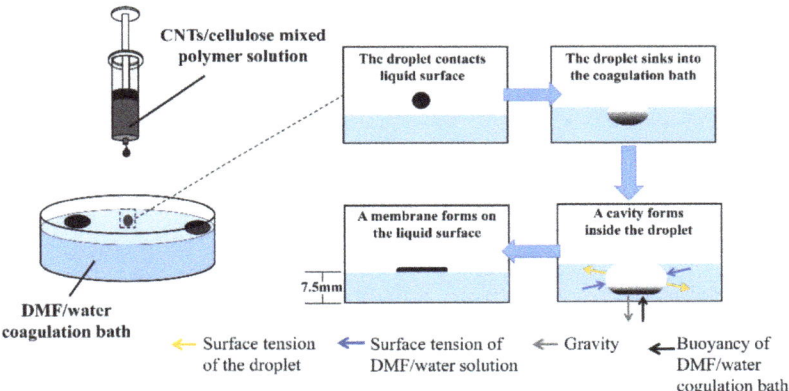

Figure 1. Schematic diagram of the preparation of the cellulose/CNTs membrane.

2.4. Characterization

The surface morphologies and cross-section morphologies of the CCM were characterized by a scanning electron microscope (Hitachi, Tokyo, Japan, S-4800). Surface chemical composition of CCM was examined by an ATR-FTIR spectra which was measured by using a Spectrum One instrument (Perkin Elmer, Waltham, MA, USA). UV-vis-NIR diffuse reflectance spectra (DRS) was measured with a Perkin-Elmer Lambda 950 UV-vis-NIR spectrophotometer (USA). The concentrations of ions were detected by inductively coupled plasma-optical emission spectrometer (ICP-OES, Optima 8300, Perkin Elmer, USA).

2.5. Solar-Driven Interfacial Evaporation Experiments

To investigate the solar-driven interfacial evaporation performance, the cellulose membrane with/without CNTs was employed as the photothermal material and placed on a transporter-assisted evaporation system. A polystyrene foam and a glass fiber filter were used as the heat insulation and water channel, respectively. In this experiment, the area of the membrane used was 4 cm^2, and the volume of water used was 20 mL. All evaporation experiments were conducted under a solar simulator (Perfect Light PLS-SXE300DUV).

The mass of the water loss is measured by an electrical balance. The *surface temperature* of the photothermal membrane was measured by an infrared thermal image (FLIR ONE). The solar-heat energy conversion efficiency (η) was calculated using the following formulas:

$$\eta = \frac{\left(m_{light} - m_{dark}\right)h_{LV}}{I} \times 100\%$$

$$h_{LV} = c(T_{surface} - T_0) + L$$

$$L = -0.00006 T_{surface}^3 + 0.0016 T_{surface}^2 - 2.36 T_{surface}^1 + 2500.8$$

where m_{dark} and m_{light} are water evaporation rates under *dark* and *light* conditions, respectively. h_{LV} is the total enthalpy, c is the specific heat capacity and L is the specific latent heat of phase change. $T_{surface}$ and T_0 represented the temperature of the evaporation surface and bulk water reservoir, respectively.

The desalination efficiencies of different ions(η_i) were calculated using the following formulas [44–46]:

$$\eta_i = \frac{C_0 - C_t}{C_0} \times 100\%$$

where C_0 is the concentrations of different ions before desalination and C_t is the concentrations of different ions after desalination.

3. Results and Discussion

When the droplet enters the DMF/water solution of coagulation bath, they will sink to the liquid under gravity. Meanwhile, [Bmim] Cl molecules in the droplet diffuses into the DMF/water solution, resulting in generating large number of tiny pores. As [Bmim] Cl content in the droplet decreases, the 3D spherical droplet becomes a two-dimensional (2D) membrane and then floats on the surface of the coagulation bath. Since [Bmim] Cl molecules diffuse relatively rapidly when the droplet contacts with DMF/water solution, the cellulose/CNTs mixed polymer solution occurred gelation, dense skin layers are formed on the liquid–liquid surfaces of the CCM (Figure 1). The dense skin layer could help resist the pulling effect of surface tension in the coagulation bath. In contrast, the interior of the CCM exhibited an obvious 3D interconnected porous structure, whose reason lied in the sufficient solvent diffusion.

The modified droplet method for preparing the CCM mainly involves phase-inversion process in the coagulation bath. There are many influencing factors of phase-inversion, and the key factors of phase-inversion are solvent type and composition, polymer type and composition, non-solvent type and composition and membrane forming conditions. In this process, cellulose/CNTs in the droplet is selected as polymer, [Bmin]Cl/DMF in the droplet as solvent, and DMF/water as nonsolvent (the coagulation medium) [47]. Therefore, attention also needs to be paid to the concentration of solvent and nonsolvent. When the droplet enters the coagulation bath, a portion of the droplet surface is in contact with the nonsolvent of the coagulation bath. Due to the gradients in density and/or interfacial energy of the polymer-nonsolvent interface, slow convective flow of nonsolvent-solvent occurs in the droplet, resulting in a large number of small pores which is the sponge-like 3D interconnected porous structure [47,48]. The membrane at liquid–liquid interface presents a dense structure due to the direct contact between the coagulation bath and the droplet. The other part of the droplet surface, that is not in direct contact with the nonsolvent of the coagulation bath, is relatively distant from the nonsolvent, so it has difficulty forming convective flow with the nonsolvent. The relatively high concentration of polymer causes it gather together, resulting in the size of pores at the gas–liquid interface are relatively large [48,49]. The modified droplet method is achieved at the interface by gravity, buoyancy and interaction between the surface tension of cellulose/CNTs mixed solution and the surface tension of the coagulating bath. The main purpose of adding DMF is to adjust the surface tension of cellulose/CNTs mixed polymer solution and coagulation bath. The addition of DMF had little effect on gravity and buoyancy, so their impact could be almost ignored in this research. When the appropriate content of DMF is added to cellulose/CNTs mixed solution, the concentration of cellulose/CNTs mixed solution at the liquid–liquid surface is higher than that at other parts of the membrane through phase separation, so that the CCM would form a dense layer at the liquid–liquid surface. The dense layer can better resist the surface tension of the coagulating bath and prevent the film from being damaged by the surface tension of the coagulating bath. Therefore, when the droplet is dropped into the coagulation bath, the solution bead is formed in the coagulation bath and, at the same time, the original stable state should be destroyed to make the cellulose/CNTs solution flow so as to make the droplet form a membrane [50]. So, there are two crucial factors affecting the preparation of CCM: cellulose/CNTs mixed polymer solution concentration and coagulation bath concentration.

As shown in Figure 2, when the DMF/water volume ratio of the coagulated bath remained unchangeable and the DMF content of droplet increased, the concentration of cellulose/CNTs mixed polymer solution decreased and the surface tension of droplet would decrease, which would also affect the state of CCM formation. It was taken as the research object that DMF/water volume ratio of the coagulation bath was 1/3 (surface tension= 0.07100 N m^{-1}). As the DMF content in the droplet increased and concentration of cellulose/CNTs mixed polymer solution decreased, the final state of the droplet in the coagulation bath gradually changed from bead-like to membrane-like and then even was split into small pieces. This was because the surface tension of the droplet was unable to resist the surface tension of the coagulating bath, therefore the droplet was split under the action of the surface tension of the coagulating bath. As can be seen from the Figure 2, due to the surface tension of the cellulose/CNTs mixed polymer droplet which added 4 mL DMF (surface tension = 0.07112 N m^{-1}) was too large to form a membrane, the droplet fell into the coagulation bath and formed like-bead float on the coagulation bath under combination of gravity, buoyancy and surface tensions. Meanwhile, the surface tension of the droplet was too small, due to the addition of 7 mL DMF (surface tension = 0.06711 N m^{-1}). When the droplet dropped into the coagulating bath, the surface tension of the droplet cannot compete with the surface tension of the coagulating bath. So, it was difficult to form a complete circular membrane and will be dispersed into small pieces or the membrane formed was too thin. With the increase of DMF content, the gelation degree of cellulose/CNTs mixed polymer solution also decreased, and it cannot resist the surface tension of coagulation bath effectively.

In Figure 2, by fixing DMF content of cellulose/CNTs mixed polymer droplet and increasing DMF/water volume ratio of coagulation bath, the surface tension of coagulation bath would decrease, but the decrement was small. If DMF cellulose/CNTs mixed polymer solution added 6 mL DMF was taken as the research object (surface tension= 0.06756 N m^{-1}), we can clearly see that different volume ratio of coagulation bath had different effect on the formation of membrane. With the decrease of the volume ratio, the formed membrane gradually changed from a thin membrane which was split easily to a circular membrane with uniform and appropriate thickness, finally a complete film cannot be formed. Considering the two variables, the membrane which was prepared from cellulose/CNTs mixed polymer solution added 6 mL DMF and water (surface tension = 0.07222 N m^{-1}) was similar to the membrane which was prepared from cellulose/CNTs mixed polymer solution added 7 mL DMF (surface tension = 0.06711 N m^{-1}) and coagulation bath with DMF/water (volume ratio = 1/3). The appearance further illustrated that CCM obtained from the interaction between the surface tension of coagulation bath and the surface tension of polymer solution. From the above analysis, it was clear that when the DMF content added to cellulose/CNTs polymer solution and DMF/water volume ratio of coagulation bath were appropriate, a circular membrane with uniform and appropriate thickness would be formed. In a word, the coagulation bath with DMF/water (volume ratio= 1/1) and the cellulose/CNTs mixed polymer solution added 6 mL DMF were the most suitable conditions to prepare membranes by the droplet method.

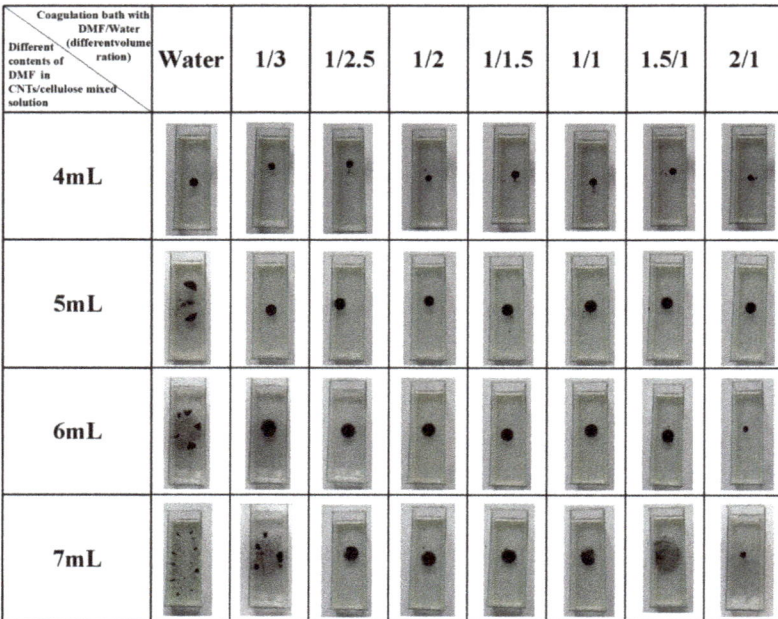

Figure 2. Digital photos of cellulose/CNTs mixed polymer solution added different DMF contents dropped into coagulating bath with DMF/water (different volume ratio).

In the coagulation bath with DMF/water (volume ratio = 2/1), less water can be gelatinized for cellulose/CNTs mixed polymer solution, the degree of gelation was lower, while more polymer solution appeared as liquid state compared to others at the same time. When the membrane area reached its maximum, non-gelatinization polymer solution returned to the droplet shape by its own surface tension, resulting in the contraction of membrane as shown in Figure 2.

In Figure 3, the digital photographs taken by the high-speed camera showed that the entire process from droplet to the CCM took 4 ms. In this process, the cellulose/CNTs mixed polymer solution was diffused and formed a complete circular membrane under the interaction of droplet surface tension and coagulation bath surface tension. At 20 ms, a cavity could be seen forming after the droplets had been dropped into the coagulation bath. Then a circular membrane was formed under the interaction of the surface tensions. In less than 0.04 s, CCM could be formed. When a complete circular film was formed, the membrane floated on the coagulation bath under the action of surface tensions. When a complete circular membrane was formed, the membrane floated on the coagulation bath by the action of surface tensions.

A mathematical model was established for analyzing membrane formation through modified droplet method. The A is the spreading coefficient of polymer solution, which is the ratio of spreading ability (δ_{spread}) to anti-spreading ability ($\delta_{anti\text{-}spread}$). The spreading ability of the polymer solution is derived from the surface tension of the coagulation bath. The spreading resistance (B_{gel}) is the tensile resistance of the instantaneous gelation product of polymer solution, which is related to the concentration of polymer solution. Meanwhile, gelation is accomplished by non-solvent in the coagulation bath. The ratio of water in the coagulation bath affects the thickness and strength of the gel layer, so k water is induced. The value of A can be used to determine whether the gel can be formed and the state of the membrane.

$$A = \frac{\delta_{spread}}{\delta_{anti\text{-}spread}} = \frac{\sigma_{cb}}{k \times B_{gel}}$$

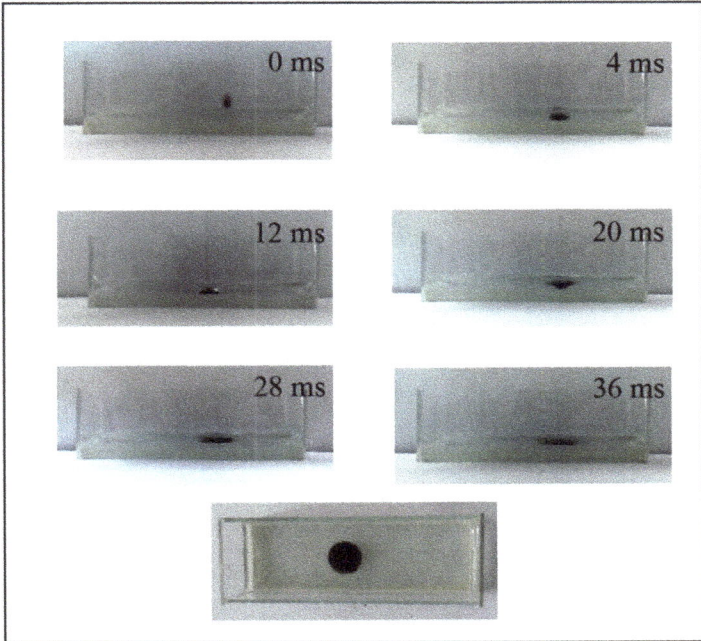

Figure 3. Digital photos of cellulose/CNTs mixed polymer solution added 6 mL DMF dropped into coagulating bath with DMF/water (volume ratio = 1/1) taken by high-speed camera.

By the measurement of the gel membrane size, the formula is obtained by fitting. Different requirements of the gel membrane can be prepared by changing the parameters.

$$r_{membrane} = A \times \frac{a}{\sigma_{polymer\ solution}} + b$$

$$h_{membrane} = \frac{V}{A} = \frac{V}{\pi r^2_{membrane}}$$

We can see from the above research, the best CCM was prepared from cellulose/CNTs mixed polymer solution added 6 mL DMF and coagulation bath with DMF/water (volume ratio = 1/1). Therefore, we chose this membrane for further tests.

SEM images in Figure 4 showed the surfaces and internal morphology of the CCM with a thickness of 400 ± 25 μm. The characteristic of asymmetric membrane can be clearly seen by Figure 4. The dense skin layer which located at liquid–liquid surface was 1~3 μm (Figure 4c). At the gas–liquid interface, pore diameters could up to 22 μm (Figure 4b). After the sufficient solvent diffusion, the interior of the CCM exhibited an obvious 3D interconnected pores structure (Figure 4c,d).

The FTIR spectra shown in Figure 5a demonstrated that the CCM exhibited characteristic peaks at 1708 cm^{-1} and 1053 cm^{-1}, which were attributed to the vibration of C=O and C-O-C pyranoid ring of PMAA and cellulose, respectively. Comparing with the cellulose and PMAA samples, a new characteristic peak appeared at 1562 cm^{-1} corresponding to the stretching vibration of COO$^-$. Besides, the carboxyl group in PMAA could generate electrostatic attraction with the hydroxyl group in cellulose to form a cross-linked network, which destroyed the intermolecular hydrogen bond of cellulose, with the characteristic peak at 3600–3000cm^{-1} strength decreased. Comparing with the CM, the CCM exhibited small optical reflectance (≈4–6%) in the 250–2500 nm wavelength range, indicating the large optical absorption of the CCM (Figure 5b).

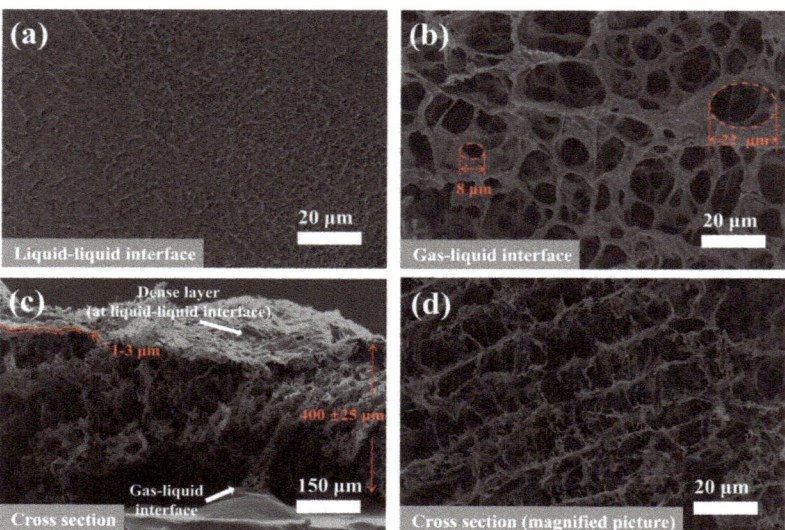

Figure 4. SEM images of (**a**) liquid–liquid interface; (**b**) gas–liquid interface; (**c**) cross section; (**d**) magnified cross section of the CCM.

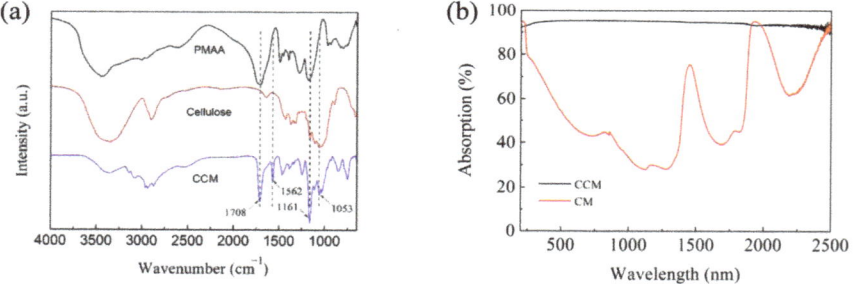

Figure 5. (**a**) FTIR spectra of the cellulose, PMAA and CCM; (**b**) UV-vis-NIR spectra of CM and CCM.

It was shown in Figure 6 that the transporter-assisted interfacial evaporation system, where the CCM was used as the photothermal membrane to absorb light heat water and generate vapor. As a result, compared with the bulk water and cellulose membrane, the surface temperature of the CCM can quickly rise to 44.4 °C within 600 s under 1 sun irradiation (Figure 7a). Additionally, excellent light absorption and heat localization of the CCM enabled the evaporation system to present higher water evaporation rates under dark and light conditions. As shown in Figure 7b, the membrane with CNTs exhibited an excellent evaporation rate of roughly 1.6 kg m^{-2} h^{-1}, and the corresponding light-to-heat energy conversion efficiency of the transporter-assisted evaporation system was calculated to be 89%, which was higher than that of reported photothermal membranes. In contrast, the evaporation rate of the evaporation system using the CM was less than 0.9 kg m^{-2} h^{-1} due to the absence of light absorber and lower surface temperature, and the evaporation rate of the evaporation system using the CMM-N was less than 1.3 kg m^{-2} h^{-1} due to the absence of 3D interconnected pore structure. The results indicated that the addition of CNTs and optimal porous structure can significantly enhance the solar-driven interfacial evaporation performance and maximize the energy conversion efficiency.

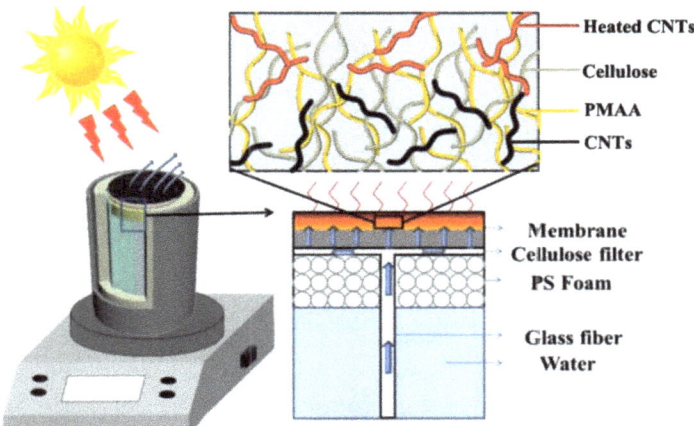

Figure 6. Schematic diagram of the transporter-assisted evaporation system.

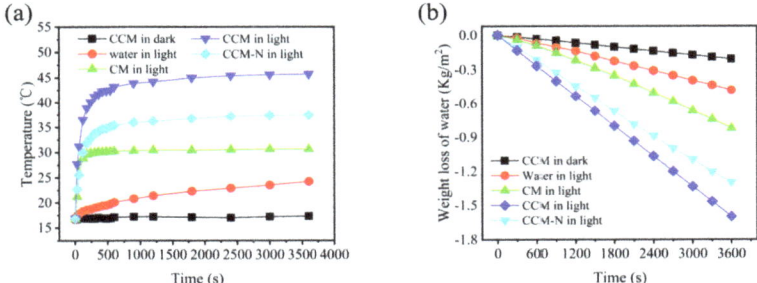

Figure 7. (**a**) Temperature changes of the water surface and membrane surface using CCM and CM under light and dark conditions; (**b**) weight loss of water using CCM and CM during the interfacial evaporation process under light and dark conditions.

We also evaluated the desalination performance of the transporter-assisted evaporation system. Figure 8a showed that evaporation rate of seawater was slightly lower than that of pure water, which was caused by more complex components such as salt ions, natural organic matters and bacteria [51]. For NaCl solutions with different salinities, the evaporation rate of 20 wt% NaCl solution can still reach about 1.4 kg m^{-2} h^{-1} after one-hour evaporation test, although the lower partial vapor pressure resulted in a slower evaporation rate as the concentration increased from 1.4 wt% to 20 wt% (Figure 8a,b). In the 10-cycle test, the evaporator can maintain a stable evaporation rate of about 1.55 kg m^{-2} h^{-1} (Figure 8c). More importantly, during the evaporation of seawater, the desalination efficiencies of Na$^+$, Mg^{2+}, K$^+$ and Ca^{2+} were 99.96%, 99.97%, 99.30% and 99.86%. Meanwhile, the concentrations of Na$^+$, Mg^{2+}, K$^+$ and Ca^{2+} in the condensed water were greatly reduced, which were lower than the salinity levels defined by World Health Organization (WHO) (Figure 8d).

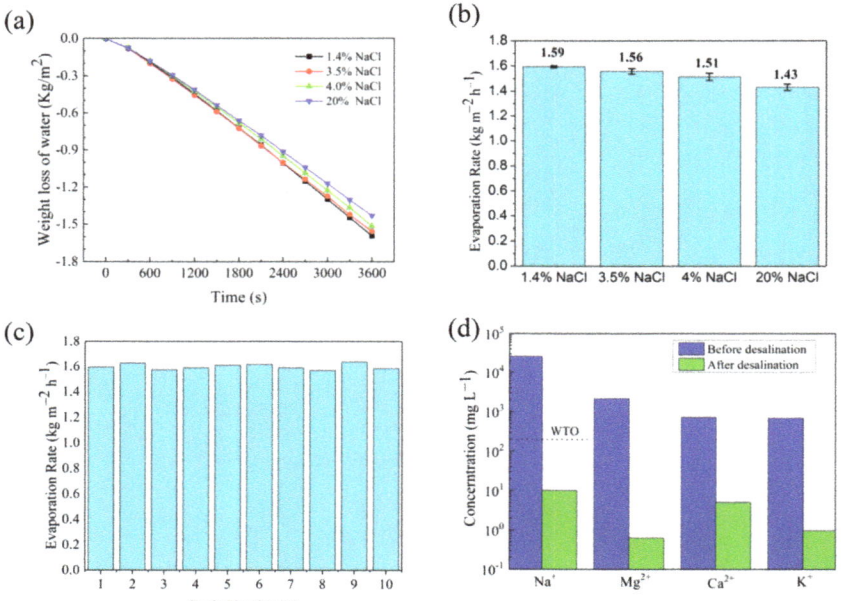

Figure 8. (**a**) Weight loss of water using the CCM when processing NaCl solutions with different salinities; (**b**) the corresponding evaporation rate of the CCM when processing NaCl solutions with different salinities; (**c**) stability test of the evaporation system using the CCM in 3.5 wt% NaCl solution; (**d**) ion concentrations using CCM before and after the solar-driven interfacial desalination.

4. Conclusions

The asymmetric CCM was fabricated by droplet method with 3D interconnected pores structure and good light absorption ability, which resulted in high evaporation rate and energy conversion efficiency when processing pure water, salt water and fresh seawater under one sun irradiation. The stable reusability and high-quality condensed water made the transporter-assisted evaporation system a potential candidate for solar-driven water purification and desalination.

Author Contributions: Conceptualization, T.D. and Z.Y.; methodology, T.D. and Z.Y.; software, Z.Y.; validation, Z.Y., T.D. and L.Z.; formal analysis, T.D. and L.S.; investigation, Z.Y.; resources, T.D.; data curation, T.D. and Z.Y.; writing—original draft preparation, Z.Y.; writing—review and editing, Y.Z., D.Z., Y.X. and L.S.; visualization, T.D.; supervision, Y.Z.; project administration, D.Z.; funding acquisition, L.S. All authors have read and agreed to the published version of the manuscript.

Funding: This research was funded by Heilongjiang Natural Science Foundation Project, grant number LH2020E107 and Innovation Training Program for College Students of China, grant number 202110212052.

Institutional Review Board Statement: Not applicable.

Informed Consent Statement: Not applicable.

Data Availability Statement: Not applicable.

Conflicts of Interest: The authors declare no conflict of interest.

References

1. Zhao, F.; Bae, J.; Zhou, X.; Guo, Y.; Yu, G. Nanostructured Functional Hydrogels as an Emerging Platform for Advanced Energy Technologies. *Adv. Mater.* **2018**, *30*, e1801796. [CrossRef] [PubMed]
2. Dao, V.-D.; Vu, N.H.; Yun, S. Recent Advances and Challenges for Solar-Driven Water Evaporation System toward Applications. *Nano Energy* **2020**, *68*, 104324. [CrossRef]
3. Tao, P.; Ni, G.; Song, C.; Shang, W.; Wu, J.; Zhu, J.; Chen, G.; Deng, T. Solar-Driven Interfacial Evaporation. *Nat. Energy* **2018**, *3*, 1031–1041. [CrossRef]
4. Zhu, L.; Gao, M.; Peh, C.K.N.; Ho, G.W. Solar-Driven Photothermal Nanostructured Materials Designs and Prerequisites for Evaporation and Catalysis Applications. *Mater. Horiz.* **2018**, *5*, 323–343. [CrossRef]
5. Tan, Z.; Chen, S.; Peng, X.; Zhang, L.; Gao, C. Polyamide Membranes with Nanoscale Turing Structures for Water Purification. *Science* **2018**, *360*, 518–521. [CrossRef]
6. Werber, J.R.; Osuji, C.O.; Elimelech, M. Materials for Next-Generation Desalination and Water Purification Membranes. *Nat. Rev. Mater.* **2016**, *1*, 16018. [CrossRef]
7. Zhu, L.; Gao, M.; Peh, C.K.N.; Ho, G.W. Recent Progress in Solar-Driven Interfacial Water Evaporation: Advanced Designs and Applications. *Nano Energy* **2019**, *57*, 507–518. [CrossRef]
8. Wang, Z.; Elimelech, M.; Lin, S. Environmental Applications of Interfacial Materials with Special Wettability. *Environ. Sci. Technol.* **2016**, *50*, 2132–2150. [CrossRef]
9. Ghim, D.; Wu, X.; Suazo, M.; Jun, Y.-S. Achieving Maximum Recovery of Latent Heat in Photothermally Driven Multi-Layer Stacked Membrane Distillation. *Nano Energy* **2021**, *80*, 105444. [CrossRef]
10. Al-Karaghouli, A.; Kazmerski, L.L. Energy Consumption and Water Production Cost of Conventional and Renewable-Energy-Powered Desalination Processes. *Renew. Sustain. Energ. Rev.* **2013**, *24*, 343–356. [CrossRef]
11. Li, Y.; Chen, J.; Cai, P.; Wen, Z. An Electrochemically Neutralized Energy-Assisted Low-Cost Acid-Alkaline Electrolyzer for Energy-Saving Electrolysis Hydrogen Generation. *J. Mater. Chem. A* **2018**, *6*, 4948–4954. [CrossRef]
12. Liu, X.; Mishra, D.D.; Wang, X.; Peng, H.; Hu, C. Towards Highly Efficient Solar-Driven Interfacial Evaporation for Desalination. *J. Mater. Chem. A* **2020**, *8*, 17907–17937. [CrossRef]
13. Kim, C.; Shin, D.; Baitha, M.N.; Ryu, Y.; Urbas, A.M.; Park, W.; Kim, K. High-Efficiency Solar Vapor Generation Boosted by a Solar-Induced Updraft with Biomimetic 3D Structures. *ACS Appl. Mater. Interfaces* **2021**, *13*, 29602–29611. [CrossRef]
14. Zhang, Y.; Xiong, T.; Nandakumar, D.K.; Tan, S.C. Structure Architecting for Salt-Rejecting Solar Interfacial Desalination to Achieve High-Performance Evaporation With In Situ Energy Generation. *Adv. Sci.* **2020**, *7*, 1903478. [CrossRef]
15. Sun, P.; Zhang, W.; Zada, I.; Zhang, Y.; Gu, J.; Liu, Q.; Su, H.; Pantelić, D.; Jelenković, B.; Zhang, D. 3D-Structured Carbonized Sunflower Heads for Improved Energy Efficiency in Solar Steam Generation. *ACS Appl. Mater. Interfaces* **2020**, *12*, 2171–2179. [CrossRef]
16. Yin, X.; Zhang, Y.; Guo, Q.; Cai, X.; Xiao, J.; Ding, Z.; Yang, J. Macroporous Double-Network Hydrogel for High-Efficiency Solar Steam Generation Under 1 Sun Illumination. *ACS Appl. Mater. Interfaces* **2018**, *10*, 10998–11007. [CrossRef]
17. Shi, Y.; Li, R.; Jin, Y.; Zhuo, S.; Shi, L.; Chang, J.; Hong, S.; Ng, K.-C.; Wang, P. A 3D Photothermal Structure toward Improved Energy Efficiency in Solar Steam Generation. *Joule* **2018**, *2*, 1171–1186. [CrossRef]
18. Wang, X.; Liu, Q.; Wu, S.; Xu, B.; Xu, H. Multilayer Polypyrrole Nanosheets with Self-Organized Surface Structures for Flexible and Efficient Solar–Thermal Energy Conversion. *Adv. Mater.* **2019**, *31*, 1807716. [CrossRef]
19. Wu, S.; Xiong, G.; Yang, H.; Tian, Y.; Gong, B.; Wan, H.; Wang, Y.; Fisher, T.S.; Yan, J.; Cen, K.; et al. Scalable Production of Integrated Graphene Nanoarchitectures for Ultrafast Solar-Thermal Conversion and Vapor Generation. *Matter* **2019**, *1*, 1017–1032. [CrossRef]
20. Liu, F.; Lou, D.; Liang, E.; Gu, Y.; Wang, Z.; Shi, X.; Bradley, R.; Zhao, B.; Wu, W. Nanosecond Laser Patterned Porous Graphene from Monolithic Mesoporous Carbon for High-Performance Solar Thermal Interfacial Evaporation. *Adv. Mater. Technol.* **2021**, *6*, 2101052. [CrossRef]
21. Zhu, Z.; Xu, Y.; Luo, Y.; Wang, W.; Chen, X. Porous Evaporators with Special Wettability for Low-Grade Heat-Driven Water Desalination. *J. Mater. Chem. A* **2021**, *9*, 702–726. [CrossRef]
22. Elma, M.; Rampun, E.L.A.; Rahma, A.; Assyaifi, Z.L.; Sumardi, A.; Lestari, A.E.; Saputro, G.S.; Bilad, M.R.; Darmawan, A. Carbon Templated Strategies of Mesoporous Silica Applied for Water Desalination: A Review. *J. Water Process Eng.* **2020**, *38*, 101520. [CrossRef]
23. Zhou, S. Novel Janus Membrane with Unprecedented Osmosis Transport Performance. *J. Mater. Chem. A* **2019**, *7*, 632–638. [CrossRef]
24. Huang, Y. Robust Preparation of Tubular PTFE/FEP Ultrafine Fibers-Covered Porous Membrane by Electrospinning for Continuous Highly Effective Oil/Water Separation. *J. Membr. Sci.* **2018**, *10*, 87–96. [CrossRef]
25. Liu, P.-F.; Miao, L.; Deng, Z.; Zhou, J.; Su, H.; Sun, L.; Tanemura, S.; Cao, W.; Jiang, F.; Zhao, L.-D. A Mimetic Transpiration System for Record High Conversion Efficiency in Solar Steam Generator under One-Sun. *Mater. Today Energy* **2018**, *8*, 166–173. [CrossRef]
26. Li, X.; Lin, R.; Ni, G.; Xu, N.; Hu, X.; Zhu, B.; Lv, G.; Li, J.; Zhu, S.; Zhu, J. Three-Dimensional Artificial Transpiration for Efficient Solar Waste-Water Treatment. *Natl. Sci. Rev.* **2018**, *5*, 70–77. [CrossRef]

27. Jiang, Q.; Gholami Derami, H.; Ghim, D.; Cao, S.; Jun, Y.-S.; Singamaneni, S. Polydopamine-Filled Bacterial Nanocellulose as a Biodegradable Interfacial Photothermal Evaporator for Highly Efficient Solar Steam Generation. *J. Mater. Chem. A* **2017**, *5*, 18397–18402. [CrossRef]
28. Athanasekou, C.; Sapalidis, A.; Katris, I.; Savopoulou, E.; Beltsios, K.; Tsoufis, T.; Kaltzoglou, A.; Falaras, P.; Bounos, G.; Antoniou, M.; et al. Mixed Matrix PVDF/Graphene and Composite-Skin PVDF/Graphene Oxide Membranes Applied in Membrane Distillation: Enhanced Flux Properties and Scaling Behavior of Mixed Matrix PVDF/Graphene and Composite-Skin PVDF/Graphene Oxide Membranes in Direct Contact Membran. *Polym. Eng. Sci.* **2019**, *59*, E262–E278. [CrossRef]
29. Li, Y.; Qi, L. Advances in Fabrication of Two-dimensionally Ordered Porous Membranes by Nanosphere Lithography at the Gas-liquid Interface. *Acta Chim. Sin.* **2015**, *73*, 869. [CrossRef]
30. Ding, X.; Cao, Y.; Zhao, H.; Wang, L. Interfacial Morphology between the Two Layers of the Dual-Layer Asymmetric Hollow Fiber Membranes Fabricated by Co-Extrusion and Dry-Jet Wet-Spinning Phase-Inversion Techniques. *J. Membr. Sci.* **2013**, *444*, 482–492. [CrossRef]
31. Wei, C.; Cheng, Q.; Lin, L. One-Step Fabrication of Recyclable Polyimide Nanofiltration Membranes with High Selectivity and Performance Stability by a Phase Inversion-Based Process. *J. Mater.Sci.* **2018**, *53*, 11104–11115. [CrossRef]
32. Fen'ko, L.A.; Semenkevich, N.G.; Bil'dyukevich, A.V. The Kinetics of Membrane Pore Structure Formation by Phase Inversion. *Pet. Chem.* **2011**, *51*, 527–535. [CrossRef]
33. Ding, Z.; Liu, Z.; Wei, W.; Li, Z. Preparation and Characterization of Plla Composite Scaffolds by ScCO$_2$-Induced Phase Separation. *Polym. Compos.* **2012**, *33*, 1667–1671. [CrossRef]
34. Chen, X.; Shi, C.; Wang, Z.; He, Y.; Bi, S.; Feng, X.; Chen, L. Structure and Performance of Poly(Vinylidene Fluoride) Membrane with Temperature-Sensitive Poly(*n*-Isopropylacrylamide) Homopolymers in Membrane Pores. *Polym. Compos.* **2013**, *34*, 457–467. [CrossRef]
35. Roy, K.J.; Anjali, T.V.; Sujith, A. Asymmetric Membranes Based on Poly(Vinyl Chloride): Effect of Molecular Weight of Additive and Solvent Power on the Morphology and Performance. *J. Mater. Sci.* **2017**, *52*, 5708–5725. [CrossRef]
36. Watanabe, T.; Yasuhara, Y.; Ono, T. Multilayer Poly(Ionic Liquid) Microcapsules Prepared by Sequential Phase Separation and Subsequent Photopolymerization in Ternary Emulsion Droplets. *ACS Appl. Polym. Mater.* **2022**, *4*, 348–356. [CrossRef]
37. Wang, F.; Altschuh, P.; Ratke, L.; Zhang, H.; Selzer, M.; Nestler, B. Progress Report on Phase Separation in Polymer Solutions. *Comput. Mater. Sci.* **2019**, *31*, e1806733. [CrossRef]
38. Ye, Q. The Formation of Regular Porous Polyurethane Membrane via Phase Separation Induced by Water Droplets from Ultrasonic Atomizer. *Mater. Lett.* **2013**, *100*, 23–25. [CrossRef]
39. Su, Y.; Beltsios, K.G.; Cheng, L. Phase Inversion in Reusable Baths (PIRBs): A New Polymer Membrane Fabrication Method as Applied to EVOH. *J. Appl. Polym. Sci.* **2019**, *136*, 48193. [CrossRef]
40. Wang, H.; Wang, L.; Liu, C.; Xu, Y.; Zhuang, Y.; Zhou, Y.; Gu, S.; Xu, W.; Yang, H. Effect of Temperature on the Morphology of Poly (Lactic Acid) Porous Membrane Prepared via Phase Inversion Induced by Water Droplets. *Int. J. Biol. Macromol.* **2019**, *133*, 902–910. [CrossRef]
41. Yang, H.; Ye, Q.; Zhou, Y.; Xiang, Y.; Xing, Q.; Dong, X.; Wang, D.; Xu, W. Formation, Morphology and Control of High-Performance Biomedical Polyurethane Porous Membranes by Water Micro-Droplet Induced Phase Inversion. *Polymer* **2014**, *55*, 5500–5508. [CrossRef]
42. Zhu, R.; Wang, D.; Liu, Y.; Liu, M.; Fu, S. Bifunctional Superwetting Carbon Nanotubes/Cellulose Composite Membrane for Solar Desalination and Oily Seawater Purification. *Chem. Eng. J.* **2022**, *433*, 133510. [CrossRef]
43. Wang, Y.; Qi, Q.; Fan, J.; Wang, W.; Yu, D. Simple and Robust MXene/Carbon Nanotubes/Cotton Fabrics for Textile Wastewater Purification via Solar-Driven Interfacial Water Evaporation. *Sep. Purif. Technol.* **2021**, *254*, 117615. [CrossRef]
44. Zhao, Y.; Wu, M.; Shen, P.; Uytterhoeven, G.; Mamrol, N.; Shen, J.; Gao, C.; Van der Bruggen, B. Composite Anti-Scaling Membrane Made of Interpenetrating Networks of Nanofibers for Selective Separation of Lithium. *J. Membr. Sci.* **2021**, *618*, 118668. [CrossRef]
45. Son, M.; Cho, K.H.; Jeong, K.; Park, J. Membrane and Electrochemical Processes for Water Desalination: A Short Perspective and the Role of Nanotechnology. *Membranes* **2020**, *10*, 280. [CrossRef]
46. Ye, B.; Liu, H.; Ye, M.; Zeng, C.; Luo, H.; Liu, G.; Zhang, R.; Huang, H. Seawater Desalination Using the Microbial Electrolysis Desalination and Chemical-Production Cell with Monovalent Selective Cation Exchange Membrane. *Desalination* **2022**, *523*, 115394. [CrossRef]
47. Guillen, G.R.; Pan, Y.; Li, M.; Hoek, E.M.V. Preparation and Characterization of Membranes Formed by Nonsolvent Induced Phase Separation: A Review. *Ind. Eng. Chem. Res.* **2011**, *50*, 3798–3817. [CrossRef]
48. Guillen, G.R.; Ramon, G.Z.; Kavehpour, H.P.; Kaner, R.B.; Hoek, E.M.V. Direct Microscopic Observation of Membrane Formation by Nonsolvent Induced Phase Separation. *J. Membr. Sci.* **2013**, *431*, 212–220. [CrossRef]
49. Menut, P.; Su, Y.S.; Chinpa, W.; Pochat-Bohatier, C.; Deratani, A.; Wang, D.M.; Huguet, P.; Kuo, C.Y.; Lai, J.Y.; Dupuy, C. A Top Surface Liquid Layer during Membrane Formation Using Vapor-Induced Phase Separation (VIPS)—Evidence and Mechanism of Formation. *J. Membr. Sci.* **2008**, *310*, 278–288. [CrossRef]
50. Gao, J.; Li, W.; Wong, J.S.-P.; Hu, M.; Li, R.K.Y. Controllable Morphology and Wettability of Polymer Microspheres Prepared by Nonsolvent Assisted Electrospraying. *Polymer* **2014**, *55*, 2913–2920. [CrossRef]
51. Zang, L.; Sun, L.; Zhang, S. Nanofibrous Hydrogel-Reduced Graphene Oxide Membranes for Effective Solar-Driven Interfacial Evaporation and Desalination. *Chem. Eng. J.* **2021**, *422*, 129998. [CrossRef]

Article

Metal-Coordinated Nanofiltration Membranes Constructed on Metal Ions Blended Support toward Enhanced Dye/Salt Separation and Antifouling Performances

Xiaofeng Fang [1,2,*], Shihao Wei [1], Shuai Liu [1], Ruo Li [1], Ziyi Zhang [1], Yanbiao Liu [1,3], Xingran Zhang [1], Mengmeng Lou [1], Gang Chen [1,3] and Fang Li [1,3,*]

[1] Textile Pollution Controlling Engineering Centre of Ministry of Ecology and Environment, College of Environmental Science and Engineering, Donghua University, Shanghai 201620, China; wei20211226@163.com (S.W.); liushuai5274@163.com (S.L.); 18379467166@163.com (R.L.); zzy13966616646@163.com (Z.Z.); yanbiaoliu@dhu.edu.cn (Y.L.); xrzhang@dhu.edu.cn (X.Z.); mengmeng_lou@outlook.com (M.L.); cheng@dhu.edu.cn (G.C.)
[2] Key Laboratory of New Membrane Materials, Ministry of Industry and Information Technology, Nanjing University of Science & Technology, Nanjing 210094, China
[3] Shanghai Institute of Pollution Control and Ecological Security, Shanghai 200092, China
* Correspondence: fxf595@dhu.edu.cn (X.F.); lifang@dhu.edu.cn (F.L.)

Abstract: Metal-phenol coordination is a widely used method to prepare nanofiltration membrane. However, the facile, controllable and scaled fabrication remains a great challenge. Herein, a novel strategy was developed to fabricate a loose nanofiltration membrane via integrating blending and interfacial coordination strategy. Specifically, iron acetylacetonate was firstly blended in Polyether sulfone (PES) substrate via non-solvent induced phase separation (NIPS), and then the loose selective layer was formed on the membrane surface with tannic acid (TA) crosslinking reaction with Fe^{3+}. The surface properties, morphologies, permeability and selectivity of the membranes were carefully investigated. The introduction of TA improved the surface hydrophilicity and negative charge. Moreover, the thickness of top layer increased about from ~30 nm to 119 nm with the increase of TA assembly time. Under the optimum preparation condition, the membrane with assembly 3 h (PES/Fe-TA3h) showed pure water flux of 175.8 L·m^{-2}·h^{-1}, dye rejections of 97.7%, 97.1% and 95.0% for Congo red (CR), Methyl blue (MB) and Eriochrome Black T (EBT), along with a salt penetration rate of 93.8%, 95.1%, 97.4% and 98.1% for Na_2SO_4, $MgSO_4$, $NaCl$ and $MgCl_2$ at 0.2 MPa, respectively. Both static adhesion tests and dynamic fouling experiments implied that the TA modified membranes showed significantly reduced adsorption and high FRR for the dye solutions separation. The PES/Fe-TA3h membrane exhibited high FRR of 90.3%, 87.5% and 81.6% for CR, EBT and MB in the fouling test, stable CR rejection (>97.2%) and NaCl permeation (>94.6%) in 24 h continuous filtration test. The combination of blending and interfacial coordination assembly method could be expected to be a universal way to fabricate the loose nanofiltration membrane for effective fractionation of dyes and salts in the saline textile wastewater.

Keywords: nanofiltration; metal-coordination; polyphenol; dye/salt separation; antifouling

Citation: Fang, X.; Wei, S.; Liu, S.; Li, R.; Zhang, Z.; Liu, Y.; Zhang, X.; Lou, M.; Chen, G.; Li, F. Metal-Coordinated Nanofiltration Membranes Constructed on Metal Ions Blended Support toward Enhanced Dye/Salt Separation and Antifouling Performances. *Membranes* **2022**, *12*, 340. https://doi.org/10.3390/membranes12030340

Academic Editor: Hongjun Lin

Received: 25 January 2022
Accepted: 24 February 2022
Published: 18 March 2022

Publisher's Note: MDPI stays neutral with regard to jurisdictional claims in published maps and institutional affiliations.

Copyright: © 2022 by the authors. Licensee MDPI, Basel, Switzerland. This article is an open access article distributed under the terms and conditions of the Creative Commons Attribution (CC BY) license (https://creativecommons.org/licenses/by/4.0/).

1. Introduction

Due to the rapid development of the textile industry, a large amount of wastewater is produced and discharged [1,2]. The textile wastewater generally consists of dyes, inorganic salts and other chemicals [3]. The discharge of such textile wastewater has negative effects on aquatic ecosystems and public health due to the features of highly toxic and bio-accumulation of dyes [4,5]. It is noteworthy that the existence of salt impedes textile wastewater from biodegrading. In addition, these inorganic salts are also a recyclable resource in textile wastewater [6]. Therefore, separating salt/dye mixture is a critical step to reuse the inorganic salts and polluted textile wastewater [7].

Membrane separation technology is deemed to be an effective way for treating textile wastewater, owing to its small footprint, low energy consumption and high selectivity [8–13]. Typically, nanofiltration (NF) has become one optimal choice for removing organic matters with a molecular weight of 200–1000 Da. However, most commercial NF membranes with a dense separation layer exhibit high salts rejection and low permeability [14,15]. Thus, it is unsatisfactory for separating dye/salt in textile wastewater to recycle the resources. To overcome this problem, the loose nanofiltration membrane (LNM) has recently drawn intense attention to achieve the effective fractionation of dye and salt [16–19]. Compared with traditional NF membranes, LNMs possess a relatively looser structure and larger pore size, which promote water and salt permeation. The organic dyes are rejected by the combination of size exclusion and electrostatic repulsion [20,21]. For the dyeing wastewater treatment, LNMs have high efficiency and economic value for separating organic pollutants from inorganic salts [3,22,23]. During the past few years, various LNMs and their separation mechanisms have been reported [24–26]. The pore size distribution and surface properties of the separation layer are pivotal factors to enhance the separation efficiency for the dye/salt mixture. In view of the academic and practical development of LNMs, the rational design and precise manipulation of the separation layer is quite challenging.

Polyphenol chemistry, including metal-phenol coordination and amine-phenol deposition, has received significant attention as effective tools for the preparation of the separation membrane [27–29]. Tannic acid (TA), a natural plant polyphenol with abundant catechol and pyrogallol groups, has been widely used in the preparation and modification of membranes, because of the presence of abundant active groups and low-cost properties [30]. The catechol and pyrogallol groups can crosslink with metal ions and organic molecules to form the complex. The application of a TA-based complex for membrane separation has been investigated in some meaningful research works [31,32]. For instance, Li et al. [33] have demonstrated the co-deposition of TA and PEI to prepare LNF membrane-selective layers. Shao et al. [34] developed the novel metal-TA network to prepare high-flux nanofiltration membranes. Wu et al. [35] have fabricated a low-pressure nanofiltration membrane via a bio-inspired one-pot assembly on the polysulfone (PSf) substrate with a tannic acid-titanium (TA-Ti) network coating as the selective layer. Fan et al. [36] have reported the preparation of an LNF membrane via coordination complexes of TA and iron (III) ions. Peinemann et al. [37] reported a facile and cost-effective co-deposition method to prepare NF membranes via the complexation of TA and copper. Shen et al. [38] used the TA and Fe as the two-phase monomers to fabricate metal-organic composite membrane via the interfacial coordination method. The obtained LNMs in the above-mentioned works exhibited both high dye rejection and salt permeation. However, these preparation methods, such as co-deposition process and biphasic interfacial coordination, are uncontrollable due to the rapid reactions in the mixed system and the aggregated particles are easy to form on the substrate surface. In addition, the stability of the separation layer should be considered, since that there is no direct chemical bonding between the TA layer and the substrate in most systems. Additionally, while most of these studies were performed at the laboratory scale, the scale-up production of NF membranes with polyphenol chemistry is still difficult. Therefore, a facile, precision controllable and widely applicable strategy for the LNM construction with precise dye/salt separation is still needed.

In this work, a novel strategy was developed to fabricate a loose separation layer via integrating blending and interfacial coordination. For the metal-phenol coordination coating, the metal ion source is a critical point. Introduction of the metal ions into the membrane substrate, as the active sites for coordinative reaction, is supposed to be a simple and efficient method. Specifically, the membrane substrate was firstly prepared by blending PES with iron acetylacetonate (Fe(acac)$_3$), as the Fe (III) source, via non-solvent induced phase separation (NIPS). The hydrophobic chain of acetylacetonate can intertwist with the PES molecular chain, which increase the stability of Fe(acac)$_3$ in the membrane matrix. Subsequently, TA was introduced on the surface and interface of PES/Fe substrate by crosslinking reaction with Fe^{3+}, forming the loose selective layer. The thickness of the selective layer was varied with the assembly time. The surface chemical properties, membrane structures and separation performance were evaluated in detail. The optimized PES/Fe-TA membrane displayed satisfactory water permeance, high dye/salt fractionation efficiency and excellent antifouling properties towards dye/salts mixtures. This work provides a facile and scalable production strategy to construct the loose NF membrane, which has great potential for industrial application.

2. Materials and Methods

2.1. Chemicals and Materials

Polyethersulfone (PES, Ultrason E6020P with M_w = 58 kDa) was bought from BASF Co., Ltd. (Shanghai, China).and dried at 110 °C for 12 h before use. Tannic acid (TA, 99%) and iron (III) acetylacetonate (Fe(acac)$_3$) were purchased from Aladdin Reagent Co. Ltd. (Shanghai, China). Polyvinylpyrrolidone (PVP, 99%) and N, N-dimethyl formamide (DMF, 99%), Congo Red (CR, 99%), Methyl blue (MB, 99%), Eriochrome Black T (EBT, 99%), Acid Orange74 (AO74, 99%), magnesium chloride (MgCl$_2$), magnesium sulfate (MgSO$_4$), sodium sulfate (Na$_2$SO$_4$) and sodium chloride (NaCl) were purchased from Sinopharm Chemical Reagent Co., Ltd. (Shanghai, China).

2.2. Membrane Fabrication

The PES/Fe membranes were prepared via non-solvent induced phase separation (NIPS). In detail, PES, PVP and a certain weight of iron acetylacetone (0%, 0.5%, 1.0%, 1.5% and 2.0 wt%) were dispersed in DMF solution and stirred at 60 °C for 6 h to obtain a uniform casting solution. The casting solutions were stored at room temperature for 12 h to ensure a complete release of bubbles and then cast on non-woven fabric using an automated film applicator with a gap of 320 μm. Subsequently, the cast films were immersed into a coagulation bath at room temperature after being exposed to the atmosphere for 20 s. Then, the prepared membranes were immersed in pure water for at least 24 h to leach out the residual solvent before using.

The PES/Fe-TA nanofiltration membranes were prepared via the coordination reaction between TA and Fe. Firstly, the cleaned PES/Fe membranes were immersed in a TA solution (10.0 g/L) and oscillated at different times (0, 1, 2, 3 and 4 h) with a shaker. The PES/Fe-TA nanofiltration membranes were then thoroughly washed with deionized water to remove the unreacted TA. Before testing, the PES/Fe-TA nanofiltration membranes were stored in deionized water.

2.3. Membrane Characterization

The chemical structures and elemental compositions of these NF membranes were analyzed by Fourier transform infrared spectroscopy (ATR-FTIR, Nicolet 6700, TMO, Waltham, MA, USA) and X-ray photoelectron spectroscopy (XPS), respectively. The field emission scanning electron microscopy (FESEM, Hitachi S4800, Tokyo, Japan) was operated to characterize the morphology of the NF membranes. The hydrophilicity of these membranes was characterized by a contact angle goniometer (SL-200C, KINO, Boston, MA, USA). The surface zeta potential of membrane was measured by a Sur-PASS electrokinetic analyzer (SurPASS, Anton Paar GmbH, Graz, Austria). Thermogravimetric analysis (TGA, METTLER TGA SF, Mettler Toled, Switzerland) was conducted with a heating rate of 10 °C/min from room temperature to 700 °C under 100 mL/min in air atmosphere.

2.4. Filtration Performance

The NF performance of these membranes was tested by a commercial laboratory scale cross-flow flat membrane module with an effective area of 7.065 cm^2 at room temperature. A schematic diagram of the experimental set up is shown in Figure 1. The membranes were initially compacted for 20 min under 0.3 MPa to obtain a steady permeation and then the pressure was lowered to 0.2 MPa. The water flux (J, $L \cdot m^{-2} \cdot h^{-1}$) was measured and calculated by the following equation:

$$J = \frac{V}{A \times \Delta t} \quad (1)$$

where V (L) is the volume of permeated water, A (m^2) is the effective membrane area and Δt (h) is the permeation time.

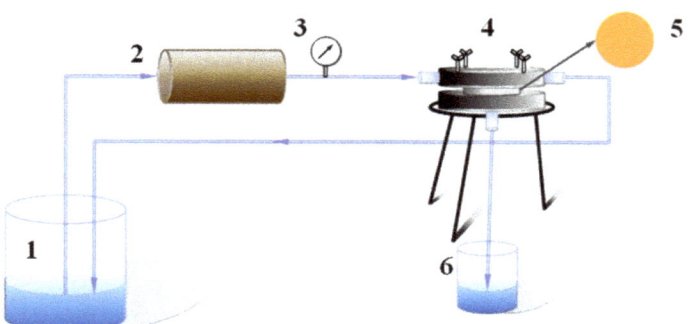

Figure 1. Schematic diagram of cross-flow experimental device: 1. Feed liquid; 2. Peristaltic pump; 3. Pressure gauge; 4. Membrane assembly; 5. Measured film; 6. Penetrating fluid.

The separation performance of these NF membranes was conducted by using 1 g/L salt solution (Na$_2$SO$_4$, MgSO$_4$, MgCl$_2$ and NaCl) and 0.1 g/L dye solution (CR, MB, EBT and AO74) as feed solution, respectively. Furthermore, the CR solution (0.1 g/L) mixed with different concentration (2, 4, 6, 8 and 10 g/L) of NaCl solution were also used as feed solution to judge the dye/salt mixture fraction ability. The rejection ratio (R) was defined by the following equation:

$$R = 1 - \frac{C_p}{C_f} \quad (2)$$

where C_p and C_f is the concentration of permeate and feed solution, respectively. Herein, the salt concentration was measured by an electrical conductivity. The dye concentration was measured by a UV-vis spectrophotometer. The maximum absorption wavelength of CR, MB, EBT and AO74 are 488 nm, 664 nm, 410 nm and 484 nm, respectively. All flux and rejection measurements were conducted using three membrane samples.

2.5. Antifouling Performance
2.5.1. Static Adsorption Tests

The antifouling measurements of the NF membrane were conducted using CR, MB and EBT as representative pollutants. For the static adsorption tests, the membrane samples (7.56 cm^2) were immersed in dye solutions (0.1 g/L, C_i) for 2 h. Equilibrium concentrations of dye (C_e) were measured by UV-vis spectrophotometry. The adsorbed mass of dye per unit area of membrane (Q, $\mu g cm^{-2}$) was calculated using Equation (3):

$$Q = \frac{(C_i - C_e)V}{A} \quad (3)$$

where A is the effective membrane area (cm^2), V is the volume of dye solution (mL) and C_i and C_e are the initial and equilibrium dye concentrations (g/L), respectively.

2.5.2. Dynamic Fouling Experiments

In the dynamic antifouling test, 0.1 g/L CR, EBT and MB solutions were used as representative pollutants, respectively. The antifouling filtration experiments were conducted at 0.2 MPa and room temperature. The antifouling test process is as follows: Firstly, the membrane sample was pressurized to reach a stable water flux before the measurement. Then, the pure water flux (J_{w1}) was continuously measured for 60 min and recorded every 10 min. Afterwards, the membrane filtration was conducted using dye solution as feed solution for another 60 min. The permeate flux of CR, EBT or MB solution (J_p) was also recorded every 10 min. Subsequently, the membrane was cleaned with distilled water for 30 min. Finally, the pure water flux (J_{w2}) was measured again for 60 min. The water fluxes were calculated by Equation (1).

The antifouling properties was further evaluated by flux recovery ratio (FRR), total fouling ratio (Rt), reversible fouling ratio (Rr) and irreversible fouling ratio (Rir). Those parameters were defined and calculated as follows:

$$\text{FRR} = \frac{J_{w2}}{J_{w1}} \times 100\% \tag{4}$$

$$\text{Rt} = \left(1 - \frac{J_p}{J_{w1}}\right) \times 100\% \tag{5}$$

$$\text{Rr} = \frac{J_{w2} - J_p}{J_{w1}} \times 100\% \tag{6}$$

$$\text{Rir} = \left(1 - \frac{J_{w2}}{J_{w1}}\right) \times 100\% \tag{7}$$

2.6. Long-Term Stability of the Membrane

To evaluate the long-term stability of optimized NF membranes, the CR (0.1 g/L) solutions mixed with NaCl (2 g/L) were used as feed solution to filtrated for 24 h. The permeate flux and rejections for CR and NaCl were monitored by the aforementioned methods.

3. Results and Discussion

3.1. Chemical Structure and Properties of Membranes

The fabrication process of LNM combined blending and interfacial coordination strategy was depicted in Figure 2. For the control, the pristine PES membrane was also fabricated using the NIPS technique. The photographs of the PES, PES/Fe and PES/Fe-TA membrane are shown in Figure 3a. It can be observed that the pristine PES membrane shows a white color and the PES/Fe membrane exhibits an orange color. The color change is ascribed to the color of iron acetylacetonate, suggesting the successful incorporation of Fe^{3+} on the membrane matrix. After the immersion of TA solution, the PES/Fe-TA membrane displays dark grey, demonstrating the TA coating is assembled on the membrane surface. In order to further confirm the existence of Fe and TA on PES membrane, the TGA and FTIR analysis were studied. Figure 3b exhibits the results of TGA analysis for the PES, PES/Fe and PES/Fe-TA membrane. The residual weights of PES/Fe and PES/Fe-TA membrane were 9.8% and 10.2%, while the pristine PES membrane was completely burned out in the air atmosphere. The increase of residual mass should correspond to the iron base compound in the PES/Fe and PES/Fe-TA membranes, providing further evidence of the iron complex loading. FT-IR was employed to analyze the surface chemical structure of PES, PES/Fe and PES/Fe-TA membranes, as shown in Figure 3c. The absorption peaks at 1150 cm^{-1} and 1296 cm^{-1} are the symmetric and asymmetric stretching vibrational peaks of the S=O functional group in PES. The stretching vibration peak between benzene ring and S in PES is located at 1100 cm^{-1}. Compared with the PES and PES/Fe membrane, an additional adsorption band at 1720 cm^{-1} can be observed in the spectrum of the PES/Fe-

TA membrane. This can be ascribed to the C=O stretching vibrations of the of TA [39], suggesting that the TA were successfully incorporated on PES/Fe membrane surface.

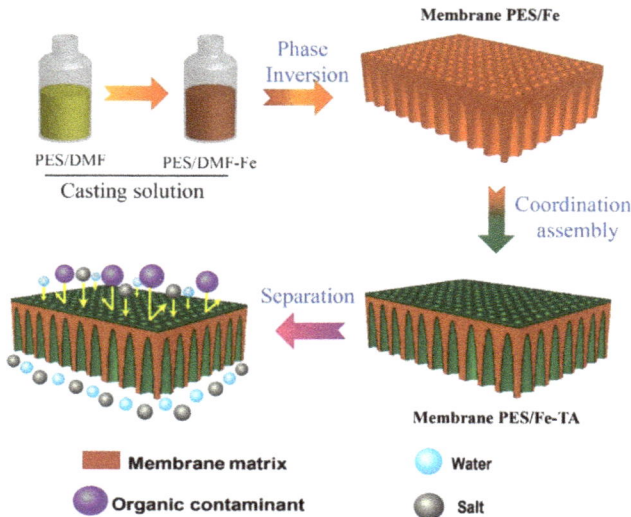

Figure 2. Schematic representation of the fabrication process of loss nanofiltration membrane and selective separation of dye and salt.

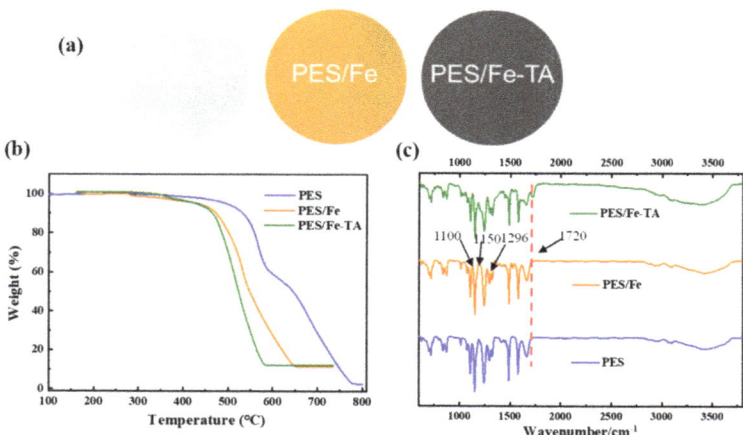

Figure 3. (**a**) Digital photographs of membrane surface, (**b**) TGA curves and (**c**) ATR-FTIR spectra of the PES, PES/Fe and PES/Fe-TA membrane.

Hydrophilicity of membrane is a vital parameter affecting membrane permeability and antifouling performance during filtration applications [40]. The surface hydrophilicity of the pristine PES, PES/Fe and PES/Fe-TA membranes was evaluated by dynamic water contact angle (WCA) measurements, and the results are shown in Figure 4a. The pristine PES membrane exhibited high hydrophobicity with the initial WCA of 75°, and almost remained unchanged within 60 s. For the PES/Fe membrane, the WCA is slightly higher than that of the PES membrane, which is ascribed to the low surface energy of Fe(acac)$_3$. However, the initial WCA of PEA/Fe-TA membrane decreased to around 49.8°

and dramatically declined to 27.5° in 60 s, indicating the improved hydrophilicity after TA assembly. This can be attributed to the hydrophilic phenolic hydroxyl groups formed on the surface of the PES/Fe-TA membrane. Moreover, the surface charge also plays a significant role in the separation properties of membranes. The surface charge of PES, PES/Fe and PES/Fe-TA membrane are studied by the surface zeta potential (Figure 4b). It can be seen that PES/Fe-TA membrane displays enhanced negative charge compared with PES and PES/Fe membranes. That is because TA had many phenolic hydroxyl groups which could release hydrogen ions to endow membrane surfaces with negative charge. As the pH increases, more phenolic hydroxyl groups deprotonate, resulting in a stronger negative charge.

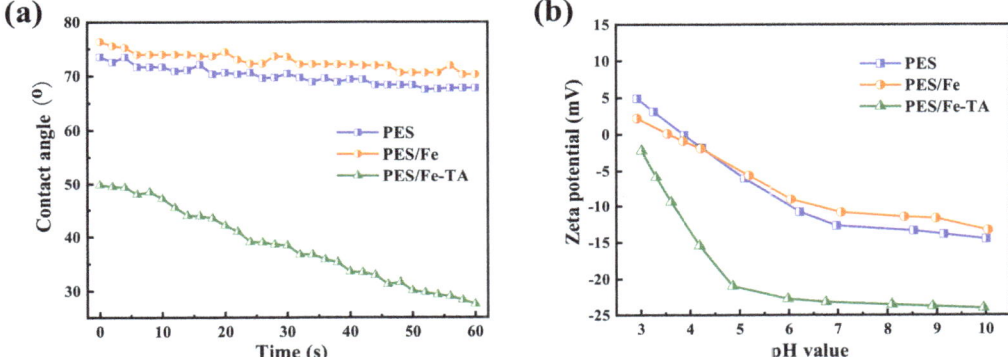

Figure 4. (a) The water contact angle and (b) zeta potentials of the PES, PES/Fe and PES/Fe-TA membranes.

3.2. Effects of TA Assembly Time on the Membrane Structure and Performance

The reaction time is an important factor for TA deposition on the PES/Fe substrate. To regulate the TA layer in a controllable thickness, the effects of assembly time on the membrane structure and performance were studied. Due to the instability of Fe(acac)$_3$ in ethanol, the PES/Fe sample has not been dried by supercritical drying apparatus, which means that its morphologies cannot be shown for control. The morphologies of PES/Fe-TA membranes treated at different TA assembly times were inspected by SEM, as shown in Figure 5. The membranes with TA coordination assembly have a flat and smooth surface. Since the interaction between Fe^{3+} in membrane matrix and TA can effectively suppress the TA-Fe complex particles on the membrane surface, a relatively smooth surface was observed on the PES/Fe-TA membranes. In addition, many pores (pore size of 10–50 nm) were observed on the surface of PES/Fe-TA1h membrane and these pores became lesser and smaller with the increase of TA assembly time (Figure 5a–d). This result was attributed to the assembly of TA on the PES/Fe substrate, forming a uniform Fe-TA complex layer on the membrane surface and reducing the pore sizes.

The cross-section images in Figure 5e–h show that a top layer formed on the PES/Fe support after TA assembly. Moreover, the thickness of the top layer increased from ~30 nm to 119 nm with the increase of the TA assembly time. The Fe^{3+} migrated to the membrane surface and coordinated with TA to form the thicker separation layer with the increase of the TA assembly time. Therefore, the interfacial coordination of polyphenolic layer can form a smooth surface and controllable thickness of separation layer by varying the TA assembly time, which might decide the separation performance of the membrane.

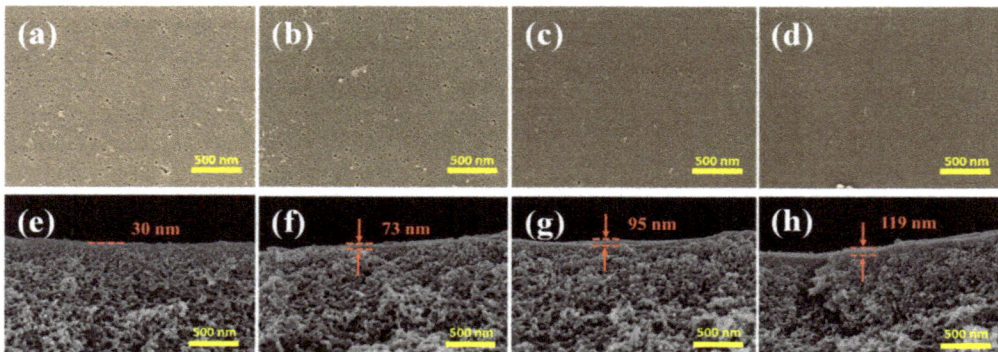

Figure 5. The surface (**a**–**d**) and cross section morphology (**e**–**h**) of PES/Fe-TA membranes with different TA assembly times.

The pure water flux and CR rejection of PES/Fe-TA membranes with different TA assembly times is shown in Figure 6. It can be seen that the pure water permeance of the PES/Fe-TA membrane gradually decreased, while the rejections of CR increased with the increase of assembly time. The water permeance decreased from 305.2 L·m^{-2}·h^{-1} to 133.6 L·m^{-2}·h^{-1}, and the CR rejection increased from 89.7% to 98.1%. These results could be attributed to the fact that a thicker separation layer was formed on the membrane surface with the increase of TA assembly time, leading to the smaller pore size and increased permeation resistance, as shown in Figure 5. Comprehensively considering the water flux and rejection, the assembly time was fixed to 3 h for the following tests. For the PES/Fe-TA3h membrane, the water permeance reached 175.8 L·m^{-2}·h^{-1} and the rejection rate of CR was 97.7% at 0.2 MPa.

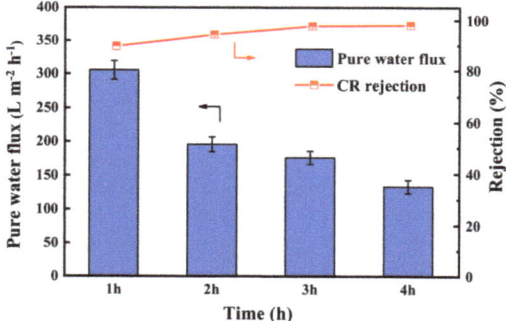

Figure 6. The pure water flux and CR rejection of PES/Fe-TA membrane at different TA assembly times.

3.3. Nanofiltration Performance

The nanofiltration properties of PES/Fe-TA3h membrane were measured by using four dyes (CR, MB, EBT and AO74) and four inorganic salts (Na$_2$SO$_4$, MgSO$_4$, MgCl$_2$ and NaCl). The flux and rejections of the PES/Fe-TA3h membrane for different dyes and inorganic salts were measured in a cross-flow filtration apparatus under 0.2 MPa. Figure 7a,b presented the results of nanofiltration performance for a single component of dye (0.1 g/L) or salt (2 g/L). The rejections to CR, MB, EBT and AO 74 was 97.7%, 97.1%, 95.0% and 58.6%, while the fluxes were 80.1, 70.0, 93.5 and 133.5 L·m^{-2}·h^{-1}, respectively. The difference of rejections for the dyes may be ascribed to the molecular size. The relative molecular weight of AO

74 was 417.35 g/mol, which is lower than 696.66, 799.80 and 461.38 g/mol for CR, MB and EBT; thus, they can more easily pass through the membrane pores when permeating the membrane. The retentions of Na_2SO_4, $MgSO_4$, NaCl and $MgCl_2$ was 6.2%, 4.9%, 2.6% and 1.9% respectively, which conforms to the typical negatively charged NF membranes (Figure 7b). The rejections of dyes and inorganic salts are decided with the coaction of steric and Donnan effects [41,42]. For this result, the PES/Fe-TA membrane is only used for dyes with a molecular weight of a least 700 Da. The SO_4^{2-} has stronger repulsive interaction with membrane surface than that with Cl^-, resulting in higher rejections for Na_2SO_4 and $MgSO_4$. The high rejections of dyes may be ascribed to the presence of a hydration shell of charged dye molecules and/or aggregates of dye molecules with a size of tens of nanometers. However, the hydrate radius (<0.5 nm) of salts is much smaller than the membrane pore size, due to the dominant role of sieve principle, and the rejection of salts was low. High dye rejection ability and weak salt rejection ability proved that the prepared PES/Fe-TA membrane can be applied to dye desalination.

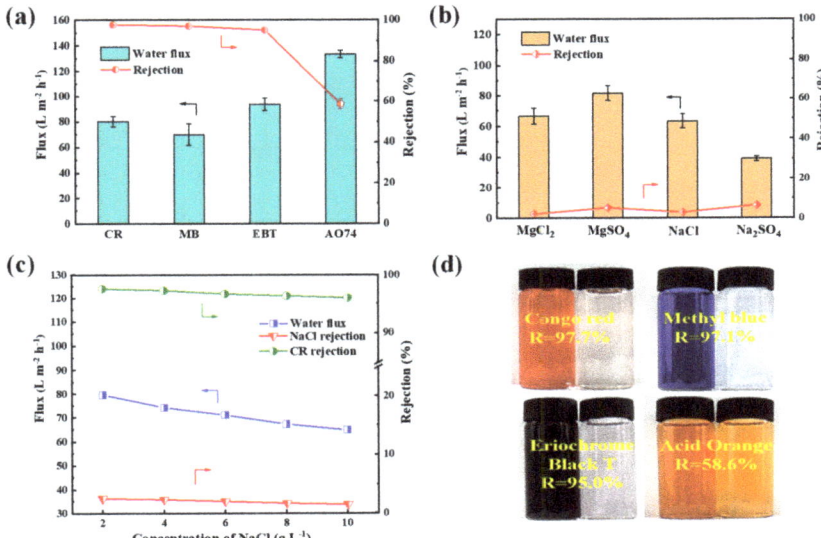

Figure 7. (**a**) Filtration performance of single dye solution, (**b**) rejections of salts, (**c**) filtration performance for the dye/salt mixture solution for the PES/Fe-TA3h membrane and (**d**) photographs of feed and penetration solutions.

It is believed that the presence of salt in the dye solution has some effects on the membrane separation performance; therefore, the separation properties of dye and salt mixture was investigated. The 0.1 g/L CR with different NaCl content were used to form different salinity feed solution. The results of permeate flux and solute rejections for the CR/NaCl mixture are shown in Figure 7c. It can be seen that the permeability of CR/NaCl mixture solution decreased from 79.6 $L \cdot m^{-2} \cdot h^{-1}$ to 64.8 $L \cdot m^{-2} \cdot h^{-1}$. Moreover, the rejections of CR and NaCl also decreased, from 97.6% to 96.0% and from 2.5% to 1.45% respectively, with the increase of NaCl concentration (from 2 g/L to 10 g/L). This was because salt tends to disperse dye molecules uniformly in the mixed solution and avoid the aggregation of dye molecules, resulting in smaller dye particles to permeate through the membrane pores. Meanwhile, the dye adsorbed on the membrane pores and the concentration polarization occurred on the membrane surface, which resulted in decreased permeability and dye rejection.

In order to highlight the prominent properties of the membranes prepared in this work, we compared the water flux, dye and salt rejections of PES/Fe-TA3h membrane with those reported in other research (Table 1). It can be seen that the LNM prepared in this study showed good dye/salt separation capability compared to the results reported in the selected literature.

Table 1. Comparison of the performance of the NF membranes in the literature.

Membranes	Permeate Flux (L m^{-2} h^{-1})	Congo Red C(g/L)	Congo Red R(%)	NaCl C(g/L)	NaCl R(%)	Pressure (MPa)	Ref.
TA/GOQDs-1	23.3	0.1	99.8	1	17.2	0.2	[43]
TiO$_2$-HMDI	30.5	0.035	97.4	1	2.7	0.2	[44]
PSF/GO	73.7–95.4	0.1	99.9	1	<5	0.2	[45]
PAN-PEI-GA	51.0	0.1	97.1	1	5	0.2	[46]
PAN-DR80	113.6	0.1	99.8	1	12.4	0.4	[47]
Fe(III)-phos-(PEI)/HPAN	12.1	0.1	99.5	1	7.5	0.2	[48]
CaCO$_3$/PEI-GA	141	0.1	99.6	1	6.9	0.3	[49]
PST-1	52.3	0.1	99.0	1	<7	0.6	[50]
TAIP M4	31.5	0.2	99.4	2	5.4	0.1	[51]
PDA/SBMA/HPAN	68.8	0.5	98.2	1	5.0	0.4	[52]
LNFM-2	212.9	0.2	99.6	1	5.6	0.4	[25]
PES/Fe-TA3h	77.0	0.1	97.7	2	2.6	0.2	This work

3.4. Antifouling Properties

Membrane fouling is one of trickiest problems in membrane processes and it results in many drawbacks, such as permeance decline, increase in operational costs and membrane degeneration. The TA-Fe complex was super-hydrophilicity and expected to overcome the fouling problem of nanofiltration membrane for the separation dye solution. The antifouling property of the PES/Fe-TA membrane was evaluated with static adsorption and dynamic filtration experiments using CR, EBT and MB as the model dye pollutants. The results of static adsorption experiments for three dyes on PES, PES/Fe and PES/Fe-TA membranes are shown in Figure 8. It can be seen that the PES/Fe-TA membrane has the smallest adsorption capacity for the three dyes, compared with PES and PES/Fe membranes. The adsorption amounts of CR, EBT and MB are, respectively, 0.018, 0.026 and 0.043 mg/cm^2 for the PES/Fe-TA membrane compared to 0.25, 0.22 and 0.33 mg/cm^2 for the pristine PES membrane. Figure 8b shows the surface colors of PES, PES/Fe and PES/Fe-TA3h membrane with static adsorption of the three dyes. We can clearly see that the membrane colors have changed to the dye color after adsorption. Moreover, the adsorption behavior for the PES and PES/Fe membrane are more serious than for the PES/Fe-TA membrane. This phenomenon was attributed to the corporation of the hydrophilicity and the charge repulsion to negative dyes. The TA on the PES/Fe-TA membrane make it more difficult for the dyes to adhere to the membrane surface, which can enhance the dye pollution resistance in the filtration dye solution.

In order to further evaluate the antifouling property of PES/Fe-TA membrane, the dynamic cyclic filtration experiment with different dye solutions was conducted and the results are revealed in Figure 9a. It can be seen that the permeate flux of the dye solution are lower than the pure water. This is probably caused by the dye fouling and concentration polarization. Moreover, the water flux recovers by the water/ethanol dilute solution cleaning treatment after each cycle of dye filtration test. In addition, the values of J_{w1}, J_p and J_{w2} measured in the two cycles were used to calculate FRR, Rt, Rr and Rir to evaluate the anti-fouling properties of the PES/Fe-TA membrane, as shown in Figure 9b. After two cycles, the FRR values of CR, EBT and MB were 90.3%, 87.5% and 81.6%, respectively, and the corresponding R_t values were 62.6%, 57.3% and 65.7%, respectively. It shows that the PES/Fe-TA3h membrane has excellent antifouling property on CR, EBT and MB. Based on the above results, the excellent antifouling ability of PES/Fe-TA3h membrane promotes its further application in dye desalination and dye wastewater treatment. These results clearly indicated that TA, indeed, acted as a strong dye adsorption resistance.

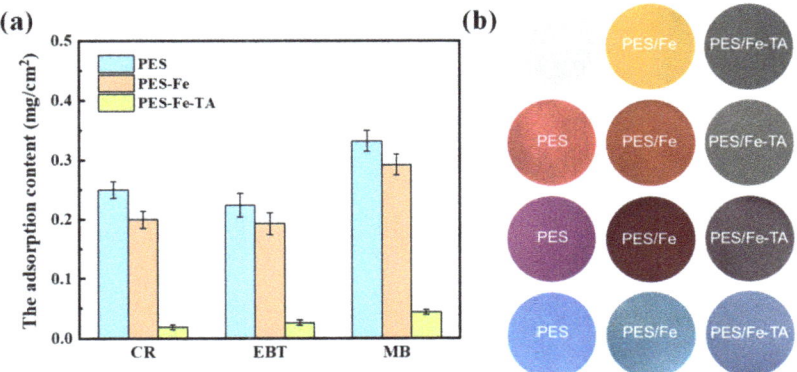

Figure 8. The dye adsorption content (**a**) and digital photos of surface color (**b**) on PES, PES/Fe and PES/Fe-TA3h membranes with static adsorption for different dyes.

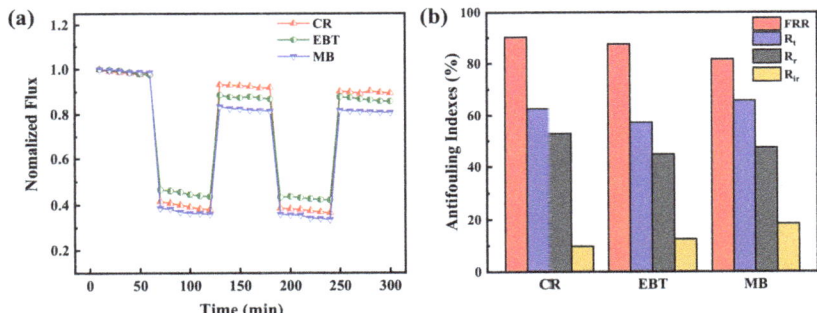

Figure 9. (**a**) The time-dependent normalized flux during the filtration of CR, EBT and MB solution and (**b**) antifouling indexes for the PES/Fe-TA3h membrane.

3.5. Long-Term Stability of the PES/Fe-TA Membrane

In order to explore the long-term stability of the membrane, the dye/salt fractionation performance of the PES/TA-Fe3h membrane was investigated with 24 h continuous filtration of the mixed solution (0.1 g/L CR and 2 g/L NaCl). As shown in Figure 10, the permeate flux slightly declined from 77.0 to 57.0 $L \cdot m^{-2} \cdot h^{-1}$ in the first few hours, which could be ascribed to the adsorption of CR and evolution of a dye cake layer on the membrane surface during the filtration. Moreover, the permeate flux stabilizes at 49.0 $L \cdot m^{-2} \cdot h^{-1}$ when the adsorption reaches dynamic equilibrium. Meanwhile, the rejections of CR and NaCl were increased slightly at the beginning, and then reached stability (>97.2% to CR and <5.4% to NaCl). The results confirmed that PES/Fe-TA3h membrane exhibited a long-term stability, which has great potential to be used in dye desalination and dye wastewater treatment.

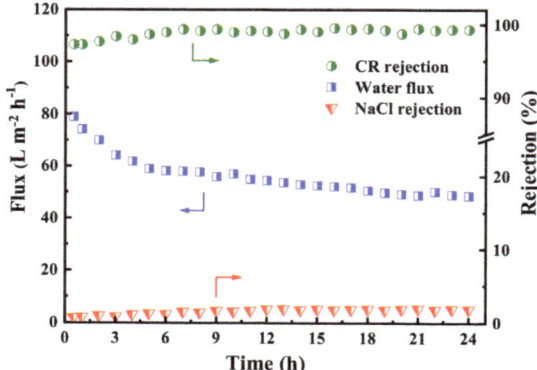

Figure 10. The long-term operation stability of the PES/Fe-TA3h membrane for the CR/NaCl mixture solution (feed: 0.1 g/L CR and 2 g/L NaCl).

4. Conclusions

In this study, a polyphenol-based loose nanofiltration membrane was successfully developed by integrating blending and interfacial coordination strategy. The iron ion complex was introduced in PES membrane via NIPS method and TA coordinated with Fe^{3+} forming the loose separation layer. The thickness of TA layer was controlled by altering the TA assembly time. The introduction of TA improved the surface hydrophilicity and negative charge. The optimized membrane with assembly 3 h (PES/Fe-TA3h) showed a pure water flux of 175.8 $L·m^{-2}·h^{-1}$, dye rejections of 97.7%, 97.1% and 95.0% for CR, MB and EBT, respectively, along with salt penetration rates of 93.8%, 95.1%, 97.4% and 98.1% for Na_2SO_4, $MgSO_4$, NaCl and $MgCl_2$, respectively, at 0.2 MPa. Moreover, the PES/Fe-TA3h membrane exhibited a stable dye rejection and salt permeation in the 24 h continuous test and high FRR of 90.3%, 87.5% and 81.6% for CR, EBT and MB, respectively. This study provides a new way to fabricate the loose nanofiltration membrane for effective fractionation of dyes and salts in the saline textile wastewater.

Author Contributions: X.F.: Conceptualization, Writing, Review, Investigation, Methodology, Visualization. S.W.: Methodology, Writing—original draft. S.L.: Investigation, Methodology, Data curation, Writing—original draft. M.L.: Methodology, Validation. R.L.: Writing—original draft. Z.Z.: Data curation. X.Z.: Visualization. G.C.: Formal analysis. Y.L.: Formal analysis. F.L.: Visualization, Validation. All authors have read and agreed to the published version of the manuscript.

Funding: This work was supported by the Shanghai Sailing Program (20YF1400100) and Fundamental Research Funds for the Central Universities (No. 2232020D-54).

Institutional Review Board Statement: Not applicable.

Informed Consent Statement: Informed consent was obtained from all subjects involved in the study.

Data Availability Statement: Not applicable.

Conflicts of Interest: The authors declare no conflict of interest.

References

1. Grant, S.B.; Saphores, J.D.; Feldman, D.L.; Hamilton, A.J.; Fletcher, T.D.; Cook, P.L.M.; Stewardson, M.; Sanders, B.F.; Levin, L.A.; Ambrose, R.F.; et al. Taking the "Waste" out of "Wastewater" for human water security and ecosystem sustainability. *Science* **2012**, *337*, 681–686. [CrossRef] [PubMed]
2. Guo, D.; You, S.; Li, F.; Liu, Y. Engineering carbon nanocatalysts towards efficient degradation of emerging organic contaminants via persulfate activation: A review. *Chin. Chem. Lett.* **2022**, *33*, 1–10. [CrossRef]
3. Lin, J.; Ye, W.; Zeng, H.; Yang, H.; Shen, J.; Darvishmanesh, S.; Luis, P.; Sotto, A.; Van der Bruggen, B. Fractionation of direct dyes and salts in aqueous solution using loose nanofiltration membranes. *J. Membr. Sci.* **2015**, *477*, 183–193. [CrossRef]

4. Cui, M.H.; Sangeetha, T.; Gao, L.; Wang, A.-J. Efficient azo dye wastewater treatment in a hybrid anaerobic reactor with a built-in integrated bioelectrochemical system and an aerobic biofilm reactor: Evaluation of the combined forms and reflux ratio. *Bioresour. Technol.* **2019**, *292*, 122001. [CrossRef] [PubMed]
5. Ren, Y.; Liu, Y.; Liu, F.; Li, F.; Shen, C.; Wu, Z. Extremely efficient electro-Fenton-like Sb(III) detoxification using nanoscale Ti-Ce binary oxide: An effective design to boost catalytic activity via non-radical pathway. *Chin. Chem. Lett.* **2021**, *32*, 2519–2523. [CrossRef]
6. Ye, W.; Liu, R.; Chen, X.; Chen, Q.; Lin, J.; Lin, X.; Van der Bruggen, B.; Zhao, S. Loose nanofiltration-based electrodialysis for highly efficient textile wastewater treatment. *J. Membr. Sci.* **2020**, *608*, 118182. [CrossRef]
7. Ji, D.; Xiao, C.; Zhao, J.; Chen, K.; Zhou, F.; Gao, Y.; Zhang, T.; Ling, H. Green preparation of polyvinylidene fluoride loose nanofiltration hollow fiber membranes with multilayer structure for treating textile wastewater. *Sci. Total Env.* **2021**, *754*, 141848. [CrossRef] [PubMed]
8. Gohil, J.M.; Ray, P. A review on semi-aromatic polyamide TFC membranes prepared by interfacial polymerization: Potential for water treatment and desalination. *Sep. Purif. Technol.* **2017**, *181*, 159–182. [CrossRef]
9. Ding, J.; Wu, H.; Wu, P. Preparation of highly permeable loose nanofiltration membranes using sulfonated polyethylenimine for effective dye/salt fractionation. *Chem. Eng. J.* **2020**, *396*, [CrossRef]
10. Zhang, X.; Ma, J.; Zheng, J.; Dai, R.; Wang, X.; Wang, Z. Recent advances in nature-inspired antifouling membranes for water purification. *Chem. Eng. J.* **2022**, *432*, 134425. [CrossRef]
11. Peng, L.E.; Yang, Z.; Long, L.; Zhou, S.; Guo, H.; Tang, C.Y. A critical review on porous substrates of TFC polyamide membranes: Mechanisms, membrane performances, and future perspectives. *J. Membr. Sci.* **2022**, *641*, 119871. [CrossRef]
12. Liu, Y.; Liu, F.; Ding, N.; Hu, X.; Shen, C.; Li, F.; Huang, M.; Wang, Z.; Sand, W.; Wang, C. Recent advances on electroactive CNT-based membranes for environmental applications: The perfect match of electrochemistry and membrane separation. *Chin. Chem. Lett.* **2020**, *31*, 2539–2548. [CrossRef]
13. Tijing, L.D.; Woo, Y.C.; Yao, M.; Ren, J. Electrospinning for Membrane Fabrication: Strategies and Applications. In *Comprehensive Membrane Science and Engineering*; Elsevier: Oxford, UK, 2017; ISBN 9780124095472.
14. Mohammad, A.W.; Teow, Y.H.; Ang, W.L.; Chung, Y.T.; Oatley-Radcliffe, D.L.; Hilal, N. Nanofiltration membranes review: Recent advances and future prospects. *Desalination* **2015**, *356*, 226–254. [CrossRef]
15. Zhang, Q.; Chen, S.; Fan, X.; Zhang, H.; Yu, H.; Quan, X. A multifunctional graphene-based nanofiltration membrane under photo-assistance for enhanced water treatment based on layer-by-layer sieving. *Appl. Catal. B Environ.* **2018**, *224*, 204–213. [CrossRef]
16. Van der Bruggen, B.; Curcio, E.; Drioli, E. Process intensification in the textile industry: The role of membrane technology. *J. Environ. Manag.* **2004**, *73*, 267–274. [CrossRef]
17. Guo, S.; Wan, Y.; Chen, X.; Luo, J. Loose nanofiltration membrane custom-tailored for resource recovery. *Chem. Eng. J.* **2021**, *409*, 127376. [CrossRef]
18. Xiao, Y.; Guo, D.; Li, T.; Zhou, Q.; Shen, L.; Li, R.; Xu, Y.; Lin, H. Facile fabrication of superhydrophilic nanofiltration membranes via tannic acid and irons layer-by-layer self-assembly for dye separation. *Appl. Surf. Sci.* **2020**, *515*, 146063. [CrossRef]
19. Guo, D.; Xiao, Y.; Li, T.; Zhou, Q.; Shen, L.; Li, R.; Xu, Y.; Lin, H. Fabrication of high-performance composite nanofiltration membranes for dye wastewater treatment: Mussel-inspired layer-by-layer self-assembly. *J. Colloid Interface Sci.* **2020**, *560*, 273–283. [CrossRef]
20. García Doménech, N.; Purcell-Milton, F.; Gun'ko, Y.K. Recent progress and future prospects in development of advanced materials for nanofiltration. *Mater. Today Commun.* **2020**, *23*, 100888. [CrossRef]
21. Ernst, M.; Bismarck, A.; Springer, J.; Jekel, M. Zeta-potential and rejection rates of a polyethersulfone nanofiltration membrane in single salt solutions. *J. Membr. Sci.* **2000**, *165*, 251–259. [CrossRef]
22. Guo, S.; Chen, X.; Wan, Y.; Feng, S.; Luo, J. Custom-tailoring loose nanofiltration membrane for precise biomolecule fractionation: New insight into post-treatment mechanisms. *Acs Appl. Mater. Interfaces* **2020**, *12*, 13327–13337. [CrossRef] [PubMed]
23. Bian, L.; Shen, C.; Song, C.; Zhang, S.; Cui, Z.; Yan, F.; He, B.; Li, J. Compactness-tailored hollow fiber loose nanofiltration separation layers based on "chemical crosslinking and metal ion coordination" for selective dye separation. *J. Membr. Sci.* **2021**, *620*, 118948. [CrossRef]
24. Liu, S.; Fang, X.; Lou, M.; Qi, Y.; Li, R.; Chen, G.; Li, Y.; Liu, Y.; Li, F. Construction of Loose Positively Charged NF Membrane by Layer-by-Layer Grafting of Polyphenol and Polyethyleneimine on the PES/Fe Substrate for Dye/Salt Separation. *Membranes* **2021**, *11*, 699. [CrossRef] [PubMed]
25. Jin, P.; Zhu, J.; Yuan, S.; Zhang, G.; Volodine, A.; Tian, M.; Wang, J.; Luis, P.; Van der Bruggen, B. Erythritol-based polyester loose nanofiltration membrane with fast water transport for efficient dye/salt separation. *Chem. Eng. J.* **2021**, *406*, 126796. [CrossRef]
26. Liu, L.; Qu, S.; Yang, Z.; Chen, Y. Fractionation of Dye/NaCl Mixtures Using Loose Nanofiltration Membranes Based on the Incorporation of WS2 in Self-Assembled Layer-by-Layer Polymeric Electrolytes. *Ind. Eng. Chem. Res.* **2020**, *59*, 18160–18169. [CrossRef]
27. Ye, W.; Ye, K.; Lin, F.; Liu, H.; Jiang, M.; Wang, J.; Liu, R.; Lin, J. Enhanced fractionation of dye/salt mixtures by tight ultrafiltration membranes via fast bio-inspired co-deposition for sustainable textile wastewater management. *Chem. Eng. J.* **2020**, *379*, 122321. [CrossRef]

28. Qiu, W.-Z.; Lv, Y.; Du, Y.; Yang, H.-C.; Xu, Z.-K. Composite nanofiltration membranes via the co-deposition and cross-linking of catechol/polyethylenimine. *RSC Adv.* **2016**, *6*, 34096–34102. [CrossRef]
29. Wang, J.; Zhu, J.; Tsehaye, M.T.; Li, J.; Dong, G.; Yuan, S.; Li, X.; Zhang, Y.; Liu, J.; Van der Bruggen, B. High flux electroneutral loose nanofiltration membranes based on rapid deposition of polydopamine/polyethyleneimine. *J. Mater. Chem. A* **2017**, *5*, 14847–14857. [CrossRef]
30. Rahim, M.A.; Ejima, H.; Cho, K.L.; Kempe, K.; Müllner, M.; Best, J.P.; Caruso, F. Coordination-driven multistep assembly of metal-polyphenol films and capsules. *Chem. Mater.* **2014**, *26*, 1645–1653. [CrossRef]
31. Fang, X.; Li, J.; Li, X.; Pan, S.; Sun, X.; Shen, J.; Han, W.; Wang, L.; Van der Bruggen, B. Iron-tannin-framework complex modified PES ultrafiltration membranes with enhanced filtration performance and fouling resistance. *J. Colloid Interface Sci.* **2017**, *505*, 642–652. [CrossRef]
32. Lou, M.; Fang, X.; Liu, Y.; Chen, G.; Zhou, J.; Ma, C.; Wang, H.; Wu, J.; Wang, Z.; Li, F. Robust dual-layer Janus membranes with the incorporation of polyphenol/Fe^{3+} complex for enhanced anti-oil fouling performance in membrane distillation. *Desalination* **2021**, *515*, 115184. [CrossRef]
33. Li, Q.; Liao, Z.; Fang, X.; Wang, D.; Xie, J.; Sun, X.; Wang, L.; Li, J. Tannic acid-polyethylenimine crosslinked loose nanofiltration membrane for dye/salt mixture separation. *J. Membr. Sci.* **2019**, *584*, 324–332. [CrossRef]
34. Zhang, Y.; Ma, J.; Shao, L. Ultra-thin trinity coating enabled by competitive reactions for unparalleled molecular separation. *J. Mater. Chem. A* **2020**, *8*, 5078–5085. [CrossRef]
35. Wu, H.; Xie, J.; Mao, L. One-pot assembly tannic acid-titanium dual network coating for low-pressure nanofiltration membranes. *Sep. Purif. Technol.* **2020**, *233*, 116051. [CrossRef]
36. Fan, L.; Ma, Y.; Su, Y.; Zhang, R.; Liu, Y.; Zhang, Q.; Jiang, Z. Green coating by coordination of tannic acid and iron ions for antioxidant nanofiltration membranes. *RSC Adv.* **2015**, *5*, 107777–107784. [CrossRef]
37. Chakrabarty, T.; Pérez-Manríquez, L.; Neelakanda, P.; Peinemann, K.V. Bioinspired tannic acid-copper complexes as selective coating for nanofiltration membranes. *Sep. Purif. Technol.* **2017**, *184*, 188–194. [CrossRef]
38. Shen, Y.J.; Fang, L.F.; Yan, Y.; Yuan, J.J.; Gan, Z.Q.; Wei, X.-Z.; Zhu, B.-K. Metal-organic composite membrane with sub-2 nm pores fabricated via interfacial coordination. *J. Membr. Sci.* **2019**, *587*, 117146. [CrossRef]
39. Gao, H.; Xue, Y.; Zhang, Y.; Zhang, Y.; Meng, J. Engineering of ag-nanoparticle-encapsulated intermediate layer by tannic acid-inspired chemistry towards thin film nanocomposite membranes of superior antibiofouling property. *J. Membr. Sci.* **2022**, *641*, 119922. [CrossRef]
40. Liu, Y.; Liu, J.; Jiang, Y.; Meng, M.; Ni, L.; Qiu, H.; Yang, R.; Liu, Z.; Liu, H. Synthesis of novel high flux thin-film nanocomposite nanofiltration membranes containing $GO-SiO_2$ via interfacial polymerization. *Ind. Eng. Chem. Res.* **2019**, *58*, 22324–22333. [CrossRef]
41. Ding, W.; Zhuo, H.; Bao, M.; Li, Y.; Lu, J. Fabrication of organic-inorganic nanofiltration membrane using ordered stacking SiO_2 thin film as rejection layer assisted with layer-by-layer method. *Chem. Eng. J.* **2017**, *330*, 337–344. [CrossRef]
42. Zheng, J.; Li, Y.; Xu, D.; Zhao, R.; Liu, Y.; Li, G.; Gao, Q.; Zhang, X.; Volodine, A.; Van der Bruggen, B. Facile fabrication of a positively charged nanofiltration membrane for heavy metal and dye removal. *Sep. Purif. Technol.* **2022**, *282*, 120155. [CrossRef]
43. Zhang, C.; Wei, K.; Zhang, W.; Bai, Y.; Sun, Y.; Gu, J. Graphene Oxide Quantum Dots Incorporated into a Thin Film Nanocomposite Membrane with High Flux and Antifouling Properties for Low-Pressure Nanofiltration. *ACS Appl. Mater. Interfaces* **2017**, *9*, 11082–11094. [CrossRef] [PubMed]
44. Zhang, L.; Guan, H.; Zhang, N.; Jiang, B.; Sun, Y.; Yang, N. A loose NF membrane by grafting TiO_2-HMDI nanoparticles on PES/β-CD substrate for dye/salt separation. *Sep. Purif. Technol.* **2019**, *218*, 8–19. [CrossRef]
45. Ji, D.; Xiao, C.; An, S.; Zhao, J.; Hao, J.; Chen, K. Preparation of high-flux PSF/GO loose nanofiltration hollow fiber membranes with dense-loose structure for treating textile wastewater. *Chem. Eng. J.* **2019**, *363*, 33–42. [CrossRef]
46. Zhao, S.; Wang, Z. A loose nano-filtration membrane prepared by coating HPAN UF membrane with modified PEI for dye reuse and desalination. *J. Membr. Sci.* **2017**, *524*, 214–224. [CrossRef]
47. Shen, L.; Li, P.; Zhang, T. Green and feasible fabrication of loose nanofiltration membrane with high efficiency for fractionation of dye/NaCl mixture by taking advantage of membrane fouling. *J. Appl. Polym. Sci.* **2019**, *136*, 47438. [CrossRef]
48. Li, P.; Wang, Z.; Yang, L.; Zhao, S.; Song, P.; Khan, B. A novel loose-NF membrane based on the phosphorylation and cross-linking of polyethyleneimine layer on porous PAN UF membranes. *J. Membr. Sci.* **2018**, *555*, 56–68. [CrossRef]
49. Zhang, J.; Yang, L.; Wang, Z.; Yang, S.; Li, P.; Song, P.; Ban, M. A highly permeable loose nanofiltration membrane prepared via layer assembled in-situ mineralization. *J. Membr. Sci.* **2019**, *587*, 117159. [CrossRef]
50. Chu, Z.; Chen, K.; Xiao, C.; Ji, D.; Ling, H.; Li, M.; Liu, H. Improving pressure durability and fractionation property via reinforced PES loose nanofiltration hollow fiber membranes for textile wastewater treatment. *J. Taiwan Inst. Chem. Eng.* **2020**, *108*, 71–81. [CrossRef]
51. Li, Q.; Liao, Z.; Fang, X.; Xie, J.; Ni, L.; Wang, D.; Qi, J.; Sun, X.; Wang, L.; Li, J. Tannic acid assisted interfacial polymerization based loose thin-film composite NF membrane for dye/salt separation. *Desalination* **2020**, *479*, 114343. [CrossRef]
52. Li, G.; Liu, B.; Bai, L.; Shi, Z.; Tang, X.; Wang, J.; Liang, H.; Zhang, Y.; Van der Bruggen, B. Improving the performance of loose nanofiltration membranes by poly-dopamine/zwitterionic polymer coating with hydroxyl radical activation. *Sep. Purif. Technol.* **2020**, *238*, 116412. [CrossRef]

Article

Simultaneous Partial Nitrification and Denitrification Maintained in Membrane Bioreactor for Nitrogen Removal and Hydrogen Autotrophic Denitrification for Further Treatment

Kun Dong [1,†], Xinghui Feng [†], Wubin Wang, Yuchao Chen, Wei Hu, Haixiang Li * and Dunqiu Wang *

College of Environmental Science and Engineering, Guilin University of Technology, 319 Yanshan Street, Guilin 541006, China; 2020005@glut.edu.cn (K.D.); xinghfeng@glut.edu.cn (X.F.); wangwubin2021@163.com (W.W.); 1020180208@glut.edu.cn (Y.C.); huwei5018@163.com (W.H.)
* Correspondence: 2011042@glut.edu.cn (H.L.); wangdunqiu@sohu.com (D.W.)
† These authors contributed to the work equally and should be regarded as co-first authors.

Abstract: Low C/N wastewater results from a wide range of factors that significantly harm the environment. They include insufficient carbon sources, low denitrification efficiency, and NH_4^+-N concentrations in low C/N wastewater that are too high to be treated. In this research, the membrane biofilm reactor and hydrogen-based membrane biofilm reactor (MBR-MBfR) were optimized and regulated under different operating parameters: the simulated domestic sewage with low C/N was domesticated and the domestic sewage was then denitrified. The results of the MBR-MBfR experiments indicated that a C/N ratio of two was suitable for NH_4^+-N, NO_2^--N, NO_3^--N, and chemical oxygen demand (COD) removal in partial nitrification-denitrification (PN-D) and hydrogen autotrophic denitrification for further treatment. The steady state for domestic wastewater was reached when the MBR-MBfR in the experimental conditions of HRT = 15 h, SRT = 20 d, 0.04 Mpa for H_2 pressure in MBfR, 0.4–0.8 mg/L DO in MBR, MLSS = 2500 mg/L(MBR) and 2800 mg/L(MBfR), and effluent concentrations of NH_4^+-N, NO_3^--N, and NO_2^--N were 4.3 ± 0.5, 1.95 ± 0.04, and 2.05 ± 0.15 mg/L, respectively. High-throughput sequencing results revealed the following: (1) The genus *Nitrosomonas* as the ammonia oxidizing bacteria (AOB) and *Denitratisoma* as potential denitrifiers were simultaneously enriched in the MBR; (2) at the genus level, *Meiothermus*, *Lentimicrobium*, *Thauera*, *Hydrogenophaga*, and *Desulfotomaculum* played a dominant role in leading to NO_3^--N and NO_2^--N removal in the MBfR.

Keywords: partial nitrification-denitrification; hydrogen autotrophic denitrification; MBR-MBfR

1. Introduction

Conventional biological nitrogen removal (BNR) includes ammonification, nitrification, and denitrification. This type of approach is considered to be a good choice for reducing nitrogenous compounds in wastewater treatment because it is economic, effective, easy to operate, and results in no secondary pollution [1,2]. However, for low C/N (chemical oxygen demand (COD)/ammonia nitrogen, NH_4^+-N) wastewater, the BNR imposes stringent restrictions on the insufficient carbon source and results in incomplete nitrogen removal, which in turn requires external organic carbon sources, high operating costs, and high aeration-associated energy consumption for nitrification [3–5]. Partial nitrification has been regarded as a sound self-sustaining biological nitrogen removal process because it can reduce aeration energy by 25%, reduce carbon dioxide emissions and sludge production, decrease organic carbon requirement by 40–100%, and reduce biomass production by 300% [1,6]. According to Equations (1) and (2), the effectiveness of partial nitrification was closely determined by the concentration of influent NH_4^+ and the precise control of dissolved oxygen (DO) concentration. In addition, NO_2^--N did not coexist with NH_4^+-N in most of the wastewater (usually needed in situ conversion to initialize the partial nitrification process), but NO_3^--N and COD were frequently found in water and wastewater:

$$NH_4^+ + 1.5O_2 \rightarrow NO_2^- + 2H^+ + H_2O, \tag{1}$$

$$NO_2^- + 1.5O_2 \rightarrow NO_3^- \tag{2}$$

Various forms of integrated techniques were implemented in the past few years to improve the ability of total nitrogen (TN) removal and to treat low C/N wastewater; novel and cost-effective partial nitrification-based BNR processes have been put forward, including partial nitrification-denitrification (PN-D), partial nitrification-simultaneous anammox and denitrification (PN-SAD), and anaerobic ammonium oxidation (Anammox). PN-D may represent a good alternative compared with conventional nitrification-denitrification because the process uses nitrite nitrogen as an electron acceptor and organic matter as an electron donor. It has the advantages of lower yields of sludge, being energy saving (low aeration consumption), reducing the carbon source, and being suited for low C/N wastewater [7]. Partial nitrification combined with denitrification achieves excellent nitrite accumulation through the accumulation of ammonia-oxidizing bacteria (AOB) and the inhibition of nitrite-oxidizing bacteria (NOB) in the reactors [8].

Considerable effort was made to control the operating conditions. However, it was still hard to avoid the effluent NO_2^--N concentration, the NO_3^--N concentration, and COD exceeding the stringent discharge standard all the time. Therefore, further improvements of the NO_2^--N concentration, NO_3^--N concentration, and COD were necessary. Furthermore, low DO aeration and NH_4^+-N residues were conducive to partial nitrification stability [4]. There were many factors affecting partial nitrification, including but not limited to DO, temperature, pH, free ammonia (FA), and free nitrous acid (FNA). The oxygen saturation concentration was 0.3 mg/L and 1.1 mg/L for the ammonia oxidation reaction and nitrite oxidation reactions, respectively, and the low DO (less than 0.5 mg/L) benefitted AOB and inhibited the NOB [9]. Li et al. [10] reported that the optimal temperature for AOB and anaerobic ammonium oxidizing bacteria (AnAOB) for partial nitrification was 35 °C. A suitable pH for the partial nitrification process was 6.5–8.5. When the pH was higher than 8.5, the partial nitrification was inhibited and the alkaline consumption was increased [11]. Soliman et al. [12] reported that the inhibition limit for NOB was 0.1–1.0 mg N/L and the inhibition limit for AOB was 10–150 mg N/L of FA. The start of the partial nitrification process was restricted by the aforementioned conditions and its industrial applications were severely hindered. Therefore, to the method of quickly starting and maintaining a stable partial nitrification reaction remains extremely challenging.

At present, the widely reported partial nitrification processes are built into sequential batch reactors (SBR). However, SBR reactors are constrained by complicated control and poor stability performance [13]. Membrane bioreactors (MBRs) are alternatives to be applied in the PN-D process. They can avoid biomass wash-out and increase biomass retention, which allows the reactor to operate at a high biomass concentration, therefore improving the stability of the PN-D process. However, poor control of oxygen and other influencing factors lead to the production of NO_2^--N and NO_3^--N, as shown in Equations (1) and (2). This can lead to the quality of the effluent exceeding the wastewater disposal standards if no further treatment is applied.

The hydrogen-based membrane biofilm reactor (MBfR), an emerging biodegradation technology based on a hydrogenotrophic reduction process, efficiently removes nitrite and nitrate from wastewater. In the MBfR, pressurized H_2 is supplied to the lumen of the hollow fiber membrane (HFM). The gas diffuses through the HFM wall through nanopores in the autotrophic biofilm formed on the outer HFM surface. Here, nitrate and nitrite diffuse from water into biofilm being reduced [14,15]. The advantages of this hydrogenotrophic denitrification compared with conventional heterotrophic denitrification technology include the utilization of nontoxic and inexpensive H_2 as electron donors, no requirement for the addition of external organic carbon, a small footprint, relatively low cost, and low production of biological sludge [16,17]. Promising outcomes have been reported with MBfR for treating nitrate- and nitrite-contaminated water for research and commercial applications. For example, Park et al. [18] applied an MBfR to treat high-strength wastewater containing 50 mg/L NO_3^--N without supplying any source of organic carbon and a maximum nitrate removal rate of 0.1 g NO_3^--N/(m^{2}d) was achieved. In addition, heterotrophs are usually also present in MBfR, which may bring further removal of COD and contribute to the denitrification in a heterotrophic pathway if COD is present in the influent. Therefore, an MBfR could be ideally suited for the treatment of nitrate and nitrite byproducts from processes of partial nitrification and excessive COD.

In this study, strategies to initiate start-up NOB suppression and to adapt the partial nitrification process to a hydrogenotrophic denitrification process were proposed. The objectives of this study were as follows: (1) to couple partial-nitrification and denitrification in MBR (MLSS = 2500 mg/L, 0.4–0.8 mg/L DO, reactor temperature was 35 °C) and use a small amount of carbon source to achieve high-efficiency nitrogen removal through a denitrification process; (2) to experimentally take advantage of MBfR (H_2 pressure was 0.04 MPa and the pH value around 7.5) to quickly remove TN (nitrate and nitrite) and low concentrations of nitrous; and (3) to explore the most suitable operating parameters for low C/N wastewater when MBR-MBfR (in the experimental conditions of HRT = 15 h, SRT = 20 d) was employed for domestic wastewater treatment.

2. Materials and Methods

2.1. Influent

The reactor was operated with both synthetic wastewater and domestic wastewater. The synthetic wastewater used in this study was composed of substrates and trace elements as influent. Ammonium was provided in the form of NH_4Cl and added as required. The pH in the reactor was maintained automatically at 7.5 ± 0.5 by addition of 0.1 M HCl and 0.1 M NaOH. An aeration device was set at the water pipe to control the concentration of DO between 0.4 and 0.8 mg/L. Peristaltic pumps (BT101L-DG-1, Lead Fluid, Baoding, China) were used to control the influent and the effluent feed rate. The synthetic domestication water was adapted from an earlier study and comprised the following components per liter [19]: 1 g $NaHCO_3$, 0.1 g $CaCl_2 \cdot 2H_2O$, 0.05 g KH_2PO_4, 0.05 g Na_2SO_3, and 1 mL of a stock solution containing trace elements. The trace element stock solution contained (per liter): 5 g $MgSO_4 \cdot 7H_2O$, 0.1 g $ZnSO_4 \cdot 7H_2O$, 6 g $FeCl_2 \cdot 4H_2O$, 0.1 g H_3BO_3, 0.88 g $CoCl_2 \cdot 6H_2O$, 0.5 g $MnCl_2 \cdot 4H_2O$, 0.036 g $NiCl_2 \cdot 6H_2O$, and 0.035 g $CuCl_2 \cdot 2H_2O$. The domestic wastewater was obtained from the effluent of a primary settling tank in a domestic wastewater treatment plant (WWTP) from Guilin University of Technology. The domestic wastewater conditions were as follows (similar to the operational conditions of the batch experiment in phase III with various C/N ratios in the MBR): COD, 140–160 mg/L; NH_4^+-N, 80–100 mg/L; NO_3^--N, 5–10 mg/L; NO_2^--N, 0 mg/L; C/N ratio, 2.

2.2. Reactor Configuration

A scheme of the MBR-MBfR used in this study is shown in Figure 1 and the physical characteristics of the reactors are listed in Table 1. The MBR set-up was made of plexiglass and consisted of four parts: (I) an inner MBR unit used for culturing PN-D sludge; (II) a thermostatic jacket filled with hot water to maintain a fixed temperature of 35 °C for PN-D; (III) a hollow fiber membrane (HFM) module, in which the HFM was made of commercially available PVC; and (IV) a programmable logic controller (PLC) system for monitoring the pH and for controlling the concentration of DO. Notably, the effluent of MBR was realized by the operation of a pump, in which the outlet at the top of the MBR was connected to a buffer tank placed lower than the MBR for further treatment.

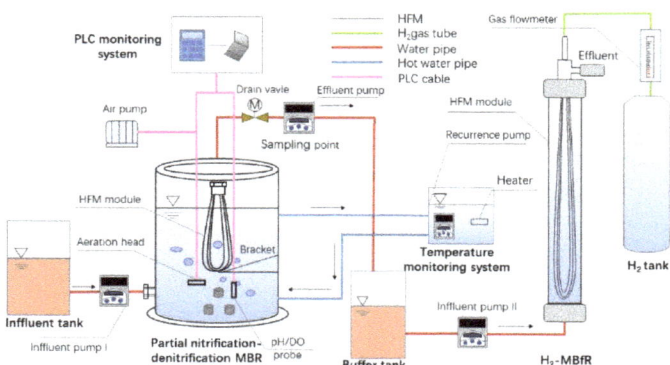

Figure 1. Schematic representation of the MBR-MBfR set-up.

Table 1. The physical characteristics of MBR-MBfR.

Reactor	Parameter	Units	Value
MBR	MBR height	cm	30
	Number of HFM		80
	HFM inner diameter	mm	1.2
	HFM outer diameter	mm	2.2
	HFM pore size	μm	0.1
	Active surface area	m^2	0.06
	Active volume	L	4.32
MBfR	MBfR height	cm	64
	Number of HFM		65
	HFM inner diameter	mm	1.0
	HFM outer diameter	mm	1.66
	HFM pore size	μm	0.02
	Active surface area	m^2	0.28
	Active volume	L	1.8

The start-up processes for MBR and MBfR were conducted under different operating conditions. The startup operation for MBR was divided into two periods. The MBR was inoculated with 500 mL of partial nitrification bacterial sludge: an initially mixed liquor suspended solids (MLSS) of 2500 mg/L from a stable operating partial nitrification reactor in the first period and 500 mL of denitrification bacterial sludge (MLSS of 2500 mg/L) from a stable operating denitrification reactor in the second period. The start-up phase began using synthetic wastewater containing NH_4^+-N under the conditions of low DO (0.4–0.8 mg/L). The SRT of the MBR was 20d. The startup operation for MBfR was HRT = 10 h (shortened to 5 h in the post-start experiment), MLSS = 2800 mg/L, and the SRT of the MBfR was 20d.

The medium consisted of tap water with the following components added: 1.386 g Na_2PO_4, 0.849 g KH_2PO_4, 0.05 g $MgSO_4 \cdot 7H_2O$, and 0.025 g $(NH_4)_2SO_4$ per liter. $NaNO_3$ and $NaHCO_3$ were used as inorganic nitrogen and carbon sources for the rapid growth of hydrogenotrophic microorganisms. A mixture of KH_2PO_4 0.128 g/L and Na_2HPO_4 0.434 g/L was used as the phosphate buffer to keep the pH value of the MBfR around 7.5 [20]. The hollow-fiber membranes were made from microporous polyethylene with a thin polyurethane core (Watercode, Guangzhou, China). The total membrane active surface area of the MBfR was 0.28 m^2. The MBfR system consisted of an HFM module with 65 HFMs located inside of a vertical plexiglass cylindrical shell and an ultrapure H_2 tank for supplying pressurized H_2 to the HFM module. The MBfR module was sealed using waterproof epoxy glue. The fiber was connected to a hydrogen-supplying manifold supplied at the top end and sealed individually at the bottom end. Smaller pore size HFMs made of PVC with a pore size of 0.02 μm were used in the MBfR to deliver bubble-less H_2 through the HFM wall. The H_2 pressure was 0.04 MPa, while the pressure was also adjusted in the range of 0.02–0.06 MPa to evaluate the effect of H_2 pressure on nitrogen removal performance.

2.3. Samping and Analytical Methods

The operating performance of the reactors was evaluated by analyzing influent and effluent samples on a daily basis. Samples were subsequently filtered through a 0.22 μm membrane filter. Influent and effluent samples were collected daily for both MBR and MBfR to analyze the concentration of NH_4^+-N, NO_3^--N, and TN according to a standard method provided by the American Public Health Association (APHA). NO_2^--N concentrations were determined using a colorimetric assay based on sulfanilamide and N-(1-naphthyl) ethylenediamine dihydrochloride. The absorbance was measured at 540 nm in a spectrophotometer (UV6100, METASH, Shanghai, China). The pH and DO were monitored in situ via a PLC system that was connected with pH and DO probes.

The biomass samples were sent out for microbial structure analysis at Novogene Co., Ltd. (Beijing, China). The relative abundance of partial nitrification bacteria, denitrification bacteria, and hydrogen autotrophic denitrifying bacteria were determined by high-throughput sequencing analysis. The V4-V5 hypervariable region of the 16s RNA gene was amplified by polymerase chain reaction (PCR). The amplified PCR used the bacterial primers 515F(5'-GTGCCAGCMGCCGCGGTAA-3') and 907R(5'-CCGTCAATTCCTTTGAGTTT-3'). The operational taxonomic units (OTUs) of the representative sequences were annotated with the Ribosomal Database Project (RDP) classifier method and the Greengene database [21] for species annotation analysis (the threshold value was set to 0.8–1). The sequence number of each sample was normalized and the trimmed sequences were grouped into OTUs using 97% identity thresholds. [22]. Kingdom, phylum, class, order, family, genus, and species were used to analyze the community composition of each sample. Taxonomic classification at the genus level was performed using the RDP algorithm to classify the representative sequences from each OTU.

3. Results and Discussion

3.1. Start-Up and Experimental Strategy of MBR

The startup operation for MBR was divided into two periods (shown in Section 2.2). In the first period, which consists of the demand for partial nitrification bacteria, the incubator was continuously fed by the peristaltic pump with an increasing nitrogen load (gradually increasing the NH_4^+-N concentration and decreasing the hydraulic retention time (HRT); the detailed operating conditions during the start-up process are shown in Table 2). After 30 days of the start-up phase for the partial nitrification operation, more than 96% removal of NH_4^+-N was achieved under a hydraulic retention time (HRT) of 10 h and the concentration of reactor NO_2^--N rose to 102 mg/L. This was accompanied by a small amount of NO_3^--N generation (10 mg/L). The start-up and stability of the partial nitrification process were successful under these conditions.

Table 2. Operational conditions during the start-up of partial nitrification in the MBR.

Phase	Time (days)	NH_4^+-N (mg/L)	NO_3^--N (mg/L)	HRT (h)
I	1–7	50.11	10.90	16
II	8–15	61.66	10.10	14
III	16–23	71.43	9.41	12
IV	24–30	80.74	9.90	10

After 31 days, the carbon source was added to the MBR relative to the phase (showed in Table 3) and 500 mL of denitrification bacteria sludge was inoculated in the MBR. A batch experiment with various C/N ratios in the range 0.5–3 was performed in the MBR to see how the performance of denitrification was affected and to promote the stability and acclimatization of denitrification with the co-existence of COD and nitrogen. This was conducted because a certain amount of COD is always present in domestic wastewater. The existence of COD promoted the denitrification process to a certain extent. At this stage, denitrifying bacteria had a certain effect on the removal of COD, NO_3^--N, and NO_2^--N, but they could not be completely removed. The operational conditions during this batch experiment are shown in Table 3. The nitrogen concentration in the influent was fixed with a concentration of NH_4^+-N of 80 mg/L, while the influent C/N was increased in a stepwise manner by adding the required volume of white sugar.

Table 3. Operational conditions during the different phases with various C/N ratios in the MBR.

Phase	Time (Days)	C/N	COD (mg/L)	NH_4^+-N (mg/L)	HRT (h)
I	31–38	0.5	40		
II	39–46	1	80	80	10
III	47–54	2	160		
IV	55–62	3	240		

An excessively high concentration of organic carbon may therefore significantly suppress the removal of nitrogen via denitrification. The effect of COD on nitrogen and COD removal was evaluated to determine the effectiveness of the denitrification period of MBR. A preliminary batch test with various C/N ratios in the range of 0.5–3 was performed in the MBR after the start-up and steady state of MBR without adding an organic carbon source. The performance of NH_4^+-N and COD removal for the MBR in the denitrification period is shown in Figure 2. Negligible change of COD removal, which was around 13.06%, was found when the C/N ratio was varied in the range of 0.5–3. Notably, the highest NH_4^+-N removal of 87.59% was obtained at a C/N ratio of two, while a dramatic decrease was found when the C/N ratio was higher than two. As shown in Figure 3, 9.9% and 5.2% higher values of NH_4^+-N removal at a C/N ratio of two were observed than those obtained at ratios of 0.5 and 1. This might be attributable to the higher contribution of NH_4^+-N removal through the heterotrophic denitrification pathway with COD concentration in the influent (C/N ratio of two). When the C/N ratio was 0.5 and the influent COD concentration decreased to 40 mg/L, the COD content was low, and the lack of carbon source led to incomplete denitrification. The effluent NO_2^--N concentration decreased and the COD removal rate decreased. As shown in Figure 2, a C/N ratio of two was suitable for NH_4^+-N, NO_2^--N, NO_3^--N, and COD removal and resulted in no apparent inhibition of denitrification activity. However, around 10 mg/L of NO_3^--N (shown in Figure 2) was detected in the effluent when the C/N ratio in the influent was two in the MBR. When the C/N ratio was three, the activated sludge system in the reactor was a complex multi-bacteria coexistence system. With the increase of C/N in the domestic wastewater, the activity of heterotrophic bacteria and denitrifiers in the reactor increased. The change in the composition of the influent matrix made PN need an adaptation process, but with the increase of organic matter, AOB and denitrifying bacteria also had a dynamic change until a new balance occurred. Increasing organic matter promoted the growth of denitrifiers. Organics promote the growth of denitrifiers, while these denitrifiers or heterotrophs would consume more oxygen, which is necessary for AOB growth, inhibiting the growth of AOB involved in partial nitrification and compete with AOB for living space and substrate [23], leading to an increase in the concentration of NH_4^+-N in the effluent. Too much carbon inhibited the growth of the AOB participating in the partial nitrification. This increased the effluent NH_4^+-N concentration at this phase (C/N ratio of three) and the partial nitrification product NO_2^--N was used by heterotrophic denitrifying bacteria to produce PN-D. This reduced the nitrite nitrogen concentration and led to an increase of sludge produced by subsequent denitrification. Therefore, a further sufficient treatment of nitrate was required.

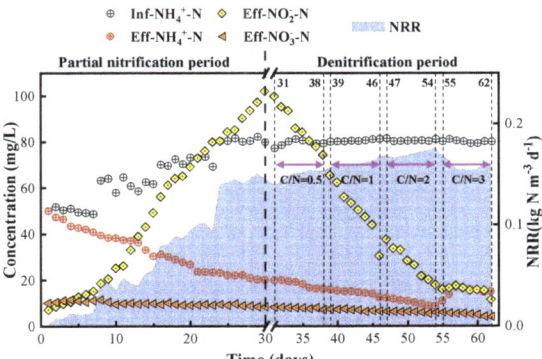

Figure 2. NH_4^+-N, NO_2^--N, and NO_3^--N in the effluent and ammonia-nitrogen removal rate (NRR) of the start-up of the MBR.

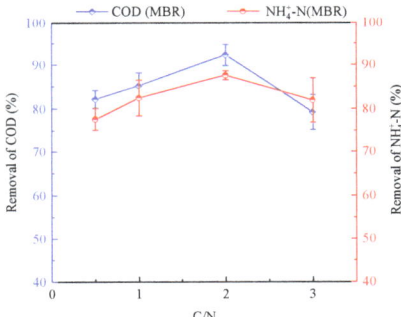

Figure 3. NH_4^+-N and COD removal in the MBR during batch experiments in the denitrification period.

3.2. Start-Up and Experimental Strategy of MBfR

Different synthetic wastewater containing 10 mg/L of NO_3^--N was used for the MBfR start-up. The inoculation seeding sludge of hydrogenotrophic bacteria was collected from a lab-scale denitrifying MBfR in our laboratory. The MBfR start-up procedure was like the procedure for the MBR, which was to continuously feed the synthetic influent with increasing loading of nitrate. After successful start-up, the MBfR was able to reach a NO_3^--N removal of more than 98% at an HRT of 10 h with an influent NO_3^--N concentration of 10 mg/L. Previous studies proved that the H_2 supplying pressure and influent nitrate loading were the two key operational factors that affected the nitrate removal performance in MBfR [24,25].

In this study, two series of experiments were conducted to investigate the performance of nitrate removal in the MBfR. These included looking at the H_2 supplying pressure and influent NO_3^--N concentration. A H_2 supplying pressure in the range of 0.02–0.08 MPa has been acknowledged to be preferable for use in most MBfRs [26]. In the H_2 series, three pressures of 0.02, 0.04, and 0.06 MPa were involved to discover the optimal H_2 pressure for our reactor, while the influent NO_3^--N concentration was fixed at 10 mg/L as the H_2 pressure varied. Most natural water in China usually has a certain amount of nitrate (e.g., 10 mg/L of NO_3^--N was detected in tap water used to make synthetic medium in our lab). Therefore, we added 10 mg/L NO_3^--N during the H_2 series. For the NO_3^--N series, the influent contained 10, 20, and 30 mg/L of NO_3^--N to discover the potential ability of nitrate removal in the MBfR and to evaluate the capability to encounter fluctuations of influent. The H_2 pressure was fixed at 0.04 MPa. The HRT was maintained at 5 h for all the experiments in the two series.

The MBfR was used to handle the left-over nitrate produced in the PN-D process. We assessed the effects of two key factors on the performance of nitrate removal, namely, H_2 pressure and influent NO_3^--N concentration, to discover the capability of the MBfR to remove nitrate. The results are shown in Figure 4. An apparent increase of nitrate removal was observed from 96% to 99%, when the H_2 pressure increased from 0.02 MPa to 0.04 MPa, compared with an increase from 0.04 MPa to 0.06 MPa (less than 1%) as shown in Figure 4a. MBfR was therefore already able to efficiently remove nitrate from the influent when the H_2 supplying pressure was set at 0.04 MPa. It was not necessary to use a higher supplying pressure, which helped with safety and preserved the life of the membranes. The influent NO_3^--N concentration was varied at 10, 20, and 30 mg/L and the effect on nitrate removal is shown in Figure 4b. The influent NO_3^--N concentration in this range covered the nitrate produced from PN-D MBR plus the fluctuation of nitrate concentration in the tap water used for synthetic wastewater or domestic wastewater. The highest effluent NO_3^--N concentration of 3.2 mg/L was found in all three phases for drinking water when the influent NO_3^--N concentration was 30 mg/L. A stepwise decrease of nitrate removal was found as the influent concentration of NO_3^--N increased.

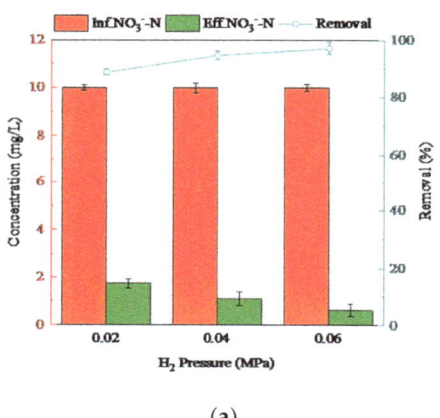
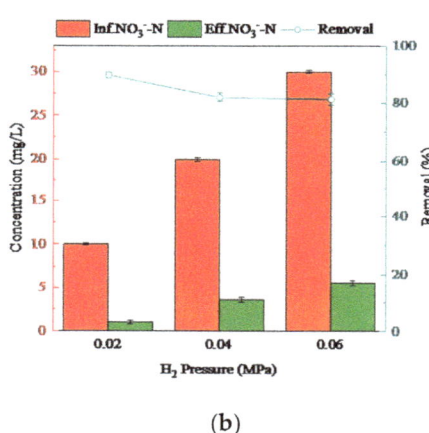

(a)   (b)

Figure 4. Effects of H_2 pressure and influent NO_3^--N concentration on NO_3^--N concentration and removal in the effluent. (a) H_2 pressure: 0.02, 0.04, and 0.06 MPa; (b) influent NO_3^--N concentration: 10, 20, and 30 mg/L.

3.3. Effects of the MBR-MBfR on Nitrogen Compound Removal at C/N = 1.5–2.5

The domestic wastewater was tested in the integrated MBR-MBfR to verify the application of the system. The domestic wastewater was obtained from the effluent of a primary settling tank in a wastewater treatment plant. Of note, 120–160 mg/L COD was added proportionally to the influent at the appropriate concentration. This addition was done to validate the PN-D process to biodegrade nitrogen and COD, and to verify the optimum C/N ratio of PN-D and hydrogen autotrophic denitrification for further treatment in the MBR-MBfR. The domestic wastewater was also fed continuously at an HRT of 15 h for both MBR and MBfR.

Extensive experiments in three phases were conducted under various C/N conditions to investigate the behaviors in both the MBR and MBfR and how the nitrogen and COD removal performed in each reactor. This was achieved based on the PN-D process and adding simulated wastewater at different C/N ratios. Each phase was operated long enough to reach steady state. The HRT of the MBR and MBfR was set to 10 h and 5 h, respectively, which resulted in a total HRT of 15 h for the integrated MBR-MBfR system. The operational conditions under different phases are summarized in Table 4. A C/N ratio of two was still suitable for the remaining PN-D process to allow high NH_4^+-N, NO_2^--N,

NO_3^--N, and COD removal in the MBR. Three phase experiments at C/N ratios of = 1.5, 2.0, and 2.5 were conducted with fixed 80 mg/L NH_4^+-N to further investigate how the nitrogen and COD removal performed in the integrated MBR-MBfR system.

Table 4. Operational conditions at various C/N ratios in the MBR.

Phase	C/N	COD (mg/L)	NH_4^+-N (mg/L)	HRT (h)
I	1.5	120		
II	2.0	160	80	15
III	2.5	200		

COD removal had no distinct change in the MBR when the C/N ratio increased from 1.5 to 2.5, as shown in Figure 5a. However, a significant decrease was found at a C/N ratio higher than two after the treatment of MBfR, owing to the increase of the influent COD concentration to 200 mg/L. The effluent COD concentrations of 8.64, 11.23, and 23.77 mg/L were detected at C/N ratios of 1.5, 2.0, and 2.5, respectively. The removal of TN in the integrated MBR-MBfR system is shown in Figure 5b. The lowest effluent TN concentration of 16.69 mg/L was detected at the influent TN concentration of 107.98 mg/L. MBfR had no significant difference in contribution to TN and COD removal among the three phase experiments at different C/N ratios.

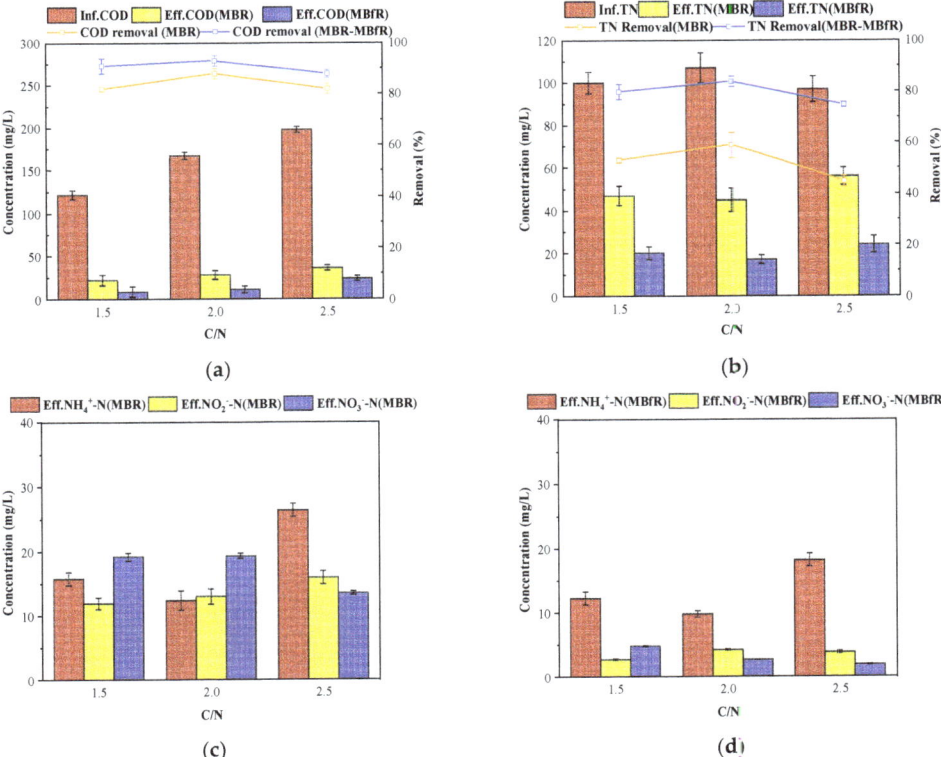

Figure 5. (a) COD removal in the MBR-MBfR during batch experiments at C/N ratios of 1.5, 2.0, and 2.5; (b) TN removal in the MBR-MBfR at C/N ratios of 1.5, 2.0, and 2.5; (c) NH_4^+-N, NO_3^--N, and NO_2^--N removal of the MBR in the MBR-MBfR at C/N ratios of 1.5, 2.0, and 2.5; (d) NH_4^+-N, NO_3^--N, and NO_2^--N removal of the MBfR in the MBR-MBfR at C/N ratios of 1.5, 2.0, and 2.5.

The performance for nitrogen removal for NH_4^+-N, NO_3^--N, and NO_2^--N in the MBR is shown in Figure 5c. The effluent concentrations of TN in the MBR were somewhat lower in phase I than phases II and III, which indicated that PN-D activity was greater in phase II. A dramatic increase of NH_4^+-N concentration in the effluent of MBR was found owing to the suppression of the growth of AOB bacteria proliferation by organic matter, and other reason is that organic matter will be absorbed by heterotrophic aerobic bacteria in large quantities, which will inhibit the growth of AOB to a certain extent. This increase is shown in Figure 5b compared with the other two phases when the C/N ratio increased to 2.5.

In addition, the effluent NO_3^--N and NO_2^--N concentrations of the MBR were lower and higher, respectively, in the third phase because of a higher COD loading that may result in partial denitrification. The PN-D activity can often be outcompeted by heterotrophic denitrification and severely inhibited at a C/N ratio greater than 2.0.

MBfR performed an efficient removal for both nitrate and nitrite that remained in the effluent of MBR in all three phases, as shown in Figure 5d. Less than 1.90 mg/L NO_3^--N and 3.83 mg/L NO_2^--N were detected in the effluent. However, there were small but insignificant differences of NH_4^+-N being reduced after the treatment of MBfR in all three phases. To our knowledge, there is little evidence in the literature that the anaerobic MBfR can effectively remove ammonium. The main contribution to TN concentration in the effluent was, therefore, that ammonium remained after the treatment of PN-D in the MBR. Therefore, in further studies, the main measure to promote TN removal could be to create a suitable condition to proliferate AOB bacteria and suppress the activity of NOB in the PN-D MBR.

3.4. Experimental Study on Treatment of Low C/N Wastewater by MBR-MBfR Reactor

In the treatment and application stage, the domestic wastewater came from the sewage treatment plant of Guilin University of Technology. The typical characteristics of this domestic wastewater are described in Section 2.1. After the stable operation of the reactor in the previous stage, the operation cycle was selected as 20 days. The operation results were shown in Figure 6.

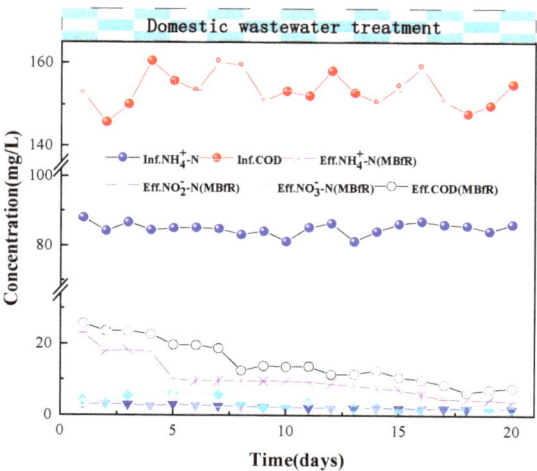

Figure 6. NH_4^+-N, NO_3^--N, NO_2^--N, and COD removal in the MBR-MBfR during the treatment and application stage.

At the beginning, the influent water of the MBR-MBfR changed for domestic wastewater under conditions of low DO and low C/N. The PN-D bacteria in the MBR reactor and the hydrogen autotrophic denitrifying bacteria in the MBfR reactor were affected by actual sewage. The start-up period was used for adaptation of the biomass and the PN-D process was targeted for treatment with the hydrogen autotrophic denitrification process. A short-term decline in AOB activity made the change in removal rate of COD and nitrogen not obvious and NH_4^+-N built up in the effluent in the MBR at the beginning.

In the first four days, a period of acclimation for the concentration of NH_4^+-N and COD was observed in the MBR-MBfR effluent. The concentration of NH_4^+-N and COD decreased from 23.00 mg/L and 25.66 mg/L to 17.80 mg/L and 22.69 mg/L, respectively. The influent NH_4^+-N concentration was maintained at about 80 mg/L from the fifth day to the end of the period of the domestic wastewater treatment. The effluent concentration of NH_4^+-N decreased significantly from 10.21 mg/L to 3.89 mg/L compared with the first four days and the average removal rate was 90.26%. The effluent concentrations of NO_3^--N and NO_2^--N were 1.95 mg/L and 1.91 mg/L, respectively.

The steady state was reached during days 16–20. Reactor and effluent concentrations of NH_4^+-N, NO_3^--N, and NO_2^--N were 4.3 ± 0.5 mg/L, 1.95 ± 0.04 mg/L, and 2.05 ± 0.15 mg/L, respectively. Here we compare the results of the proposed method with the previous study about an experiment of using a membrane bioreactor to treat actual wastewater; when the nitrogen loading rate was similar to that reported before, MBR-MBfR system could remove the excess NO_2^--N and NO_3^--N remaining after the partial nitrification-denitrification process, and the total nitrogen removal rate and COD removal rate could reach 84.75% and 90.57%, higher than the 43% and 87% mentioned in the previous study [27]. Compared with another study on the treatment of municipal wastewater with low C/N ratios by the A^2O-MBR process, the total nitrogen removal rate of the MBfR in this study was greater than the 79% mentioned in previous studies (under conditions of MLSS = 3000 mg/L, close to the actual wastewater MLSS in this study) [28]. The DO concentration was at a low level between 0.4 and 0.8 mg/L. Under these conditions, the activity of AOB was enhanced and completely outcompeted NOB from the reactor. This result indicated that the PN-D process and hydrogen autotrophic denitrification in the MBR were the main processes taking place. NOB inhibition was effective while maximizing the activity of AOB, although some residual nitrite oxidation was still present. The effluent water of the MBfR met Class 1A level for Chinese discharge standards from municipal wastewater treatment plant (GB18918-2002) for NH_4^+-N, NO_3^--N, and NO_2^--N concentrations.

3.5. Microbial Community Analysis in Different Phases of MBR-MBfR

Sludge samples were collected on MR1 (the start-up period of MBR), MR2 (treatment stage of low C/N wastewater of MBR), MR3 (treatment stage of domestic wastewater of MBR), FR1 (the start-up period of MBfR), FR2 (treatment stage of low C/N wastewater of MBfR), and FR3 (treatment stage of domestic wastewater of MBfR). In the taxonomic analyses, the samples collected from MBR (MR1, MR2, and MR3) were grouped into 304, 324, and 363 OTUs, and the samples collected from MBfR (FR1, FR2, and FR3) were grouped into 613, 619, and 684 OTUs, respectively. The genera and phyla with relative abundance rates greater than 0.1% are shown in Figure 7.

In the sample from MBR (MR1), nine phyla with relative abundance greater than 0.1% were detected. Proteobacteria (53.54%), Bacteroidetes (23.49%), Planctomycetes (7.68%), and Chloroflexi (9.56%) were the dominant phyla in the resultant bacterial community. Proteobacteria were the predominant phyla with an occurrence of 53.54%. The genus *Brevundimonas* within this phylum had the greatest relative abundance of 14.80%. It has been reported to have a process of promoting ammonia oxidation [29] and boosting the development of the PN-D process [30]. The denitrifying bacteria of the genus *Denitratisoma*, which perform denitrification via nitrite, were present in the MR1, MR2, and MR3, with 0.80%, 1.84%, and 3.63% relative abundance. It should be noted that the main reason for the *Denitratisoma* abundance increase was that MBR-MBfR had a denitrification process, and the denitrification process was gradually enhanced after incubation with nitrogen wastewater containing carbon sources. High-throughput analysis revealed that an abundance of the genera *Denitratisoma*, which are potential denitrifiers, improved TN removal efficiency.

The continuous increase of the relative abundance of *Denitratisoma* was closely related to the COD concentration rising from MR1 to MR3. This result was consistent with the findings of Ge [31] and Tao [32]. The genus *Nitrosomonas*, which belongs to the AOB, has been reported to be the first step in partial nitrification, and also found in diverse aquatic and terrestrial environments [33]. Proteobacteria and *Nitrosomonas* were the dominant phyla and genus, respectively, for MR1 to MR3. The Proteobacteria increased to 62.56% in MR3; *Nitrosomonas* increased from 0.38% to 29.53% and performed ammonium oxidation to nitrite. In the case of low DO, the NOB activity was inhibited. DO correlated with AOB and NOB abundance [34].

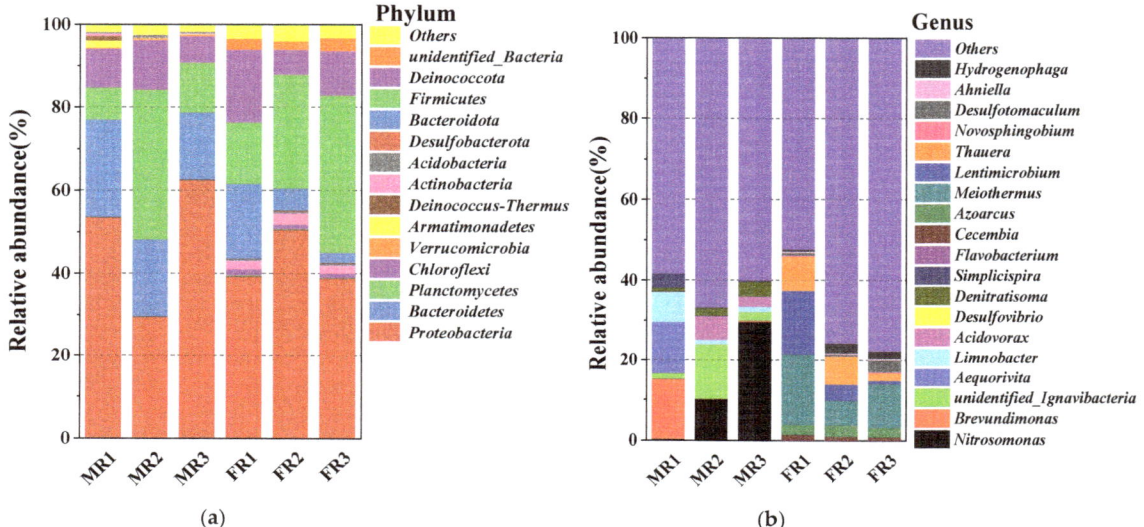

Figure 7. (**a**) Relative abundance of the phyla found in samples from the reactors MBR-MBfR; (**b**) relative abundance of genera found in samples from the reactors MBR-MBfR.

The abundance of partial nitrification microorganisms (*Nitrosomonas* bacteria) and denitrifying microorganisms (*Denitratisoma* bacteria) in the MBR shed light on their remarkable performance in the combined partial oxidation of ammonium and the denitrification of nitrite and nitrate. In the MBR, the species richness increased with the three phases of operation of MBR (MR1, MR2, and MR3), as evidenced by the OTUs and the Chao1 indexes. This increase was possibly because of the change from synthetic wastewater to domestic wastewater that contained complex organic matter and nitrogen compounds. The data summarized above clearly showed that partial nitrifying bacteria and denitrifying bacteria were simultaneously present in the MBR, which further demonstrates the fusion of the PN-D process in the MBR.

In the MBfR, the abundance of the ten most found bacteria at the genus and phylum level were investigated in the different phases (FR1, FR2, and FR3). The qualified sequence reads of the biological samples (FR1, FR2, and FR3) were 83270, 90419, and 80113. The most abundant genera of the operation for FR1 were *Meiothermus* and *Lentimicrobium* with a relative abundance of 17.69% and 15.96%, respectively, as shown in Figure 7b. *Meiothermus*, which has been reported as a denitrifying bacteria, played an important role of reducing Cr (VI), BrO_3^-, and NO_3^- in several MBfRs [35,36]. The genus *Lentimicrobium*, known as a potential denitrifier [37], has been reported to be indispensable for the successive removal of high concentrations of nitrate [38]. The genus *Thauera*, with a relative abundance of 8.4% in MR1, could also not be ignored. *Thauera* was deemed to be the most active denitrifying bacteria in a sewage treatment system and was the most dominant and major contributor to the denitrification of nitrogen wastewater [39]. The occurrence of the genus *Hydrogenophaga* and *Desulfotomaculum* at the beginning of the operation (FR1) was 0.4% and 0.5%, respectively. They occurred as smaller populations, increasing to 1.8% and 3%, respectively, at the end (FR3). *Hydrogenophaga*, an autotrophic denitrifier, belongs to the autotrophic genera. It was a known genus of hydrogen-oxidizing bacteria dominant in the microbial community of MBfR [40]. *Hydrogenophaga* had the characteristic of dominating the biofilm and was responsible for the reduction of NO_3^- [41]. The population of *Hydrogenophaga* exhibited low relative abundances in our study compared with previous work [42,43]. The possible explanations for the lack of *Hydrogenophaga* in the present study are as follows: (1) the successful PN-D process brings about the accumulation of NO_2^- in the MBR effluent; the presence of NO_2^- has a toxic effect on *Hydrogenophaga*, and the reproduction of *Hydrogenophaga* is sensitive to its presence [17]. (2) The MBfR influent contained synthetic wastewater and domestic wastewater with COD, NO_3^-, and NO_2^-, and other nitrogen compounds. The loaded influent and the components were different in the MBR of each phase from FR1–FR3, which led to the limitation of the activity of *Hydrogenophaga*. Additionally, in the treatment and application stage, the treated water after the partial nitrification-denitrification process was MBfR influent, which had a more complex composition and was quite different from the experimental water in the previous study (synthetic groundwater with additives) [43]. The influent of MBfR contained a small amount of organic matter from the MBR reactor, and the presence of a small amount of residual organic matter promoted the proliferation of heterotrophic denitrifying bacteria and competed with hydrogen autotrophic denitrifying bacteria (living space and nutrients). As a result, the dominant species of hydrogen autotrophic denitrifying bacteria possessed a low stable relative abundance for a long time.

Meanwhile, the entry of organic matter in the influent made heterotrophic denitrifying bacteria compete more effectively with the autotrophic denitrifying bacteria. The heterotrophic denitrifying bacteria therefore had an advantage, seizing the electron donor, which resulted in inhibition of the growth of the autotrophic denitrifying bacteria.

4. Conclusions

A two-stage system was applied for nitrogen removal from a wastewater treatment plant processing wastewater in an MBR-MBfR reactor. The proper functioning of the system was achieved by coupling the PN-D process in an MBR with further treatment in an MBfR. More than 96% of NH_4^+-N was removed via PN-D in MBR. In experiments with C/N (ratios of 1.5, 2.0, and 2.5) for the MBR-MBfR, the PN-D process was often outcompeted by heterotrophic denitrification and severely inhibited at a C/N greater than 2.0. MBfR performed a further treatment for both nitrate and nitrite that remained in the effluent of MBR, which obtained the average NO_3^--N removal of 89.3% when the influent NO_3^--N concentration was 30 mg/L and the HRT was 5 h. The effluent water of MBfR met Class 1A level for Chinese discharge standards after the stable operation of the MBR-MBfR (in the experimental conditions of HRT = 15 h, SRT = 20 d) used for domestic wastewater treatment. Microbial community analysis revealed that a successful AOB-proliferation stage was achieved with denitrifying bacteria (*Denitratisoma* genus), which performed denitrification in MBR at the same time. In the MBfR, the dominant bacteria were *Meiothermus*, *Lentimicrobium*, *Thauera*, *Hydrogenophaga*, and *Desulfotomaculum*, which proved the success of the hydrogenotrophic denitrification process in MBfR and showed the characteristics of efficient nitrate and nitrite removal.

Author Contributions: Conceptualization, K.D. and X.F.; methodology, H.L.; formal analysis, K.D. and X.F.; investigation, D.W., W.W., Y.C. and W.H.; resources, H.L.; data curation, D.W., W.W., Y.C. and W.H.; writing—original draft preparation, K.D. and X.F.; writing—review and editing, K.D. and X.F.; visualization, W.W. and W.H.; supervision, H.L.; project administration, K.D. and H.L.; funding acquisition, H.L. and D.W. All authors have read and agreed to the published version of the manuscript.

Funding: This research was funded by the National Natural Science Foundation of China (grant number 51638006; 51878197 and 51768012); the Basic Ability Enhancement Program for Young and Middle-aged Teachers of Guangxi (grant number 2021KY0265).

Institutional Review Board Statement: Not applicable.

Informed Consent Statement: Not applicable.

Data Availability Statement: The data that support the findings of this study are available from the corresponding author upon reasonable request.

Conflicts of Interest: The authors declare no conflict of interest.

References

1. Liu, X.; Kim, M.G.; Nakhla, G.; Andalib, M.; Yuan, F. Partial nitrification-reactor configurations, and operational conditions: Performance analysis. *J. Environ. Chem. Eng.* **2020**, *8*, 103984. [CrossRef]
2. Yang, Y.; Chen, Z.; Wang, X.; Zheng, L.; Gu, X. Partial nitrification performance and mechanism of zeolite biological aerated filter for ammonium wastewater treatment. *Bioresour. Technol.* **2017**, *241*, 473–481. [CrossRef]
3. Wang, Y.; Li, B.; Xue, F.; Wang, W.; Wang, Z. Partial nitrification coupled with denitrification and anammox to treat landfill leachate in a tower biofilter reactor (TBFR). *J. Water Process Eng.* **2021**, *42*, 102155. [CrossRef]
4. Xiao, H.; Peng, Y.; Zhang, Q.; Liu, Y. Pre-anaerobic treatment enhanced partial nitrification start-up coupled with anammox for advanced nitrogen removal from low C/N domestic wastewater. *Bioresour. Technol.* **2021**, *337*, 125434. [CrossRef]
5. Zhang, M.; Wang, S.; Ji, B.; Liu, Y. Towards mainstream deammonification of municipal wastewater: Partial nitrification-anammox versus partial denitrification-anammox. *Sci. Total Environ.* **2019**, *692*, 393–401. [CrossRef]
6. Ge, S.; Wang, S.; Yang, X.; Qiu, S.; Li, B.; Peng, Y. Detection of nitrifiers and evaluation of partial nitrification for wastewater treatment: A review. *Chemosphere* **2015**, *140*, 85–98. [CrossRef]
7. Li, J.; Du, Q.; Peng, H.; Zhang, Y.; Liu, T. Optimization of biochemical oxygen demand to total nitrogen ratio for treating landfill leachate in a single-stage partial nitrification-denitrification system. *J. Clean. Prod.* **2020**, *266*, 121809. [CrossRef]
8. Chen, Z.; Wang, X.; Yang, Y.Y.; Mirino, M.W.; Yuan, Y. Partial nitrification and denitrification of mature landfill leachate using a pilot-scale continuous activated sludge process at low dissolved oxygen. *Bioresour. Technol.* **2016**, *218*, 580–588. [CrossRef]
9. Dong, H.; Zhang, K.; Han, X.; Du, B.; Wei, Q.; Wei, D. Achievement, performance and characteristics of microbial products in a partial nitrification sequencing batch reactor as a pretreatment for anaerobic ammonium oxidation. *Chemosphere* **2017**, *183*, 212–218. [CrossRef] [PubMed]
10. Li, X.; Lu, M.Y.; Huang, Y.; Yuan, Y.; Yuan, Y. Influence of seasonal temperature change on autotrophic nitrogen removal for mature landfill leachate treatment with high-ammonia by partial nitrification-Anammox process. *J. Environ. Sci.* **2021**, *102*, 291–300. [CrossRef] [PubMed]
11. Duan, Y.; Liu, Y.; Zhang, M.; Li, Y.; Zhu, W.; Hao, M.; Ma, S. Start-up and operational performance of the partial nitrification process in a sequencing batch reactor (SBR) coupled with a micro-aeration system. *Bioresour. Technol.* **2020**, *296*, 123321. [CrossRef]
12. Soliman, M.; Eldyasti, A. Development of partial nitrification as a first step of nitrite shunt process in a Sequential Batch Reactor (SBR) using Ammonium Oxidizing Bacteria (AOB) controlled by mixing regime. *Bioresour. Technol.* **2016**, *221*, 85–95. [CrossRef]
13. Li, J.; Zhang, Q.; Li, X.; Peng, Y. Rapid start-up and stable maintenance of domestic wastewater nitritation through short-term hydroxylamine addition. *Bioresour. Technol.* **2019**, *278*, 468–472. [CrossRef]
14. Nerenberg, R. The membrane-biofilm reactor (MBfR) as a counter-diffusional biofilm process. *Curr. Opin. Biotechnol.* **2016**, *38*, 131–136. [CrossRef]
15. Wu, J.; Yin, Y.; Wang, J. Hydrogen-based membrane biofilm reactors for nitrate removal from water and wastewater. *Int. J. Hydrogen Energy* **2018**, *43*, 1–15. [CrossRef]
16. Jiang, M.; Zheng, J.; Perez-Calleja, P.; Picioreanu, C.; Lin, H.; Zhang, X.; Zhang, Y.; Li, H.; Nerenberg, R. New insight into CO2-mediated denitrification process in H2-based membrane biofilm reactor: An experimental and modeling study. *Water Res.* **2020**, *184*, 116177. [CrossRef] [PubMed]
17. Jiang, M.; Zhang, Y.; Yuan, Y.; Chen, Y.; Lin, H.; Zheng, J.; Li, H.; Zhang, X. Nitrate Removal and Dynamics of Microbial Community of A Hydrogen-Based Membrane Biofilm Reactor at Diverse Nitrate Loadings and Distances from Hydrogen Supply End. *Water* **2020**, *12*, 3196. [CrossRef]
18. Park, J.H.; Choi, O.; Lee, T.H.; Kim, H.; Sang, B.I. Pyrosequencing analysis of microbial communities in hollow fiber-membrane biofilm reactors system for treating high-strength nitrogen wastewater. *Chemosphere* **2016**, *163*, 192–201. [CrossRef]
19. Chi, Y.; Zhang, X.; Shi, X.; Ren, T.; Jin, P. Quick start-up of partial nitrification in a novel anaerobic/pulse washout (APW) process for treating municipal wastewater. *J. Clean. Prod.* **2020**, 124850. [CrossRef]
20. Xia, S.; Zhong, F.; Zhang, Y.; Li, H.; Yang, X. Bio-reduction of nitrate from groundwater using a hydrogen-based membrane biofilm reactor. *J. Environ. Sci.* **2010**, *22*, 257–262. [CrossRef]
21. Zhou, X.; Zhang, Z.; Zhang, X.; Liu, Y. A novel single-stage process integrating simultaneous COD oxidation, partial nitritation-denitritation and anammox (SCONDA) for treating ammonia-rich organic wastewater. *Bioresour. Technol.* **2018**, *254*, 50–55. [CrossRef] [PubMed]

22. Hu, T.; Peng, Y.; Yuan, C.; Zhang, Q. Enhanced nutrient removal and facilitating granulation via intermittent aeration in simultaneous partial nitrification endogenous denitrification and phosphorus removal (SPNEDpr) process. *Chemosphere* **2021**, *285*, 131443. [CrossRef] [PubMed]
23. Liu, W.; Hao, S.; Ma, B.; Zhang, S.; Li, J. In-situ fermentation coupling with partial-denitrification/anammox process for enhanced nitrogen removal in an integrated three-stage anoxic/oxic (A/O) biofilm reactor treating low COD/N real wastewater. *Bioresour. Technol.* **2021**, 126267. [CrossRef] [PubMed]
24. Chung, J.; Nerenberg, R.; Rittmann, B.E. Bio-reduction of soluble chromate using a hydrogen-based membrane biofilm reactor. *Water Res.* **2006**, *40*, 1634–1642. [CrossRef]
25. Lee, K.C.; Rittmann, B.E. Applying a novel autohydrogenotrophic hollow-fiber membrane biofilm reactor for denitrification of drinking water. *Water Res.* **2002**, *36*, 2040–2052. [CrossRef]
26. Xia, S.; Zhang, Z.; Zhong, F.; Zhang, J. High efficiency removal of 2-chlorophenol from drinking water by a hydrogen-based polyvinyl chloride membrane biofilm reactor. *J. Hazard. Mater.* **2011**, *186*, 1367–1373. [CrossRef]
27. Vo, T.-K.-Q.; Dang, B.-T.; Ngo, H.H.; Nguyen, T.-T.; Nguyen, V.-T.; Vo, T.-D.-H.; Ngo, T.-T.-M.; Nguyen, T.-B.; Lin, C.; Lin, K.-Y.A.; et al. Low flux sponge membrane bioreactor treating tannery wastewater. *Environ. Technol. Innov.* **2021**, *24*, 101989. [CrossRef]
28. Hu, X.; Xie, L.; Shim, H.; Zhang, S.; Yang, D. Biological Nutrient Removal in a Full Scale Anoxic/Anaerobic/Aerobic/Pre-anoxic-MBR Plant for Low C/N Ratio Municipal Wastewater Treatment. *Chin. J. Chem. Eng.* **2014**, *22*, 447–454. [CrossRef]
29. Gao, Y.X.; Li, X.; Zhao, J.R.; Zhang, Z.X.; Fan, X.Y. Response of microbial communities based on full-scale classification and antibiotic resistance genes to azithromycin and copper combined pollution in activated sludge nitrification laboratory mesocosms at low temperature. *Bioresour. Technol.* **2021**, *341*, 125859. [CrossRef]
30. Dong, L.; Luo, Y.; Cai, Y.; Zeng, H.; Jie, Z. Bacterial composition and nutrient removal with a novel PIA-A2/O sewage treatment. *Water Sci. Technol.* **2016**, *73*, 2722–2730. [CrossRef]
31. Ge, G.; Zhao, J.; Chen, A.; Hu, B.; Ding, X. Nitrogen Removal and Nitrous Oxide Emission in an Anaerobic/Oxic/Anoxic Sequencing Biofilm Batch Reactor. *Environ. Eng. Sci.* **2017**, *35*, 19–26. [CrossRef]
32. Tao, X.A.; Hong, L.B.; Peng, W.A.; Dga, B. Insights into two stable mainstream deammonification process and different microbial community dynamics at ambient temperature—ScienceDirect. *Bioresour. Technol.* **2021**, *331*, 125058.
33. Sedlacek, C.J.; McGowan, B.; Suwa, Y.; Sayavedra-Soto, L.; Laanbroek, H.J.; Stein, L.Y.; Norton, J.M.; Klotz, M.G.; Bollmann, A. A Physiological and Genomic Comparison of Nitrosomonas Cluster 6a and 7 Ammonia-Oxidizing Bacteria. *Microb. Ecol.* **2019**, *78*, 985–994. [CrossRef]
34. Sui, Q.; Liu, C.; Zhang, J.; Dong, H.; Zhu, Z.; Wang, Y. Response of nitrite accumulation and microbial community to free ammonia and dissolved oxygen treatment of high ammonium wastewater. *Appl. Microbiol. Biotechnol.* **2016**, *100*, 4177–4187. [CrossRef] [PubMed]
35. Zhong, L.; Lai, C.Y.; Shi, L.D.; Wang, K.D.; Dai, Y.J.; Liu, Y.W.; Ma, F.; Rittmann, B.E.; Zheng, P.; Zhao, H.P. Nitrate effects on chromate reduction in a methane-based biofilm. *Water Res.* **2017**, *115*, 130–137. [CrossRef]
36. Lai, C.Y.; Lv, P.L.; Dong, Q.Y.; Yeo, S.L.; Rittmann, B.E.; Zhao, H.P. Bromate and Nitrate Bioreduction Coupled with Poly-beta-hydroxybutyrate Production in a Methane-Based Membrane Biofilm Reactor. *Environ. Sci. Technol.* **2018**, *52*, 7024–7031. [CrossRef]
37. Wang, H.; Sun, Y.; Zhang, L.; Wang, W.; Guan, Y. Enhanced nitrogen removal and mitigation of nitrous oxide emission potential in a lab-scale rain garden with internal water storage—ScienceDirect. *J. Water Process Eng.* **2021**, *42*, 102147. [CrossRef]
38. Wang, H.; Chen, N.; Feng, C.; Deng, Y.; Gao, Y. Research on efficient denitrification system based on banana peel waste in sequencing batch reactors: Performance, microbial behavior and dissolved organic matter evolution. *Chemosphere* **2020**, *253*, 126693. [CrossRef]
39. Patel, R.J.; Patel, U.D.; Nerurkar, A.S. Moving bed biofilm reactor developed with special microbial seed for denitrification of high nitrate containing wastewater. *World J. Microbiol. Biotechnol.* **2021**, *37*, 1–13. [CrossRef] [PubMed]
40. Park, H.I.; Choi, Y.J.; Pak, D. Autohydrogenotrophic Denitrifying Microbial Community in a Glass Beads Biofilm Reactor. *Biotechnol. Lett.* **2005**, *27*, 949–953. [CrossRef] [PubMed]
41. Zhao, H.P.; Ginkel, S.V.; Tang, Y.; Kang, D.W.; Rittmann, B.; Krajmalnik-Brown, R. Interactions between Perchlorate and Nitrate Reductions in the Biofilm of a Hydrogen-Based Membrane Biofilm Reactor. *Environ. Sci. Technol.* **2011**, *45*, 10155–10162. [CrossRef] [PubMed]
42. Zhou, L.; Xu, X.; Xia, S. Effects of sulfate on simultaneous nitrate and selenate removal in a hydrogen-based membrane biofilm reactor for groundwater treatment: Performance and biofilm microbial ecology. *Chemosphere* **2018**, *211*, 254–260. [CrossRef] [PubMed]
43. Xia, S.; Xu, X.; Zhou, L. Insights into selenate removal mechanism of hydrogen-based membrane biofilm reactor for nitrate-polluted groundwater treatment based on anaerobic biofilm analysis. *Ecotoxicol. Environ. Saf.* **2019**, *178*, 123–129. [CrossRef] [PubMed]